Experimental approaches
to mammalian embryonic development

Experimental approaches to mammalian embryonic development

Edited by
Janet Rossant and Roger A. Pedersen

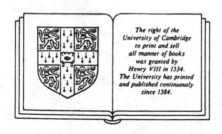

The right of the
University of Cambridge
to print and sell
all manner of books
was granted by
Henry VIII in 1534.
The University has printed
and published continuously
since 1584.

CAMBRIDGE UNIVERSITY PRESS

Cambridge
New York   New Rochelle
Melbourne   Sydney

Published by the Press Syndicate of the University of Cambridge
The Pitt Building, Trumpington Street, Cambridge CB2 1RP
32 East 57th Street, New York, NY 10022, USA
10 Stamford Road, Oakleigh, Melbourne 3166, Australia

First published 1986
First paperback edition 1988

Library of Congress Cataloging-in-Publication Data

Experimental approaches to mammalian embryonic development.
1. Embryology – Mammals.   2. Embryology, Experimental.
3. Mammals – Genetics.   I. Rossant, Janet.   II. Pedersen,
Roger A.
QL959.E86   1986   599′.03′3   86–9655

British Library Cataloguing in Publication Data

Experimental approaches to mammalian
embryonic development.
1. Embryology – Mammals
I. Rossant, Janet   II. Pedersen, Roger A.
599.03′34   QL959

ISBN 0 521 30991 3 hard covers
ISBN 0 521 36891 X paperback

Transferred to digital printing 2005

But actually it is the whole (four dimensional) pattern of the "phenotype," the visible and manifest nature of the individual, which is reproduced without appreciable change for generations . . . and borne at each transmission by the material structure of the two cells which unite to form the fertilized egg cell. That is a marvel — than which only one is greater . . . I mean the fact that we, whose total being is entirely based on a marvellous interplay of this very kind, yet possess the power of acquiring considerable knowledge about it.

E. Schrödinger *What Is Life?* Cambridge University Press, p. 31.

Contents

Preface

During the decade since the publication of McLaren's *Mammalian Chimaeras* (Cambridge University Press, 1976), the field of mammalian embryology has undergone tremendous expansion and diversification. There is now a substantial body of information available on cellular, biochemical, and molecular aspects of the early embryo, and exciting new approaches to the genetics of development are now possible through recombinant DNA techniques. Yet during this period of rapid growth there has been no single volume summarizing these advances. This book is our attempt to remedy this deficiency.

The book is intended for any serious student of the mammalian embryo and should be useful at many levels. Advanced undergraduates and graduate students can learn from the authors' emphasis on general principles; established investigators of mammalian development will benefit from the attention to detail. Developmental biologists working in other systems will enjoy the opportunity for cross-species comparisons. For the many molecular biologists who have discovered the mouse embryo in their quest for transgenic mice it will provide an introduction to the wider issues of mammalian embryogenesis. Scientists working in the more applied areas of embryology, such as human in vitro fertilization and livestock embryo manipulation, can contemplate applications of new experimental techniques to their systems.

The goal of a readable, affordable book has required some compromises. We have omitted the topics of fertilization and embryo transfer because these have been covered recently in other reviews. We have also omitted discussion of aspects of later embryonic development, such as lineage analysis of the nervous system and sex differentiation, so that we could concentrate on events in the early stages of development before and just after implantation. The literature reviewed here includes studies published through 1985 and early 1986. We have included approaches and results that we hope will provide insights and applications throughout the next decade.

The volume is divided into three general sections as follows: cellular aspects, including studies of potency, allocation, differentiation, and fate in the preimplantation and early postimplantation embryo; molecular and biochemical aspects, including gene expression during gametogenesis and early development, the physiology of the embryo, cell surface and cytoskeletal differentiation and X-chromosome inactivation; and finally a section dealing with new approaches toward a genetic understanding of development, including nuclear transfer, analysis of genetic mutations, and introduction of new genetic material into the embryo via stem cell lines or production of transgenic mice.

Why study the mammalian embryo? One obvious answer is that we ourselves are mammals, and analysis of embryogenesis in other mammals can fulfill our need to understand more about our own development. Moreover, work with laboratory mammals has proved of direct benefit in laying the groundwork for the current rapid growth of human in vitro fertilization programs. Another answer that the adventurous developmental biologist might give is that mammals represent the last embryological frontier. The small size, fragility, and inaccessibility of early viviparous life had kept mammalian embryos beyond the reach of any but the most persevering of explorers. But the advent of techniques of superovulation, embryo culture, and, most important, embryo manipulation opened up the area for settlement. The sheer detail of the body of knowledge about early mammalian development − cellular, molecular, and genetic − now rivals or surpasses that available in other favored embryological systems. There is now a thriving community of mammalian embryologists, and new immigrants are arriving all the time, bringing with them new ideas and techniques to expand our knowledge of early embryogenesis. We look forward to another exciting decade.

We want to acknowledge the many people who have assisted and encouraged us in transforming this book from an after-dinner idea to a final published product. First, the contributors have unanimously shared our excitement about assembling the major concepts of the field in a single volume in order to make them widely accessible. They have willingly complied with urgent deadlines and responded politely to our sometimes drastic editorial comments. They deserve the credit for the success of their chapters. We have also benefited from the informal reviews and advice provided by our colleagues, particularly including John Biggers, Frank Costantini, Hans Denker, Mitch Eddy, Robert Erickson, Richard Gardner, Brigid Hogan, Gerald Kidder, Kirstie Lawson, Susan Lewis, Elwood Linney, Michael McBurney, Clement Markert, Catherine Racowsky, Tony Searle, and students at the Cold Spring Harbor course on Molecular Embryology of the Mouse, 1985. Special thanks are due to Jim DeMartino and Louise Calabro Gruendel of Cambridge University Press for their patience and care at all stages of production and for their efforts to achieve a speedy publication schedule. Finally, we wish to thank our close associates and family members for putting up with the reams of paper and weeks of distraction essential to the completion of this project. We are particularly grateful to our spouses, Alex Bain and Carmen Arbona, who shared with us during this period the mystery of the conception and birth of our own children. Perhaps someone who reads this will find a way to ease the path to life.

Janet Rossant,
Toronto, Ontario

Roger A. Pedersen,
Durham, North Carolina

Contributors

Eileen D. Adamson
La Jolla Cancer Research Foundation
10901 North Torrey Pines Road
La Jolla CA 92037, U.S.A.

Rosa Beddington
ICRF Developmental Biology Unit
Department of Zoology
South Parks Road
Oxford OX1 3PS, England

Allan Bradley
Department of Genetics
University of Cambridge
Downing Street
Cambridge CB2 3EH, England

Verne M. Chapman
Department of Molecular Biology
Roswell Park Memorial Institute
Buffalo, NY 14263, U.S.A.

Karl M. Ebert
Department of Anatomy and Cellular
  Biology
Tufts University School of Medicine
136 Harrison Avenue
Boston, MA 02111, U.S.A.

Norman B. Hecht
Department of Biology
Tufts University
Medford, MA 02155, U.S.A.

Martin H. Johnson
Department of Anatomy
University of Cambridge
Downing Street
Cambridge CB2 3DY, England

Peter L. Kaye
Department of Physiology and
  Pharmacology
University of Queensland
St. Lucia
Queensland 4067, Australia

Terry Magnuson
Department of Developmental
  Genetics and Anatomy
Case Western Reserve University
Cleveland, OH 44106, U.S.A.

Bernard Maro
Départment de Biologie Cellulaire
Centre de Génétique Moléculaire
91190 Gif sur Yvette, France

Virginia E. Papaioannou
Department of Pathology
Tufts University School of Medicine
136 Harrison Avenue
Boston, MA 02111, U.S.A.

Roger A. Pedersen
Laboratory of Radiobiology
University of California at
  San Francisco
San Francisco, CA 94143, U.S.A.

Jean Richa
Wistar Institute of Anatomy and
  Biology
Thirty-Sixth Street at Spruce Street
Philadelphia, PA 19104, U.S.A.

Elizabeth J. Robertson
Department of Genetics
University of Cambridge
Downing Street
Cambridge CB2 3EH, England

Janet Rossant
Research Institute
Division of Molecular and
   Developmental Biology
Mount Sinai Hospital
600 University Avenue
Toronto, Ontario M5G 1X5, Canada

Gilbert A. Schultz
Department of Medical Biochemistry
Univeristy of Calgary Health Sciences
   Centre
3330 Hospital Drive N.W.
Calgary, Alberta T2N 4N1, Canada

Richard M. Schultz
Department of Biology
University of Pennsylvania
Philadelphia, PA 19104, U.S.A.

Davor Solter
Wistar Institute of Anatomy and
   Biology
Thirty-Sixth Street at Spruce Street
Philadelphia, PA 19104, U.S.A.

Colin L. Stewart
European Molecular Biology
   Laboratory
Postfach 10.2209
D-6900 Heidelberg, Federal Republic
   of Germany

M. Azim H. Surani
AFRC Institute of Animal Physiology
Animal Research Station
307 Huntingdon Road
Cambridge CB3 0JQ, England

Erwin F. Wagner
European Molecular Biology
   Laboratory
Postfach 10.2209
D-6900 Heidelberg, Federal Republic
   of Germany

Cellular aspects

1   Potency, lineage, and allocation in preimplantation mouse embryos

ROGER A. PEDERSEN

CONTENTS

I. Introduction

Although early mammalian development has fascinated embryologists for centuries (Bodemer 1971), our detailed mechanistic understanding of the mammalian embryo has been accumulated only during the past several decades. We now know a great deal about the timing of the early events of development and the potency and the fate of early embryonic cells, particularly in rodent embryos. In this chapter we shall examine the temporal sequence of early developmental events as a basis for understanding the cellular processes in the formation of the blastocyst; this information also provides a useful context for the genetic and molecular aspects of development covered in other chapters.

Cell potency during cleavage and blastocyst formation has been studied extensively in mammalian embryos, particularly in the mouse. The individual cells or blastomeres of the cleaving embryo retain their developmental totipotency until late in cleavage, even at the blastocyst stage in the case of the inner cell mass (Figure 1.1).

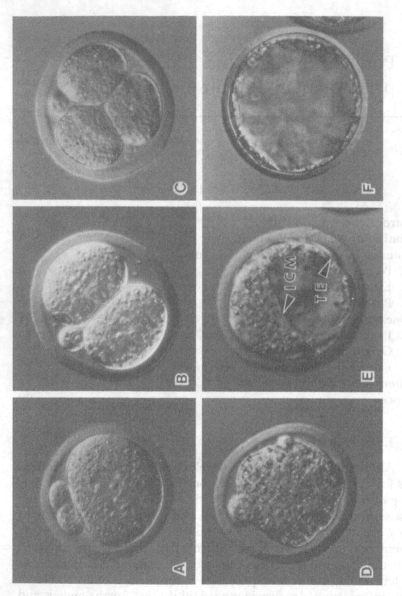

Figure 1.1. Preimplantation stages of mouse embryo development: (A) zygote; (B) 2-cell stage; (C) 4-cell stage; (D) compacted morula stage (~16 cells); (E) early blastocyst stage; TE, trophectoderm (~32 cells); (F) expanded blastocyst stage; ICM is out of focal plane (~64 cells). Magnification approximately 300×

Virtually all our information on mammalian cell lineage pertains to the preimplantation and early postimplantation mouse embryo. Postimplantation cell fate, potency, and lineage are discussed by Rossant in Chapter 4 and by Beddington in Chapter 5. Analysis of preimplantation embryos indicates that the early blastomeres have descendants that occupy both the inner cell mass and the trophectoderm and that clonal partitioning (allocation) of descendants to either of these blastocyst tissues occurs only late in cleavage or at the blastocyst stage. Furthermore, consideration of the relationship between potency and lineage suggests that the persistent totipotency of cleavage blastomeres is utilized in cell allocation in normal preimplantation development.

My major objective in reviewing what is known about cell potency and lineage in preimplantation embryos is to deduce some of the "rules" the mammalian embryo follows in forming the blastocyst. Without this body of knowledge about preimplantation embryos it would be impossible to predict when important cellular and molecular events occur or to formulate a concept of decision making in the early embryo.

II. Timing of early development

There are several conventions for measuring time in mammalian embryonic development. Absolute time after conception can be measured in embryos fertilized in vitro, and in embryos derived by artificial insemination or by brief mating, because the time of insemination is then known precisely. When conception occurs by natural mating, gestational age can be estimated by assuming that fertilization occurs immediately after ovulation, the middle of the dark phase (i.e., between 2400 and 0200 hours). Most investigators studying spontaneously ovulated embryos used this convention, referring to embryonic age in gestational days (d.g.) and fractions thereof.

A rather precise estimate of time is also possible in embryos from gonadotropin-induced ovulations. In this case, ovulation (and fertilization) is known to occur 10–14 h after administration of human chorionic gonadotropin (hCG) to mature mice primed with pregnant mare serum gonadotropin (Edwards and Gates 1959). Time after hCG is thus approximately 0.5 day greater than time after conception (d.g.).

Because gestational age in rodents is generally measured as time after copulation, the first day after mating is referred to as the plug day, referring to the appearance of the copulatory plug. The plug day can be counted either as day 0 or day 1; so it is important to note the convention used in each study. Embryonic ages given in this chapter are expressed as absolute time after conception, correcting for time after hCG or other conventions (noon of plug day = 0.5 d.g.).

The timing of fertilization events has been studied both in vitro and in vivo. The time from insemination to sperm-egg fusion is approximately 1 h in vitro (Gaddum-Rosse et al. 1982) and 1.75 h in vivo (Edwards and Gates 1959; Krishna and Generoso 1977). The second polar body is extruded between 1 and 2 h post insemination in vitro (Howlett and Bolton 1985). The maternal and paternal pronuclei form just after complete sperm penetration, at 4.5 to 6.5 h after insemination; the maternal pronucleus lies closer to the second polar body and is smaller than the male pronucleus (Luthardt and Donahue 1973; Howlett and Bolton 1985). Pronuclear DNA synthesis begins at 8 to 10 h after fertilization in vivo and has a general duration of 4–8 h, depending on the strain (summarized by Molls, Zamboglou, and Streffer 1983; Howlett and Bolton 1985).

The length of the first cell cycle appears to have a substantial genetic component. In the mouse strains studied by Shire and Whitten (1980a, 1980b), the median estimated cleavage times ranged from 22 to 26 h, depending on both maternal and paternal genotypes. Such observations, however, may represent several effects, including times of mating, ovulation, meiotic maturation, and fertilization. When direct cell-cycle analysis has been carried out, $G_2 + M$ has ranged from 3 to 8 h; combined with the variation in length of the S phase, the estimated length of the first cell cycle has been as short as 14 h (Molls et al. 1983) or as long as 20 h in the strains studied (Krishna and Generoso 1977; Bolton, Oades, and Johnson 1984; Howlett and Bolton 1985).

The length of the second cell cycle also varies between strains, owing to differences in the S phase (range 4–7 h) and in $G_2 + M$ (range 12–15 h). In general, the S phase is slightly longer and G_2 markedly longer than in the first cell cycle (Luthardt and Donahue 1975; Molls et al. 1983), and the total length of the second cell cycle among most strains studied ranges from 18 to 22 h. The second cell cycle is particularly interesting from a molecular viewpoint, because much of the maternal messenger RNA is degraded by the early 2-cell stage and replaced with transcripts of the embryonic genome (see Chapter 8).

The second cleavage division is invariably asynchronous in mice, yielding a 3-cell embryo as a transitory intermediate. Three-cell embryos are also observed in other mammalian species; this feature seems to be a general mammalian trait, differing from the situation for some vertebrates, in which virtually complete early synchrony occurs (Newport and Kirschner 1982a). Despite an asynchrony of up to 2 h in the time of second cleavage, descendants of both 2-cell blastomeres have similar fourth-cell-cycle lengths in mouse embryos (Kelly, Mulnard, and Graham 1978). Indeed, cell-cycle length is relatively invariant between the 4-cell and 64-cell stages. Embryos of all strains studied in vivo have cycle lengths of approximately 10 h. Culturing embryos delays growth between 2 and

7 h per cycle (Bowman and McLaren 1970; McLaren and Bowman 1973; Streffer et al. 1980; Harlow and Quinn 1982; Molls et al. 1983). Thus, embryos cultured from the 2-cell stage may lag a full day behind in vivo development by the blastocyst stage. A correlation between H−2 haplotype and cleavage rate (k and r haplotypes being associated with slow development, and b, d, q, s, and u haplotypes with fast development) has been interpreted as evidence for a gene in the H−2 complex affecting preimplantation developmental rate (Goldbard and Warner 1982).

The flattening of blastomeres known as compaction begins at the 8-cell stage in mouse embryos and is characteristic of all placental mammals (Lewis and Wright 1935). By contrast, embryos of marsupials do not appear to undergo compaction (Selwood and Young 1983). The complex cell surface and cytoskeletal modifications that accompany compaction are described in detail in chapters 2 and 10. During this same period (late morula, >20-cell stage), mouse embryos become resistant to the effects of α-amanitin, an inhibitor of RNA polymerase, and subsequently undergo blastocyst formation, with characteristic changes in the patterns of polypeptides synthesized (Braude 1979). The late morula accumulates sufficient mRNA for fluid accumulation 5−7 h before this process actually occurs, as shown by α-amanitin resistance. Translation products necessary for the onset of cavitation accumulate nearly concomitantly (i.e., 1−3 h earlier) with the process itself, as shown by cycloheximide treatment (Kidder and McLachlin 1985). In view of the cell-cycle dynamics described previously, the mouse morula thus acquires competence to differentiate into the blastocyst during the S phase or G_2 + M of the fifth cell cycle (late 16-cell stage).

Although the entire process of compaction presages trophectoderm differentiation, formation of the blastocoel cavity itself in the mouse embryo occurs on completion of the fifth cleavage division. Nascent blastocysts have average cell numbers of 30 to 34 (Smith and McLaren 1977; Chisholm et al. 1985). When two or three embryos were aggregated as chimeras, the resultant nascent blastocysts had twofold or threefold the numbers of cells in single embryos. Conversely, half-embryos and tetraploid embryos had half the normal cell numbers when they became nascent blastocysts. The most obvious explanation for these results is a requirement for five nuclear divisions before blastocoel formation (Smith and McLaren 1977). Blastocoel formation in other placental mammals also seems closely tied to the number of cell cycles, but this number varies. According to the sparse observations reviewed by Lewis and Wright (1935), rat and pig embryos also form blastocysts with about 30 cells, whereas rabbit blastocysts form at the 128-cell stage. Hamster embryos, on the other hand, cavitate at the 16-cell stage (Y. Kato, unpublished observations), and some primitive placental mammals cavitate at even earlier stages; see Wimsatt (1975) for review. A process resembling cavitation can be induced in 2-cell mouse embryos by treatment with wheat germ agglutinin (Johnson 1985), showing that fluid accumulation is dissociable from cleavage.

What is the nature of the clock governing events in the early mammalian embryo? Despite the correlation between nuclear divisions and blastocoel formation, this may not be the way the embryo counts time. Perturbing the cell cycle with phorbol esters, aphidicolin (an inhibitor of DNA polymerase alpha), or inhibitors of spermine and spermidine synthesis results in embryos that cavitate with fewer cells and at earlier times than normal (Alexandre 1979, 1982; Sawicki and Mystkowska 1981; Dean and Rossant 1984; Spindle, Nagano, and Pedersen 1985). These observations raise the possibility that blastocyst formation depends on the interaction between the nucleus and the cytoplasm (classically, the nucleocytoplasmic ratio), or perhaps on the total amount of DNA in relationship to some limited cytoplasmic factor(s), as in the mid-blastocyst transition of the amphibian embryo (Newport and Kirschner 1982b), rather than on the number of nuclear divisions per se, as in ascidian differentiation (Satoh and Ikegami 1981; Mita-Miyazawa, Ikegami, and Satoh 1985). The persistent contraction of activated, enucleated Xenopus eggs and the autonomous cortical activity of anucleate fragments of mouse zygotes and parthenogenetically activated eggs provide direct evidence for a cytoplasmic clock in these species (Hara, Tydeman, and Kirschner 1980; Newport and Kirschner 1984; Waksmundzka et al. 1984).

The role of cytoplasmic and membrane components in regulating the cell cycle is just beginning to be understood. Maturation-promoting factor, which is present in unfertilized amphibian eggs (Masui and Markert 1971), in mature mouse eggs (Kishimoto et al. 1984; Sorenson, Cyert, and Pedersen 1985), and in numerous other cell types at the time of mitosis, can induce nuclear breakdown and chromosomal condensation of meiotic as well as mitotic nuclei (Newport and Kirschner 1984). In addition, membrane phospholipids, oncogene protein kinases, and cytosolic calcium are linked to the activity of calcium- and phospholipid-dependent kinase C; all are strongly implicated in regulation of the cell cycle, DNA replication, and differentiation in somatic cells (Berridge 1984; Faletto, Arrow, and Macara 1985; Macara 1985). Some of these components appear to act through a change in cytoplasmic pH, which oscillates during the cell cycle in Dictyostelium and other cell types (Aerts, Durston, and Moolenaar 1985); see Busa and Nuccitelli (1984) for review. Finally, polyamines are known to be required for cell proliferation in cultured cells and in numerous embryonic systems, including the mouse (Alexandre 1979; Zwierzchowski, Czlonkowska, and Guszkiewicz 1986); see Heby (1981) for review. In sum, the simple geometry of the cleaving mouse embryo may belie its complexity; an understanding of the clock that regulates cleavage and cavitation will require substantially more information about the cell biology of proliferation in general.

During blastocyst formation and subsequent growth of the inner cell mass and trophectoderm populations there is increasing asynchrony in the mouse embryo. It was found, however, that all blastocyst cells became labeled with [^{3}H]thymidine after long labeling times (>6 h), implying that all cells are dividing during this period. Inner cells had a higher labeling index than outer cells; this might imply a faster rate of division in inner cells, but only if one assumes that the S phases are the same length in both cell populations (Barlow, Owen, and Graham 1972). At the late blastocyst stage, 105–108 h after fertilization (4.5 d.g.), nuclei with >4C DNA content could be detected, indicating the endoreduplication of trophectoderm cell DNA. This correlates with the appearance in abembryonic mural trophectoderm of giant cells. In half of the embryos of the strain studied by Dickson (1966), giant cell transformation was complete to the edge of the inner cell mass by noon at 4.5 d.g. This also is the time at which decidual uterine swellings can first be recognized, and the number of embryos that can be recovered by uterine flushing decreases. Implantation has begun.

III. Potency in preimplantation embryos

Prospective potency can be defined as the full range of developmental capacities of an embryonic cell under any circumstances (Rossant and Papaioannou 1977). This definition implies a requirement for experimental manipulation to test the potency of the cell. This has been accomplished with mammalian embryos by micromanipulative procedures: extirpation, disaggregation, and aggregation. Simply culturing the intact or zona-free embryo does not appear to alter the developmental fate of its cells.

A. Potency during cleavage stages

The potency of cleavage-stage blastomeres was first examined by extirpating one of the 2-cell-stage blastomeres (by convention, a $\frac{1}{2}$ blastomere). The surviving $\frac{1}{2}$ blastomere was capable of developing normally (Tarkowski 1959). Taking advantage of this totipotency, Epstein and associates (1978) have developed an elegant approach to genetic analysis of early embryos using the half-embryos for either karyotypic or molecular analysis. Embryos of other species have been isolated at the 2-cell stage, with both blastomeres being transferred separately to foster mothers, in which they have developed into identical twins (see Chapter 3). Thus, the 2-cell blastomeres appear to be developmentally totipotent in mammalian embryos in general.

Blastomeres isolated from 4-cell mouse embryos ($\frac{1}{4}$ blastomeres), however, frequently give vesicular structures in culture, and isolated $\frac{1}{8}$ blastomeres predominantly give trophectodermal vesicles lacking inner cells (Tarkowski and Wroblewska 1967). The low incidence of morula/inner-cell-mass-like forms in those experiments led to the idea that the inner cell mass (ICM) developed from cells that were enclosed at the morula stage, and trophectoderm from cells that were outside. Tarkowski and Wroblewska (1967) concluded that "different environmental conditions play a decisive role in the differentiation of cells in one of the two directions (trophectoderm vs. inner cell mass)." This idea has become known as the inside-outside or epigenetic hypothesis for early mouse embryo differentiation. As its major prediction, the epigenetic hypothesis requires that blastomeres remain totipotent until they acquire an inner or outer position in the morula. Furthermore, the experimentally altered position of a totipotent blastomere should determine its fate. To test the epigenetic hypothesis, subsequent investigators have examined the potency and fate of isolated blastomeres.

The potency of early morula (16-cell) outer blastomeres was examined using disaggregation-reaggregation techniques. In one study, single isolated 16-cell blastomeres were injected under the zona pellucida of 8-cell embryos, which were then transferred to foster mothers and analyzed for their glucose phosphate isomerase (GPI) patterns at midgestation (Rossant and Vijh 1980). Although there was a preponderance of embryos with donor descendants in trophectodermal tissues, several had descendants in ICM-derived tissues as well as trophectodermal tissues, demonstrating the totipotency of 16-cell blastomeres. This result was confirmed in another study (Ziomek, Johnson, and Handyside 1982) in which the polar, outer cells and apolar, inner cells of disaggregated 16-cell embryos were recombined as homogeneous or mixed aggregates (see subsequent discussion and Chapter 2 for details of this polarized phenotype). Despite significant differences in the timing of blastocyst formation, there were no differences in postimplantation development for pure inner, pure outer, or mixed aggregates, showing that at least some cells of each population are totipotent.

The potency of late morula-stage embryos (approximately 30 cells) was also determined by Rossant and Vijh (1980), who microsurgically removed individual cells from the outside of experimentally decompacted embryos, then reconstituted morulae by injecting 20 or more of these cells into empty zonae for transfer to foster mothers. When analyzed at early postimplantation stages, half of these had normal embryonic development, with ICM as well as trophectoderm derivatives, whereas the remainder had trophectodermal tissues only. Thus, even late morulae often include at least some totipotent cells in the outer population. The potency of late morula inner cells was determined by isolating inner cells with the immunosurgical technique of Solter and Knowles (1975), which involves treatment with heat-inactivated antiserum, then with active complement to lyse the outer cells of late morulae and blastocysts. When the isolated inner cells of late morulae are cultured overnight, they develop into small blastocysts or trophectodermal vesicles (Handyside 1978; Hogan and Tilly 1978; Spindle 1978). These results show that the mouse embryo has totipotent inner cells until the formation of the blastocoel cavity (i.e., until late in the fifth cell cycle).

B. Effect of position on fate

The consequences of experimentally altering the positions of blastomeres were also examined by determining the contribution of labeled blastomeres in reaggregates cultured to the blastocyst stage. In these experiments, donor blastomeres from 4-cell or 8-cell embryos were aggregated with unlabeled morulae. Alternatively, labeled entire 8- to 16-cell embryos were aggregated in various combinations with unlabeled morulae. As predicted, labeled cells placed in outer positions tended to contribute descendants to trophectoderm, whereas cells placed inside contributed predominantly to the ICM when these aggregation chimeras were examined as blastocysts. In a few decisive cases, entire labeled embryos placed in inner positions contributed all their descendants to the ICM, showing that most, if not all, blastomeres form ICM when placed in inside positions (Hillman, Sherman, and Graham 1972).

In another approach, 4-cell embryos were disaggregated and then allowed to divide once, and each pair of $\frac{1}{8}$ descendants was aggregated with unlabeled blastomeres (quartets); alternatively, the $\frac{1}{8}$ pairs were separated before they were aggregated (octets) (Figure 1.2). The pattern of GPI isozymes was used to distinguish donor and host embryos' tissues at 10.5 d.g. Occasionally all four members of a quartet had donor descendants in either the trophectoderm or the ICM, depending on whether the donor was placed outside or inside of the reaggregate (Hillman et al. 1972; Kelly 1977). These experiments demonstrate that individual 4- and 8-cell blastomeres can give rise either to trophectoderm or to ICM, depending on their position in the embryo. However, quartets with inner $\frac{2}{8}$ blastomeres and octets with inner $\frac{1}{8}$ blastomeres frequently had marked descendants in trophectodermal as well as ICM-derived tissues, which indicates either that these reaggregates do not entirely enclose the labeled blastomere or that there are rearrangements during cleavage or blastocyst formation that expose inner cells.

The maintenance of totipotency until the late morula stage and the responsiveness of totipotent blastomeres' fate to their position confirm the central predictions of the inside-outside hypothesis. They provide no information, however, about the nature of environmental cues that cause differentiation into either trophectoderm or ICM. All that is known

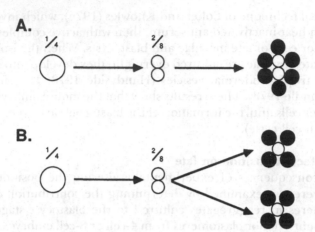

Figure 1.2.   Quartet/octet experiment to determine the effect of position on blastomere fate: (A) quartets with 2 paired, labeled $\frac{1}{8}$ descendants located on inside of aggregate; (B) octets with 2 separated, labeled $\frac{1}{8}$ descendants on inside of aggregate. [Adapted from Kelly (1977).]

is that the soluble components of the blastocoel fluid do not direct the differentiation of totipotent cells into ICM by suppressing trophectoderm differentiation. This was shown by injecting entire 8-cell or morula-stage embryos (presumably consisting of totipotent cells) into the blasto-coel of giant, chimeric blastocysts (Pedersen and Spindle 1980). Rather than forming pure ICM, the enclosed embryos developed into blastocysts (Figure 1.3). Other possible mechanisms (besides diffusible extracellular substances) for transmitting positional information include cell surface interactions involving the cytoskeleton, gap-junctional communication, cell shape, and cell polarization (see Chapter 2). No experiments have been reported to exclude any of these as sources of epigenetic information.

C. Potency in blastocysts

Using immunosurgery to isolate inner cells, several investigators have studied the potency of the ICM during blastocyst growth. The majority of ICMs isolated from early blastocysts (~32-cell stage) form blastocyst-like structures on subsequent culture (Handyside 1978; Hogan and Tilly 1978; Spindle 1978). Such ICM-derived embryos are capable of forming trophoblast giant cells in culture and of inducing a normal decidual response when transferred to foster mothers (Rossant and Lis 1979; Nichols and Gardner 1984). Thus, at least some inner cells of the early blastocyst are also totipotent. As assessed by immunosurgery and the capacity to form trophectodermal vesicles, this ICM totipotency persists

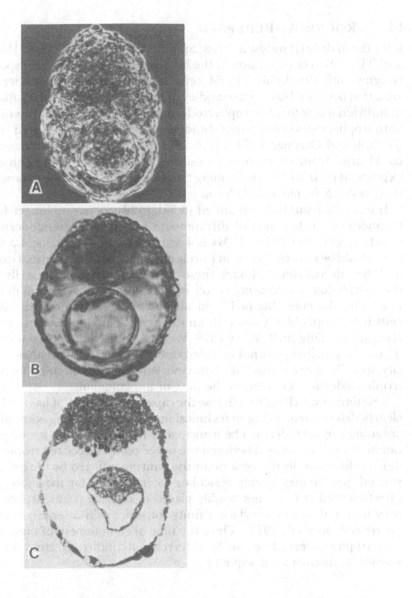

Figure 1.3.   Eight-cell- and morula-stage embryos formed blastocysts when placed in the blastocoel cavity of giant chimeric blastocysts. A: Giant blastocyst recovering from microsurgery, containing morula in blastocyst cavity. B: Giant blastocyst 24 h after surgery, showing development of morula into blastocyst. C: Section of embryos shown in B. [Reproduced from Pedersen and Spindle (1980), with permission of the publisher.]

until the middle (Handyside 1978; Spindle 1978) or later phases (Hogan and Tilly 1978) of expansion of the blastocoel cavity, which corresponds roughly with the division to 64 cells. However, there is considerable variation between blastocysts—and even in a single ICM—in the capacity of individual cells to form trophectoderm, as evidenced by the presence of both trophectoderm and parietal endoderm cells in single cultured ICMs (Nichols and Gardner 1984; Chisholm et al. 1985). This heterogeneity could arise from asynchrony in cell cycles within the ICM, with cells exposed at their sixth cycle forming trophectoderm and those exposed at their seventh forming endoderm.

Inner cells from fully expanded (3.5-d.g.) blastocysts no longer form trophectoderm, but instead differentiate into primitive endoderm or ectoderm (Rossant 1975). ICMs isolated from late blastocysts (4.5 d.g.) thus would be expected to form purely endodermal or ectodermal tissues and their derivatives. Although these derivatives predominate, cells with the morphology of extraembryonic ectoderm—a trophectoderm derivative—have also been obtained from late blastocyst ICMs; moreover, these cells form trophoblast giant cells on disaggregation and further culture (Hogan and Tilly 1978; Wiley 1978; Wiley, Spindle, and Pedersen 1978). Thus, the possibility cannot be ruled out that inner cells lose their capacity to differentiate into the extraembryonic ectoderm component of trophectoderm later, around the time of implantation.

The time at which outer cells lose the capacity to form ICM has not been clearly determined, owing to technical limitations in dealing with differentiating trophectoderm. The principal difficulty lies in the strong junctional complexes formed between the outer cells. Dissociated trophectoderm cells generally die because of the damage inflicted by the isolation procedures, or they simply vesiculate (Stern 1972). Furthermore, trophectodermal cells are not readily placed in internal positions, because they have reduced cell surface affinity for other cells or embryos (Burgoyne and Ducibella 1977). Thus, the time of commitment of outer cells to a trophectoderm fate can be inferred only indirectly from lineage studies, as discussed subsequently.

IV. Lineage and cell allocation

Because cell lineage describes the relationships between progenitors and their descendants, lineage analysis in embryos is a particularly valuable approach for understanding how cells become allocated to diverse organ rudiments during development (Gardner 1978; Rossant 1984). Cell lineage analysis requires an experimental means for marking individual progenitors and for recognizing their descendants at a subsequent stage of development. This has been accomplished in mammalian embryos either by aggregating genetically distinct or physically marked blastomeres as chimeras or by labeling cells in intact embryos by microinjection. In chimeras, donor cells are removed from their native environment and placed in a host embryo. As a consequence, the chimera approach to lineage analysis confounds potency and fate. Both are being tested in any given experiment, because the results reflect both the capacity of the donor cell to accommodate to its new environment and its fate at its new position with the host. The alternative approach, labeling individual cells in the intact embryo, avoids this problem, thus assessing fate exclusively, but requires the assumption that the labeling procedure does not distort the fate of the progenitor cell. This assumption is difficult to test rigorously, because usually there is no way to determine the behavior or fate of cells in the intact embryos except by marking them. The results of both approaches are discussed because together they give a relatively complete view of cell allocation to clonally distinct trophectoderm and ICM populations.

Because the ICM gives rise to the fetus and its membranes, whereas trophectoderm contributes strictly to placental structures (see Chapter 4), the location of a progenitor cell in the preimplantation embryo may ultimately determine whether its descendants will survive only until birth or will contribute to the somatic and germ cell lineages of the new individual. It is thus important to determine how cells of the preimplantation embryo acquire their positions.

A. Is allocation random or orderly?

The limited extent of cell mixing during preimplantation development eliminates any mechanism requiring extensive cell sorting as a basis for allocation to trophectoderm and ICM lineages. Blastocysts that developed from 8-cell chimeras containing a pair of [³H]thymidine-labeled cells, or containing an entire labeled embryo, had labeled descendants grouped together as a coherent clone, with relatively little mixing (Garner and McLaren 1974; Kelly 1979). Thus, it is unlikely that either the trophectoderm or the ICM arises from randomly distributed precursor cells that subsequently migrate to appropriate positions.

Further evidence for an orderly, rather than random, origin of blastocyst tissues comes from an analysis of the relationship between early cell divisions and allocation to trophectoderm and ICM lineages. Descendants of the earlier-dividing 2-cell blastomere (the AB cell) were labeled with [³H]thymidine and reaggregated at the 8-cell stage with host blastomeres (⅔ labeled). Similar constructions were made from descendants of the later-dividing 2-cell blastomere (the CD cell). A comparison of numbers of labeled cells in the trophectoderm and ICM of sectioned blastocysts showed a significantly greater contribution of AB descendants to the ICM

(Kelly et al. 1978). This result has been confirmed by using fluorescein isothiocyanate to label the CD blastomere at the 2-cell state and then aggregating this blastomere with an unlabeled AB cell. In this case, outer cells were identified by labeling with rhodamine-conjugated antiserum, and then embryos were dissociated as late morulae for scoring. There was a significant excess of unlabeled inner cells, as compared with controls that had both labeled and unlabeled CD blastomeres, thus confirming the "advantage" of the (unlabeled) AB cell in generating ICM descendants (Surani and Barton 1984). Yet, in both studies, the differences in numbers of AB versus CD descendants among inner cells were small (~2–3 cells). A more dramatic result was obtained by aggregating entire labeled 8-cell embryos with unlabeled 4-cell embryos, and vice versa; in this approach, the ICM at the blastocyst stage consisted predominantly of 8-cell descendants (Spindle 1982). This phenomenon has recently been exploited to make sheep-goat chimeras by aggregating embryos of different cleavage stages (Meinecke-Tillmann and Meinecke 1984).

In an attempt to define how earlier-dividing blastomeres contribute more cells to the ICM, Graham and associates studied the cell contact patterns between blastomeres derived from AB and CD cells. The earlier-dividing cell and its descendants had more contacts with other cells and were closer to the visual center of the embryo. When these same earlier-dividing descendants were injected with oil droplets at the 8-cell stage, most were found to contribute to the ICM (Graham and Deussen 1978; Graham and Lehtonen 1979). These investigators proposed that microvilli were primarily responsible for the preferential enclosure of AB descendants, but this has not been substantiated experimentally. An intriguing observation suggests that early division of the AB blastomere may itself be related to prior events of fertilization. Sperm tails that were labeled with fluorescein isothiocyanate during fertilization, marking the site of the sperm entry, were associated with the early dividing blastomere at the 3-cell stage (J. Bennett and R.A. Pedersen, unpublished observations). The lack of stable landmarks in the mammalian egg makes it difficult to establish any relationship between the egg's polarity and the polarity of the blastocyst (Gulyas 1975; Graham and Deussen 1978). This area nonetheless deserves further study because of the relationship between egg polarity or sperm entry site and the embryonic axis in other developing systems (Davidson 1976).

Despite the lack of axial landmarks in the mouse egg, lineage analysis clearly shows a preferential contribution of cortical regions of cleavage-stage blastomeres to trophectoderm. When oil droplets were injected into outer portions of 2-cell or 8-cell embryos, they appeared in trophectoderm cells (Wilson, Bolton, and Cuttler 1972). This result implies that this portion of the cell becomes predisposed toward a trophectodermal fate early in cleavage and that early cells pass this property to their descendants. The role of heritable phenotype in trophectoderm differentiation has been studied extensively by Johnson and associates, who observed that 8-cell embryos develop an asymmetric distribution of microvilli on the free surfaces of blastomeres at the time of compaction (see Chapter 2). These and accompanying cytoskeletal and cortical features form opposite to the point of contact with another cell and become stable a few hours after formation. Moreover, the polarized phenotype, together with a consistent size differential between outer (larger, polar) and inner (smaller, apolar) blastomeres, can be used to distinguish progenitors or their descendants after dissociation at the 8- and 16-cell stages (Handyside 1980; Ziomek and Johnson 1980; Johnson and Ziomek 1981a, 1981b). When dissociated 8-cell polarized blastomeres divide, they form predominantly polar-apolar pairs (80%), the remainder being polar-polar pairs (Johnson and Ziomek 1981b). In an approach designed to assess the fate of these cells in the fifth cleavage, single polar or apolar blastomeres were labeled with fluorescein isothiocyanate, then aggregated with 15 unlabeled $\frac{1}{16}$ blastomeres in various combinations (pure and mixed types) or with a single unlabeled $\frac{1}{16}$ blastomere; they were scored for the positions of fluorescent-labeled descendants after development to the 32-cell stage. Labeled polar cells invariably produced at least one trophectoderm descendant; labeled apolar cells contributed predominantly to ICM when placed inside of mixed aggregates, but sometimes (15–20%) contributed also to trophectoderm when they were on the outside of mixed aggregates or were in pure apolar aggregates (Ziomek and Johnson 1982; Johnson and Ziomek 1983). Similar results were obtained by aggregating labeled polar or apolar cells with partially decompacted morulae (Randle 1982; Surani and Handyside 1983). These results imply that the polarized apical surface of an outer cell is a phenotypic constraint that causes a cell possessing it to produce at least one outer descendant in subsequent divisions. Apolar cells, lacking this constraint, are more promiscuous. In view of the strong predisposition of polarized (outer) cells to contribute descendants to the trophectoderm, we must conclude that origin of the trophectoderm is both nonrandom and strongly influenced by lineage.

This view of trophectoderm origin was first proposed as the "polarization" hypothesis (Johnson, Pratt, and Handyside 1981). How does it compare with the inside-outside hypothesis? Both focus on the importance of position as the decisive information for differentiation, and in this sense both are epigenetic hypotheses. According to the inside-outside hypothesis, the positional information is acquired at the morula stage, after the ICM progenitors have been internalized. In the polarization hypothesis, this information acts on the 8-cell embryo, and on uncommitted cells at subsequent stages. The inside-outside hypothesis presumes that positional information takes the form of microenvironmental signals (although their nature can be interpreted broadly), whereas the polarization hypothesis requires the polarizing response of outer cells to asymmetric cell contacts, together with their inherent tendency to flatten (Johnson 1985) (see Chapter 2). We can further examine these hypotheses on the basis of their predictions about the origin of the ICM. The inside-outside hypothesis predicts that any cell that is totally enclosed at the morula stage will, by virtue of its microenvironment, differentiate into ICM. The polarization hypothesis predicts that any inner cells that remain apolar will form ICM, but it does not preclude the possibility that they or their descendants could emerge and differentiate into trophectoderm, provided that they retained the ability to polarize and flatten (Johnson and Ziomek 1983).

B. Origin of the ICM

The origin of the ICM has been studied descriptively as well as experimentally. The first cells are allocated to the ICM lineage in the division from 8 to 16 cells. The absence of inner cells at earlier stages is readily observable before compaction and can be demonstrated by sectioning compacted embryos (Barlow et al. 1972). However, the number of inside cells at the 16-cell stage of mouse embryos is controversial. Estimates range from a low of two (based on serial paraffin sections) to four to six (based on polar/apolar phenotype) (Barlow et al. 1972; Handyside 1981; Johnson and Ziomek 1981b). A careful analysis of 16-cell embryos after glutaraldehyde fixation and plastic embedding would provide a better estimate of the range of cell numbers in normal embryos at this stage.

Estimates of cell numbers in early blastocysts also vary, owing to the methods used and perhaps other variables. A lower estimate of approximately 8 inner cells was obtained by paraffin sectioning in one study (Barlow et al. 1972), with a higher estimate of 13 cells in another (Copp 1978). Isolating inner cells by immunosurgery or disaggregation of embryos after labeling outer cells gave estimates of 14 to 16 cells (Handyside 1978; Surani and Barton 1984). The lower estimates after paraffin sectioning at this and the 16-cell stage may be attributable to shrinkage artifact, which exposes internal cells. A recent study of two strains using four methods to determine cell numbers in nascent blastocysts may provide the best estimates (Chisholm et al. 1985). Early cavitating blastocysts (34 ± 7 cells) had a range of 7 to 18 inner cells, with a mean of 10 to 13 cells.

Comparing these descriptions of the inner-cell lineage at the fifth and sixth cell cycles, two scenarios for the origin of inner cells are apparent. Low estimates of inner-cell numbers at the 16-cell stage imply that cells are added to the inner population at the fifth cleavage division (Barlow et al. 1972; Graham and Deussen 1978; Rossant and Vijh 1980). Given high estimates of inner-cell numbers at the 16-cell stage, no additional allocation to the inner-cell lineage would be expected (Johnson and Ziomek 1981b). Following the 32-cell stage, the ratio of inner cells to total blastocyst cell numbers gradually declines (Barlow et al. 1972); Handyside 1978; Copp 1979), which suggests, but does not prove, that no further recruitment of inner cells takes place after the blastocoel forms.

In an experimental approach to the origin of the ICM, Balakier and Pedersen (1982) microinjected horseradish peroxidase as a lineage tracer, attempting to determine when trophectoderm and ICM become clonally distinct. The results of this approach are notable because they are derived from the intact embryo and could differ in principle with the results from isolated cells or reaggregates. Almost all labeled 2-cell blastomeres contributed descendants to both trophectoderm and ICM lineages (Figure 1.4). Labeled blastomeres of intact 8-cell embryos always contributed to trophectoderm. They often contributed also to the ICM, but not at the high frequency to be expected if inner-cell allocation were occurring solely at the fourth cleavage division. Moreover, labeled outer blastomeres of intact 16-cell embryos frequently had descendants not only in trophectoderm but also in the ICM. This suggested that a second allocation event was occurring at the fifth cleavage division (Balakier and Pedersen 1982).

To resolve the issue of when the ICM originates, we subsequently studied cell lineage in the fourth, fifth, and sixth cell cycles using microinjected lineage tracers (Pedersen, Wu, Balakier, in press). In the first approach (prospective analysis) we injected single outer 8-cell, 16-cell, and 32-cell blastomeres with a mixture of horseradish peroxidase and rhodamine-conjugated dextran produced as described by Gimlich and Braun (1985). The fluorescent dextran served as a vital indicator of the marked progenitor cell's position in the embryo. In the second approach (retrospective analysis) we studied the locations of sister-cell pairs labeled after microinjecting single outer blastomeres with horseradish peroxidase at the 16-cell, 32-cell, or 4-cell stage. [As originally observed by Lo and Gilula (1979), horseradish peroxidase and other lineage markers readily diffuse between recently divided cells through the cytoplasmic bridge that connects them, thus passively labeling the sister of the injected cell (Pedersen et al. in press).] Both approaches showed that cells are allocated to inner positions during the fourth cleavage (38−50% of divisions yield inner cells) and fifth cleavage (20−33% of outer-cell divisions yield inner cells). These results differ from those with disaggregated $\frac{1}{8}$ blastomeres, which yield outer/inner pairs more frequently than in the intact embryo, and from those with disaggregated $\frac{1}{16}$ blastomeres, which yield outer/inner pairs less frequently than in the intact embryo (Johnson and Ziomek 1981b, 1983).

In the population of embryos that we studied, statistical estimates of the

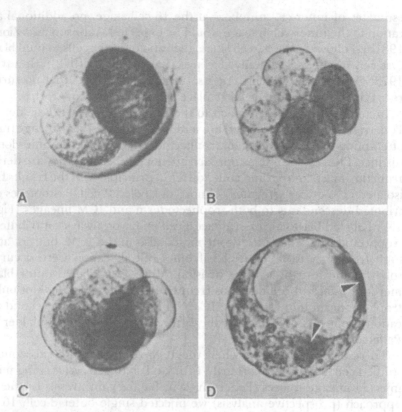

Figure 1.4. Cell lineage analysis of preimplantation mouse embryos by microinjection of horseradish peroxidase into single blastomeres. A: Two-cell embryo stained after injection into one blastomere. B: Two labeled descendants stained after culture. C: Early compacting embryo with several stained blastomeres. D: Blastocyst with stained trophectoderm and inner cells (arrows). [Reproduced from Balakier and Pedersen (1982), with permission of the publisher.]

numbers of allocated inner cells at the 16-cell stage (3−4 cells) and 32-cell stage (2−4 additional cells, plus the descendants of those previously allocated) were in the range of previous estimates of inner cell numbers at these stages (Figure 1.5). Our results confirm the findings from chimeric analysis that outer cells at the 8- and 16-cell stages invariably have at least one outer descendant. However, in view of the variability in inner-cell number between individual embryos at the same stage, inner-cell allocation should not be considered a quantitatively rigid event. Indeed, studies of chimeric aggregates containing fluorescein-marked cells suggest that embryos regulate the number of inner cells allocated in the fifth cleavage, depending on the number of inner cells already present (John-

MODEL

8-Cell Embryo   **16-Cell Embryo**   **32-Cell Embryo**
                                                             (early blastocyst)

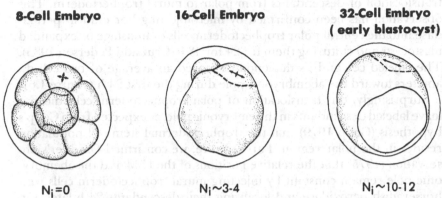

$N_i = 0$ $N_i \sim 3\text{-}4$ $N_i \sim 10\text{-}12$

Figure 1.5. Model for allocation of cells to the trophectoderm and inner-cell populations of the mouse embryo. N_i is the estimated number of inner cells at each stage; arrows represent the relative probabilities of outer/outer and outer/inner pairs forming at the next division.

son and Ziomek 1983; Surani and Barton 1984; Chisholm et al. 1985). Allocations of inner cells at the fourth and fifth divisions might therefore be regarded as complementary events, yielding a number of cells sufficient to establish the ICM, but varying from embryo to embryo in the precise number allocated at each division. The number of inner cells also depends on the total protoplasmic volume of the embryo (Rand 1985) and thus will be influenced by cell death or other blastomere loss occurring during cleavage.

Does the process of inner-cell allocation continue beyond the fifth cleavage division? We carried out a lineage analysis of early ($\sim$32-cell) and expanded ($\sim$64-cell) blastocysts and found that their outer cells had only trophectodermal descendants (Cruz and Pedersen 1985; Pedersen et al. in press). Cells of the trophectoderm lineage may thus become committed at the 32-cell stage, according to the definition of Johnson and Ziomek (1983), because they no longer appear capable of the differentiative divisions that produce inner cells. This conclusion about their potency should be viewed with caution, however, because it is based on behavior in the intact embryo, rather than a direct test in which cell position is experimentally altered.

C. Fate of the polar trophectoderm

A major aspect of blastocyst cell lineage is the fate of the polar trophectoderm during blastocyst growth. After labeling polar trophectoderm cells with melanin granules, Copp (1979) studied the locations of marked descendants in vitro and in utero. There was no evidence for labeled trophectoderm descendants in the ICM, but there was a consistent translocation of descendants from polar to mural trophectoderm. This observation has been confirmed by microinjecting horseradish peroxidase into the central polar trophectoderm cells of midstage or expanded blastocysts and culturing them for 24 for 48 h (Cruz and Pedersen 1985). The labeled cell and its descendants moved an average of two cell diameters toward the abembryonic pole during the first 24 h (Figure 1.6).

Surprisingly, this translocation of polar trophectoderm cells did not leave labeled descendants in the embryonic pole, as expected from Copp's hypothesis (Copp 1979) that the trophectodermal stem cell population resides in the polar region. Furthermore, we confirmed Gardner's observation (1975) that the relative positions of the ICM and the abembryonic pole remain constant by injecting mural trophectoderm cells with horseradish peroxidase and localizing their descendants 24 h later. On this basis, we postulated that the ICM contributes cells to the polar trophectoderm during blastocyst growth (Cruz and Pedersen 1985).

Evidence confirming an ICM contribution to polar trophectoderm has been obtained by microinjecting ICM cells in early/mid-blastocysts (32–64-cell stage) with a mixture of horseradish peroxidase and rhodamine-conjugated dextran, then culturing them for 24 or 48 h (G. Winkel and R.A. Pedersen, unpublished observations). In 35 percent of embryos injected in a single inner cell by a mural injection route (traversing mural trophectoderm), labeled descendants were found in the polar trophectoderm after culture; no polar trophectoderm cells were directly labeled in the controls injected by the mural route. In the same injection series, single labeled inner cells also had descendants in the ICM (58% of embryos) and in endoderm (46% of embryos). We thus concluded that the ICM is a stem cell population for the entire blastocyst at early stages of blastocyst growth.

This view differs from that derived by studying ICM fate in blastocysts reconstituted from ICM and trophectodermal vesicles of expanded (3.5-d.g.) blastocysts (Gardner 1978). In these reconstituted blastocysts, with few exceptions, the ICM descendants are restricted to the fetus and its membranes, plus the mesodermal component of the placenta (Gardner

Figure 1.6.   Fate of polar trophectoderm during blastocyst growth. A: Midstage blastocyst prepared for microinjection into central polar trophectoderm cell. B: Fluorescence of rhodamine-conjugated dextran immediately after injection. C: Staining of injected cell with horseradish peroxidase immediately after injection. D: Staining of descendant cells with horseradish peroxidase after 48 h culture, showing predominantly mural localization of stained trophectoderm cells. [Reproduced from Cruz and Pedersen (1985), with permission of the publisher.]

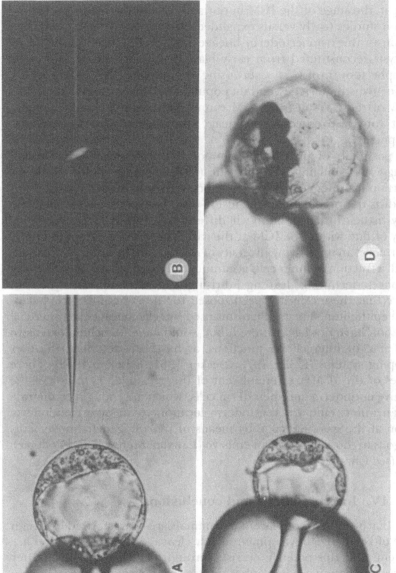

1978; Papaioannou 1982). The main reason for this difference, however, may lie in the stage of the ICM in our studies and the blastocyst reconstruction studies (early versus expanded blastocysts). An inner-cell contribution to the trophectoderm lineage has recently been observed in blastocysts reconstituted from early blastocyst inner cells of *Mus caroli* and trophectoderm of *M. musculus*, using recombinant DNA probes or GPI isozymes to distinguish descendant populations (Rossant and Croy 1985). In view of these results, the ICM cannot be considered to be clonally distinct from the trophectoderm lineage until sometime later in blastocyst development.

The possibility of an ICM contribution to the trophectoderm lineage was suggested initially by Handyside (1978) because of the declining proportion of inner cells during blastocyst growth. Such a contribution contradicts the prediction of the inside-outside hypothesis that all inner cells by virtue of their position will differentiate into ICM. Such heterogeneity of fate within the ICM at the early blastocyst stage could imply, furthermore, an underlying diversity of cells within the inner population. Indeed, a fraction of inner cells accumulate glycogen in a manner similar to trophectoderm cells, leading Edirisinghe, Wales, and Pike (1984) to propose that these cells are predisposed to enter the polar trophectodermal epithelium. The mechanisms used by cells to leave the inner-cell population have not been studied, but would have to include extensive remodeling of intercellular junctions, as has been described in other developing systems (Keller 1978; Decker 1981; Fristrom 1982). These features of the ICM are reminiscent of the presumed behavior of the definitive endoderm and mesoderm cells, which are thought to migrate from an inner, embryonic ectoderm location to an outer, endoderm location at the gastrula stage by means of morphogenetic movements analogous to those in the avian embryo (Lawson, Meneses, and Pedersen 1986) (see Chapter 5).

IV. Interpretations and conclusions

Early embryos of placental mammals are characterized by a high degree of regulative development. Early cleavage-stage blastomeres are totipotent, and this potency persists through late cleavage. Even after blastocyst formation, when trophectoderm cells behave as if committed to their fate, inner cells remain totipotent for an additional cell cycle, or perhaps longer. What is the relationship between potency and lineage in the early mouse embryo? How does an orderly and consistent allocation to ICM and trophectoderm occur amidst extensive lability? Finally, what are the implications of these results for further research on decision-making processes in early development?

The well-documented potency of cleavage-stage mouse blastomeres can now be recognized as having a necessary role in early cell allocation events, because progenitor cells frequently have descendants in both ICM and trophectoderm populations, even at late cleavage stages. Without totipotency, this would be impossible. The same functional relationship between totipotency and lineage in the ICM is implicit in the observation that these cells contribute to polar trophectoderm during growth of the early blastocyst. Thus, the totipotency of the early mouse embryo is utilized in its normal development and cell allocation processes.

We can deduce certain rules about cell allocation in early mouse development from the results of cell lineage analysis. Allocation of inner cells first occurs when about half of the 8-cell blastomeres undergo a delaminatory fourth cleavage division (Soltynska 1982), and this allocation is completed when one-fourth to one-third of outer 16-cell blastomeres undergo a delaminatory fifth cleavage division. As a consequence of this pattern of divisions, 8-cell blastomeres and outer 16-cell blastomeres are each destined to have at least one outer descendant; the other descendant is either an outer or an inner descendant. After division to the 32-cell stage, outer cells have only outer descendants. Outer blastomeres complete their differentiation into trophectoderm and begin fluid accumulation shortly after the division to the 32-cell stage. These rules seem to ensure that the mouse embryo will generate populations of both trophectoderm and ICM precursors. Similar rules presumably apply to early embryos of other species of placental mammals. In these cases, however, their differing numbers of cells at the blastocyst stage will determine the number of cleavage divisions over which allocation to the ICM occurs.

A quite different set of rules must apply to the marsupial embryo, which forms a unilaminar blastocyst devoid of an ICM (Selwood and Young 1983). The unilaminar marsupial blastocyst has compelling implications for understanding the nature of epigenetic information in early steps of differentiation, because all cells are initially equivalent in their positions (Denker 1976). In these embryos, blastocyst polarity is first evident when endoderm cells are budded into the cavity by outer cells at the prospective embryonic pole; outer cells at this pole differentiate directly into embryonic ectoderm (Wimsatt 1975). Thus, outer cells of the marsupial embryo are not destined to differentiate into trophectoderm, as they are in placental mammals, but remain capable of totipotent stem cell development. The time of their commitment to either embryonic or extraembryonic lineage is unknown. By comparison, the trophectodermal lineage of the mouse embryo contributes no further descendants to the embryo proper after the early blastocyst stage. Because of recent evidence from two laboratories cited earlier for an ICM contribution to trophectoderm at least during the transition from early (32-cell)

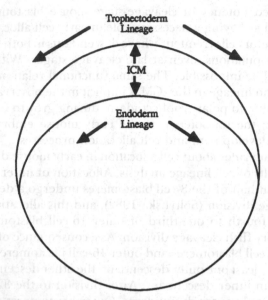

Figure 1.7.   Diagram of mouse blastocyst lineage and allocation, showing ICM as stem cell population for entire embryo during blastocyst growth.

to expanded (64-cell) stages, these two lineages cannot be considered clonally distinct at the blastocyst stage. Moreover, in view of the "one-way" direction of descent from ICM to trophectoderm, but not in the reverse direction, the ICM can be viewed as a stem cell population for the entire embryo during blastocyst growth (Figure 1.7). Unlike the marsupial embryo with its outer layer of presumably pluripotent stem cells, the mouse embryo thus appears to have a sequestered inner stem cell population, with precocious differentiation of outer cells into trophectoderm. Given this propensity of outer cells to differentiate into trophectoderm, embryos of placental mammals could not generate the fetus and its membranes without a mechanism for establishing a sequestered inner stem cell population during cleavage.

How are these inner cells allocated? The polarity of cell divisions in the fourth and fifth cleavages is undoubtedly crucial, with delaminatory divisions giving inner cells. Of course, this does not preclude important roles for other intercellular forces, particularly differential cell adhesion and contractility (Harris 1976; Surani and Handyside 1983; Sobel 1984). All of these mechanisms focus our attention on the cytoskeleton, including the spindle and centrosomes, and its interactions with the cell surface. Compaction is clearly the central manifestation of all the cellular events leading to allocation of inner cells. Without the close relationship between cells established at compaction, allocation of inner cells does not occur (see Chapter 2). A circumstantial clue to the importance of compaction in inner-cell allocation is the absence of compaction, and of the ICM, in the marsupial embryo (Selwood and Young 1983). Thus, an understanding of the cell biology and evolution of compaction should be a priority for future efforts in mammalian development.

Several additional areas deserve particular attention if we are to achieve a complete understanding of the relationships among potency, lineage, and cell allocation in early mammalian development. We need to know more about the nature of the cytoplasmic and nuclear timekeeping mechanisms in the early embryo. What are the processes that regulate the cell cycle at the completion of meiosis and throughout cleavage and blastocyst formation? Are they unique, or are they shared with somatic cells, as proposed by Newport and Kirschner (1984)? We also need to know more about the molecular basis of potency. What does potency consist of at the level of the chromosome and the gene, and how are these conditions altered at the time of commitment? Finally, what is the physical information about a cell's position in the embryo that is transduced into a differentiated pattern of gene expression? On the basis of the evidence presented here, it would be naive to assume that a simple diffusible molecule conveys this epigenetic signal. It is much more likely that the cell interacts with its fluid environment and its neighbors at many levels of organization, and in doing so alters both itself and its environment.

The preimplantation mouse embryo remains one of the best-characterized differentiating systems available for experimental analysis. And despite its peculiarity (among vertebrate embryos) of precocious trophectoderm differentiation and inner stem cell sequestration, the blastocyst of placental mammals offers a superb model for early decision-making processes in embryogenesis.

Acknowledgments

The work performed in my laboratory was supported by the U. S. Department of Energy, contract no. DE-AC03-76-SF01012. This chapter was written while I was a guest of the Laboratory of Reproductive and Developmental Toxicology (LRDT), National Institute of Environmental Health Sciences, Research Triangle Park, North Carolina, where I was partially supported under the terms of the Interagency Personnel Act. I am grateful to the LRDT staff for their hospitality, and particularly to Susan DeBrunner-Bass and Ms. Vickie Englebright for their assistance with the manuscript. I thank Carmen Arbona for the computer graphic illustrations. I also thank Drs. Gerald Kidder and Sabina Sobel for their comments.

References

Aerts, R. J., Durston, A. J., and Moolenaar, W. H. (1985) Cytoplasmic pH and the regulation of the *Dictyostelium* cell cycle. *Cell* 43, 653−7.

Alexandre, H. (1979) The utilization of an inhibitor of spermidine and spermine synthesis as a tool for the study of determination of cavitation in the preimplantation mouse embryo. *J. Embryol. Exp. Morphol.* 53, 145−62.

−(1982) Effet de l'inhibition specifique de la replication de l'ADN par l'aphidicoline sur la differentiation primarie de l'oeuf de souris en preimplantation. *C. R. Acad. Sci. (Paris)* III 294, 1001−6.

Balakier, H., and Pedersen, R. A. (1982) Allocation of cells to inner cell mass and trophectoderm lineages in preimplantation mouse embryos. *Dev. Biol.* 90, 352−62.

Barlow, P., Owen, D. A. J., and Graham, C. (1972) DNA synthesis in the preimplantation mouse embryo. *J. Embryol. Exp. Morphol.* 27, 431−45.

Berridge, M. J. (1984) Inositol triphosphate and diacylglycerol as second messengers. *Biochem. J.* 220, 345−60.

Bodemer, C. W. (1971) The biology of the blastocyst in historical perspective. In *Biology of the Blastocyst*, ed. R. J. Blandau, pp. 1−26. University of Chicago Press.

Bolton, V. N., Oades, P. J., and Johnson, M. H. (1984) The relationship between cleavage, DNA replication and activation of transcription in the mouse 2-cell embryo. *J. Embryol. Exp. Morphol.* 79, 139−63.

Bowman, P., and McLaren, A. (1970) Cleavage rate of mouse embryos *in vivo* and *in vitro*. *J. Embryol. Exp. Morphol.* 24, 203−7.

Braude, P. R. (1979) Time-dependent effects of *a*-amanitin on blastocyst formation in the mouse. *J. Embryol. Exp. Morphol.* 52, 193−202.

Burgoyne, P. S., and Ducibella, T. (1977) Changes in the properties of the developing trophoblast of preimplantation mouse embryos as revealed by aggregation studies. *J. Embryol. Exp. Morphol.* 40, 143−57.

Busa, W. B., and Nuccitelli, R. (1984) Metabolic regulation via intracellular pH. *Am. J. Physiol.* 246, R409−38.

Chisholm, J. C., Johnson, M. H., Warren, P. D., Fleming, T. P., and Pickering, S. J. (1985) Developmental variability within and between mouse expanding blastocysts and their ICMs. *J. Embryol. Exp. Morphol.* 86, 311−36.

Copp, A. J. (1978) Interaction between inner cell mass and trophectoderm of the mouse blastocyst. I. A study of cellular proliferation. *J. Embryol. Exp. Morphol.* 48, 109−25.

−(1979) Interaction between inner cell mass and trophectoderm of the mouse blastocyst. II. Fate of the polar trophectoderm *J. Embryol. Exp. Morphol.* 51, 109−20.

Cruz, Y. P., and Pedersen, R. A. (1985) Cell fate in the polar trophectoderm of mouse blastocysts as studied by microinjection of cell lineage tracers. *Dev. Biol.* 112, 73−83.

Davidson, E. (1976) *Gene Activity in Early Development*, pp. 245−318. New York: Academic Press.

Dean, W. L., and Rossant, J. (1984) Effect of delaying DNA replication on blastocyst formation in the mouse. *Differentiation* 26, 134−7.

Decker, R. S. (1981) Disassembly of the zonula occludens during amphibian neurulation. *Dev. Biol.* 81, 12–22.

Denker, H. W. (1976) Formation of the blastocyst: determination of trophoblast and embryonic knot. *Curr. Top. Pathol.* 62, 59–78.

Dickson, A. D. (1966) The form of the mouse blastocyst. *J. Anat.* 100, 335–48.

Edirisinghe, W. R., Wales, R. G., and Pike, I. L. (1984) Studies of the distribution of glycogen between the inner cell mass and trophoblast cells of mouse embryos. *J. Reprod. Fertil.* 71, 533–8.

Edwards, R. G., and Gates, A. H. (1959) Timing of the stages of the maturation divisions, ovulation, fertilization and the first cleavage of eggs of adult mice treated with gonadotrophins. *J. Endocrinol.* 18, 292–304.

Epstein, C. J., Smith, S., Travis, B., and Tucker, G. (1978) Both X chromosomes function before visible X-chromosome inactivation in female mouse embryos. *Nature* 274, 500–3.

Faletto, D. L., Arrow, A. S., and Macara, I. G. (1985) An early decrease in phosphatidylinositol turnover occurs on induction of Friend cell differentiation and precedes the decrease in c-myc expression. *Cell* 43, 315–25.

Fleming, T. P., Warren, P. D., Chisholm, J. C., and Johnson, M. H. (1984) Trophectodermal processes regulate the expression of totipotency within the inner cell mass of the mouse expanding blastocyst. *J. Embryol. Exp. Morphol.* 84, 63–90.

Fristrom, D. K. (1982) Septate junctions in imaginal discs of *Drosphila*: a model for the redistribution of septa during cell rearrangement. *J. Cell Biol.* 94, 77–87.

Gaddum-Rosse, P., Blandau, R. J., Langley, L. B., and Sato, K. (1982) Sperm tail entry into the mouse egg in vitro. *Gamete Res.* 6, 215–23.

Gardner, R. L. (1975) Analysis of determination and differentiation in the early mammalian embryo using intra- and inter-specific chimaeras. In *The Developmental Biology of Reproduction* (33rd symposium, Society for Developmental Biology), ed. C. L. Markert, pp. 207–38. New York: Academic Press.

Gardner, R. L. (1978) The relationship between cell lineage and differentiation in the early mouse embryo. In *Genetic Mosaics and Cell Differentiation*, ed. W. J. Gehring, pp. 205–41. Berlin: Springer-Verlag.

Garner, W., and McLaren, A. (1974) Cell distribution in chimaeric mouse embryos before implantation. *J. Embryol. Exp. Morphol.* 32, 495–503.

Gimlich, R. L., and Braun, J. (1985) Improved fluorescent compounds for tracing cell lineage. *Dev. Biol.* 109, 509–14.

Goldbard, S. B., and Warner, C. M. (1982) Genes affect the timing of early mouse embryo development. *Biol. Reprod.* 27, 419–24.

Graham, C. F., and Deussen, Z. A. (1978) Features of cell lineage in preimplantation mouse development. *J. Embryol. Exp. Morphol.* 49, 277–94.

Gulyas, B. J. (1975) A reexamination of cleavage patterns in eutherian mammalian eggs: rotation of blastomere pairs during second cleavage in the rabbit. *J. Exp. Zool.* 193, 235–48.

Handyside, A. H. (1978) Time of commitment of inside cells isolated from preimplantation mouse embryos. *J. Embryol. Exp. Morphol.* 45, 37–53.

–(1980) Distribution of antibody- and lectin-binding sites on dissociated blastomeres from mouse morulae: evidence of polarization at compaction. *J. Embryol. Exp. Morphol.* 60, 99–116.

30   ROGER A. PEDERSEN

−(1981) Immunofluorescence techniques for determining the numbers of inner and outer blastomeres in mouse morulae. *J. Reprod. Immunol.* 2, 339−50.

Hara, K., Tydeman, P., and Kirschner, M. W. (1980) A cytoplasmic clock with the same period as the division cycle in *Xenopus* eggs. *Proc. Natl. Acad. Sci. U.S.A.* 77, 462−6.

Harlow, G. M., and Quinn, P. (1982) Development of pre-implantation mouse embryos in vivo and in vitro. *Aust. J. Biol. Sci.* 35, 187−93.

Harris, A. K. (1976) Is cell sorting caused by differences in the work of intercellular adhesion? A critique of the Steinberg hypothesis. *J. Theor. Biol.* 61, 267−85.

Heby, O. (1981) Role of polyamines in the control of cell proliferation and differentiation. *Differentiation* 19, 1−20.

Hillman, N., Sherman, M. I., and Graham, C. F. (1972) The effect of spatial arrangement on cell determination during mouse development. *J. Embryol. Exp. Morphol.* 28, 263−78.

Hogan, B., and Tilly, R. (1978) In vitro development of inner cell masses isolated immunosurgically from mouse blastocysts. II. Inner cell masses from 3.5- to 4.0-day p.c. blastocysts. *J. Embryol. Exp. Morphol.* 45, 107−21.

Howlett, S. K., and Bolton, V. N. (1985) Sequence and regulation of morphological and molecular events during the first cell cycle of mouse embryogenesis. *J. Embryol. Exp. Morphol.* 87, 175−206.

Johnson, L. V. (1985) Wheat germ agglutinin induces compaction- and cavitation-like events in two-cell mouse embryos. *Dev. Biol.* 113, 1−9.

Johnson, M. H. (in press). Manipulation of early mammalian development: What does it tell us about cell lineages? In *Manipulation of Mammalian Development*, ed. R. B. L. Gwatkin. New York: Plenum.

Johnson, M. H., Pratt, H. M. P., and Handyside, A. H. (1981) The generation and recognition of positional information in the preimplantation mouse embryo. In *Cellular and Molecular Aspects of Implantation,* eds. S. R. Glasser and D. W. Bullock, pp. 55−74. New York: Plenum.

Johnson, M. H., and Ziomek, C. A. (1981a) Induction of polarity in mouse 8-cell blastomeres: specificity, geometry, and stability. *J. Cell Biol.* 91, 303−8.

−(1981b) The foundation of two distinct cell lineages within the mouse morula. *Cell* 24, 71−80.

−(1983) Cell interactions influence the fate of mouse blastomeres undergoing the transition from the 16- to the 32-cell stage. *Dev. Biol.* 95, 211−18.

Keller, R. E. (1978) Time-lapse cinemicrographic analysis of superficial cell behavior during and prior to gastrulation in *Xenopus laevis*. *J. Morphol.* 157, 223−48.

Kelly, S. J. (1977) Studies of the developmental potential of 4- and 8-cell stage mouse blastomeres. *J. Exp. Zool.* 200, 365−76.

−(1979) Investigations into the degree of cell mixing that occurs between the 8-cell stage and the blastocyst stage of mouse development. *J. Exp. Zool.* 207, 121−30.

Kelly, S. J., Mulnard, J. G., and Graham, C. F. (1978) Cell division and cell allocation in early mouse development. *J. Embryol. Exp. Morphol.* 48, 37−51.

Kidder, G. M., and McLachlin, J. R. (1985) Timing of transcription and protein synthesis underlying morphogenesis in preimplantation mouse embryos. *Dev. Biol.* 112, 265–75.

Kirschner, M., Newport, J., and Gerhart, J. (1985) The timing of early developmental events in *Xenopus. Trends in Genetics* 1, 41–6.

Kishimoto, T., Yamazaki, K., Kato, Y., Koide, S. S., and Kanatani, H. (1984) Induction of starfish oocyte maturation by maturation-promoting factor of mouse and surf clam oocytes. *J. Exp. Zool.* 231, 293–5.

Krishna, M., and Generoso, W. M. (1977) Timing of sperm penetration, pronuclear formation, pronuclear DNA synthesis, and first cleavage in naturally ovulated mouse eggs. *J. Exp. Zool.* 202, 245–52.

Lawson, K. A., Meneses, J. J., and Pedersen, R. A. (1986) Cell fate and cell lineage in the endoderm of the presomite mouse embryo, studied with an intracellular tracer. *Dev. Biol.* 115, 325–39.

Lewis, W. H., and Wright, E. S. (1935) On the early development of the mouse egg. *Carnegie Institute Contrib. Embryol.* 25, 113–43.

Lo, C. W., and Gilula, N. B. (1979) Gap junctional communication in the preimplantation mouse embryo. *Cell* 18, 399–409.

Luthardt, F. W., and Donahue, R. P. (1973) Pronuclear DNA synthesis in mouse eggs: an autoradiographic study. *Exp. Cell Res.* 82, 143–51.

–(1975) DNA synthesis in developing two-cell mouse embryos. *Dev. Biol.* 44, 210–16.

Macara, I. G. (1985) Oncogenes, ions and phospholipids. *Am. J. Physiol. (Cell Physiol.* 17) 248:C3–C11.

McLaren, A., and Bowman, P. (1973) Genetic effects on the timing of early development in the mouse. *J. Embryol. Exp. Morphol.* 30, 491–8.

Masui, Y., and Markert, C. L. (1971) Cytoplasmic control of nuclear behavior during meiotic maturation of frog oocytes. *J. Exp. Zool.* 117, 129–46.

Meinecke-Tillmann, S., and Meinecke, B. (1984) Experimental chimaeras – removal of reproductive barrier between sheep and goat. *Nature* 307, 637–8.

Mita-Miyazawa, I., Ikegami, S., and Satoh, N. (1985) Histospecific acetylcholinesterase development in the presumptive muscle cells isolated from 16-cell-stage ascidian embryos with respect to the number of DNA replications. *J. Embryol. Exp. Morphol.* 87, 1–12.

Molls, M., Zamboglou, N., and Streffer, C. (1983) A comparison of the cell kinetics of pre-implantation mouse embryos from two different mouse strains. *Cell Tissue Kinet.* 16, 277–83.

Newport, J., and Kirschner, M. (1982a) A major developmental transition in early *Xenopus* embryos: I. Characterization and timing of cellular changes at the midblastula stage. *Cell* 30, 675–86.

–(1982b) A major developmental transition in early *Xenopus* embryos: II. Control of the onset of transcription. *Cell* 30, 687–96.

–(1984) Regulation of the cell cycle during early *Xenopus* development. *Cell* 37, 731–42.

Nichols, J., and Gardner, R. L. (1984) Heterogeneous differentiation of external cells in individual isolated early mouse inner cell masses in culture. *J. Embryol. Exp. Morphol.* 80, 225–40.

Papaioannou, V. E. (1982) Lineage analysis of inner cell mass and trophectoderm using microsurgically reconstituted mouse blastocysts. *J. Embryol. Exp. Morphol.* 68, 199–209.

Pedersen, R. A., and Spindle, A. I. (1980) Role of the blastocoel microenvironment in early mouse embryo differentiation. *Nature* 284, 550–2.

Pedersen, R. A., Wu, K., and Balakier, H. (in press) Origin of the inner cell mass in mouse embryos: cell lineage analysis by microinjection. *Dev. Biol.*

Rand, G. F. (1985) Cell allocation in half- and quadruple-sized preimplantation mouse embryos. *J. Exp. Zool.* 236, 67–70.

Randle, B. J. (1982) Cosegregation of monoclonal antibody reactivity and cell behavior in the mouse preimplantation embryo. *J. Embryol. Exp. Morphol.* 70, 261–78.

Rossant, J. (1975) Investigation of the determinative state of the mouse inner cell mass. II. The fate of isolated inner cell masses transferred to the oviduct. *J. Embryol. Exp. Morphol.* 33, 991–1001.

–(1984) Somatic cell lineages in mammalian chimeras. In *Chimeras in Developmental Biology*, eds. N. LeDouarin and A. McLaren, pp. 89–109. New York: Academic Press.

Rossant, J., and Croy, B. A. (1985) Genetic identification of tissue of origin of cellular populations within the mouse placenta. *J. Embryol. Exp. Morphol.* 86, 177–89.

Rossant, J., and Lis, W. T. (1979) Potential of isolated mouse inner cell masses to form trophectoderm derivatives *in vivo. Dev. Biol.* 70, 255–61.

Rossant, J., and Papaioannou, V. E. (1977) The biology of embryogenesis. In *Concepts in Mammalian Embryogenesis*, ed. M. I. Sherman, pp. 1–36. Cambridge: M.I.T. Press.

Rossant, J., and Vijh, K. M. (1980) Ability of outside cells from preimplantation mouse embryos to form inner cell mass derivatives. *Dev. Biol.* 76, 475–82.

Satoh, N., and Ikegami, S. (1981) On the "clock" mechanism determining the time of tissue-specific enzyme development during ascidian embryogenesis. II. Evidence for association of the clock with the cycle of DNA replication. *J. Embryol. Exp. Morphol.* 64, 61–71.

Sawicki, W., and Mystkowska, E. T. (1981) Phorbol ester-mediated modulation of cell proliferation and primary differentiation of mouse preimplantation embryos. *Exp. Cell Res.* 136, 455–8.

Selwood, L., and Young, G. J. (1983) Cleavage in vivo and in culture in the dasyurid marsupial *Antechinus stuartii* (Macleay). *J. Morphol.* 176, 43–60.

Shire, J. G. M., and Whitten, W. K. (1980a) Genetic variation in the timing of first cleavage in mice: effect of paternal genotype. *Biol. Reprod.* 23, 363–8.

–(1980b) Genetic variation in the timing of first cleavage in mice: effect of maternal genotype. *Biol. Reprod.* 23, 369–76.

Smith, R., and McLaren, A. (1977) Factors affecting the time of formation of the mouse blastocoel. *J. Embryol. Exp. Morphol.* 41, 79–92.

Sobel, J. S. (1984) Myosin rings and spreading in mouse blastomeres. *J. Cell Biol.* 99, 1145–50.

Solter, D., and Knowles, B. B. (1975) Immunosurgery of mouse blastocyst. *Proc. Natl. Acad. Sci. U.S.A.* 72, 5099–102.

Soltynska, M. S. (1982) The possible mechanism of cell positioning in mouse morulae: an ultrastructural study. *J. Embryol. Exp. Morphol.* 68, 137–47.

Sorensen, R. A., Cyert, M. S., and Pedersen, R. A. (1985) Active maturation-promoting factor is present in mature mouse oocytes. *J. Cell Biol.* 100, 1637–40.

Spindle, A. I. (1978) Trophoblast regeneration by inner cell masses isolated from cultured mouse embryos. *J. Exp. Zool.* 203, 483–9.

–(1982) Cell allocation in preimplantation mouse chimeras. *J. Exp. Zol.* 219, 361–7.

Spindle, A. I., Nagano, H., and Pedersen, R. A. (1985) Inhibition of DNA replication in preimplantation mouse embryos by aphidicolin. *J. Exp. Zol.* 235, 289–95.

Stern, M. S. (1972) Experimental studies on the organization of the preimplantation mouse embryo. II. Reaggregation of disaggregated embryos. *J. Embryol. Exp. Morphol.* 28, 255–61.

Streffer, C., Van Beuningen, D., Molls, M., Zamboglou, N., and Schultz, S. (1980) Kinetics of cell proliferation in the preimplanted mouse embryo in vivo and in vitro. *Cell Tissue Kinet.* 13, 135–43.

Surani, M. A. H., and Barton, S. C. (1984) Spatial distribution of blastomeres is dependent on cell division order and interactions in mouse morulae. *Dev, Biol.* 102, 335–43.

Surani, M. A. H., and Handyside, A. H. (1983) Reassortment of cells according to position in mouse morulae. *J. Exp. Zool.* 225, 505–12.

Tarkowski, A. K. (1959) Experiments on the development of isolated blastomeres of mouse eggs. *Nature* 184, 1286–7.

Tarkowski, A. K., and Wroblewska, J. (1967) Development of blastomeres of mouse eggs isolated at the 4- and 8-cell stage. *J. Embryol. Exp. Morphol.* 18, 155–80.

Waksmundzka, M., Krysiak, E., Karasiewicz, J., Czolowska, R., and Tarkowski, A. K. (1984) Autonomous cortical activity in mouse eggs controlled by a cytoplasmic clock. *J. Embryol. Exp. Morphol.* 79, 77–96.

Wiley, L. M. (1978) Apparent trophoblast giant cell production in vitro by core cells isolated from cultured mouse inner cell masses. *J. Exp. Zool.* 206, 13–16.

Wiley, L. M., Spindle, A. I., and Pedersen, R. A. (1978) Morphology of isolated mouse inner cell masses developing in vitro. *Dev. Biol.* 63, 1–10.

Wilson, I. B., Bolton, E., and Cuttler, R. H. (1972) Preimplantation differentiation in the mouse egg as revealed by microinjection of vital markers. *J. Embryol. Exp. Morphol.* 27, 467–79.

Wimsatt, W. A. (1975) Some comparative aspects of implantation. *Biol. Reprod.* 12, 1–40.

Ziomek, C. A., and Johnson, M. A. (1980) Cell surface interaction induces polarization of mouse 8-cell blastomeres at compaction. *Cell* 21, 935–42.

–(1982) The roles of phenotype and position in guiding the fate of 16-cell mouse blastomeres. *Dev. Biol.* 91, 440–7.

Ziomek, C. A., Johnson, M. H., and Handyside, A. H. (1982) The developmental potential of mouse 16-cell blastomeres. *J. Exp. Zool.* 221, 345–55.

Zwierzchowski, L., Czlonkowska, M., and Guszkiewicz, A. (1986) Effect of polyamine limitation on DNA synthesis and development of mouse preimplantation embryos *in vitro. J. Reprod. Fertil.* 76, 115–21.

2   Time and space in the mouse early embryo: a cell biological approach to cell diversification

MARTIN H. JOHNSON AND BERNARD MARO

CONTENTS

I. Introduction

The study of developmental biology seeks to explain how a diversity of cell types arises from a single fertilized egg and does so in the appropriate sequence yielding appropriate spatial relationships. Traditionally, developmental studies have concentrated on two aspects of this problem. A molecular approach has sought to understand how similar complements of genes in all the diverse cell types of the body are regulated and expressed differentially. Active genes, or their flanking sequences, have been analyzed for evidence of physical rearrangement or chemical change, and conclusions have then been drawn concerning the possible nature of the developmental decision-making process. In contrast, the experimental embryological approach to development has followed the subsequent fate of progeny by marking cells in the early

35

embryo, thus building up a descriptive picture of the paths by which cell diversification occurs. This latter approach, when used in conjunction with spatial relocation of cells, has also yielded information on the plasticity of development. As a result, hypotheses have been formulated in which the relative importance for development of a cell's recent history versus its current positional relationships can be evaluated. There is, however, a considerable gulf between molecular analysis of gene expression and description and categorization of developmental pathways. Between lies the cell biology of development: how cells move, measure the passage of time, and recognize relative position. This chapter represents an attempt to interpret the developmental biology of the early mammalian embryo in terms of cell biological mechanisms and thus to focus the questions of molecular control more precisely.

It has been generally acknowledged for nearly 20 years that cells within the early mouse embryo are responsive to their relative positions on the inside or outside of a cellular aggregate. They then tend to differentiate along different lineages according to their positions, but they retain some plasticity for at least two or three cell cycles such that a change in relative position can still be accommodated by a change in developmental fate (see Chapter 1) (Johnson 1985). By the 64- to 128-cell stage, a blastocyst with two committed cell subpopulations has formed. An outer layer of epithelial trophectoderm cells, derived largely from outer cells of earlier stages, surrounds an inner cluster of cells, the inner cell mass (ICM), located eccentrically within the blastocoelic cavity and derived largely from the inner cells of earlier stages. How do cells recognize and respond to position in order to initiate lineage divergence, and what form does their plasticity take?

It has long been suspected that events occurring at the 8-cell stage of development may constitute the earliest phases of positional recognition. At this stage, the embryo undergoes a major physical reorganization called compaction (Figure 2.1), in which cells flatten upon one another to maximize intercellular contact, establish gap-junctional communication for the first time, and reorder the components of each cell from nonpolar into a polarized array. The problem has been to determine which, if any, of these features of compaction are critical for recognition of position and to understand how cell diversification results. Recent work has revealed that the emergence of the trophectodermal and ICM cell lineages and the expression of plasticity within each are indeed influenced by the events of compaction and can be described in terms of three simple cell biological mechanisms. First, individual cells at the 8-, 16-, and 32-cell stages of development seem to be "programmed" to change their morphological and functional phenotype from radially symmetric to polarized (Figure 2.2A). Second, the pattern of contacts that a cell makes influences whether or not this change occurs, with polarization being suppressed

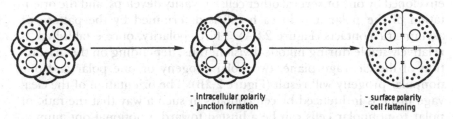

COMPACTION

- intracellular polarity
- junction formation

- surface polarity
- cell flattening

Figure 2.1. Schematic view of the changes undergone by 8-cell mouse embryos at compaction. Individual blastomeres flatten upon each other to maximize cell contact, establish gap-junctional coupling, and develop a polar phenotype both cytocortically and cytoplasmically.

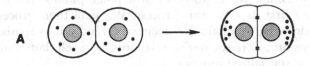

Cell Polarization

A

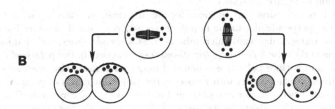

Cell Division

B

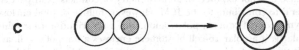

Cell Adhesion

C

Figure 2.2. Diagrammatic summary of the three mechanisms principally involved in the generation of cell diversity in the mouse blastocyst (Johnson 1985b). A: Cell polarization first becomes evident at the 8-cell stage, but nonpolar 16- and 32-cell blastomeres seem to obey similar rules (Ziomek and Johnson 1981; Fleming et al. 1984). Cells that are exposed to an asymmetry

(continued)

when a cell is surrounded totally by other cells (Figure 2.2C). However, when contact with other cells is incomplete (that is, a cell is not totally enveloped by one or several other cells), polarity develops, and the orientation of the polar axis in each cell is determined by the pattern of asymmetric contacts (Figure 2.2A). Third, polarity, once established, is relatively stable during mitotic division; thus, depending on the orientation of the cleavage plane, two polar progeny or one polar and one nonpolar progeny will result (Figure 2.2B). The orientation of the cleavage plane is influenced by cell contacts in such a way that the ratio of polar to nonpolar cells can be adjusted toward a notional optimum at successive division cycles. Results from recent lineage studies demonstrate quite clearly that such a regulative adjustment actually operates in situ (T. P. Fleming, personal communication). The detailed evidence supporting these conclusions (summarized in Figure 2.2) has recently been presented (Johnson in press).

Central to this description of the cell biology of mouse early development is the acquisition by blastomeres of a polar phenotype. In this chapter we shall first describe this process. We shall then proceed to analyze the underlying regulation of polarization in both its spatial and temporal dimensions, in particular its relationship to cell flattening and the formation of specialized junctions.

Caption to Figure 2.2 (*cont.*)
of cell contact, form gap junctions for the first time, and develop an axis of polarity perpendicular to the region of contact, as reflected in the redistribution of intracellular actin, microtubules, MTOCs, endosomes, and clathrin. B: At the end of the 8-cell stage (or 16-cell stage, in part C), the polarized cell divides, and surface polarity is retained throughout mitosis. The orientation of the division plane varies with respect to the longitudinal axis of polarity, and thus either two polar or one polar and one nonpolar cell can result. Cytoplasmic polarity, which, unlike surface polarity, is lost during mitosis, is restored in only those progeny cells that have inherited all or part of the surface pole, and this cytoplasmic pole locates beneath the surface pole. C: Polar and nonpolar cells differ in their adhesive properties. Polar cells tend to enclose nonpolar cells. If total enclosure is achieved, whether as illustrated in a pair of cells or within the intact embryo, then the nonpolar cell remains nonpolar, because it is not exposed to an asymmetry of contacts as in part A. It will then tend to contribute to the ICM. However, failure of total enclosure leads to asymmetry of contact, polarization, and a contribution to trophectoderm. Similarly, if a polar 16-cell blastomere encloses a nonpolar cell, as illustrated, it tends to divide so as to generate two polar 32-cells and thereby to contribute to trophectoderm only. However, in the absence of a nonpolar cell, it remains rounded up and can divide across its axis of polarity (as illustrated in the right-hand panel of B) to generate polar and nonpolar progeny and thus contributes to both trophectoderm and ICM (Ziomek and Johnson 1982; Johnson and Ziomek 1983).

II. Polarity

Most types of stably differentiated cells manifest regular asymmetry in their organization; this is most obvious in epithelial cells. The organization of polarity in at least some mature epithelial cells appears to be hierarchical, with some features being cell-autonomous and persisting after isolation of single cells, whereas other features are contingent on the continuing integration of cells within an epithelium and its associated matrix (Ziomek, Schulman, and Edidin 1980). The polarized nature of mature epithelial cells relates to their role in defining the boundaries and composition of tissue spaces by regulating the passage of materials between these spaces. Embryonic epithelial sheets also serve these functions, but in addition function as proliferative sources. Within the expanding blastocyst, the trophectoderm conforms to this prototypic embryonic epithelium. The trophectodermal sheet defines the boundary between the maternal and embryonic/fetal compartments. Its constituent cells are polarized both structurally and functionally. Zonular tight junctions delimit the apical, microvillous domain of the plasma membrane from the relatively smooth basolateral domain and thereby constitute a barrier to free paracellular diffusional exchange (Calarco and Brown 1969; Calarco and Epstein 1973; Ducibella et al. 1975; Magnuson, Demsey, and Stackpole 1977; Wiley and Eglitis 1981; Johnson and Ziomek 1982). Desmosomes and cytokeratin intermediate filaments, typical of epithelial cells, are present only in the cells of the trophectoderm (Brulet et al. 1980; Jackson et al. 1980; Lehtonen et al. 1983; Oshima et al. 1983). In the cytoplasm, endosomes and coated vesicles are found near the apex of the cell, whereas secondary lysosomes, mitochondria and cytoplasmic droplets tend to localize basally (Calarco and Brown 1969; Wiley and Eglitis 1981; Fleming et al. 1984; Fleming and Pickering 1985; Maro et al. in press, b). Not surprisingly, this physical organization is associated with vectorial transport of materials (Borland, Biggers, and Lechene 1977; Dizio and Tasca, 1977; Vorbrodt et al. 1977; Kaye et al. 1982; Fleming and Goodall in press) (see Chapter 9). The trophectoderm, in addition to its role as barrier and selective transporter, serves as a proliferative source of cells in response to signals from the underlying ICM (Gardner, Papaioannou, and Barton 1973; Copp, 1979).

Trophectodermal cells do not achieve this highly organized polar phenotype in a single developmental transition. Rather, it is arrived at progressively in three phases: a nonpolar phase, a phase during which individual cell polarity is established, and a true epithelial phase.

A. Nonpolar phase

The earliest time during development at which asymmetry in the organization of an individual embryonic cell can be observed is at fertilization, when two domains can be detected, each characterized by distinct cytocortical and cytoskeletal elements associated with and regulated by the chromosomes of the zygote (see Maro 1985 for a review). This asymmetry is evidently associated with polar-body formation (Maro et al. 1984; Maro et al. 1986; Maro, Howlett, and Johnson in press, a), and no evidence is available to link it to the ensuing cell diversification.

The earliest cell asymmetries, organized about a radial axis through the embryo and thus potentially of importance in the generation of differences between the inside and the outside of the embryo, are detected at the 2-cell stage and involve both the cell surface and the cytoskeleton.

Myosin, actin, clathrin, and an antigen associated with the endoplasmic reticulum show heterogeneous intracellular organization at the 2-, 4-, and early 8-cell stages, being excluded from the subcortical zone of cytoplasm adjacent to areas of membrane apposition (Sobel 1983; Johnson and Maro 1984; Maro et al. in press, b). This zonal clearing represents a transitory or reversible process, because separation of cells, and their examination either in isolation or after reaggregation in a new orientation, removes or relocates, respectively, the zone free of cytoplasmic elements.

Use of the detergent probes filipin and tomatin, in conjunction with freeze-fracture, has revealed that the organization of intramembranous cholesterol varies between central areas of membrane apposition and the more peripheral, nonapposed areas (Pratt 1985). This heterogeneity persists throughout subsequent cleavage to the blastocyst stage. It is not known if the membrane heterogeneity is retained on isolation of blastomeres, nor if it is reoriented on changing the pattern of blastomere contacts.

B. Establishment of polarity

Early during the 8-cell stage, cells develop a stable axis of polarity that develops radially in situ and in isolated pairs of cells develops perpendicular to the area of cell-cell contact (Ziomek and Johnson 1980; Johnson and Ziomek 1981a). If contact patterns are changed at any time up to 5 h into the 8-cell stage, there is a corresponding change in the orientation of the axis of polarity. Thus, polarization, like the cytoplasmic clearing described in the previous section, is evidently not stable and is influenced by cell contact at this stage of development.

The developing polar axis is manifested at various levels. The earliest visible manifestations of polarity are detectable within the cytoskeletal and cytoplasmic components of the blastomeres. Over the first half of the fourth cell cycle, actin, clathrin, and endosomes accumulate in an apical focus, and the cell nucleus tends to migrate basally (Figure 2.3) (Reeve 1981a; Reeve and Kelly 1983; Johnson and Maro 1984; Fleming and

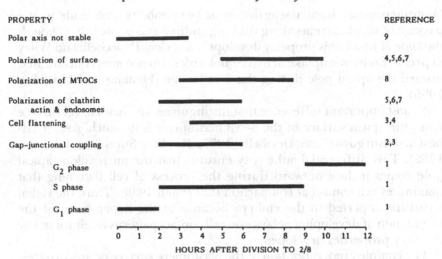

Figure 2.3. The range of time over which a population of pairs of 2/8 blastomeres undergoes various of the developmental transitions is recorded on the left of the figure. The relevant references are indicated on the right as follows: 1, Smith and Johnson (1985a); 2, Goodall and Johnson (1982); 3, Goodall and Johnson (1984); 4, Ziomek and Johnson (1980); 5, Johnson and Ziomek (1981a); 6, Johnson and Maro (1984); 7, Maro et al. (1985b); 8, Fleming and Pickering (1985); 9, B. Maro and S. J. Pickering (unpublished data).

Pickering 1985; Maro et al. in press, b). Microtubule organizing centers (MTOCs), which in blastomeres take the form of multiple foci of an amorphous pericentriolar material, also aggregate apically but appear to do so with a slightly more protracted time course. This polar aggregation of MTOC activity is associated with a polar accumulation of microtubules (Figure 2.3) (B. Maro and S. J. Pickering, unpublished data).

Polarity becomes detectable at the cell surface some 3 to 4 h after the first signs of intracellular polarization. Surface polarity may be visualized by use of a range of molecular probes, including lectins and antibodies, all of which reveal increased concentrations of their target molecules at the apical (outward-facing) pole of the cell (Handyside 1980). This "pole" of ligand-binding sites reflects the concentration of long microvilli at the apex of the cell (Reeve and Ziomek 1981), and there is currently no clear evidence to suggest specific enrichment of particular molecules at this site. The proposal that alkaline phosphatase activity might be depleted specifically in the apical membrane (Izquierdo and Marticorena 1975; Mulnard and Huygens 1978) has proved controversial and difficult to confirm, and it seems more likely that alkaline phosphatase activity is in fact also concentrated at the cell apex (Vorbrodt et al. 1977; Ziomek et al. in press). There is also evidence that the apical region in polarized blastomeres may have a selective ionic permeability that leads to the passage of small currents along the longitudinal axis of the cell, although the time at which this property develops is not clear (Nuccitelli and Wiley in press). Endocytic uptake of macromolecules also becomes concentrated toward the apical pole during the 8-cell stage (Fleming and Pickering 1985).

A most important difference that distinguishes the basolateral surface from the apical surface of the 8-cell blastomere is its much greater adhesivity (Burgoyne and Ducibella 1977; Kimber, Surani, and Barton 1982). This differential adhesivity ensures that the microvillous apical pole comes to face outward during the process of cell flattening that maximizes cell contact at compaction (Lehtonen 1980). Thus, the radial structure imparted to the embryo because of the dependence of the orientation of developing polarity on cell contact patterns is reinforced by the very properties developed.

The complex reorganization of the blastomere surface occurs progressively, with the pole at first occupying most of the cell surface not engaged in intercellular apposition, but then "shrinking" in size to form a discrete circular pole of longer microvilli. It is important to stress that the restriction of microvilli to a polar area is not merely a consequence of the cell flattening "squeezing" them apically, because (a) poles can form in non-flattened embryos (Johnson, Maro, and Takeichi 1986), (b) polar cells that are isolated from intact compact embryos retain their polar phenotype in the absence of continued intercellular flattening (Ziomek and Johnson 1981), and (c) the poles that develop in intact embryos, especially in aggregates of only 2 or 3 cells, are not merely confined to the region of free, nonapposed membrane but are restricted to a limited area of that membrane (Figure 2.2A). Within the apical pole there is evidence of heterogeneity of microvilli, the pole being bounded by a ring of microvilli that are structurally distinct and more resistant to disturbance (Johnson and Maro 1984, 1985; Smith and Johnson 1985). The sequence in which polarity and the other related events of compaction develop at the 8-cell stage is summarized in Figure 2.3. The events span the fourth cell cycle, and although many of the visible manifestations of polarity do not develop until relatively late in the cell cycle, it is clear that the primary axis of polarity is set up within 3 to 5 h of entry into the 8-cell stage. Once this axis has been laid down, isolation of a cell, or alteration of its position or contacts relative to other cells, does not appear to disturb the axis (Johnson and Ziomek 1981a). In consequence, the relocated or rotated blastomere can integrate successfully into the embryo only if it moves or rerotates to regain its appropriate alignment within the embryo with its apical region outward-facing and its basolateral region internally (Johnson 1985). This process of relocation presumably is achieved as a result of the differential adhesivity of the apical and basolateral surfaces of the polarized cells (Kimber et al. 1982; Johnson 1985).

Stability in any biological system is a relative concept and means merely that the half-life of the property under consideration is relatively long. In the context of the polarity established during the 8-cell stage, a most important feature for development is that its stability is sufficient to survive the subsequent cleavage to the 16-cell stage. Cells sampled at intervals throughout this mitotic division retain elements of polarity (Johnson and Ziomek 1981b). In consequence, the daughter progeny cells may inherit different domains of the parental polar cell. If the cleavage plane is oriented along the axis of polarity, two similar cells result, each possessing both apical and basolateral domains, whereas cleavage across the axis of polarity yields cells with differing phenotypes, one incorporating only the basolateral domain (therefore termed "apolar") and the other inheriting elements of both domains (termed "polar") (Figure 2.2B). Cell heterogeneity is first achieved within the embryo by this route. The relative frequency with which cells divide along or across the axis of polarity will determine the ratio of polar to apolar cells achieved, and there is now evidence that this ratio is influenced by cell interaction. Isolated polar 8-cells tend to divide mainly (ca. 85%) across the axis of polarity, whereas polar 8-cells in aggregates of two or more do so in only about 50 percent of cases. Thus, within the embryo, about 4 to 5 apolar cells will result, on average (B. Maro, S. J. Pickering, and T. P. Fleming, unpublished results).

The stability of polarity during cleavage is restricted to only certain elements. As might be expected from their obligatory involvement in cleavage, microfilamentous actin, microtubular tubulin, and the pericentriolar material of the MTOCs lose their polar organization and are distributed to both progeny cells regardless of phenotype (Johnson and Maro 1985; B. Maro and S. J. Pickering, unpublished results). Similarly, those cytoplasmic features that are polar, namely clathrin and endosomes, disperse and are distributed to both progeny (Reeve 1981b; Fleming and Pickering 1985; Maro et al. in press, b), an observation that is not surprising, because their polar distribution at the 8-cell stage is known to be dependent on the continuing integrity of the cytoskeletal system (Johnson and Maro 1985; Fleming, Cannon, and Pickering 1986). The surface pole of ligand binding sites and microvilli is more resistant during division, but its general form changes, with the pole expanding to cover more of the cell's surface (Johnson and Ziomek 1981b). Those features of polarity that develop most rapidly (cytoskeletal and cytoplasmic) are those that are lost first on entry into division, whereas the slow-developing surface polarity is retained, albeit showing a tendency to reverse by polar expansion. With the completion of mitotic cleavage, a discrete, compact surface pole reforms rapidly in those cells inheriting all or part of the polar cortical domain. The actin, clathrin, and endosomes are repolarized, but *only* in those 16-cell blastomeres that inherit all or part of the apical domain of the polar 8-cell. Moreover, the position of the cytoplasmic pole appears to be determined by the position in which that cytocortical domain is located in the progeny cells (Fleming and Pickering 1985; Maro et al. in press, b).

The polar and apolar cells evident at the 16-cell stage are distinguishable by a number of features (Ziomek and Johnson 1981; Reeve 1981b; Kimber et al. 1982; Randle 1982; MacQueen and Johnson 1983; Edirisinghe, Wales, and Pike 1984). The degree and variety of these differences increase as the cells proceed through the fifth cell cycle. Thus, the endocytic processing system in both cell types matures to the point at which definitive lysosomes and residual bodies first become evident, but whereas these elements are randomly located in the apolar cells, they occupy a basal position in the polar cells. Moreover, in the polar cells, endocytic uptake becomes even more heavily concentrated in the region of the apical pole, and the whole polar endocytic apparatus shows evidence of a higher degree of organization (Fleming and Pickering 1985; Fleming et al. 1986; Fleming and Goodall in press; Maro et al. in press, b).

Perhaps the most important distinction developmentally between polar and apolar cells at the 16-cell stage, however, is their surface adhesive and junctional properties. As was pointed out earlier, the apical and basolateral domains of polar 8-cells differ markedly in their adhesiveness, and this difference is also transmitted to their progeny. As a result, the apolar cells are uniformly adhesive, whereas the polar cells have adhesive basolateral surfaces and nonadhesive microvillous apical poles (Ziomek and Johnson 1981; Kimber et al. 1982). Both cell types attempt to maximize contact between adhesive surfaces, and the result is that the apolar cells tends to become internalized, whereas the polar cells expose only their microvillous poles. We believe that this interaction has important consequences for the segregation and regulation of the two lineages of the blastocyst, because it tends to reinforce the inside and the outside positions along the radial axis through the embryo of basolaterally derived and apically derived cells, respectively. This process of positional reinforcement also leads to the next phase in the hierarchy of polar development, namely, the integration of the outer polar cells into an epithelial array.

C. Epithelial phase

The intercellular flattening that occurs during the 8-cell stage is mediated by a homotypic, Ca^{2+}-dependent, cell-cell adhesion system (CDS) in which the active molecule has been identified as E-cadherin (also known as uvomorulin, L-CAM, cell CAM 120/80, and gp123) (Kemler et al. 1977; Hyafil et al. 1980; Hyafil, Babinet, and Jacob 1981; Damsky et al. 1983; Gallin, Edelman, and Cunningham 1983; Shirayoshi, Okada, and Takeichi 1983; Peyrieras et al. 1983; Vestweber and Kemler 1984; Yoshido-Noro, Suzuki, and Takeichi 1984; Johnson et al. 1986) (see Chapter 10). Removal of calcium, use of antibodies to E-cadherin, or exposure to trypsin under calcium-free conditions inactivates the CDS, and as a result the cells round up. From the middle to late 16-cell stage onward, these treatments are less effective at reducing intercellular flattening. Moreover, use of various other treatments to modify cell contact patterns and use of probes to assess the glycosylation status of the cell surface suggest a progressive increase in the complexity of intercellular adhesion systems during the 16- and 32-cell stages (Surani, Kimber, and Handyside 1981; Kimber and Surani 1982; Surani, Kimber, and Osborne 1983; Bird and Kimber 1984; Sato, Muramatsu, and Berger 1984; Rastan et al. 1985). Direct observation of embryos over this period reveals that they are forming zonular tight junctions and point desmosomes (Ducibella et al. 1975; Magnuson et al. 1977) and that a system of intracellular intermediate filaments is developing (Jackson et al. 1980; Lehtonen et al. 1983). Distinct membrane domains, vectorial transport of materials, and a permeability seal between the inside and the outside of the embryo develop (Wiley and Eglitis 1981). The basis for a true epithelium is established and the production and retention of blastocoelic fluid begins.

D. Polarity and developmental lability

We have described the progressive changes in cell organization that lead to formation of a blastocyst and have indicated how important the polar or nonpolar nature of the blastomeres is for their subsequent developmental fate. However, the two cell subpopulations formed by differential cleavage of polarized 8-cell blastomeres are not rigidly determined to become trophectoderm and ICM, respectively. It is commonly but erroneously assumed (see, for example, Wiley 1984) that a mosaic embryo must be a determinate embryo. Mosaic organization and determinacy are *not* necessarily linked in any developmental system and certainly are not linked in the mouse early embryo (Johnson and Pratt 1983; Johnson et al. 1984). The observed developmental lability of the 16-cell blastomeres is perfectly compatible with the known properties of polar and apolar cells. We know that a polar mouse blastomere always gives rise to at least one polar progeny but can also produce apolar progeny, depending on the plane of cleavage. The ability to generate both trophectoderm and ICM cells from polar 16-cell blastomeres is not, therefore, surprising (Randle 1982; Ziomek and Johnson 1982; Ziomek, Johnson, and Handyside 1982; Surani and Handyside 1983). When apolar 16- or 32-cell blastomeres cease to be totally enclosed within the center of an embryo, they also start to polarize (Ziomek and Johnson

1981; Fleming et al. 1984). As we observed for de novo polarization, the acquisition of a polar phenotype by erstwhile apolar cells is also progressive (Fleming et al. 1984) and can explain the ability of apolar cells to generate ICM alone or trophectoderm alone or a mixture of both, depending on the experimental situation. Thus, both primary and regulative cell diversifications involve the same mechanisms, and central to these events are the process of polarization and cleavage of the polarized cell.

E. Summary

We have described the sequence of events whereby cells diverge during development to form a blastocyst and have commented briefly on the fact that the same mechanisms can explain the expression of cell lability. It is clear that if we are to gain a comprehensive understanding of how a blastocyst forms, we need to understand how the development of polarity is initiated and organized, how the orientation of the axis of polarity is determined, and how polar stability is achieved. This attempt to get at the detailed mechanics of polarization is the subject of the remainder of this chapter.

III. Mechanisms underlying polarization

Polarization involves a cascade of events resulting in a major reorganization of the cell triggered at a characteristic time during development, namely, the early part of the fourth cell cycle. Although it may appear convenient to dissect the process of polarization into spatial and temporal components, in practice this is difficult because the two are intimately linked. Thus, spatial reorganization is progressive and involves sequential ordering of various cellular elements (Figure 2.3). The process of differentiative reorganization is imposed on underlying proliferative cell cycles, and many of the elements manifestly involved in spatial reordering also show cell-cycle-dependent variation. Our task is to dissect apart those elements of spatial organization that are cell-cycle-related from those that endure beyond the confines of an individual cell cycle, and then determine how these two temporally distinct elements of spatial organization interact to achieve progressive differentiation.

One approach is to examine the effects on the sequence of events described in Section II (and summarized in Figure 2.3) of the application of selective inhibitors at discrete points prior to or during the process of differentiation. From such an approach we can gain insights into how the process of polarization is initiated and which of its sequential components are related causally. We shall review recent work based on this approach using two general categories of inhibitors, namely, those affecting the cytoskeleton and its relationship to the cytocortex, and those affecting macromolecular biosynthetic activity. The results illuminate the spatial and temporal aspects, respectively, of compaction.

A. Spatial aspects

The cytoskeleton is one of the most obvious candidates for control over changes in spatial organization, and it could also provide a measure of the passage of time via its cyclical modulation at interphase-mitotic transitions (Maro et al. in press, a). In early 8-cell embryos, only microfilaments and microtubules are present, with polymerized intermediate filaments first being detected at later stages, as described earlier. The sequence of events observed during polarization and summarized in Figure 2.3 reveals that the apical polarization of actin and, associated with it, clathrin and endosomes is a very early event. The elaboration of a cytoplasmic and cytoskeletal axis, originating from the region of intercellular contact and penetrating toward the presumptive apical region, might therefore be an early causative element in the chain of events leading to a stable polar phenotype. This axis could then serve as a basis both for further cytoplasmic polarization and for redistribution of microvilli to yield cytocortical polarity (Johnson and Maro 1984; Fleming et al. 1986; Maro et al. in press, b). Drugs that interfere with the cytoskeletal system of the cell can be used to test this hypothesis.

I. Induction of polarity. Many studies have attempted to analyze the role of cytoskeletal elements during compaction, and almost all have used specific inhibitors such as colchicine, colcemid, or nocodazole, which lead to depolymerization of microtubules, taxol, which stabilizes microtubules, and the cytochalasins (B and D), which disrupt the microfilament network and inhibit the formation of new filaments. Unfortunately, many of these studies were compromised by a failure to distinguish the effects of the drugs on interphase cells from their effects on mitotic events (Ducibella and Anderson 1975; Surani, Barton, and Burling 1980; Pratt et al. 1982; Sutherland and Calarco-Gillam 1983). However, when these drugs were applied at the beginning of interphase of the fourth cell cycle and compaction was allowed to proceed in their presence (Ducibella 1982; Kimber and Surani 1982; Johnson and Maro 1984, 1985; Maro and Pickering 1984; Fleming et al. 1986), useful information was obtained (Figure 2.4 and Table 2.1).

It is clear that the development of cytocortical polarity is not inhibited by drugs that destroy the cytoskeletal system of the cell. In the presence of microtubule inhibitors, microvillous poles form, although they are spread over a larger area of the exposed apical domain, and their boundary with the nonmicrovillous region of the cell is not as sharp as in control blastomeres. The role of microfilaments is more difficult to assess, because they

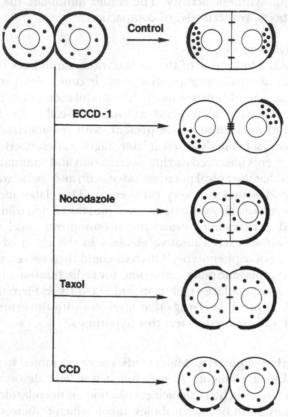

Figure 2.4. Summary of effects of incubating pairs of 8-cell blastomeres in monoclonal antibody ECCD-1, nocodazole, taxol, or cytochalasin D (CCD). Features analyzed include cell flattening, gap-junction formation (bars linking cells), surface polarity (surface hatching), and cytoplasmic polarity (dots). For references, see Table 2.1.

are structural features of the microvilli themselves (albeit relatively resistant to disruption by cytochalasins, presumably because of their relatively slow turnover, to be described later). However, it is clear that although the precise form of the surface pole is disturbed by cytochalasin D, poles attempt to form in the presence of the drug. In contrast to surface polarity, cytoplasmic polarity, however assayed, is very sensitive to the actions of both microtubule and microfilament inhibitors. These experiments do not, therefore, support the proposition that the polarization of cytoplasmic and cytoskeletal elements is one link in a causal chain leading to cytocortical polarization. Indeed, in experiments in which cytocortical polarity develops in the absence of cytoplasmic polarity, and then conditions are changed to permit the development of cyto-

Table 2.1. *Effects of various agents applied during interphase on compaction of 8-cell mouse embryo*

Treatment	Effect	Cytoplasmic polarity[a]	Surface polarity[b]	Junctional communication[c]	Intercellular flattening[d]
Nocodazole	Microtubule depolymerization	Absent	Broad poles	Normal	Normal but completed more rapidly
Taxol	Microtubule stabilization	Absent	Broad poles	Reduced	Reduced and delayed
CCD	Microfilament disruption	Absent	Abnormal morphology and off axis	Absent	Absent
ECCD-1	Cell adhesion inhibition	Normal morphology off axis and delayed	Normal morphology off axis and delayed	Normal	Absent

Note:
[a] Johnson and Maro 1984, 1985; Fleming et al. 1986; Johnson et al. 1986.
[b] Ducibella 1982; Maro and Pickering 1984; plus references in footnote a.
[c] Goodall 1986; Goodall and Maro 1986.
[d] Shirayoshi et al. 1983; Surani et al. 1980; plus references in footnotes a and b.

plasmic polarity, the cytoplasmic pole always seems to underlie the cytocortical pole. The same phenomenon is observed after mitotic division, when the pole of clathrin or of endosomes develops only subjacent to the cytocortical pole of microvilli, as described earlier. Thus, the cytocortex seems to provide the focus for cytoplasmic polarity, rather than the converse, as was suggested originally.

Two main conclusions follow from this interpretation. First, because during normal development cytoplasmic polarity is detected in advance of surface polarity (Figure 2.3), the earliest cytocortical changes in the presumptive apical region of the polarizing cell must have gone undetected. We have termed these putative changes "covert polarity" (Johnson and Maro 1985). They presumably must be completed within 3 to 5 h of entry into the fourth cell cycle, because by that time a stable axis of polarity has been laid down. These proposed changes may also serve as the basis for the cytocortical "memory trace" of the axis of polarity that persists during mitotic division.

Second, the setting up of covert polarity in the cytocortical domain during the early part of the fourth cell cycle would seem to be remarkably insensitive to major disorganization within the cytoplasmic matrix. How, then, is the contact region at the presumptive basal region of the blastomere translated into a cytocortical memory trace at its presumptive apical region? There seem to be two broad categories of explanations. Cell-cell interaction in the basal region could lead to generation of an internal cytoplasmic signal that is independent of an organized cytoplasmic matrix for its transmission to apical regions. For example, a change in ionic permeability in the region of cell contact might lead to the setting up of an ionic gradient through the cell and to the generation of small local currents along the presumptive axis of polarity. Such currents might then effect changes in the polar cytocortex leading to establishment of a stable covert axis in the cell. Such a model is supported by the observation that small local currents may indeed exist in fully polarized cells (Nuccitelli and Wiley in press), but we do not yet know whether these currents are the causes or consequences of polarization. However, such a mechanism need not necessarily be restricted to ions; it could involve any low-molecular-weight, diffusable molecule that is biologically active. Alternatively, independence from an organized cytoplasmic matrix could be achieved if the signal mediated by intercellular contact traveled from the basal region in the plane of the cytocortex itself, via spread of a catalytic chemical change such as protein phosphorylation or lipid modification, or lateral movement of intramembranous molecules of sufficiently high lateral mobility (Pratt 1978; Wolf 1983).

II. Elaboration of polarity. Although neither microtubules nor microfilaments appear to be involved in the very early events of covert polarization, they do seem to be involved in later cell reorganization. Their involvement is indicated by two types of observations. First, the cytocortical poles that do develop in the continuing presence of the drugs have a modified morphology. Second, when drugs are applied after the process of polarization has been completed, the effects on polarization differ somewhat from those observed after their addition early in the fourth cell cycle. Use of the evidence from each of these types of observations allows us to suggest a possible role for microfilaments and microtubules in the later events of compaction.

Microtubule-disrupting drugs applied to fully polarized cells cause (a) a broadening of the cytocortical pole of microvilli over most of the exposed cell surface to give a less distinct boundary and (b) dispersal of most polar aggregates of endosomes and clathrin. Removing the drugs leads to restoration of endosomal clustering in a polar location (Johnson and Maro 1985; Fleming et al. 1986).

These results suggest that once polarity is established, microtubules are required for maintenance of endosomal clustering subjacent to the cortical pole; the results also suggest that the requirement for microtubules in the setting up of cytoplasmic polarity, and during repolarization following a pulse of drugs, might involve direction of the movement of clustered endosomes toward the apical region. A similar role for microtubules has been deduced from studies on a range of other cell types (Freed and Lebowitz 1970; Phaire-Washington, Silverstein, and Wang 1980; Imhof et al. 1983; Collot, Louvard, and Singer 1984; Herman and Albertini 1984). We know that microtubules are redistributed from a general subcortical to a polar location during polarization, and we suspect that they form a matrix on which endosomes cluster and that must, directly or indirectly, be linked to the overlying polar cytocortex, influencing its detailed morphology.

Microfilaments are present in both the cytocortical and cytoplasmic poles. The microvilli of the pole, particularly those forming its boundary, seem to be comparatively resistant to the action of cytochalasin D (Handyside 1980; Sutherland and Calarco-Gillam 1983; Johnson and Maro 1984, 1985; Fleming et al. 1986). The process of cytocortical polarization seems to involve an alteration in the spatial distribution of these stable microvilli. Early in the 8-cell stage, microvilli containing stable microfilaments are distributed in patches over the whole surface of the blastomere, as is revealed, for example, by placing the blastomeres in cytochalasin D and examining them shortly thereafter for evidence of residual microvilli. As cytocortical polarization progresses, the stable cytochalasin-resistant microvilli become increasingly concentrated in the apical pole until the definitive pole is formed (Fleming et al. 1986). It seems possible that polarization may involve restriction of a stabilizing system for microfilaments to the polar region coupled with a global rise in the turnover of actin. In this way, only polar microvilli will be retained. Such localized changes in the stability of microvillous and subcortical actin have been reported in unfertilized and newly fertilized mouse eggs (Maro et al. 1984, 1986). Although cytochalasin D does not destroy the fully formed microvillous poles, it does modify their morphology, causing the poles to become sharply defined, smaller, and denser. Cytochalasin D acts to sever the network associations of microfilaments (Schliwa 1982), and so one might explain the observed result as reflecting the rupturing of the connection between the resistent mesh of microfilaments at the base of the stabilized microvilli and those dispersed elsewhere around the cortex. The pole would in essence contract in on itself, unopposed by the global cytocortical tension.

Cytochalasin D, although it prevents the development of cytoplasmic polarity, does *not* destroy it once a polar cell has formed. Thus, it is possible that stabilization of microvilli is accompanied by stabilization of the subjacent cytoplasmic actin. Moreover, nocodazole-induced dispersion of polar endosomal clusters is *prevented* by concurrent treatment with cytochalasins (Fleming et al. 1986). These results suggest that microfilaments are required for both the microtubule-dependent clustering of endosomes and their movement toward an apical position during polarization; microfilaments also seem to be involved in the displacement and dispersal of endosomal clusters in the absence of microtubules. Such a role for microfilaments is not confined to mouse blastomeres (Goldberg et al. 1980; Solomon and Magendantz 1981; Tomasek and Hay 1984; Maro et al. 1986).

We need to ask how it is that both cytoskeletal elements become concentrated to this region of the cell. In principle, concentration could be achieved by local nucleation and/or by local stabilization (or by destabilization elsewhere). MTOCs are indeed found to be localized in the same region, but they arrive there relatively late in the fourth cell cycle (B. Maro and S. J. Pickering, unpublished data). This result, together with the fact that areas of intercellular contact in early 8-cell embryos are devoid of cytoplasmic and cytoskeletal elements, suggests that it may be a local destabilization of microtubules by the cytocortex in the basolateral domain that initiates events. The relative stability of microtubules in the more apical domain may then lead to localization of MTOCs in the same area, thereby reinforcing local stability by local nucleation.

III. Flattening. We have concentrated attention on the events of polarization because we believe they are central to cell diversification. However, both intercellular coupling and flattening also occur at compaction, and as we saw earlier, the latter, at least, is implicated in the expression or suppression of plasticity in the two developing cell lineages (Figure 2.2). The use of drugs has also been informative about its possible role in the foundation of the two cell lineages. The data in Table 2.1 do suggest that there is some relationship between intercellular flattening and polarization. Under the two types of conditions in which flattening is completely inhibited, namely, inactivation of the CDS and disruption of the microfilament network together with suppression of new microfilament formation, polarization nonetheless occurs, but it is *not* oriented perpendicular to the points of residual contact between the nonflattened cells, but rather appears to be oriented randomly (Figure 2.4) (Johnson and Maro 1984; Fleming and Pickering 1985; Johnson et al. 1986). This result appears to be explicable as follows: Isolated, newly formed 8-cell blastomeres cultured in isolation, and therefore in the absence of any cell contacts at all, do polarize, but, as a population, they do so over a much more protracted time course than do cells in pairs or clusters (Ziomek and Johnson 1980; Johnson et al. 1986). When the time course of polarization in pairs of cells exposed to ECCD-1 is compared with that for control pairs and that for singletons, it is found to resemble that for singletons (Johnson et al. 1986). Thus, "blinded" by the antibody, the cells, as a population, polarize more slowly and with an axis of polarity that is random with respect to the companion cells that they can no longer "see." These results with single blastomeres, and with pairs and clusters of blastomeres prevented from flattening on each other, suggest that whatever the nature of the signal between cells might be, its primary function is to ensure synchronized and appropriately oriented polarization. Flattening will favor the appearance of the covert changes leading to polarity in the area of intercellular contact and thus speed up and orient the process of compaction. Finally, these experiments provide an insight into the integration of time and space in the development of the early embryo, because cell interaction influences not only the spatial orientation of polarization but also the timing of its occurrence. The problem of how the timing of the events of compaction might be regulated will be considered later.

IV. Coupling. From the results in Table 2.1 it is clear that the available evidence does not indicate a role for the newly developed gap junctions in mediating the setting up of a polar axis. A popular but unproven theme in the literature of developmental biology is the possible role of gap-junctional communication in the transmission of developmental information among cells. The temporal coincidence of the first appearance of gap junctions in mouse development (Lo and Gilula 1979; Goodall and Johnson 1982, 1984; McLachlin, Caveney, and Kidder 1983) with the cell-contact-mediated signal orienting and synchronizing polarity has provoked the suggestion of causality (Lo 1982). However, it seems unlikely that this is the case, because polarity can be induced by cell types that do not establish gap-junctional communication with each other (Johnson and Ziomek 1981a; Goodall and Johnson 1982, 1984; McLachlin et al. 1983). Moreover, polarity can develop under conditions in which gap-junctional communication has been blocked or impaired, and polarization may be impaired under conditions in which junctional coupling is nonetheless present (Goodall 1986; Goodall and Maro 1986).

B. Temporal aspects

There are three temporal aspects to consider in connection with the process of compaction: What initiates compaction? What regulates the rate at which it progresses once initiated? What determines whether or not the process goes to completion? These distinctions, although easy to make in principle, may be more difficult to achieve in practice. Thus, cells cultured in isolation or in the presence of ECCD-1 do polarize, but, as a population, they do so over a more protracted period. It is not clear whether this phenomenon results from a delay in the signal to polarize, from a delay in implementing the response to that signal, or from a combination of these two. Likewise, a blastomere that is totally surrounded by other cells does flatten and establish gap junctions, but it does not complete compaction, showing no evidence of polarity. In both of these situations, spatial and temporal interactions are occurring. Can we dissect out more cleanly the levels at which temporal controls operate by use of selective inhibitory drugs?

Use of nocodazole provides an example of the way in which the rate at which compaction occurs can be modulated while leaving its time of initiation unaffected. Thus, application of nocodazole to 4- or 8-cell embryos does not advance the onset of compaction, but it does advance the completion of the process among a population of embryos (Maro and Pickering 1984; Levy et al. in press). Conversely, in the presence of taxol, the flattening process is reduced and slowed down (Maro and Pickering 1984; Goodall and Maro 1986). These results have led to the suggestion that an intact network of microtubules might actually slow down implementation of the changes initiated at compaction.

What controls the initiation of compaction? A number of timing devices have been proposed to explain the ordered occurrence of developmental events, and most implicate a role for the cell cycle. It is known that passage through the cell cycle involves local or global changes in microenvironment that lead to restriction to characteristic points in the cell cycle of a whole variety of cellular enzyme activities, changes in molecular conformation, biosynthetic patterns, and supramolecular assembly. These cyclic restrictions could lead to the appearance of "developmental windows," which, coinciding with a positional signal, for example, might effect a change that would endure beyond the cell cycle. In addition, the cell cycles during cleavage are atypical in that interphase growth does not occur, which means that cell volume is reduced by approximately half

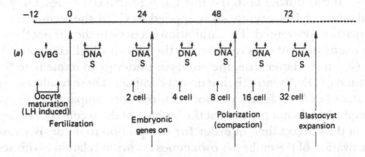

Figure 2.5. Schematic view of early mouse development showing relationships between major developmental transitions and the cell cycle. [Adapted from Johnson (1981).]

at each division. Such a situation offers opportunities for temporal control by dilution. For example, a constant level of cytoplasmic factor might be titrated against an increasing number of nuclei or quantity of chromatin. If the titration of such a factor should coincide with a developmental window, differentiative change could occur. The observations that major transitions in early mouse development are associated characteristically with particular developmental cell cycles (Figure 2.5), when taken together with the types of considerations outlined earlier, have led to the quest for possible causal relationships between specific elements of an individual cell cycle (especially cytokinesis, karyokinesis, and DNA replication) and a particular developmental transition.

In the early mouse embryo there is little evidence to suggest that cytokinesis plays any critical role in measuring the passage of time. Thus, when cytokinesis is prevented, but DNA replication is allowed to proceed, the pattern and pace of development to the blastocyst remain unaffected (except, of course, for any properties contingent on an increasing cell number) (Surani et al. 1980; Pratt, Chakraborty, and Surani 1981; Petzoldt et al. 1983). It is more difficult to assess the impact on development of inhibiting karyokinesis in the mouse embryo, because there is as yet no way of specifically isolating this component of the cell cycle. Thus, although application of drugs such as nocodazole arrests cells at metaphase, the metaphase state is global, involving as it does a characteristic pattern of biosynthesis and modification of proteins, and a characteristic stabilization of proteins and RNA species. The cell arrests by all criteria examined, whether they be cell cycle or developmental in nature (Goodall and Maro 1986; Howlett 1986; Maro et al. in press, a).

In contrast, specific deletion of DNA replication can be achieved in the early mouse embryo, and the events of compaction are found to be remarkably independent of proximate replicative activity (Smith and Johnson 1985). Thus, the events of compaction occur in the absence of

DNA synthesis during both the third and fourth cell cycles. Only when the inhibitor of DNA synthesis is applied in G_1 of the second cell cycle is compaction prevented. This inhibition occurs whether or not the inhibitor remains present or is removed at the time at which control embryos enter G_2. In the latter case, the embryos undergo an immediate "catch-up" round of DNA synthesis, but do not compact. These results mean that proximate DNA replication is not required for compaction. They may also imply, but do not prove, that the S phase of the second cell cycle must occur at the correct time in order for compaction to occur. Because the first activation of the embryo's own genes occurs in relation to this second round of replication (see Chapter 8 for discussion and references), the results hint at a role in compaction for embryonic transcripts and the proteins that they encode. Use of inhibitors of protein synthesis provides an interesting insight into this possibility.

It has long been known that inhibition of protein synthesis impairs cleavage (Thomson and Biggers 1966; Monesi et al. 1970; Tasca and Hillman 1970) and a more precise timing of the requirement for protein synthesis has recently been determined for the first, second, and third mitotic divisions (Kidder and McLachlin 1985; Howlett 1986; Levy et al. 1986). The most detailed of these studies suggests that there is an absolute requirement for protein synthesis if the transition from interphase to mitosis is to occur successfully, but a protein biosynthetic requirement for transitions between other phases of the cell cycle cannot be excluded at present (Howlett 1986). The arresting effect of protein biosynthetic inhibitors on cell cycle events is in marked contrast to their effect on developmental events and allows us to dissect the two sets of parameters. Thus, application of inhibitors to the very late 2-cell or early 4-cell embryo results in *advancement* of both intercellular flattening and surface polarization in many embryos (Levy et al. in press). Gap junctions do not form unless the application of the drug is delayed to the middle of the 4-cell stage (McLachlin et al. 1983; Levy et al. in press), and cytoplasmic polarization is sensitive up to the early 8-cell stage (Levy et al. in press). These results suggest two conclusions. First, it seems that most of the proteins required for compaction are already present from as early as the late 2- and early 4-cell stages. Second, the timing of the onset of compaction may be controlled by releasing an inhibitory restraint on otherwise competent cells, using presumably posttranslational mechanisms. Because the protein biosynthetic activity during the late 2-cell stage is achieved largely on embryonic templates, it is tempting to speculate that the earliest task of the embryonic genes is to produce the proteins required for compaction together with a system for regulating the timing of their use.

The notion that a key developmental transition might be regulated posttranslationally is not entirely novel. Thus, a similar situation applies at fertilization, at which sperm activation functions at an entirely posttranslational level to effect a complex sequence of major structural changes and does so by neutralization of a restraining factor (reviewed by Levy et al. 1986). Such a regulatory device has certain advantages. It will aid the achievement of relative synchrony among blastomeres within an embryo, will diminish the chance that a slight deficiency in one component could prejudice an elaborate spatial and temporal program of cell reorganization, and, most important, will allow environmental or positional signals ready access to the developmental program. As we saw earlier, such access constitutes an important component of normal development in the mouse embryo. We now need to understand the nature of the putative restraining factor that holds the process of compaction in check, the way in which the activity of this factor declines naturally early in the fourth cell cycle, and whether the accelerating effect of cell interaction on polarization operates via an influence on the decline of that factor or on the consequences of that decline.

IV. Conclusion

A blastocyst is a deceptively simple structure. Only slowly are we gaining some clues as to how it develops. Can we draw together some strands from the mass of data now accumulated on the earliest events in this process and speculate on how compaction is achieved? The process of compaction consists of separable but interrelated events. The expression of these events is evidently regulated by some global change within the cells of the embryo that may involve a release from a restraint. At present there is little evidence to suggest that gap-junctional coupling constitutes an important part of the developmental regulation at the time of compaction. However, both intercellular flattening and cell polarization are important in the initiation of cell diversification and in the expression of lability by the cells within each diversing lineage.

Cell polarization gives spatial heterogeneity to each blastomere at the 8-cell stage. The surface component of this heterogeneity, namely, the differential adhesiveness of apical and basolateral regions, ensures that the polar cells integrate to give an embryo that is itself spatially heterogeneous, with distinctive inside and outside domains. The embryo capitalizes on this heterogeneity at cleavage to found two cell lineages. Cell polarization can occur spontaneously, but in situ it is advanced and oriented by a cell interaction that involves the cell-cell adhesion system mediating intercellular flattening. The consequences of intercellular flattening could be some relatively nonspecific local intracellular change in, for example, an ion or a small metabolite, which could in turn lead to the observed local destabilization of the cytoskeleton, with loss of microvilli, microfilaments, microtubules, and myosin in the area of cell apposition. We can envisage that when intercellular flattening coincides with a global release from restraint, these local subcortical changes may be able to spread beyond the confines of the area of cell apposition. Then, gradually, the microtubules and their associated organelles (including microfilaments) will be excluded from these areas, leading to the retention of an apical cluster of microtubules, microfilaments, and associated organelles to which the MTOCs will be drawn subsequently. The spread of these subcortical changes will also lead, more slowly, to the disappearance of microvilli in the basolateral part of the cytocortex, resulting in the formation of a surface pole of microvilli and allowing an extension of intercellular flattening. This latter effect on microvilli and flattening will be facilitated by the absence of microtubules in these areas. Rather than being the driving force in compaction, microtubules may help to coordinate the various changes taking place during the process of compaction and reinforce the asymmetry set up in the cell cortex. The missing links in this sequence are the nature of the restraining factor, the nature of the local changes beneath areas of cell apposition, and the way these two unknowns interact to yield a progressive spread around the cytocortex of the cell.

Acknowledgments

We thank our research colleagues for their stimulation and their permission to cite unpublished work, Raith Overhill and Ian Edgar for their technical assistance, and Francis Woodman for assistance in the preparation of the manuscript. The work reported here was supported by grants from the Medical Research Council to Drs. M. H. Johnson and P. R. Braude, from the Fondation pour la Recherche Medicale to Dr. B. Maro, and from the Cancer Research Campaign to Drs. M. H. Johnson, H. P. M. Pratt, and R. T. Hunt. Dr. Maro is an EMBO Fellow.

References

Bird, J. M., and Kimber, S. J. (1984) Oligosaccharides containing fucose linked alpha(1−3) and alpha(1−4) to N-acetyl glucosamine cause decompaction of mouse morulae. Dev. Biol. 79, 449−60.

Borland, R. M., Biggers, J. D., and Lechene, C. P. (1977) Studies on the composition and formation of mouse blastocoele fluid using electron probe microanalysis. Dev. Biol. 55, 1−8.

Brulet, P., Babinet, C., Kemler, R., and Jacob, F. (1980) Monoclonal antibodies against trophectodermal-specific markers during mouse blastocyst formation. Proc. Natl. Acad. Sci. U.S.A. 77, 4113−17.

Burgoyne, P. S., and Ducibella, T. (1977) Changes in the properties of the developing trophoblast of preimplantation mouse embryos as revealed by aggregation studies. *J. Embryol. Exp. Morphol.* 40, 143–57.

Calarco, P. G., and Brown, E. H. (1969) An ultrastructural and cytological study of preimplantation development of the mouse. *J. Exp. Zool.* 171, 253–84.

Calarco, P. G., and Epstein, C. J. (1973) Cell surface changes during preimplantation development in the mouse. *Dev. Biol.* 32, 208–13.

Collot, M., Louvard, D., and Singer, S. J. (1984) Lysosomes are associated with microtubules and not with intermediate filaments in cultured fibroblasts. *Proc. Natl. Acad. Sci. U.S.A.* 81, 788–92.

Copp, A. J. (1979) Interaction between inner cell mass and trophectoderm of the mouse blastocyst. II. The fate of the polar trophectoderm. *J. Embryol. Exp. Morphol.* 51, 109–20.

Damsky, C. H., Richa, C., Solter, D., Knudsen, K., and Buck, C. A. (1983) Identification and purification of a cell surface glycoprotein mediating intercellular adhesion in embryonic and adult tissue. *Cell* 34, 455–6.

Dizio, S. M., and Tasca, R. J. (1977) Sodium-dependent amino-acid transport in preimplantation mouse embryos. III. Na^+-K^+-ATPase-linked mechanism in blastocysts. *Dev. Biol.* 59, 198–203.

Ducibella, T. (1982) Depolymerization of microtubules prior to compaction. *Exp. Cell. Res.* 138, 31–8.

Ducibella, T., Albertini, D. F., Anderson, E., and Biggers, J. (1975) The preimplantation mammalian embryo: characterization of intracellular junctions and their appearance during development. *Dev. Biol.* 45, 231–50.

Ducibella, T., and Anderson, E. (1975) Cell shape and membrane changes in the eight-cell mouse embryo. Prerequisites for morphogenesis of the blastocyst. *Dev. Biol.* 47, 45–58.

Edirisinghe, W. R., Wales, R. G., and Pike, I. L. (1984) Studies of the distribution of glycogen between the inner cell mass and trophoblast cells of mouse embryos. *J. Reprod. Fertil.* 71, 533–8.

Fleming, T. P., Cannon, P., and Pickering, S. J. (1986) The cytoskeleton, endocytosis and cell polarity in the mouse preimplantation embryo. *Dev. Biol.* 113, 406–19.

Fleming, T. P., and Goodall, H. (in press) Endocytic traffic in trophectoderm and polarized blastomeres of the mouse preimplantation embryo. *Anat. Rec.*

Fleming, T. P., and Pickering, S. J. (1985) Maturation and polarization of the endocytotic system in outside blastomeres during mouse preimplantation development. *J. Embryol. Exp. Morphol.* 89, 175–208.

Fleming, T. P., Pickering, S. J., Qasim, F., and Maro, B. (1986) The generation of cell surface polarity in mouse 8-cell blastomeres: the role of cortical microfilaments analysed using cytochalasin D. *J. Embryol. Exp. Morphol.* 95, 169–91.

Fleming, T. P., Warren, P., Chisholm, J. C., and Johnson, M. H. (1984) Trophectodermal processes regulate the expression of totipotency within the inner cell mass of the mouse expanding blastocyst. *J. Embryol. Exp. Morphol.* 84, 63–90.

60   MARTIN H. JOHNSON AND BERNARD MARO

Freed, J. J., and Lebowitz, M. M. (1970) The association of a class of saltatory movements with microtubules in cultured cells. *J. Cell Biol.* 45, 334–54.

Gallin, W. J., Edelman, G. M., and Cunningham, B. A. (1983) Characterization of L-CAM, a major cell adhesion molecule from embryonic liver cells. *Proc. Natl. Acad. Sci. U.S.A.* 80, 1038–42.

Gardner, R. L., Papaioannou, V. E., and Barton, S. C. (1973) Origin of the ectoplacental cone and secondary giant cells in mouse blastocysts reconstituted from isolated trophoblast and inner cell mass. *J. Embryol Exp. Morphol.* 30, 561–72.

Goldberg, D. J., Harris, D. A., Lubit, B. W., and Schwartz, J. H. (1980) Analysis of the mechanism of fast axonal transport by intracellular injection of potentially inhibitory macromolecules: evidence for a possible role of actin filaments. *Proc. Natl. Acad. Sci. U.S.A.* 77, 7448–52.

Goodall, H. (1986) Manipulation of gap junctional communication during compaction of the mouse early embryo. *J. Embryol. Exp. Morphol.* 91, 283–96.

Goodall, H., and Johnson, M. H. (1982) Use of carboxyfluorescein diacetate to study formation of permeable channels between mouse blastomeres. *Nature* 295, 524–6.

–(1984) The nature of intercellular coupling within the preimplantation mouse embryo. *J. Embryol. Exp. Morphol.* 79, 53–76.

Goodall, H., and Maro, B. (1986) Loss of junctional coupling during mitosis in early mouse embryos. *J. Cell Biol.* 102, 568–75.

Handyside, A. H. (1980) Distribution of antibody and lectin binding sites on dissociated blastomeres from mouse morulae: evidence for polarization at compaction. *J. Embryol. Exp. Morphol.* 60, 99–116.

Herman, B., and Albertini, D. F. (1984) A time-lapse video image intensification analysis of cytoplasmic organelle movement during endosome translocation. *J. Cell Biol.* 98, 565–76.

Howlett, S. K. (1986) A set of proteins showing cell cycle-dependent behaviour in the one-cell mouse embryo. *Cell.* 45, 387–96.

Hyafil, F., Babinet, C., and Jacob, F. (1981) Cell-cell interactions in early embryogenesis: a molecular approach to the role of calcium. *Cell* 26, 447–54.

Hyafil, F., Morello, D., Babinet, C., and Jacob, F. (1980) A cell surface glycoprotein involved in the compaction of embryonal carcinoma cells and cleavage stage embryos. *Cell* 21, 927–34.

Imhof, B. A., Marti, U., Boller, K., and Birchmeier, W. (1983) Association between coated vesicles and microtubules. *Exp. Cell Res.* 145, 199–207.

Izquierdo, L., and Marticorena, P. (1975) Alkaline phosphatase in preimplantation mouse embryos. *Exp. Cell Res.* 92, 399–402.

Jackson, B. W., Grund, C., Schmid, E., Burki, K., Franke, W. W., and Illmensee, K. (1980) Formation of cytoskeletal elements during mouse embryogenesis. Intermediate filaments of the cytokeratin type and desmosomes in preimplantation embryos. *Differentiation.* 17, 161–179.

Johnson, M. H. (1981) The molecular and cellular basis of preimplantation mouse development. *Biol. Rev.* 56, 463–98.

–(1985) Manipulation of early mammalian development: What does it tell us about cell lineages? In *Developmental Biology: A Comprehensive Synthesis*, ed. L. Browder, pp. 27–48. New York: Plenum.

–(in press) Three types of cell interaction regulate the generation of cell diversity in the mouse blastocyst. In *The Cell in Contact: Adhesions and Junctions as Morphogenetic Determinants*, eds. G. Edelman and J. P. Thiery, New York: Wiley.

Johnson, M. H., and Maro, B. (1984) The distribution of cytoplasmic actin in mouse 8-cell blastomeres. *J. Embryol. Exp. Morphol.* 82, 97–117.

–(1985) A dissection of the mechanisms generating and stabilizing polarity in mouse 8- and 16-cell blastomeres: the role of cytoskeletal elements. *J. Embryol. Exp. Morphol.* 90, 311–34.

Johnson, M. H., Maro, B., and Takeichi, M. (1986) The role of cell adhesion in the synchronisation and orientation of polarization in 8-cell mouse blastomeres. *J. Embryol. Exp. Morphol.* 93, 239–55.

Johnson, M. H., and Pratt, H. P. M. (1983) Cytoplasmic localisations and cell interactions in the formation of the mouse blastocyst. In *Time, Space and Pattern in Embryonic Development*, eds. W. R. Jeffery and R. A. Raff, pp. 287–312. New York: Alan R. Liss.

Johnson, M. H., and Ziomek, C. A. (1981a) Induction of polarity in mouse 8-cell blastomeres: specificity, geometry and stability. *J. Cell Biol.* 91, 303–8.

–(1981b) The foundation of two distinct cell lineages within the mouse morula. *Cell* 24, 71–80.

–(1982) Cell subpopulations in the late morula and early blastocyst of the mouse. *Dev. Biol.* 91, 431–9.

–(1983) Cell interactions influence the fate of blastomeres undergoing the transition from the 16- to the 32-cell stage. *Dev. Biol.* 95, 211–18.

Johnson, M. H., Ziomek, C. A., Reeve, W. J. D., Pratt, H. P. M., Goodall, H., and Handyside, A. H. (1984) The mosaic organisation of the preimplantation mouse embryo. In *Ultrastructure of Reproduction*, eds. J. Van Blerkom and P. M. Motta, pp. 205–17. The Hague: Martinus Nijhoff.

Kaye, P., Schultz, G., Johnson, M. H., Pratt, H. P. M., and Church, R. B. (1982) Amino acid transport and exchange in preimplantation mouse embryos. *J. Reprod. Fertil.* 367–80.

Kemler, R., Babinet, C., Eisen, H., and Jacob, F. (1977) Surface antigen in early differentiation. *Proc. Natl. Acad. Sci. U.S.A.* 74, 4449–52.

Kidder, G. M., and McLachlin, J. R. (1985) Timing of transcription and protein synthesis underlying morphogenesis in preimplantation mouse embryos. *Dev. Biol.* 112, 265–75.

Kimber, S. J., and Surani, M. A. H. (1982) Spreading of blastomeres from eight-cell mouse embryos on lectin-coated beads. *J. Cell Sci.* 56, 191–206.

Kimber, S. J., Surani, M. A. H., and Barton, S. C. (1982) Interactions of blastomeres suggest changes in cell surface adhesiveness during the formation of inner cell mass and trophectoderm in the preimplantation mouse embryo. *J. Embryol. Exp. Morphol.* 70, 133–52.

Lehtonen, E. (1980) Changes in cell dimensions and intracellular contacts during cleavage-stage cell cycles in mouse embryonic cells. *J. Embryol. Exp. Morphol.* 58, 231–49.

Lehtonen, E., Lehto, U. P., Vartio, T., Badley, R. A., and Virtanen, I. (1983) Expression of cytokeratin polypeptides in mouse oocytes and preimplantation embryos. *Dev. Biol.* 100, 158–65.

Levy, J. B., Johnson, M. H., Goodall, H., and Maro, B. (1986) Control of the timing of compaction: a major developmental transition in mouse early development. *J. Embryol. Exp. Morphol.* 95, 213–37.

Lo, C. W. (1982) Gap junctional communication compartments and development. In *The Functional Integration of Cells in Animal Tissues*, eds. J. D. Pitts and M. E. Finbow, pp. 167–79. Cambridge University Press.

Lo, C. W., and Gilula, N. B. (1979) Gap junctional communication in the preimplantation mouse embryo. *Cell* 18, 399–409.

McLachlin, J. R., Caveney, S., and Kidder, G. M. (1983) Control of gap junction formation in early mouse embryos. *Dev. Biol.* 98, 155–64.

MacQueen, H. A., and Johnson, M. H. (1983) The fifth cell cycle of the mouse embryo is longer for small cells than for larger cells. *J. Embryol. Exp. Morphol.* 77, 297–308.

Magnuson, T., Demsey, A., and Stackpole, C. W. (1977) Characterization of intercellular junctions in the preimplantation mouse embryo by freeze-fracture and thin-section electron microscopy. *Dev. Biol.* 61, 252–61.

Maro, B. (1985) Fertilization and the cytoskeleton in the mouse. *Bioessays* 3, 18–21.

Maro, B., Howlett, S. K., and Johnson, M. H. (in press, a) A cellular and molecular interpretation of mouse early development: the first cell cycle. In *Gametogenesis and the Early Embryo*, ed. J. G. Gall, New York: Alan R. Liss.

Maro, B., Johnson, M. H., Pickering, S. J., and Flach, G. (1984) Changes in actin distribution during fertilization of the mouse egg. *J. Embryol. Exp. Morphol.* 81, 211–37.

Maro, B., Johnson, M. H., Pickering, S. J., and Louvard, D. (in press, b) Changes in the distribution of membranous organelles during mouse early development. *J. Embryol. Exp. Morphol.*

Maro, B., Johnson, M. H., Webb, M., and Flach, G. (1986) Mechanism of polar body formation in the mouse oocyte: an interaction between the chromosomes, the cytoskeleton and the plasma membrane. *J. Embryol. Exp. Morphol.* 92, 11–32.

Maro, B., and Pickering, S. (1984) Microtubules influence compaction in preimplantation mouse embryos. *J. Embryol. Exp. Morphol.* 84, 217–32.

Monesi, V., Molinaro, M., Spalletta, E., and Davoli, C. (1970) Effect of metabolic inhibitors on macromolecular synthesis and early development in the mouse embryo. *Exp. Cell Res.* 59, 197–206.

Mulnard, J., and Huygens, R. (1978) Ultrastructural localisation of nonspecific alkaline phosphatase during cleavage and blastocyst formation in the mouse. *J. Embryol. Exp. Morphol.* 44, 121–31.

Nuccitelli, R., and Wiley, L. (in press) Polarity of isolated blastomeres from mouse morulae: detection of transcellular ion currents. *Dev. Biol.*

Oshima, R. G., Howe, W. E., Klier, F. G., Adamson, E. D., and Shevinsky, L. H. (1983) Intermediate filament protein synthesis in preimplantation murine embryos. *Dev. Biol.* 99, 447–55.

Petzoldt, U., Burki, K., Illmensee, G. R., and Illmensee, K. (1983) Protein synthesis in mouse embryos with experimentally produced asynchrony between chromosome replication and cell division. *Roux's Arch. Dev. Biol.* 192, 138–44.

Peyrieras, N., Hyafil, F., Louvard, D., Ploegh, H. D., and Jacob, F. (1983) Uvomorulin: a nonintegral membrane protein of early mouse embryo. *Proc. Natl. Acad. Sci. U.S.A.* 80, 6274−7.

Phaire-Washington, L., Silverstein, S. C., and Wang, E. (1980) Phorbol myristate acetate stimulates microtubule and 10 nm filament extension and lysosome redistribution in mouse macrophages. *J. Cell Biol.* 86, 641−55.

Pratt, H. P. M. (1978) Lipids and transitions in embryos. In *Development in Mammals*, Vol. 3, ed. M. H. Johnson, pp. 83−130. Amsterdam: Elsevier.

−(1985) Membrane organisation in the preimplantation mouse embryo. *J. Embryol. Exp. Morph.* 90, 101−21.

Pratt, H. P. M., Chakraborty, J., and Surani, M. A. H. (1981) Molecular and morphological differentiation of the mouse blastocyst after manipulations of compaction with cytochalasin D. *Cell* 26, 279−92.

Pratt, H. P. M., Ziomek, C. A., Reeve, W. J. D., and Johnson, M. H. (1982) Compaction of the mouse embryo: an analysis of its components. *J. Embryol. Exp. Morphol.* 70, 113−32.

Randle, B. (1982) Cosegregation of monoclonal reactivity and cell behaviour in the mouse preimplantation embryo. *J. Embryol. Exp. Morphol.* 70, 261−78.

Rastan, S., Thorpe, S. J., Scudder, P., Brown, S., Gooi, H. C., and Feizi, T. (1985) Cell interactions in preimplantation embryos: evidence for involvement of saccharides of the poly-N-acetyllactosamine series. *J. Embryol. Exp. Morphol.* 87, 115−28.

Reeve, W. J. D. (1981a) Cytoplasmic polarity develops at compaction in rat and mouse embryos. *J. Embryol. Exp. Morphol.* 62, 351−67.

−(1981b) The distribution of ingested horseradish peroxidase in the 16-cell mouse embryo. *J. Embryol. Exp. Morphol.* 66, 191−207.

Reeve, W. J. D., and Kelly, F. (1983) Nuclear position in cells of the mouse early embryo. *J. Embryol. Exp. Morphol.* 75, 117−39.

Reeve, W. J. D., and Ziomek, C. A. (1981) Distribution of microvilli on dissociated blastomeres from mouse embryos: evidence for surface polarisation at compaction. *J. Embryol. Exp. Morphol.* 62, 339−50.

Sato, M., Muramatsu, T., and Berger, E. G. (1984) Immunological detection of cell surface galactosyltransferase in preimplantation mouse embryos. *Dev. Biol.* 102, 514−18.

Schliwa, M. (1982) Action of cytochalasin D on cytoskeletal networks. *J. Cell Biol.* 92, 79−91.

Shirayoshi, Y., Okada, T. S., and Takeichi, M. (1983) The calcium-dependent cell-cell adhesion system regulates inner cell mass formation and cell surface polarisation in early mouse development. *Cell* 35, 631−8.

Smith, R. K. W., and Johnson, M. H. (1986) Analysis of the third and fourth cell cycles of early mouse development. *J. Reprod. Fertil.* 76, 393−9.

−(1985) DNA replication and compaction in the cleaving embryo of the mouse. *J. Embryol. Exp. Morphol.* 89, 133−48.

Sobel, J. S. (1983) Cell-cell contact modulation of myosin organization in the early mouse embryo. *Dev. Biol.* 100, 207−13.

Solomon, F., and Magendantz, M. (1981) Cytochalasin separates microtubule disassembly from loss of asymmetric morphology. *J. Cell Biol.* 89, 157−61.

Surani, M. A. H., Barton, S. C., and Burling, A. (1980) Differentiation of 2-cell and 8-cell mouse embryos arrested by cytoskeletal inhibitors. *Exp. Cell Res.* 125, 275–86.

Surani, M. A. H., and Handyside, A. H. (1983) Reassortment of cells according to position in mouse morulae. *J. Exp. Zool.* 225, 505–11.

Surani, M. A. H., Kimber, S. J., and Handyside, A. H. (1981) Synthesis and role of cell surface glycoproteins in preimplantation mouse development. *Exp. Cell Res.* 133, 331–9.

Surani, M. A. H., Kimber, S. J., and Osborne, J. C. (1983) Mevalonate reverses the developmental arrest of preimplantation mouse embryos by Compactin, an inhibitor of HMG Co A reductase. *J. Embryol. Exp. Morphol.* 75, 205–23.

Sutherland, A. E., and Calarco-Gillam, P. G. (1983) Analysis of compaction in the preimplantation mouse embryo. *Dev. Biol.* 100, 328–38.

Tasca, R. J., and Hillman, N. (1970) Effects of actinomycin D and cycloheximide on RNA and protein synthesis in cleavage stage mouse embryos. *Nature* 225, 1022–5.

Thomson, J. L., and Biggers, J. D. (1966) Effect of inhibitors of protein synthesis on the development of preimplantation mouse embryos. *Exp. Cell Res.* 41, 411–27.

Tomasek, J. J., and Hay, E. D. (1984) Analysis of the role of microfilaments and microtubules in acquisition of bipolarity and elongation of fibroblasts in hydrated collagen gels. *J. Cell Biol.* 99, 536–49.

Vestweber, D., and Kemler, R. (1984) Rabbit antiserum against purified surface glycoprotein decompacts mouse preimplantation embryos and reacts with specific adult tissues. *Exp. Cell Res.* 152, 169–78.

Vorbrodt, A., Konwinski, M., Solter, D., and Koprowski, H. (1977) Ultrastructural cytochemistry of membrane-bound phosphatases in preimplantation mouse embryos. *Dev. Biol.* 55, 117–34.

Wiley, L. M. (1984) The cell surface of the mammalian embryo during early development. In *Ultrastructure of Reproduction*, eds. J. Van Blerkom and P. M. Motta, pp. 190–204. The Hague: Martinus Nijhoff.

Wiley, L. M., and Eglitis, M. A. (1981) Cell surface and cytoskeletal elements: cavitation in the mouse preimplantation embryo. *Dev. Biol.* 86, 493–501.

Wolf, D. E. (1983) The plasma membrane in early embryogenesis. In *Development in Mammals*, Vol. 5, ed. M. H. Johnson, pp. 187–208, Amsterdam: Elsevier.

Yoshido-Noro, C., Suzuki, N., and Takeichi, M. (1984) Molecular nature of the calcium-dependent cell-cell adhesion system in mouse teratocarcinoma and embryonic cells studied with a monoclonal antibody. *Dev. Biol.* 101, 19–27.

Ziomek, C. A., and Johnson, M. H. (1980) Cell surface interactions induce polarization of mouse 8-cell blastomeres at compaction. *Cell* 21, 935–42.

–(1981) Properties of polar and apolar cells from the 16-cell mouse morula. *Roux's Arch. Dev. Biol.* 190, 287–96.

–(1982) The roles of phenotype and position in guiding the fate of 16-cell mouse blastomeres. *Dev. Biol.* 91, 440–7.

Ziomek, C. A., Johnson, M. H., and Handyside, A. H. (1982) The developmental potential of mouse 16-cell blastomeres. *J. Exp. Zool.* 221, 345–55.

Ziomek, C. A., Lapire, M. L., Wiley, L. M., and Kim, J. (in press) Fluorescent histochemical and immunofluorescent localisation of cell surface alkaline phosphatase on mouse preimplantation embryos. *J. Embryol. Exp. Morphol.*

Ziomek, C. A., Schulman, S., and Edidin, M. (1980) Redistribution of membrane proteins in isolated mouse intestinal epithelial cells. *J. Cell Biol.* 86, 849–57.

3 Comparative aspects of embryo manipulation in mammals

VIRGINIA E. PAPAIOANNOU AND KARL M. EBERT

CONTENTS

I. Introduction

For many, mammalian embryology is synonymous with mouse embryology. This docile laboratory rodent is so inexpensive and convenient for research in experimental embryology that our concepts of the mechanisms of early mammalian development have largely been shaped by research on this single species. Notwithstanding, a vast literature and long tradition of comparative embryology stretching back to the time of Aristotle attest to the differences and similarities between embryos of different classes and species. Even among mammals there have been many fruitful studies of comparative embryology using a wide range of species. Only recently, however, have experimental techniques been applied to mammalian embryos with the intent of elucidating the underlying mechanisms guiding development. With the growth of this field, the advantages of laboratory rodents, particularly mice, became evident. More and more work was concentrated on this animal, making it a better-defined tool for continued study. For example, genetic studies of mice, which have their roots in the old mouse fancy, led to the present situation in which more is known about the genetic makeup of mice than any other mammal apart from humans. Clearly this is an advantage to the experimental embryologist who can use genetic identity to decrease experimental variation and genetic differences to provide markers to identify the origins of cells. Thus, controlled genetic variation on an otherwise uniform genetic background has become an integral part of experimental embryology, and mice provide the most flexibility for the use of this tool. Similarly, with other aspects of mouse development, the more information there is available, the more inclined researchers are to use this information to approach specific problems. As this volume attests, the advantages of the mouse have been successfully exploited, and a wealth of information is now available for this species.

In spite of this emphasis on the mouse, other species have not been totally ignored. Early development has been described for many species, and practical aims, such as improving reproductive performance in economically important animals, have led to the development of techniques for working with embryos of domestic species. Studies of the embryo in connection with the reproduction of species have yielded information, albeit incomplete, on such aspects as maternal-fetal interactions, timing of early development and implantation, and requirements for early embryogenesis. In the human, apart from descriptive information that has been obtained largely by chance acquisition of specimens, a great deal of information has been accumulated in the course of the development of techniques designed to control fertility. Contraceptive research and more recent clinical applications of in vitro fertilization and embryo transfer have yielded considerable information on the early stages of human development, in addition to information on human reproduction per se. The limitation, of course, is that ethical considerations prevent application of many experimental techniques to human embryos, and thus information bearing directly on mechanisms of development is difficult to obtain. Similarly, in other nonhuman species besides the mouse, mechanistic details are sketchy because of the sporadic application of the methods of experimental embryology.

There are many reasons why the study of a variety of species is advantageous. If hypotheses or theories of development are based on studies of one species, it is desirable to test their general applicability by comparing other species. Differences may indeed reveal unexpectedly diverse mechanisms for similar developmental events that will support or invalidate general theories. Among mammals, the variety of placentation types alone (Wimsatt 1975) necessitates the study of representative species, but a thorough understanding of the development of one type will greatly assist in understanding the variations in others. Among economically important agricultural animals, understanding differences in the timing or details of development may be essential for successful practical application of a variety of manipulative techniques.

In many respects the mouse must be recognized as an imperfect and incomplete model for humans or other mammals. For example, human and mouse placentas, although similar in later stages, have very different developmental sequences. The mouse is a polytocous species, whereas humans are monotocous, suggesting a host of differences in maternal-fetal interactions that might be better understood by studying another animal model system, such as nonhuman primates. Other limitations of the laboratory mouse as a model, such as limitations in the available genetic variability, might also be overcome by more extensive use of a variety of species.

Recently, spurred on by the promise of biotechnology for improvements in health and reproductive performance in humans and animals, experimental techniques are being applied to a wide range of species, and data on basic embryology are beginning to round out our picture of early mammalian development. In this chapter we shall review some of the experimental approaches being used in species other than *Mus musculus* and highlight a few of the species differences that have been shown to be of practical or theoretical importance. These studies will be considered within a framework designed to implement practical application of experimental procedures and biotechnology, as outlined in Figure 3.1. When the conditions for successful completion of this framework are available, it not only will meet the goals of applied science but also will result in systematic study of early embryos that will greatly enhance our understanding of species differences and our ability to formulate general principles for early mammalian development.

II. Microsurgical manipulations on preimplantation embryos

Cleavage is the specialized cell division without growth that transforms the large, single-celled zygote into a multicellular structure with cells closer in size and nuclear:cytoplasmic ratio to those characteristic of the adult organism. The embryo travels down the oviduct as it cleaves and enters the uterus, usually toward the end of this process, although the precise stage at which it enters the uterus varies in different mammals from early cleavage, as in marsupials (Lyne & Hollis 1977), to morula or early blastocyst, as in the dog (Holst & Phemister 1971). Late in cleavage, cells of the embryo — which up to this time have been fairly loosely

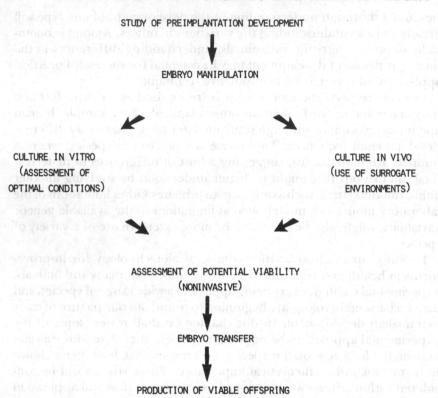

Figure 3.1. Framework for systemic study of early embryos and application of biotechnology to a variety of species.

associated spheres – begin to flatten onto one another and compact as a prelude to formation of the cystic, fluid-filled blastocyst. With blastocyst formation in eutherian mammals comes the first overt sign of morphological differentiation of two cell types, the inner cell mass (ICM) or embryonic disk and the outer layer of trophectoderm. By contrast, marsupial blastocysts consist of an unilaminar vesicle with no ICM. Following hatching from the zona pellucida, the blastocyst in some species, such as horse and mouse, remains spherical, whereas in others, such as pig and cow, it expands and elongates to a considerable extent before implanting in the uterus. The precise time and stage at implantation vary in different species, as do maternal-fetal interactions resulting in distinctively different placentation types. Marsupial embryos remain in a shell membrane until relatively late, generally two-thirds of the length of gestation, before hatching and establishing contact with the uterus.

In addition to the morphological changes during cleavage and blastulation, the zygotic genome is beginning to function and to direct the development of the organism, probably as early as the 2-cell stage in the mouse. There is molecular evidence and evidence from early lethal gene mutations in the mouse that the zygotic genome plays an essential role in development by the morula stage (Johnson 1981; Magnuson & Epstein 1981). At the blastocyst stage, differentiation has occurred as the result of a divergence in gene expression in two cell populations, the ICM and trophectoderm. This first differentiation into distinct cell types also heralds the earliest commitment of cells to specific cell lineages and results from the restriction of the potential of the previously totipotent or multipotent blastomeres.

These early events of preimplantation development have been well studied in the mouse from both morphological and experimental perspectives, and a reasonably full picture is available. But for other species of mammals such information is still very sketchy. Morphologically, the embryos of most eutherians resemble the mouse, although the timing of cleavage divisions and the relationship between morphological events and cell numbers vary. It is becoming evident from experimental work that differences in timing are not the only distinctions and that embryos of different species with the same cell numbers or at the same morphological stage may be fundamentally different in the degree of determination that has taken place, the potential of blastomeres, and the fate of different cell types.

Several types of studies in other species, particularly twinning and the formation of chimeras, bear on these issues, but comparative analysis of these studies is complicated because often it is not possible to determine the precise age of an embryo. Assumptions must be made as to the timing of ovulation and fertilization, and frequently embryos are timed from an arbitrary point such as the onset of estrus. For example, in the pig, the time of onset of estrus and first mating is conventionally used to time pregnancy, but ovulation is generally assumed to take place anywhere between 24 and 40 h after this time. This means that there will be considerable variation in the fertilization ages of embryos for a given time after the onset of estrus, particularly because ovulation in this polytocous species may occur over an extended period. With these limitations, Table 3.1 gives an indication of estimated cleavage rates for several mammals and illustrates the different rates of tubal transport and early development using estimated fertilization time as the starting point. Because of uncertainties regarding the exact ages of embryos, it is perhaps most useful in experimental studies to make species comparisons between embryos of similar cell numbers or morphological stage.

A. Cell potential during cleavage and blastulation: isolation and regulation studies

Cleavage in the mammal produces uniform, morphologically similar blastomeres. The first overt sign of differentiation is seen at compaction of the morula, just prior to blastulation, when the cells on the

Table 3.1. *Comparative cell numbers and timing (days after fertilization) of developmental events in some experimental species*

	Cleavage		Compaction of morula		Blastulation		Entry into uterus (days)	Start of implantation		References[c]
	2c (days)	8c (days)	Days	Cell no.	Days	Cell no.		Days	Stage[a]	
Marsupials	1–2	—	—[b]	—	3	32–80	1[c]	9	ES	1
Laboratory species										
Mouse	1	2	2.5	8–16	3	32	3	4.5	BC	2
Rat	1.5	3.5			4.5		3.5	5.5	BC	3
Rabbit	0.5	1.5	2	16–32	3	128	3	6.5	BC	3
Domestic species										
Pig	0.5–1	2–3	3.5	8–16	3.5–5	16	2.5–3.5	14–15	ES	4,5
Sheep	1	2	3		4.5	64	3	15–16	ES	5,6
Cow	1	2–4	4–5	16	5–6		3	17–19	ES	5
Horse	1	3	4–5	>15	7–8		5–6	34	16 mm	7
Primates										
Rhesus monkey	1	2–3	>4	>26	By 7		4	8–9	BC	8
Baboon			5		5–8		4–5	8–9	BC	9
Human	1.5–2.5	2.5–3[d]	3–5	16	4.5–5[d]	64–107	3–4	6–7	BC	9,10

Note:

[a] ES = early somite; BC = blastocyst; 16 mm = crown-rump length.

[b] No morula stage is present in marsupials.

[c] Early cleavage; 1- to 4-cell stage.

[d] In vitro.

1, Lyne and Hollis (1977); 2, Mayer and Fritz (1974); 3, Davis and Hesseldahl (1971); 4, Perry and Rowlands (1962); 5, King, Atkinson, and Robertson (1982); 6, Willadsen (1980); 7, Hamilton and Day (1945); Steven and Morriss (1975); Allen (1982); 8, Heuser and Streeter (1941); Lewis and Hartman (1941); Enders and Schlafke (1981); 9, Edwards (1980); 10, Brackett (1978).

outside of the morula undergo a series of changes along the pathway of differentiation to trophectoderm. The point at which this process is irreversible marks the time that cell potential is restricted. Until then, in the indeterminant cleavage characteristic of mammals, each blastomere theoretically has the potential to form any and all cell types. As development continues, other determination events further restrict the potential of different cells and regions of the embryo. When and how these restrictions in potential occur are topics of considerable experimental effort (see Chapters 1, 2, and 4). The most stringent test of cell potential is whether or not an isolated cell or part of an embryo is totipotent, that is, whether or not it is capable of regulating its development such that it can compensate for an overall reduction in embryonic cell numbers and produce a viable fetus and offspring. One must determine if any or all of the cells or fractions of an early embryo have this capacity. For example, development of one or even three blastomeres from a 4-cell embryo is an indication of individual blastomere totipotency, but all four blastomeres from a single embryo must be shown to develop normally before complete totipotency of 4-cell blastomeres is proven. Likewise, the production of even a single pair of identical twins from bisection of an embryo is proof that two halves can be totipotent, but a further consideration is the plane of bisection with respect to any visible or invisible asymmetry in the embryo and the determination whether or not any two halves are equally competent regardless of the plane of bisection. For the polarized blastocyst stage, it seems evident that only manipulations that bisect the ICM or embryonic disk, giving balanced half-embryos, can result in successful development of both halves (Gardner 1972, 1974). However, bilateral asymmetry has not been ruled out in the blastocyst or even in earlier stages, and if it does exist, it is possible that only particular planes of section will yield two totipotent halves. Because of this possibility, a large number of randomly bisected pairs must be analyzed. Marsupial blastocysts, which are unilaminar, with no ICM, present yet another interesting and completely unexplored problem in cell and embryonic potency.

i. Laboratory species. In the mouse, the potential of single isolated blastomeres to form embryos appears to be restricted relatively early (Table 3.2). It has been found that single blastomeres from the 2-cell stage, but not later stages, are capable of developing into complete, viable fetuses or offspring (Tsunoda and McLaren 1983). Although individual blastomeres from embryos up to at least the 8-cell stage appear to be totipotent in terms of the tissues they can produce in a chimera (Kelly 1975), in isolation they are not capable of organizing into viable fetuses (Rossant 1976). Little work has been done with rat blastomeres, but a complete egg cylinder was reported to have developed from one blastomere of the 2-cell stage, although development stopped shortly there-

Table 3.2. *Percentage normal development of manipulated partial embryos into viable fetuses or offspring*

	Single blastomere from			Half-embryos from					Both halves (twins) from				
	2c	4c	8c	2c	4c	8–16c	Compact morula	BC	2c	4c	8–16c	Compact morula	BC
Mouse	65	0	0	30	46	30	42	46	40	0	30	27	0
Rabbit	30	19	11	—	—[a]	—	—	—	—	—	—	—	—
Sheep	52	16	6	80	100	80	58	—	31	>0[b]	100[c]	41	—
Cow	—	—	—	16	—	16	46	48	—	—	9	23	28
Horse	—	36	—	16	—	16	—	—	—	—	—	—	—

Note: Results from different studies are combined (see text for references).

[a] Not done.

[b] Two pairs of twins have been produced following freezing of one half of each pair (Willadsen 1980).

[c] This figure represents only two twin pairs (Willadsen 1980).

after, and no offspring were born (Nicholas and Hall 1942). By contrast, single blastomeres from 4- and 8-cell stages of the rabbit have shown totipotency in the limited work that has been devoted to this species (Seidel 1952, 1956; Moore, Adams, and Rowson 1968).

If mouse embryos are halved, it is found that survival of half-embryos to midgestation or term can be as high as 65 percent for the 2-cell stage (Tsunoda and McLaren 1983), 46 percent for the 4-cell stage (Rossant 1976), 30–42 percent for 8- to 16-cell stages (Tsunoda and McLaren 1983; Nagashima et al. 1984), and 46 percent for the blastocyst (Gardner 1974). However, survival of both halves of the same embryo (i.e., production of identical twins) has been demonstrated only for the 2-cell stage (40% twins) and 8- to 16-cell stages (30% twins) (Moustafa and Hahn 1978; Gartner and Baunack 1981; Tsunoda and McLaren 1983; Nagashima et al. 1984), not for the blastocyst stage (Table 3.2).

ii. Domestic species. Because of the potential economic and experimental value of genetically identical individuals, attempts have been made to produce twins in several domestic species, and the experimental data have provided basic information on early embryonic development. In these studies, considerable variations in technique have been employed, even within a species, and that makes strict comparisons of success rates from one stage or study to another difficult. Consequently, the combined results presented in Table 3.2 represent the total effort in each species, rather than the best success rates obtained in any given study.

In swine, single blastomeres of 2–8-cell stages have been isolated and grown in vitro to the blastocyst stage (Moore, Polge, and Rowson 1969; Menino and Wright 1983), but as yet the further potential of these manipulated embryos has not been tested in vivo, nor have attempts to produce identical twins been reported. One reason for this may be the limitations imposed by certain features of reproduction in the pig. More than two fetuses are necessary to maintain a pregnancy, and transmigration of embryos from one uterine horn to the other is the rule. Thus, in order to recognize an experimental twin pair, control embryos from a separate stock containing a distinctive genetic marker must be transferred to recipients along with each twin pair in order to sustain the pregnancy and still provide a means of recognizing both halves of individual pairs.

Totipotency of isolated, individual sheep blastomeres is maintained up to the 8-cell stage (Trounson and Moore 1974a; Willadsen 1981), but no more than three of the blastomeres from a 4-cell embryo and one of the blastomeres from an 8-cell embryo have been proved totipotent. This could be a technical limitation, because relatively few attempts have been made to accomplish the difficult experimental procedure of testing all four or all eight blastomeres from a single embryo. Alternatively, it could represent a restriction in the potential of some of the blastomeres by these stages and indicate that the morphologically similar cells have already undergone restriction in cell fate and potential. In another experiment, four offspring (identical quadruplets) were born from a single sheep embryo that had been quartered at the 8-cell stage (Willadsen 1981). If these quarters consisted of daughter-cell pairs, as seems likely from the procedure used (Willadsen 1980), then this result might argue that all four blastomeres at the 4-cell stage do indeed retain totipotency. A high rate of survival of half-embryos to term has been demonstrated for the 2-cell stage through the blastocyst stage in sheep (Table 3.2), and both halves of all of these stages have produced identical-twin offspring (Willadsen 1979, 1980; Gatica et al. 1984).

The potential of single blastomeres of the cow embryo has not been investigated, but quarter-embryos from the 8-cell stage and 32–64-cell morula stage have produced one set of identical triplets (Willadsen and Polge 1981) and four sets of twins, respectively (Willadsen et al. 1981), indicating that at least some subsets of blastomeres at these stages are totipotent. Bisection of cow embryos has resulted in successful production of identical twins at the early morula stage through the blastocyst stage (Ozil, Heyman, and Renard 1982; Lambeth et al. 1983; Ozil 1983; Williams, Elsden and Seidel 1984) and in high overall rates of half-embryo survival (Table 3.2). In these studies, even higher rates of pregnancy were reported at midgestation for bisected embryos, but these figures can be misleading unless a complete assessment of embryonic normality is made at these earlier times. Reported losses later in gestation clearly indicate that not all embryos that implant are capable of full development. Determining the causes of fetal mortality in these cases might well provide insight into the possible restriction of potential or regulative capacity in preimplantation embryos.

Much less work has been published regarding manipulation of preimplantation horse embryos, but three offspring have been reported from bisected 2–8-cell stages (no twin pairs), and four offspring comprising two sets of twins have been obtained from isolated 4-cell blastomeres (Allen 1982; Allen and Pashen 1984). Thus, at least two of the blastomeres at the 4-cell stage are totipotent, but no information is available for later stages.

From this as yet incomplete picture of domestic and laboratory species, it appears that the capacity for independent development of a single blastomere is retained to a later stage in cleavage in rabbit, sheep, and horse embryos than in the mouse. Survival of half-embryos may be similar or somewhat better in sheep and cows than in mice, and the survival of both halves or even quarters to produce identical offspring has been successful in domestic species with significantly fewer attempts, indicating a greater capacity for independent development of partial embryos in these species, compared with the mouse. The occurrence of blastulation at a later cleavage division and thus a higher total cell number (Table 3.1) may allow greater flexibility in embryonic organization. If there is a critical number of ICM cells below which normal development of the fetus will not occur, embryos that blastulate with a higher total cell number, and thus presumably a higher proportion of ICM cells (Buehr and McLaren 1974), may be able to tolerate a proportionately greater decrease in cell number. However, the relationship among total cell number, proportion of cells in the ICM, and embryonic viability needs to be determined to test this idea.

Apart from the production of genetically identical offspring, another procedure that would be valuable for research and animal husbandry is routine sexing of preimplantation embryos. Epstein and associates (1978) have successfully sexed mouse embryos by separating blastomeres at the 2-cell stage, growing the twin halves to the blastocyst stage, and using one blastocyst to sex the pair. This procedure for the production of offspring of known sex would be simplified in species with an even greater regulative capacity than the mouse. Likewise, the procedure of sampling the trophectoderm that was devised to sex living rabbit blastocysts (Edwards and Gardner 1967; Gardner and Edwards 1968) could be extended to other species that have a good regulative capacity and abundant trophectoderm before implantation. The difficulty in obtaining biopsies from trophectoderm of the spherical blastocysts of farm animals may limit practical application of this approach. However, Hare and associates (1976) have been successful in sexing two-week-old bovine embryos and producing a high percentage of correctly sexed offspring. These older embryos that have elongated are easier to biopsy, but the success of pregnancy after transfer is much lower than following transfer of the early blastocyst. The sexing method used, identification of sex chromosomes in metaphase spreads, requires a relative large amount of tissue. Greater efficiency may result if it is possible to detect sex chromatin in biopsies from earlier blastocysts, as was done in the rabbit.

B. Cell fate and potential: chimera studies

A recent volume on the use of chimeras attests to the extensive and successful use of this experimental approach in developmental biology (LeDouarin and McLaren 1984). Among mammals, the mouse has been used almost exclusively in cell potential and cell lineage studies (reviewed by Rossant 1984), although chimeras in a few other species, such as rabbit, rat, and sheep, and between species, such as rat and mouse and different species of mouse, have also been constructed (reviewed by Papaioannou and Dieterlen-Lièvre 1984) and are potentially of value in confirming or extending information gained from mouse studies. Recently, such techniques have been applied to the production of cow chimeras (Summers, Shelton, and Bell 1983; Brem, Tenhumberg, and Krausslich 1984) and interspecies sheep-goat chimeras (Fehilly, Willadsen, and Tucker 1984a; Meinecke-Tillmann, and Meinecke 1984). The small numbers of chimeras thus far produced in species other than mouse and the lack of adequate cell markers to trace component contributions have limited the information obtainable from these studies, particularly regarding cell fate and potential. Nevertheless, some new information has emerged, for the most part indicating close similarities with the mouse.

In all chimera studies involving aggregation of preblastocyst stages, as well as in the embryo-splitting studies referred to earlier, the timing of the developmental events of compaction and blastulation appears to be unaffected by a change in cell number, as has been observed by many workers using the mouse. The capacity of embryos to regulate for cell numbers greater than normal has been demonstrated by successful development of chimeras following aggregation of cleavage stages and/or injection of a second ICM into blastocysts of the rabbit (Gardner and Munro 1974), rat (Mayer and Fritz 1974), cow (Summers et al. 1983; Brem et al. 1984), and sheep (Tucker, Moor, and Rowson 1974; Fehilly et al. 1984b), but in no case does this regulation appear to occur before the blastocyst stage. In all species studied, aggregates form blastocysts larger than normal. In one study of sheep chimeras, the claim was made that the proportion of cells allocated to the ICM was greater in blastocysts formed by aggregation (Fehilly et al. 1984b), although no evidence of cell numbers was presented. If this observation can be substantiated, it will be in agreement with mouse studies (Buehr and McLaren 1974; Rand 1985) confirming the prediction of the inside/outside hypothesis of ICM and trophectoderm determination that double-size embryos will have more than twice as many inside cells.

In the same study of sheep chimeras, aggregates of four 8-cell sheep embryos were made. Among 12 chimeric offspring in this group, inadvertent production of one set of chimeric twins from a single composite chimeric blastocyst may indicate that the limits of downward size regulation were being approached with four times the normal cell number. Two blastocysts with eight times the normal cell number failed to develop as chimeras. As in the mouse (Markert and Petters 1978), more than two embryos can make cellular contributions to offspring in sheep: Three embryos (6 parents) have been implicated in contributing to each of three chimeras (Fehilly et al. 1984b).

In the cow, a single calf chimeric only for internal tissues has been produced by blastocyst injection (Summers et al. 1983), and a single calf chimeric for coat color has been produced following aggregation of four halves from two bisected morulae (Brem et al. 1984). In the latter study, no chimeras were obtained from 16 attempts to aggregate only two halves of bisected morulae, a failure attributed to a greater possibility of one or the other half forming the entire ICM in these normal-size aggregates. However, because of the small number of offspring and the absence of markers other than coat color, it is possible that some of the other six calves that went to term following this manipulation were in fact covert chimeras. In addition, although it is likely that cell lineages will be similar among mammals, it has not been formally shown for any species other than mouse that outside morula cells become trophectoderm and that the ICM and trophectoderm do in fact have distinct fates (Papaioannou 1982), an assumption inherent in the interpretations made by these workers.

Interspecies sheep-goat chimeras have been produced by blastomere aggregation and by injection of ICM plus polar trophectoderm into blastocysts (Fehilly et al. 1984a). The results indicate that chimeras will survive if the species genotype of the trophectoderm is the same as that of the host uterus, a result compatible with evidence from mouse inter-species chimeras (Rossant, Mauro, and Croy 1982). Again, the assumption is made that the mechanism of determination of trophectoderm, as well as the later cell lineage of trophectoderm, is the same in sheep-goat chimeras as in the mouse, an assumption that has yet to be verified. The development in a sheep mother of a putative goat offspring from a sheep-goat chimeric embryo (Meinecke-Tillmann and Meinecke 1984) was also assumed possible because of preferential allocation of sheep cells to the trophectoderm of the aggregate blastocyst. This animal developed from an aggregate of one 4-cell sheep blastomere and two 8-cell goat blastomeres. In similar asynchronous aggregates of mouse blastomeres, Spindle (1982) found a preferential, although not exclusive, allocation of the more advanced cells to the ICM of chimeric blastocysts, a result that could be explained by increased cell adhesiveness at the 8-cell stage (Kimber, Surani, and Barton 1982). In interspecies rat-mouse aggregation chimeras, Tachi and Tachi (1980) found a tendency for mouse cells to contribute preferentially to ICM and rat cells to contribute to trophectoderm of blastocysts, but in no case was an exclusive allocation of cells seen. In the sheep-goat aggregate there was no direct evidence of the species origin of the trophectoderm, and the lack of suitable markers makes it impossible to rule out the possibility that the "goat" offspring was in fact a chimera. In light of the potential uses for interspecies transfers (i.e., rescue of exotic species), it is essential that this work be corroborated under more controlled conditions.

The available information from other mammals indicates that their embryos share with the mouse a considerable flexibility in regulative systems and may share mechanisms of determination, although hard data are still scarce. For this reason it is important to critically evaluate experiments to avoid basing interpretations on assumed similarities that have not been proved. One need only look at obviously exceptional animals such as the nine-banded armadillo, which regularly produces monozygotic quadruplets, or marsupials, in which the blastocyst has no ICM, to be reminded that smaller variations among more closely related species might well exist.

What has been discovered about the development of the mouse embryo beyond blastulation, including specific cell interactions and lineage relationships, may not hold for all species, especially those with very different types of placentation. The polar trophectoderm, which in the mouse proliferates rapidly under the direct influence of the ICM during implantation (Gardner, Papaioannou, and Barton 1973), disappears completely before implantation in certain other species such as pig and sheep. The hypothesis that abembryonic trophectoderm cells cease mitotic division and endoreduplicate their DNA as a result of their distance from the ICM clearly cannot hold in unmodified form for species in which the blastocyst expands into an enormous filamentous structure (up to 1 m long in the pig and 100 mm in the sheep) consisting of thousands of trophectoderm cells, both with an only slightly enlarged embryonic disk. Morphological differences in the formation of the amnion may be indicative of variations in the cell lineage relationships that have been worked out for the mouse. The amnion can form by folding of a bilaminar membrane consisting of trophectoderm and mesoderm, as described for sheep and pig, by cavitation between the polar trophectoderm and ICM, followed by separation and differentiation of "amniogenic" cells presumably derived from the trophectoderm, as in primates, or by a combination of cavitation and folding, as seen in rodents. Although the broad outline of cell lineage relationships can be extrapolated using the mouse as a model, the details can be ascertained only by experimental studies on representative species.

III. Genetic manipulation in various mammalian species

Because of developments in biotechnology, single gene modifications can now be made prior to the first cell division of an organism. Thus, alteration or integration of a gene can be established prior to cellular differentiation. It is the purpose of this section to explore some of the experimental manipulations of the mammalian embryo that result in genetic engineering of the gamete or zygote. Specifically, we shall address methods of controlling the genetic makeup of mammalian species by embryo manipulation.

A. Parthenogenesis

Parthenogenetic activation of an ovum and production of a diploid homozygote is a natural form of cloning that exists in many lower forms of life, including avian species (i.e., turkey). However, this asexual or vegetative reproduction has not been shown to be a means of propagation in mammalian species. Although parthenogenetic development can be experimentally induced in mice (Tarkowski 1975) and has been shown to occur spontaneously in ovaries of some strains (Eppig 1978), no live young have ever been produced by parthenogenetic activation of a mammalian embryo. Parthenogenetic mouse embryos can nonetheless contribute viable cells to chimeras, as discussed in Chapters 13 and 15. The reader is directed to the monograph by Kaufman (1983) for a detailed review of parthenogenesis.

B. Microinjection of embryos

i. RNA and DNA injection. The ability to change or alter the genetic makeup of an individual was significantly advanced with the development of micromanipulative and microsurgical techniques for the early embryo. The work of Markert and Ursprung (1963) and Gurdon (1963) showed that amphibian eggs can be manipulated and microinjected with various substances without affecting the viability of the developing embryo. Lin (1966) was successful in microinjecting bovine gamma globulin into the mouse embryo and, on subsequent transfer of these embryos to recipient females, produced living fetuses. Injection of foreign mRNA into the cytoplasm of the amphibian oocyte and fertilized ovum showed that foreign mRNA could be accurately translated with some peptides being secreted by the cell (Gurdon, Lane, and Woodland, 1971; Colman and Morser 1979). These successes challenged mammalian embryologists and developmental biologists to extend these techniques to mammalian embryos and oocytes and to attempt alteration of the mammalian genome.

The fertilized mouse ovum can translate exogenous mRNA that is microinjected into the cytoplasm (Brinster et al. 1980; Brinster, Chen, and Trumbauer 1981; Ebert and Brinster 1983). This ability, along with the development of techniques to identify and quantify message products, allows for direct study of mRNA translation during early development. Questions such as how the oocyte maintains a relatively constant amount of mRNA and how it maintains protein synthesis without detectable transcriptional activity may be addressed by injection of exogenous mRNA, because it can be easily detected in a pool of endogenous mRNA and monitored for its stability and translational activity. Experiments have detected significant differences between oocytes and fertilized eggs in regard to stability of mRNA; however, translation of foreign mRNA appears to be similar in both cell types (Ebert et al. 1984). This type of experiment can be useful in understanding the mechanisms of message translation in the mammalian oocyte and zygote. The relatively small amounts of material that can be injected and the large numbers of cells needed for accurate measurements of mRNA and peptide products limit the extent to which this method is being pursued. However, there may be differences among mammals that could be profitably explored using the somewhat larger eggs of domestic species.

The nuclear milieu itself can now be invaded directly by microinjection of genes that can be integrated and expressed by the individual. Recently, selected sequences of DNA consisting of a eukaryotic gene with or without a segment of a bacterial plasmid have been microinjected into the pronucleus of the mouse fertilized ovum (see Chapter 16). Transfer of the injected ova to recipient females has produced transgenic animals capable of transcribing the introduced DNA molecules, producing authentic protein products such as herpes thymidine kinase (Brinster et al. 1981), rat growth hormone (Palmiter et al. 1982), transferrin (McKnight et al. 1983), and immunoglobulin (Brinster et al. 1983). The results of these experiments clearly show that injection of foreign genes into the pronuclear region of the fertilized egg results in gene integration and expression in adult tissues. In some cases, the integrated genes were passed through the germ line from one generation to another, generating transgenic lines of animals with similar genotypic and phenotypic alterations. The most dramatic example of this genetic engineering approach was reported by Palmiter and associates (1982). They showed that mice injected with a fusion gene containing the metallothionein promoter and rat growth-hormone gene resulted in dramatic increases in the size of the animals expressing this gene. Results such as this have spearheaded a major effort to produce transgenic animals in domestic species and to explore the possibility of improving the efficiency of food production and animal health.

Although the microinjection of genes into the pronucleus of the mouse ovum is relatively easy even with bright-field microscope optics, the dense cytoplasm within the fertilized embryo in domestic species poses a distinct problem; that is, the pronucleus of the fertilized 1-cell ovum is not easily visualized by bright-field, phase, or interference-contrast optics, with the possible exception of the sheep ovum (Hammer et al. 1985). A vital DNA dye (Hoechst 33258) has been used to locate the pronuclei of mouse embryos while still maintaining viability of the treated embryos (Ebert, Hammer, and Papaioannou 1985) and is promising for use with fertilized eggs of domestic species (K. M. Ebert, unpublished data). Another vital nuclear DNA dye (DAPI) has also been used (Minhas et al. 1984). However, these dyes do not define the boundaries of pronuclei with a great degree of contrast in embryos containing a large amount of lipid within the cytoplasm. Efforts to enhance visualization of pronuclei turned to use of the centrifugation technique that was introduced by Morgan (1927) and Conklin (1931) using sea urchin and ascidian embryos. This technique has been successful in visualizing pronuclei of pig and cow embryos

(Wall et al. 1985) and has been used for microinjection of genes into the fertilized embryo of the pig (Hammer et al. 1985). The method could be a major breakthrough for production of transgenic livestock of other domestic species, providing their embryos can withstand centrifugation.

Several laboratories are presently involved in experiments designed to produce transgenic livestock, but only one success has been reported. The metallothionein–growth-hormone fusion plasmid constructs that effectively produced phenotypic changes in transgenic mice (Palmiter et al. 1982) have been injected and integrated into the genome in rabbit, pig, and sheep embryos. The incidences of gene integration in these species were lower than in mouse, but factors other than the actual DNA integration, such as manipulation, embryo transfer, and survival, can have significant effects on the overall success of the procedure. The integrated gene was expressed, leading to production of human growth hormone in one rabbit and 11 pigs, but as yet no dramatic increase in body weight has been detected (Hammer et al. 1985).

ii. Injection of sperm components. The microinjection procedure also offers an opportunity to study the various contributions that the sperm makes to the process of embryogenesis. Microinjection of sperm or sperm components can be used to genetically alter the mammalian embryo as well as to assess the importance of sperm components in the interaction of the sperm and egg during development. Another possibility is the production of normal offspring from selected or treated sperm.

At present, there have been only a few reports on sperm injection. Uehara and Yanagimachi (1977) showed that the nucleus of hamster sperm injected into the hamster egg developed into a male pronucleus, but they did not test the developmental potential of these eggs. Heterospecific sperm-egg interactions were also studied using the microinjection technique, and although male pronuclei were formed in heterospecific egg cytoplasm, further developmental potential was not tested (Thadani 1980). Markert (1983) reported that microinjection of sperm into the ovum resulted in activation of the mouse ovum and development of blastocysts in vitro; however, the ultimate viability test, that is, transfer of these blastocysts to recipient females and production of living offspring, was not attempted.

iii. Injection of mitochondria. Microinjection of mitochondria may be a way of studying nonnuclear maternal inheritance. This concept is based on the idea that the mitochondrial genome (mtDNA) of an individual is derived solely from the maternal parent (Dawid and Blackler 1972; Hutchison et al. 1974). In the mouse, the process of sperm penetration results in introduction of approximately 75 paternal mitochondria into the oocyte (Bahr and Engler 1970), which contains 100,000 mitochondria (Piko and Matsumoto 1976). Failure to detect paternal mitochondria may be due to the large dilution of paternal mtDNA. Studies designed to inject significantly higher numbers of sperm mitochondria than normally enter by fertilization and to test different sources of mitochondria (somatic and testicular) should determine whether or not sperm mitochondria become selectively defective in their ability to replicate (K. Ebert and N. Hecht, unpublished data). This approach may also be used to produce offspring that are mitochondrial hybrids, that is, possess a mixture of mitochondrial genomes.

iv. Nuclear transfer. The desire to produce genetically identical organisms by cloning has met with success in some organisms but not others. For the purposes of this discussion, cloning will be understood as the production of two or more genetically identical individuals from a single sexually derived zygote. A naturally occurring yet limited form of cloning in mammals is the formation of identical twins from a separation of embryonic cells, sometime during early development. The phenomenon can be artificially produced by the experimental means discussed in the previous section. Despite the limitations imposed by inadequate methods, it is obvious that cloning by dividing preimplantation embryos is technically feasible and can provide a means of controlling variation between animals and thus reducing the total number of animals required for experimentation.

One of the most direct means of altering the genotype of an individual is removal of the nucleus from the fertilized egg and its replacement with a foreign nucleus. This approach has been an experimental tool in developmental biology since 1952, when Briggs and King transferred nuclei from blastula cells to enucleated frog eggs to produce composite eggs that could give rise to complete tadpoles. This method has been used to study the totipotency of nuclei from early mammalian embryos in both rabbit (Bromhall 1975) and mouse (Modlinski 1978; Illmensee and Hoppe 1981), as well as nuclei from parthenogenetic mouse embryos (Hoppe and Illmensee 1982). Recently, McGrath and Solter (1983a, 1983b) have altered the technique so that egg-membrane penetration for removal and reinsertion of the nuclei is replaced by fusion of a karyoplast with the egg membrane. Their higher success rate with this modified procedure has offered the opportunity for reinvestigating earlier experiments (see Chapter 13).

The potential of a nucleus to produce a complete organism following transfer to an enucleated host zygote appears to be inversely correlated with the number of nuclear divisions the nuclear donor cell had undergone. No organism has yet developed from a nucleus taken from an adult cell, whereas nuclei from cells at early stages of development, such as the frog blastula (Briggs and King 1952), have developed into complete organisms. Similar results have been reported for nuclei of the morula and ICM of a mouse blastocyst (Illmensee and Hoppe 1981). However, experiments by McGrath and Solter (1984) have failed to produce blastocysts from nuclei taken from any stage later than the 8-cell stage, and thus the potential for cloning from the mammalian embryo seems to be limited. A nucleus from a differentiated amphibian cell, however, can clearly give rise to cell types different from the tissue of its origin (DiBerardino 1980). The greatest extent of reprogramming apparently can take place within the cytoplasm of the oocyte (Leonard, Hoffner, and DiBerardino 1982). If this process of nuclear reprogramming were better understood, then cloning in mammals might become feasible for a broader range of developmental stages.

Nuclear-transfer experiments with embryos of domestic species are just beginning, and there is some evidence indicating that this approach will be successful (Willadsen 1986). Although significant developmental differences might be discovered, it still appears unlikely that cloning of mammals using transfer of early embryonic nuclei will become economically feasible, because of the difficulty of the procedure and probable limitations in nuclear or cell developmental potential.

IV. Culture and transfer of manipulated embryos

A. Importance of the zona pellucida

Most embryo-manipulation procedures involve removal of or damage to the zona pellucida (ZP). It has been suggested that the ZP provides an encasement that maintains the position of the blastomeres during cleavage and also that it maintains a barrier between the oviduct and uterine epithelium during transport and trophectoderm differentiation. Although most studies suggest that removal of the ZP from early developmental stages of the mouse embryo does not affect subsequent development in vitro (Tarkowski 1961; Mintz 1962; Konwinski, Solter, and Koprowski 1978; cf. O'Brien, Critser, and First 1984), it has a significant effect on movement of the embryo through the oviduct following transfer to pseudopregnant recipients and thus on subsequent developmental capacity. Bronson and McLaren (1970) transferred ZP-free 8-cell embryos and blastocysts to the oviduct and showed a significant decline in pregnancy rate among females with denuded cleavage-stage embryos, compared with denuded blastocysts. This did not appear to be because of loss of embryo viability due to the pronase treatment that was used to remove the ZP, because denuded 8-cell embryos developed normally when transferred to the uterus in pseudopregnant females. They concluded that the ZP is necessary for oviduct transit of cleavage stages, but not of later embryos. However, a differential effect of pronase on embryos of different ages cannot be ruled out. In a similar experiment, Modlinski (1970) studied the fate of ZP-free early cleavage-stage mouse embryos in the oviduct and showed in histological sections that eggs stuck to each other or were trapped within the folds of the oviduct and possibly stuck to the oviductal epithelium. The denuded embryos were deformed and elongated within a few hours of transfer, and no intact embryos could be found a day after transfer – observations that support the idea that the ZP holds the cleaving blastomeres together and prevents them from adhering to the oviduct.

The ZP of rabbit embryo may be critical in maintaining blastomere contact because denuded 1- and 2-cell rabbit eggs continue to cleave in culture, but most of the blastomeres separate following cleavage (Edwards 1964). Additionally, the separate blastomeres do not undergo differentiation of trophectoderm-like cells: No adhesion or outgrowth occurred in culture, as would normally happen if embryos between the 4-cell stage and blastocyst stage were denuded and cultured under the same conditions. Assessment of viability following the removal of the ZP from later-stage rabbit embryos (morula) indicates that, unlike the situation for the mouse, denuded embryos will not implant following transfer to recipient females (Rottmann and Lampeter 1981), although in this study there was no control for the effect of pronase on embryo viability. It is obvious that the importance of the ZP for development and oviduct transit of the embryo will have to be evaluated for each species separately. In addition, manipulated embryos may be differentially sensitive to removal of the ZP or to different methods of ZP removal. Tsunoda and McLaren (1983) found an adverse effect on in vivo development of bisected 2-cell embryos denuded by pronase, as compared with mechanically denuded embryos, and O'Brien and associates (1984) found that the decreased viability of pronase-denuded, bisected 2-cell embryos could not be improved by reinsertion into another ZP.

Removal of the ZP of sheep eggs with pronase is a slow process and can result in blastomere separation and cessation of development in vitro and in vivo (Moor and Cragle 1971). Trounson and Moore (1974b) compared removal of the ZP of fertilized sheep eggs by "protease" and by mechanical means. They found no difference in proportions of embryos that developed to normal fetuses in recipient females between late morula-stage and early blastocyst-stage embryos denuded by either means. When embryos at earlier stages (8-cell to early morula) were treated with protease, however, the ZP was only partially digested, and these embryos did significantly better than similar-stage embryos completely denuded by mechanical means. Unfortunately, this study did not include controls with untreated, intact ZP, which means that the effect of ZP removal per se cannot be evaluated.

Recent experiments designed to study the developmental potential of early blastomeres of the pig indicated that in vitro development of pig blastomeres with and without the ZP was similar, at least up to the early blastocyst stage (Menino and Wright 1983). However, subsequent transfer of these blastocysts to recipient females was not performed, and thus the potential of these embryos to produce living young has not been tested.

In the cow, there is limited evidence that the ZP is not necessary for continued viability of morulae or blastocysts transferred to recipient females (Hoppe and Bavister 1983). In spite of this, most investigators still routinely place manipulated cow embryos back into a ZP prior to transfer (Ozil 1983; Lambeth et al. 1983). The importance of this aspect of domestic embryo-transfer protocols needs to be further tested, especially in experiments with manipulated embryos. For earlier stages in which reinsertion of manipulated embryos into a ZP is necessary, the species origin of the ZP appears to be unimportant, at least for the development of cow and horse embryos (Willadsen et al. 1981; Willadsen and Polge 1981; Allen and Pashen 1984).

Willadsen (1979, 1982) pioneered a method of agar embedding of manipulated embryos, followed by transfer to the oviduct in temporary recipients, often of another species, as an alternative to culture of manipulated early embryos in domestic species. The agar coating not only seals imperfections in the ZP that result from the manipulation procedures and thus provides the embryo added protection but also allows groups of embryos, for example, twin pairs, to be kept together for evaluation before final transfer to suitable hosts.

In summary, cleavage-stage embryos probably require a ZP for oviduct transit, and some species may require it to maintain the integrity of the cleaving embryo. The method of ZP removal, particularly the type and length of enzyme treatments, can affect the further development of intact and/or manipulated embryos, and later-stage embryos of some species such as the rabbit may require an intact ZP to complete preimplantation development.

B. Use of surrogate culture environments

Because manipulation of early embryos necessarily entails heavy losses, it is extremely useful to have an evaluation period before the embryos are committed to a final host, for reasons of economy and also for more accurate experimental evaluation. In vitro culture of early mammalian embryos has been used as a means of directly observing the effects on embryonic development of certain microenvironments that may occur in vivo. Culture systems also enable the investigator to systematically dissect the microenvironment surrounding the embryo in order to discover the minimal requirements necessary for normal development. A variety of media from simple basic salt solutions to complex nondefined media have been used to provide conditions for viable embryonic development. Unfortunately, the bulk of studies on species other than the mouse stress normal morphological appearance of the embryos as compared to embryos removed from the reproductive tract at the same chronological age. This means that assessment of good culture conditions often does not take into consideration the ultimate goal of embryo culture, that is, maintenance of the embryo's ability to develop into a normal living young. It should be obvious that any culture conditions that will not support normal development of the embryos after transfer to recipient females may be artifactual.

Unfortunately, to date, in vitro culture of domestic species has had limited success in supporting embryonic viability. The ovine embryo has been shown to maintain viability, as evidenced by living young from transferred embryos when morulae and blastocysts were cultured in Dulbecco enriched PBS medium for up to 27 h (Bunch et al. 1984). In the absence of in vitro culture conditions for longer periods of time, a variety of surrogate in vivo conditions have been explored for use with domestic species.

Many studies (recently summarized by Adams 1982) have dealt with interspecies embryo transfers to oviducts, ligated oviducts, or uteri of mature females either synchronous or asynchronous with the embryos. In general, embryos from early cleavage stages to early blastocysts are fairly tolerant of a heterospecific environment, but decline rapidly beyond this stage. In most of these studies, embryos were not retransferred to suitable hosts of their own species to test their capacity for development after a sojourn in another species. However, successes with domestic species (Lawson, Rowson, and Adams 1972; Boland 1984) have been substantial. In studies with manipulated embryos, Willadsen and associates (Willadsen 1982) have found the anestrous ewe to be a convenient temporary host for cow, pig, and horse embryos as an aid in evaluating experimental procedures. Clearly, the use of a laboratory species as a temporary host, if this were possible with domestic-species embryos, would offer an even more economical means of monitoring development of manipulated embryos.

We have recently been experimenting with the use of the immature mouse oviduct as a surrogate culture environment for the embryos of mice, rabbits, and pigs (V. E. Papaioannou and K. M. Ebert, unpublished observations). We have found that the immature reproductive tract is as effective as the pseudopregnant tract in maintaining the viability and developmental capacity of mouse embryos between the 1-cell stage and blastocyst stage and is more effective than in vitro culture (Papaioannou and Ebert 1986). In extending this work, we have found that 1-cell rabbit embryos will develop for two days in an immature mouse oviduct, attaining cell numbers comparable to those of in vitro cultured embryos while maintaining developmental potential. One-cell pig embryos will progress at least to the 4-cell stage in this environment. These preliminary results suggest that the immature, nonpregnant mouse oviduct might be useful as a temporary environment for manipulated embryos of other species.

Mouse blastocysts placed in the immature oviduct appear to enter a period of embryonic diapause while maintaining viability for a further four days with little morphological development (Papaioannou 1986). This state resembles the implantation delay normally brought about by lactation in the mouse. Embryonic diapause or implantation delay is a normal feature of reproduction in a wide variety of mammals and is controlled by environmental factors and the hormonal status of the pregnant mother. It is possible that the immature mouse oviduct might also be used for temporary storage of embryos of other species if they could be induced to enter diapause in this location. However, this might be impossible in species that do not normally have implantation delay as a feature of their reproduction, limiting its use in domestic species.

V. Conclusions

The application of manipulation techniques to the mouse embryo has resulted in rapid expansion of knowledge and generation of concepts of development. The resilience of the mouse embryo to this invasive experimental approach has encouraged extension of manipulation procedures to other species, including agricultural animals. The incentive for application of these techniques to agricultural animals is not primarily for basic knowledge but rather for potential economic benefits. Techniques now being applied to domestic species could eventually result in increased food production, improved efficiency of feed conversion, and rapid expansion of genetically valuable animals such as animals with specific disease resistance. In addition, techniques of embryo transfer and manipulation offer the possibility of rescuing endangered exotic species from extinction. The pioneering work reviewed here is encouraging for these goals, although it is still premature to predict the extent of the economic viability of these procedures.

A full evaluation of experimentation on preimplantation embryos cannot be made unless adequate methods become available to test the viability and potential for continued development of control and experimentally treated embryos (Figure 3.1). Although evaluation of viability is frequently made by short-term culture in vitro using morphological criteria or biochemical tests of cell viability, these tests by themselves are not fully indicative of the long-term potential of the embryos. The problem is particularly acute in domestic species, because the basic culture requirements that will sustain cleavage and blastulation for more than short periods of time have not yet been determined. Even in mice, where complete development in vitro of the 1-cell stage to the blastocyst can be easily accomplished with a high degree of success, the potential of these morphologically normal cultured embryos for continued development is severely hampered (Papaioannou and Ebert in press).

The use of temporary surrogate hosts may overcome the inadequacies of culture for domestic-species embryos until conditions are found that will support their viability. At present, the production of viable offspring following embryo transfer is the only valid test of embryonic potential. Innovative methods for noninvasive assessment of embryos that can be shown to have predictive value for developmental potential are badly needed.

Acknowledgments

This work was supported by the Charlton Fund of Tufts University, the Monsanto Group of Companies, and NIH grant AR 1R24 RR02510−01 (V.E.P.). We wish to thank Cindy Welch for help in preparation of the manuscript.

References

Adams, C. E. (1982) Egg transfer in carnivores and rodents, between species, and to ectopic sites. In *Mammalian Egg Transfer,* ed. C. E. Adams, pp. 49−61. Boca Raton: CRC Press.

Allen, W. R. (1982) Embryo transfer in the horse. In *Mammalian Egg Transfer,* ed. C. E. Adams, pp. 135−54. Boca Raton: CRC Press.

Allen, W. R., and Pashen, R. L. (1984) Production of monozygotic (identical) horse twins by embryo micromanipulation. *J. Reprod. Fertil.* 71, 607−13.

Bahr, G. F., and Engler, W. F. (1970) Consideration of volume, mass, DNA, and arrangement of mitochondria in the midpiece of bull spermatozoa. *Exp. Cell Res.* 60, 338−40.

Boland, M. P. (1984) Use of the rabbit oviduct as a screening tool for the viability of mammalian eggs. *Theriogenology* 21, 126−37.

Brackett, B. G. (1978) Experimentation involving primate embryos. In *Methods in Mammalian Reproduction,* ed. J. C. Daniel, Jr., pp. 333−57, New York: Academic Press.

Brem, G., Tenhumberg, H., and Krausslich, H. (1984) Chimerism in cattle through microsurgical aggregation of morulae. *Theriogenology* 22, 609−13.

Briggs, R., and King, T. J. (1952) Transplantation of living nuclei from blastula cells into enucleated frog's eggs. *Zoology* 38, 455−63.

Brinster, R. L., Chen H. Y., and Trumbauer, M. (1981) Somatic expression of herpes thymidine kinase in mice following injection of a fusion gene into eggs. *Cell* 27, 223−31.

Brinster, R. L., Chen, H. Y., Trumbauer, M., and Avarbock, M. R. (1980) Translation of globin messenger RNA by the mouse ovum. *Nature* 283, 499–501.

Brinster, R. L., Ritchie, K. A., Hammer, R. E., O'Brien, R. L., Arp, B., and Storb, U. (1983) Expression of a microinjected immunoglobulin gene in the spleen of transgenic mice. *Nature* 306, 332–6.

Bromhall, J. D. (1975) Nuclear transplantation in the rabbit egg. *Nature* 258, 719–22.

Bronson, R. A., and McLaren A. (1970) Transfer to the mouse oviduct of eggs with and without the zona pellucida. *J. Reprod. Fertil.* 22, 129–37.

Buehr, M., and McLaren, A. (1974) Size regulation in chimaeric mouse embryos. *J. Embryol. Exp. Morphol.* 31, 229–34.

Bunch, T. D., Foote, W. C., Call, J. W., and Clark, W. F. (1984) *In vitro* culture and maintenance of ovine embryos transported in Dulbecco's enriched phosphate-buffered saline. *Theriogenology* 21, 981–7.

Colman, A., and Morser, J. (1979) Export of proteins from oocytes of *Xenopus laevis. Cell* 17, 517–26.

Conklin, E. G. (1931) The development of centrifuged eggs of ascidians. *J. Exp. Zool.* 60, 1–119.

Davis, J., and Hesseldahl, H. (1971) Comparative embryology of mammalian blastocysts. In *The Biology of the Blastocyst*, ed. R. J. Blandau, pp. 27–48. University of Chicago Press.

Dawid, I. B., and Blackler, A. W. (1972) Maternal and cytoplasmic inheritance of mitochondrial DNA in *Xenopus. Dev. Biol.* 29, 152–61.

DiBerardino, M. A. (1980) Genetic stability and modulation of metazoan nuclei transplanted into eggs and oocytes. *Differentiation* 17, 17–30.

Ebert, K. M., and Brinster, R. L. (1983) Rabbit α-globin messenger RNA translation by the mouse ovum. *J. Embryol. Exp. Morphol.* 74, 159–68.

Ebert, K. M., Hammer, R. F., and Papaioannou, V. E. (1985) A simple method for counting nuclei in the preimplantation mouse embryo. *Experientia* 41, 1207–9.

Ebert, K. M., Paynton, B. V., McKnight, G. S., and Brinster, R. L. (1984) Translation and stability of ovalbumin messenger RNA injected into growing oocytes and fertilized ova of mice. *J. Embryol. Exp. Morphol.* 84, 91–103.

Edwards, R. G. (1964) Cleavage of one- and two-cell rabbit eggs *in vitro* after removal of the zona pellucida. *J. Reprod. Fertil.* 7, 413–15.

–(1980). *Conception in the Human Female.* New York: Academic Press.

Edwards, R. G., and Gardner, R. L. (1967) Sexing of live rabbit blastocysts. *Nature* 214, 576–7.

Enders, A. C., and Schlafke, S. (1981) Differentiation of the blastocyst of the rhesus monkey. *Am. J. Anat.* 162, 1–21.

Eppig, J. J. (1978) Developmental potential of LT/Sv parthenotes derived from oocytes matured *in vivo* and *in vitro. Dev. Biol.* 65, 244–9.

Epstein, C. J., Smith, S., Travis, B., and Tucker, G. (1978) Both X chromosomes function before visible X-chromosome inactivation in female mouse embryos. *Nature* 274, 500–3.

Fehilly, C. B., Willadsen, S. M., and Tucker, E. M. (1984a) Interspecific chimaerism between sheep and goat. *Nature* 307, 634–6.

–(1984b) Experimental chimaerism in sheep. *J. Reprod. Fertil.* 70, 347–51.

Gardner, R. L. (1972) An investigation of inner cell mass and trophoblast tissues following their isolation from the mouse blastocyst. *J. Embryol. Exp. Morphol.* 28, 279–312.

–(1974) Microsurgical approaches to the study of early mammalian development. In *Birth Defects and Fetal Development, Endocrine and Metabolic Factors,* ed. K. S. Moghissi, pp. 212–33. Springfield, Ill.: Thomas.

Gardner, R. L., and Edwards, R. G. (1968) Control of the sex ratio at full term in the rabbit by transferring sexed blastocysts. *Nature* 218, 346–8.

Gardner, R. L., and Munro, A. J. (1974) Successful construction of chimaeric rabbit. *Nature* 250, 146–7.

Gardner, R. L., Papaioannou, V. E., and Barton, S. C. (1973) Origin of the ectoplacental cone and secondary giant cells in mouse blastocysts reconstituted from isolated trophoblast and inner cell mass. *J. Embryol. Exp. Morphol.* 30, 561–72.

Gartner, K., and Baunack, E. (1981) Is the similarity of monozygotic twins due to genetic factors alone? *Nature* 292, 646–7.

Gatica, R., Boland, M. P., Crosby, T. F., and Gordon, I. (1984) Micromanipulation of sheep morulae to produce monozygotic twins. *Theriogenology* 21, 555–60.

Gurdon, J. B. (1963) Nuclear transplantation in amphibia and the importance of stable nuclear changes in promoting cellular differentiation. *Biol. Rev.* 38, 54–78.

Gurdon, J. B., Lane, C. D., and Woodland, H. R. (1971) Use of frog eggs and oocytes for the study of messenger RNA and its translation in living cells. *Nature* 233, 177–82.

Hamilton, W. J., and Day, F. T. (1945) Cleavage stages of the ova of the horse, with notes on ovulation. *J. Anat.* 79, 127–30.

Hammer, R. E., Pursel, V. G., Rexroad, C. E., Jr., Wall, R. J., Bolt, D. J., Ebert, K. M., Palmiter, R. D., and Brinster, R. L. (1985) Production of transgenic rabbits, sheep and pigs by microinjection. *Nature* 315, 680–3.

Hare, W. C. D., Mitchell, D., Betteridge, K. J., Eaglesome, M. D., and Randall, G. C. B. (1976) Sexing 2-week-old bovine embryos by chromosomal analysis prior to surgical transfer: preliminary methods and results. *Theriogenology* 5, 243–53.

Heuser, C. H., and Streeter, G. L. (1941) Development of the macaque embryo. *Contributions to Embryology,* No. 181, Vol. 29, publication 525, pp. 15–55. Washington, D.C.: Carnegie Institute of Washington.

Holst, P. A., and Phemister, R. D. (1971) The prenatal development of the dog: preimplantation events. *Biol. Reprod.* 5, 194–206.

Hoppe, R. W., and Bavister, B. D. (1983) Effect of removing the zona pellucida on development of hamster and bovine embryos *in vitro* and *in vivo.* *Theriogenology* 19, 391–404.

Hoppe, P. C., and Illmensee, K. (1982) Full-term development after transplantation of parthenogenetic embryonic nuclei into fertilized mouse eggs. *Proc. Natl. Acad. Sci. U.S.A.* 79, 1912–16.

Hutchison, C. A., Newbold, J. E., Potter, S. S., and Edgell, M. H. (1974) Maternal inheritance of mammalian mitochondrial DNA. *Nature* 251, 536–8.

Illmensee, K., and Hoppe, P. C. (1981) Nuclear transplantation in *Mus musculus*: developmental potential of nuclei from preimplantation embryos. *Cell* 23, 9–18.

Johnson, M. H. (1981) The molecular and cellular basis of preimplantation mouse development. *Biol. Rev.* 56, 463–98.

Kaufman, M. H. (1983) *Early Mammalian Development: Parthenogenetic Studies*. Cambridge University Press.

Kelly, S. J. (1975) Studies of the potency of the early cleavage blastomeres of the mouse. In *The Early Development of Mammals*, British Society for Development Biology, Symposium 2, eds. M. Balls and A. E. Wild, pp. 97–105. Cambridge University Press.

Kimber, S. J., Surani, M. A. H., and Barton, S. C. (1982) Interactions of blastomeres suggest changes in cell surface adhesiveness during the formation of inner cell mass and trophectoderm of the preimplantation mouse embryo. *J. Embryol. Exp. Morphol.* 70, 133–52.

King, G. J., Atkinson, B. A., and Robertson, H. A. (1982) Implantation and early placentation in domestic ungulates. *J. Reprod. Fertil.* Suppl. 31, 17–30.

Konwinski, M., Solter, D., and Koprowski, H. (1978) Effect of removal of the zona pellucida on subsequent development of mouse blastocysts *in vitro*. *J. Reprod. Fertil.* 54, 137–43.

Lambeth, V. A., Looney, C. R., Vrelkel, S. A., Jackson, D. A., Hill, K. G., and Godke, R. A. (1983) Microsurgery on bovine embryos at the morula stage to produce monozygotic twin calves. *Theriogenology* 20, 85–95.

Lawson, R. A. S., Rowson, L. E. A., and Adams, C. E. (1972) The development of cow eggs in the rabbit oviduct and their viability after re-transfer to heifers. *J. Reprod. Fertil.* 28, 313–15.

LeDouarin, N., and McLaren, A. (1984) *Chimeras in Developmental Biology*. New York: Academic Press.

Leonard, R. A., Hoffner, N. J., and DiBerardino, M. A. (1982) Induction of DNA synthesis in amphibian erythroid nuclei in *Rana* eggs following conditioning in meiotic oocytes. *Dev. Biol.* 92, 343–55.

Lewis, W. H., and Hartman, C. G. (1941) Tubal ova of the rhesus monkey. In *Contributions to Embryology*, No. 180, Vol. 29, publication 525, pp. 7–15. Washington, D. C.: Carnegie Institution of Washington.

Lin, T. P. (1966) Microinjection of mouse eggs. *Science* 151, 333–7.

Lyne, A. G., and Hollis, D. E. (1977) The early development of marsupials, with special reference to bandicoots. In *Reproduction and Evolution*, eds. J. A. Calaby and C. H. Tyndale-Biscoe, pp. 293–302, Canberra: Australian Academy of Science.

McGrath, J., and Solter, D. (1983a) Nuclear transplantation in the mouse embryo by microsurgery and cell fusion. *Science* 220, 1300–2.

–(1983b) Nuclear transplantation in mouse embryos. *J. Exp. Zool.* 228, 355–62.

–(1984) Inability of mouse blastomere nuclei transferred to enucleated zygotes to support development in vivo. *Science* 226, 1317–19.

McKnight, G. S., Hammer, R. E., Kuenzel, E. A., and Brinster, R. L. (1983) Expression of the chicken transferrin gene in transgenic mice. *Cell* 34, 335–41.

Magnuson, T., and Epstein, C. J. (1981) Genetic control of very early mammalian development. *Biol. Rev.* 56, 369–408.

Markert, C. L. (1983) Fertilization of mammalian eggs by sperm injection. *J. Exp. Zool.* 228, 195–201.

Markert, C. L., and Petters, R. M. (1978) Manufactured hexaparental mice show that adults are derived from three embryonic cells. *Science* 202, 56–8.

Markert, C. L., and Ursprung, H. (1963) Production of replicable persistent changes in zygote chromosomes of *Rana pipiens* by injected proteins from adult liver nuclei. *Dev. Biol.* 7, 560–77.

Mayer, J. F., and Fritz, H. I. (1974) The culture of preimplantation rat embryos and the production of allophenic rats. *J. Reprod. Fertil.* 39, 1–9.

Meinecke-Tillman, S., and Meinecke, B. (1984) Experimental chimaeras – removal of reproductive barrier between sheep and goat. *Nature* 307, 637–8.

Menino, A. R., Jr., and Wright, R. W., Jr. (1983) Effect of pronase treatment, microdissection and zona pellucida removal on the development of porcine embryos and blastomeres *in vitro*. *Biol. Reprod.* 28, 433–46.

Minhas, B. S., Capehart, J. S., Bowen, M. J., Womack, J. E., McCrady, J. D., Harms, P. G., Wagner, T. E., and Kraemer, D. C. (1984) Visualization of pronuclei in living bovine zygotes. *Biol. Reprod.* 30, 687–91.

Mintz, B. (1962) Experimental study of the developing mammalian egg: removal of the zona pellucida. *Science* 138, 594–5.

Modlinski, J. (1970) The role of the zona pellucida in the development of mouse eggs *in vitro*. *J. Embryol. Exp. Morphol.* 23, 539–47.

–(1978) Transfer of embryonic nuclei to fertilized mouse eggs and development of tetraploid blastocysts. *Nature* 273, 466–7.

Moor, R. M., and Cragle, R. G. (1971) The sheep egg: Enzymatic removal of the zona pellucida and culture of eggs *in vitro*. *J. Reprod. Fertil.* 27, 401–9.

Moore, N. W., Adams, C. E., and Rowson, L. E. A. (1968) Developmental potential of single blastomeres of the rabbit egg. *J. Reprod. Fertil.* 17, 527–31.

Moore, N. W., Polge, C., and Rowson, L. E. A. (1969) The survival of single blastomeres of pig eggs transferred to recipient gilts. *Aust. J. Biol. Sci.* 22, 979–82.

Morgan, T. H. (1927) *Experimental Embryology*. New York: Columbia University Press.

Moustafa, V. L. A., and Hahn, J. (1978) Experimentelle Erzengung von identischen Mausezwillingen. *Dtsch. Tieraerztl. Wochenschr.* 85, 242–4.

Nagashima, H., Matsui, K., Sawaski, T., and Kano, Y. (1984) Production of monozygotic mouse twins from microsurgically bisected morulae. *J. Reprod. Fertil.* 70, 357–62.

Nicholas, J. S., and Hall, B. V. (1942) Experiments on developing rats. II. The development of isolated blastomeres and fused eggs. *J. Exp. Zool.* 90, 441–59.

O'Brien, M. J., Critser, E. S., and First, N. L. (1984) Developmental potential of isolated blastomeres from early murine embryos. *Theriogenology* 22, 601-7.

Ozil, J. -P. (1983) Production of identical twins by bisection of blastocysts in the cow. *J. Reprod. Fertil.* 69, 463–8.

Ozil, J. -P., Heyman, Y., and Renard, J. P. (1982) Production of monozygotic twins by micromanipulation and cervical transfer in the cow. *Vet. Rec.* 110, 126–7.

Palmiter, R. D., Brinster, R. L., Hammer, R. E., Trumbauer, M. E., Rosenfeld, M. G., Birnberg, N. C., and Evans, R. M. (1982) Dramatic growth of mice that develop from eggs microinjected with metallothionein–growth hormone fusion genes. *Nature* 300, 611–15.

Papaioannou, V. E. (1982) Lineage analysis of inner cell mass and trophectoderm using microsurgically reconstituted mouse blastocysts. *J. Embryol. Exp. Morphol.* 68, 199–209.

Papaioannou, V. E. (1986) Diapause of mouse blastocysts transferred to oviducts of immature mice. *J. Reprod. Fertil.* 76, 105–13.

Papaioannou, V. E., and Dieterlen-Lièvre, F. (1984) Making chimeras. In *Chimeras in Developmental Biology,* eds. N. LeDouarin and A. McLaren, pp. 3–37. Orlando: Academic Press.

Papaioannou, V. E., and Ebert, K. M. (in press) Development of fertilized embryos transferred to oviducts of immature mice. *J. Reprod. Fertil.*

Perry, J. S., and Rowlands, I. W. (1962) Early pregnancy in the pig. *J. Reprod. Fertil.* 4, 175–88.

Piko, L., and Matsumoto, L. (1976) Number of mitochondria and some properties of mitochondrial DNA in the mouse egg. *Dev. Biol.* 49, 1–10.

Rand, G. F. (1985) Cell allocation in half- and quadruple-sized preimplantation mouse embryos. *J. Exp. Zool.* 236, 67–70.

Renfree, M. B., and Tyndale-Biscoe, C. H. (1978) Manipulation of marsupial embryos and pouch young. In *Methods in Mammalian Reproduction,* ed. J. C. Daniel, Jr., pp. 307–31. New York: Academic Press.

Rossant, J. (1976) Postimplantation development of blastomeres isolated from 4- and 8-cell mouse eggs. *J. Embryol. Exp. Morphol.* 36, 283–90.

–(1984) Somatic cell lineages in mammalian chimeras. In *Chimeras in Developmental Biology,* eds. N. LeDouarin and A. McLaren, pp. 89–109. Orlando: Academic Press.

Rossant, J., Mauro, V. M., and Croy, B. A. (1982) Importance of trophoblast genotype for survival of interspecific murine chimaeras. *J. Embryol. Exp. Morphol.* 69, 141–9.

Rottmann, O. J., and Lampeter, W. W. (1981) Development of early mouse and rabbit embryos without zona pellucida. *J. Reprod. Fertil.* 61, 303–6.

Seidel, F. (1952) Die Entwicklungspotenzen einer isolierten Blastomere des Zweizellenstadiums im Saugetierei. *Naturwissenschaften* 39, 355–6.

–(1956) Nachweis eines Zentrums zur Bildung der Keinscheibe in Saugetierei. *Naturwissenschaften* 43, 306–7.

Spindle, A. (1982) Cell allocation in preimplantation mouse chimeras. *J. Exp. Zool.* 219, 361–7.

Steven, D., and Morriss, G. (1975) Development of foetal membranes. In *Comparative Placentation, Essays in Structure and Function,* ed. D. H. Steven, pp. 58–86. New York: Academic Press.

Summers, P. M., Shelton, J. N., and Bell, K. (1983) Synthesis of primary *Bostaurus-Bos indicus* chimeric calves. *Anim. Reprod. Sci.* 6, 91–102.

Tachi, S., and Tachi, C. (1980) Electron microscopic studies of chimeric blastocysts experimentally produced by aggregating blastomeres of rat and mouse embryos. *Dev. Biol.* 80, 18–27.

Tarkowski, A. K. (1961) Mouse chimaeras developed from fused eggs. *Nature* 190, 857.

−(1975) Induced parthenogenesis in the mouse. In *The Developmental Biology of Reproduction,* eds. C. L. Markert and J. Papaconstantinou, pp. 107–29. New York: Academic Press.

Thadani, V. M. (1980) A study of hetero-specific sperm-egg interactions in the rat, mouse, and deer mouse using *in vitro* fertilization and sperm injection. *J. Exp. Zool.* 212, 435–53.

Trounsoun, A. O., and Moore, N. W. (1974a) Attempts to produce identical offspring in the sheep by mechanical division of the ovum. *Aust. J. Biol. Sci.* 27, 505.

−(1974b) The survival and development of sheep eggs following complete or partial removal of the zona pellucida. *J. Reprod. Fertil.* 41, 97–105.

Tsunoda, Y., and McLaren, A. (1983) Effect of various procedures on the viability of mouse embryos containing half the normal number of blastomeres. *J. Reprod. Fertil.* 69, 315–22.

Tucker, E. M., Moor, R. M., and Rowson, L. E. A. (1974) Tetraparental sheep chimaeras induced by blastomere transplantation. Changes in blood type with age. *Immunology* 26, 613–21.

Uehara, T., and Yanagimachi, R. (1977) Behavior of nuclei of testicular, caput, and cauda epididymal spermatozoa injected into hamster eggs. *Biol. Reprod.* 16, 315–21.

Wall, R. J., Pursel, V. G., Hammer, R. E., and Brinster, R. L. (1985) Development of porcine ova that were centrifuged to permit visualization of pronuclei and nuclei. *Biol. Reprod.* 32, 645–51.

Willadsen, S. M. (1979) A method for culture of micromanipulated sheep embryos and its use to produce monozygotic twins. *Nature* 277, 298–300.

−(1980) The viability of early cleavage stages containing half the normal number of blastomeres in the sheep. *J. Reprod. Fertil.* 59, 357–62.

−(1981) The developmental capacity of blastomeres from 4- and 8-cell sheep embryos. *J. Embryol. Exp. Morphol.* 65, 165–72.

−(1982) Micromanipulation of embryos of the large domestic species. In *Mammalian Egg Transfer,* ed. C. E. Adams, pp. 185–210. Boca Raton: CRC Press.

Willadsen, S. M. (1986) Nuclear transplantation in sheep embryos. *Nature* 320, 63–6.

Willadsen, S. M., Lehn-Jensen, H., Fehilly, C. B., and Newcomb, R. (1981) The production of monozygotic twins of preselected parentage by micromanipulation of non-surgically collected cow embryos. *Theriogenology* 15, 23–9.

Willadsen, S. M., and Polge, C. (1981) Attempts to produce monozygotic quadruplets in cattle by blastomere separation. *Vet. Rec.* 108, 211–13.

Williams, T. J., Elsden, R. P., and Seidel, G. E., Jr. (1984) Pregnancy rates with bisected bovine embryos. *Theriogenology* 22, 521–31.

Wimsatt, W. A. (1975) Some comparative aspects of implantation. *Biol. Reprod.* 12, 1–40.

4 Development of extraembryonic cell lineages in the mouse embryo

JANET ROSSANT

CONTENTS

I. Introduction

Cell lineage analysis has been a particularly valuable approach for understanding the relationship between cell fate and cell determination in a variety of species. In some invertebrates, such as *Caenorhabditis elegans* and the leech, a combination of visual observation and injected lineage tracers has allowed the fate of every cell in the embryo to be followed through development (Sulston et al. 1983; Weisblat and Blair 1984). These studies have shown that lineage is almost invariant in such species; a rigid pattern of cell division generates a distinct set of progeny of defined cell fate. This invariance is associated with fairly rigid cell determination. Although cell-cell interactions may affect cell fate in some stages of lineage development (Sulston et al. 1983; Weisblat, Kim, and Stent 1984; Zackson 1984), various experimental manipulations have shown that most cells of *C. elegans* become heritably restricted to their fate early in development (Sulston and White 1980; Kimble 1981). In other species, such as *Drosophila*, cell lineage cannot necessarily be defined at the single-cell level, but groups of cells, or polyclones (Crick and Lawrence 1975), are set aside early in development and give rise to discrete "compartments" in later development (Garcia-Bellido, Ripoll, and Morata 1973; Morata and Lawrence 1976). The compartment boundaries established early in embryogenesis also represent boundaries of heritable cell commitment or determination. It is important to note in all these cases that lineage determination does not necessarily imply commitment to a given cell type, but rather to a given pattern of development. Different compartments in *Drosophila*, for example, contain very similar cell types, but the pattern in which cells and tissues are assembled into distinct structures varies between compartments.

When vertebrate embryos in general, and mammalian embryos in particular, are examined, an apparently different type of development is observed. First, it has not proved possible to predict what the fate of every cell in the early embryo will be in later development. Part of this failure may be technical, because suitable long-term cell lineage tracers have not yet been developed for mammalian embryos. Second, extensive study of development in genetic chimeras indicates that there is a great deal of cell mixing during development of the embryo itself, making rigid patterns of coherent clonal development from single cells unlikely. Other experimental studies have shown that development is very flexible and regulative in the mammalian embryo, with cell commitment to different patterns of development being difficult to define at the single-cell level. However, groups of identifiable progenitors to distinct cell lineages are set aside early in development (Gardner and Rossant 1976; Rossant 1984). The ones about which most is known are those established during preimplantation development that give rise to the extraembryonic structures of the later conceptus, namely, the trophectoderm and primitive endoderm lineages. In this chapter, I shall review our knowledge of the characteristics of the progenitors of these lineages in the mouse embryo, outline information available on the exact boundaries of the lineages in later development, discuss evidence that progenitors are committed to these lineages in preimplantation development, and describe the special features of these lineages that may relate to their unique roles in development. I shall contend that the detailed experimental information available on both the progenitors and derivatives of these lineages makes them ideal model systems for investigating the process of lineage determination in mammalian development.

II. Characterization of lineage progenitors

A. Trophectoderm

i. Morphology. Trophectoderm cells are first unequivocally identifiable at the 3.5-day blastocyst stage in the mouse, where they form the outer cell layer enclosing the blastocoelic cavity. At this stage they are morphologically quite distinct from the enclosed cells of the inner cell mass (ICM); at the light-microscope level, they are flattened epitheloid cells that often stain in a manner different from that of the more nearly round ICM cells. At the electron-microscope (EM) level, several differences can be distinguished between the two populations: Trophectoderm cells are joined together by tight junctional complexes at the external surface, endowing them with the properties of an epithelium (Enders and Schalfke 1965; Calarco and Brown 1969; Ducibella et al. 1975). No such complexes are found between ICM cells. Trophectoderm cells also contain lysosomes localized in the basal region of the cell (Fleming et al. 1984). The most obvious difference between trophectoderm and ICM is physiological, namely, the ability to secrete blastocoel fluid. Trophectoderm cells pump fluid first into intracellular spaces and later into extracellular spaces to create the blastocoel (Borland 1977). The exact mechanism of fluid accumulation is still not clear, but involvement of a Na^+-K^+ ATPase perhaps specific to trophectoderm cells has been proposed (Borland 1977; Wiley 1984) (see Chapter 9).

ii. Isolation. An important prerequisite for any further characterization of the differences between trophectoderm and ICM cells is that they must be separable as pure populations. This can be achieved at the blastocyst stage by microsurgical and immunosurgical approaches. ICM cells can be isolated by microdissection from blastocysts (Gardner 1972) or by immunological destruction of the trophectoderm cells (Solter and Knowles 1975). Both these mechanisms destroy the trophectoderm, which can be isolated only by cutting the blastocyst in half, removing the polar trophectoderm plus underlying ICM (Gardner 1972; Papaioannou 1982). This approach is tedious and has the disadvantage that polar trophectoderm cells are always excluded from any study. No method has yet been found to isloate pure polar cells.

iii. Biochemical markers. Biochemical markers specific to the trophectoderm of the blastocyst are few in number. Differences in glycogen metabolism between ICM and trophectoderm have recently been described; glycogen accumulates differentially in the trophectoderm cells (Edirisinghe, Wales, and Pike 1984). Also, trophectoderm cells show less extensive alkaline phosphatase activity than do ICM cells (Johnson, Calarco, and Siebert 1977a). Polyclonal antisera raised against some teratocarcinoma cell lines bind preferentially to the surface of ICM but not trophectoderm cells (Gooding, Hsu, and Edidin 1976; Gachelin et al. 1977), as does the monoclonal antibody anti-SSEA-1 (Solter and Knowles 1978).

Several groups have used two-dimensional electrophoresis to determine whether or not there are differences in protein synthetic profiles between the two tissues (Van Blerkom, Barton, and Johnson 1976; Levinson et al. 1978; Brulet et al. 1980). Most of the abundant polypeptides detectable by such methods are common to both tissues, but some are trophectoderm-specific. Brulet and associates (1980) have shown that some of these trophectoderm-specific proteins are intermediate filaments. Extraction of labeled intermediate filaments from blastocysts but not from ICMs produced an enrichment for some of the trophectoderm-specific polypeptides. Further, a monoclonal antibody, TROMA-1 (Kemler et al. 1981), raised against a supposed trophoblastoma cell line (Nicolas et al. 1976), reacted with a network of intracellular filaments in trophectoderm cells (but not ICMs) and immunoprecipitated polypeptides identical with the trophectoderm-specific polypeptides on two-dimensional electrophoresis. Other studies showed that anticytokeratin antibodies bind specifically to trophectoderm cells (Jackson et al. 1980; Paulin et al. 1980). Extensive intermediate filament networks thus characterize trophectoderm cells. However, these filaments are not limited to trophectoderm cells but seem to typify embryonic epithelial-type cells (Jackson et al. 1981; Kemler et al. 1981; Oshima et al. 1983). The protein recognized by TROMA-1 appears to be identical with Endo A, a cytokeratin produced by extraembryonic endoderm cells (Oshima 1981). No attempts have been made to characterize further any of the other proteins that appear to differ between ICM and trophectoderm.

iv. Molecular markers. Extension of protein synthetic studies to the level of specific mRNA expression has been achieved for only the intermediate filament protein recognized by TROMA-1. Screening a cDNA library from the trophoblastoma cell line used to generate TROMA-1 has identified a gene whose message can be translated in vitro into proteins recognized by TROMA-1 (Brulet and Jacob 1982; Vasseur et al. 1985). No other trophectoderm-specific probes have yet been reported from this library, and the library may not be truly representative of trophectoderm gene expression, because the exact relationship between the trophoblastoma cell line and the trophectoderm layer of the blastocyst is unclear. To date, it has not proved possible to establish a good in vitro model for trophectoderm differentiation in the mouse, which clearly would facilitate these kinds of molecular studies. Teratocarcinoma cells, which seem to resemble pluripotent cells of the early embryo in many respects (Martin 1980), rarely differentiate into cells resembling trophectoderm. Apart from the trophoblastoma cell line mentioned earlier, the only other reported trophectoderm cell line is E6496, a cell line originally derived from a spontaneous ovarian teratocarcinoma in a C3H mouse (Damjanov, Damjanov, and Andrews 1985). This cell line has many of the properties of trophectoderm and can differentiate into cells resembling trophoblast as well as endodermal elements. Although this cell line has interesting properties, its relation to the normal embryo is again unclear, because it was maintained by passage in mice for over 30 years before its recent clonal characterization. In the absence of good in vitro model systems for trophectoderm differentiation, the only possible approach to identifying trophectoderm-specific gene expression is to make cDNA libraries from both trophectoderm and ICM. This has not yet been achieved in the mouse.

B. Primitive endoderm

i. Morphology. The primitive endoderm cells first appear as a layer of cells possibly more than one layer thick (Gardner 1985) on the blastocoelic surface of the ICM around the end of the fourth day or early on the fifth day of pregnancy in the mouse. The remaining ICM cells are termed primitive ectoderm after this differentiative event. At the light-microscope level, primitive endoderm cells are distinguishable by their loose cell contacts, unlike the closely packed cells of the earlier ICM and the primitive ectoderm. At the EM level, primitive endoderm cells are readily distinguishable by extensive rough endoplasmic reticulum, which is often swollen with apparent secretory material. Junctional complexes joining endoderm cells are similar to but not so extensive as those joining trophectoderm cells (Enders, Given, and Schalfke 1978).

ii. Isolation. Separation of primitive ectoderm and primitive endoderm cells is not as readily achieved as is separation of trophectoderm and ICM. Microdissection can enrich for one or the other cell type (Gardner and Papaioannou 1975) and with care can give pure preparations (Gardner 1985). Immunosurgery of isolated 4.5-day ICMs has been reported to destroy primitive endoderm and leave primitive ectoderm intact (Pedersen, Spindle, and Wiley 1977; Dziadek 1979), but some doubt has been cast on the efficiency of removal of the primitive endoderm by this technique (Gardner 1982, 1985). To date, the only clean separation of the two cell types has been achieved by disaggregation of the entire 4.5-day ICM. Experiments have shown that disaggregated cells with rough surface morphology as seen in the light microscope are primitive endoderm, and smooth cells are primitive ectoderm (Gardner and

Rossant 1979). This is clearly not a technique that can be adapted for large-scale isolation, especially because it is not possible to classify every cell as rough or smooth.

iii. Biochemical markers. The search for biochemical differences between primitive endoderm and ectoderm has concentrated on components of the extracellular matrix that can be detected by antibody staining directly on embryos. Type IV collagen, laminin (Adamson and Ayers 1979; Leivo et al. 1980), entactin (Wu et al. 1983), and fibronectin (Wartiovaara, Leivo, and Vaheri 1979) have all been detected on primitive endoderm cells, but not in primitive ectoderm, and presumably are the precursors of the basement membrane that is laid down between the two layers as development proceeds. Primitive endoderm cells, like other embryonic epithelia, also react with the TROMA-1 anticytokeratin antibody (Kemler et al. 1981). They are also alkaline-phosphatase-negative (Johnson, Handyside, and Braude 1977b, and personal observations). The difficulty in isolating large quantities of pure primitive endoderm or ectoderm has deterred attempts to examine the differences in protein synthetic profiles of the two tissues.

iv. Molecular markers. The same problems of isolation have impaired analysis of differences in mRNA expression between primitive endoderm and ectoderm. However, unlike the situation with trophectoderm, there is a valid model system available for studying primitive endoderm formation, namely, the differentiation of teratocarcinoma cells into endoderm following retinoic acid treatment (Strickland and Mahdavi 1978). This system has been extensively reviewed elsewhere (Hogan, Barlow, and Tilly 1983). It is important to emphasize that many of the later events of differentiation within the primitive endoderm lineage seem to be mimicked by the differentiation of F9 embryonal carcinoma (EC) cells under various conditions. Although the similarities are not complete, this system seems to be the only viable way to address the molecular differences between primitive endoderm and ectoderm at this time. Recently there have been various reports on isolation of cDNA clones that are specific to either untreated or the retinoic-acid-treated EC cells (Levine, La Rosa, Gudas 1984; Stacey and Evans 1984; Wang, La Rosa, and Gudas 1985). Most of the clones specific to cells after treatment were isolated from cells several days after differentiation and so may not have reflected properties of primitive endoderm per se. Recently, cDNA libraries have been obtained closer to the time of initiation of differentiation (D. Skup, personal communication), and these may prove more valuable in defining the gene expression patterns of primitive endoderm in the intact embryo. Some of the clones specific to differentiated EC have been characterized and have been shown to encode specific differentia-

Table 4.1. *Markers to distinguish lineage progenitors in preimplantation development*

	Trophectoderm	ICM
3.5-day blastocyst		
Junctional complexes	+	−
Blastocoelic fluid secretion	+	−
Glycogen accumulation	+	−
Alkaline phosphatase	−	+
SSEA-1	−	+
TROMA-1	+	−

	Primitive endoderm	Primitive ectoderm
4.5-day ICM		
Rough ER	+	−
Junctional complexes	+	−
Type IV collagen	+	−
Laminin	+	−
Entactin	+	−
Fibronectin	+	−
TROMA-1	+	−
Alkaline phosphatase	+	−

tion products of later endoderm derivatives, such as type IV collagen and laminin (Gudas and Wang 1983). To date, none of the other cloned sequences specific to differentiated or undifferentiated EC cells has been tested for expression in the embryo itself, with the exception of the ETn transposon-like sequence (Brulet, Condamine, and Jacob 1985). In situ hybridization revealed that this sequence was expressed in embryonic and extraembryonic ectoderm, but not the endoderm, of the postimplantation egg cylinder. The same approach should be used to analyze the embryonic expression of other cloned sequences before concluding that they play any role in primitive endoderm differentiation in the embryo.

Some of the distinctive properties of trophectoderm and primitive endoderm cells are summarized in Table 4.1.

III. Lineages derived from trophectoderm

The cell lineages derived from the trophectoderm layer of the blastocyst have been well defined by studying the fate of genetically marked cells in chimeras. The clearest results have come from reconstituted blastocysts, in which the trophectoderm and ICM are of different genotypes. Original studies of this sort were carried out by Gardner and his colleagues using isozymes of glucose phosphate isomerase (GPI) as genetic markers (Gardner, Papaioannou, and Barton 1973). The results showed that the ectoplacental cone and trophoblast giant cells of the 7.5-day egg cylinder were derived solely from the trophectoderm layer of the blastocyst and that there was also a contribution from the trophectoderm to some other extraembryonic tissues. Later studies using the same markers suggested that this contribution was to the extraembryonic ectoderm (Papaioannou 1982), whose origin had been the subject of some dispute in the descriptive literature. Complete confirmation of the exact derivatives of the trophectoderm could not be obtained using destructive markers such as GPI, but required the development of a ubiquitous in situ marker system for following cell lineage in chimeras (Siracusa et al. 1983). When blastocysts were reconstituted with trophectoderm from *Mus musculus* and ICM from *Mus caroli* and transfered back to the *M. musculus* uterus, the cellular composition of sections of 7.5-day conceptuses could be determined by in situ hybridization with a cloned probe to *M. musculus* satellite DNA (Rossant et al. 1983). *M. musculus* cells hybridizing to the probe were found in the ectoplacental cone and trophoblast giant cells as well as the entire extraembryonic ectoderm. All the rest of the conceptus was derived from *M. caroli* and did not hybridize to the probe. The boundaries between the cell lineages were very sharp; no cells crossed between the different tissues.

From these studies it is clear that all of the trophectoderm derivatives at 7.5 days are extraembryonic; no cells contribute to the fetus itself. At this stage, pure trophectoderm derivatives can be isolated for further study. Ectoplacental cone and giant cells can be isolated by simple dissection, and extraembryonic ectoderm can be separated from overlying endoderm by enzymic treatment (Rossant and Ofer 1977). However, the amount of tissue available at this stage is still limited, particularly for molecular analysis, and it would be useful to identify and isolate trophectoderm derivatives later in development. Descriptive studies show that both ectoplacental cone and extraembryonic ectoderm form part of the mature chorioallantoic placenta. This structure is not totally trophoblast in origin, because both maternal decidual cells and the fetal allantois are also known to contribute to its formation. We recently carried out a detailed analysis of the trophoblast components of the placenta, using reconstitute blastocysts and both GPI and the *M. musculus/M. caroli* marker system (Rossant and Croy 1985). This study showed that approximately 60 percent of the 12–14-day placenta was of trophectoderm origin (Figure 4.1) and that the trophectoderm contribution was to the cell layers classically described as trophoblast in the literature. However, maternal cells were found to penetrate the outer spongiotrophoblast layer, and ICM-derived cells contributed mesenchymal elements, fetal blood capillaries, and end-

LINEAGE COMPOSITION
OF THE CHORIOALLANTOIC PLACENTA

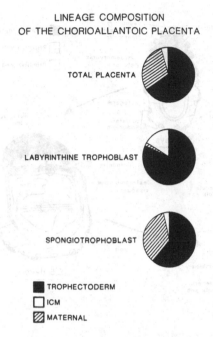

TOTAL PLACENTA

LABYRINTHINE TROPHOBLAST

SPONGIOTROPHOBLAST

■ TROPHECTODERM
☐ ICM
▨ MATERNAL

Figure 4.1. Lineage composition of the 12−14-day chorioallantoic placenta as determined by GPI analysis of reconstituted blastocysts and embryo transfer.

odermal sinuses to the labyrinthine region. Various attempts were made to separate the trophoblast cells of the placenta, but none has thus far proved successful; indeed, most procedures designed to enrich for trophoblast cells actually enriched for the other cell types.

The derivatives of the trophectoderm layer of the blastocyst are thus well defined throughout pregnancy (Figure 4.2) and show a restricted number of developmental options. Interestingly, the potential to form polyploid giant cells appears to be retained by different diploid derivatives of the trophectoderm for several days after implantation. Extraembryonic ectoderm and diploid ectoplacental cone cells from the 7.5-day conceptus differ in their protein synthetic profiles (Johnson and Rossant 1981), and yet each is capable of producing trophoblast giant cells in vitro or in ectopic sites (Rossant and Lis 1981). These and other data (Copp 1979) have led us to propose a stem cell model for the trophoblast lineage in which extraembryonic ectoderm acts as a stem cell pool for all trophoblast cell types (Rossant and Lis 1981). It is not known how long this developmental lability persists within the trophoblast lineage, mostly because of the previously mentioned problems of isolating pure trophoblast populations beyond seven or eight days of

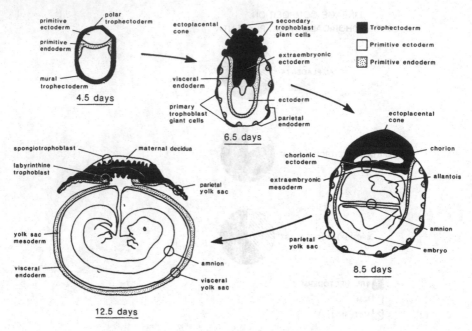

Figure 4.2. Lineage derivatives of trophectoderm, primitive ectoderm, and primitive endoderm during mouse embryonic development.

pregnancy. However, it is not inconsistent with the data to suggest that developmental lability may be a general feature of the diploid cells of this lineage.

IV. Lineages derived from primitive endoderm

The derivatives of primitive endoderm in later development have not been assessed directly by following the fate of genetically marked cells in the 4.5-day blastocysts, because such blastocysts cannot be operated on successfully and returned to the uterus for completion of gestation. Instead, the fate of primitive endoderm cells has been inferred indirectly from studies of the development of such cells injected into earlier 3.5-day blastocysts. Using GPI markers, it was shown that primitive endoderm cells dissociated from 4.5-day ICMs could contribute to the extraembryonic endoderm of the visceral yolk sac but did not make a detectable contribution to the definitive endoderm of the fetus (Gardner and Rossant 1979). More careful analysis revealed that primitive endoderm cells also contributed to the parietal endoderm and that this was in fact their preferred fate after blastocyst injection (Gardner 1982). The ability of primitive endoderm to contribute to these two forms of extraembryonic endoderm has recently been confirmed by another in situ marker system based on differences in malic enzyme activity between normal and $Mod-1^n$ /$Mod-1^n$ null-enzyme mutant cells (Gardner 1984).

It has not yet been shown conclusively in any of these studies, however, that primitive endoderm has an exclusively extraembryonic fate. A minor contribution of less than 10 percent from the injected cells to the definitive endoderm of the fetus would not be detected by GPI. Furthermore, the marked propensity of primitive endoderm cells injected into the 3.5-day blastocyst to contribute to parietal rather than visceral endoderm (Gardner 1982) suggests that cells may be shunted away from the developing embryonic region after injection and thus may not be in a position to contribute to the definitive endoderm even if they have this potential. The possibility of a minority contribution from the primitive endoderm to the definitive gut has not yet been examined using either of the in situ markers described. Marking of endoderm cells in the intact postimplantation embryo by injection of horseradish peroxidase has, however, supported the concept that the definitive endoderm arises from the embryonic ectoderm at the primitive streak stage (Lawson, Meneses, and Pedersen 1984, and personal communication) (see Chapter 5).

Although a role for primitive endoderm in the development of the fetus itself is thus not completely ruled out, it seems clear that the major role of the primitive endoderm, like the trophectoderm, is to form extraembryonic structures (Figure 4.2). Also, as with trophectoderm, the number of developmental options open to primitive endoderm cells is limited, apparently to only two: parietal and visceral endoderm. These two cell types have been extensively characterized (reviewed by Hogan et al. 1983) (see Chapter 11), and they can be separated as pure populations at various stages of development in fairly large amounts. Parietal endoderm can be dissected mechanically from the surface of trophoblast giant cell layer (Gardner 1982), and visceral endoderm can be separated enzymically from the mesodermal layers of the visceral yolk sac (Gardner and Rossant 1979). As in the trophectoderm lineage, there appears to be some lability of cell phenotype within the primitive endoderm lineage. Visceral endoderm cells from early postimplantation egg cylinders have been shown to be able to transform into parietal cells in vitro (Hogan and Tilly 1981) or after injection into the blastocyst (Gardner 1982), but it is not known how long this lability persists in development. The reverse transformation from parietal to visceral endoderm has not been observed. It is not yet clear if visceral endoderm itself acts as a stem cell line, giving rise to more visceral cells or new parietal cells, depending on its environment, or if there is a pool of primitive endoderm-like stem cells within the visceral endoderm layer (Hogan and Newman 1984).

V. Commitment of cells to extraembryonic lineages

In the foregoing sections it has been shown that the first two cell types that can be recognized by their differentiated properties and isolated from the preimplantation embryo give rise to distinct sets of extraembryonic derivatives in later development. These lineage derivatives are invariant from one embryo to another, so that trophectoderm and primitive endoderm can legitimately be considered lineage progenitors. By the time that trophectoderm and primitive endoderm cells can be recognized and separated in the embryo, it is also clear that the cells have become irreversibly committed to their particular pathways of development. Because of our inability to follow the fate of cells in the intact embryo, many of the studies on cell fate cited in the previous section are more correctly considered as studies on cell commitment. In particular, the inability of primitive endoderm cells to contribute to the fetus after blastocyst injection is often interpreted as indicating commitment of these cells (Slack 1983). This is, of course, a somewhat tautological argument, because one cannot claim that the same experiment gives information on both normal fate and potential outside of the normal environment. However, isolated primitive endoderm cells never regenerate primitive ectoderm-like cells in vitro (J. Rossant, unpublished data), and so it is reasonable to conclude that primitive endoderm cells are indeed committed to their future fate by 4.5 days. Likewise, isolated trophectoderm cells never generate ICM derivatives in vitro or in vivo (Gardner 1972), suggesting that the attainment of specialized properties is associated with a heritable change in cell potential in this lineage also.

Although it is fairly clear that trophectoderm and primitive endoderm cells are committed by the time they can be recognized as differentiated cell populations, it is less clear whether or not commitment begins before final morphological differentiation. This subject has been extensively reviewed in the literature (Johnson et al. 1977b; Rossant 1977; Johnson 1981; Rossant 1984) and in other chapters in this book. Suffice it to say that there is some evidence that commitment of cells to the extraembryonic lineages may occur prior to their morphological differentiation. Outside cells from the late morula stage of development can regenerate an entire embryo (Rossant and Vijh 1980; Ziomek, Johnson, and Handyside 1982), but there is evidence to suggest that such cells always generate at least one outside cell at each cell division (Ziomek and Johnson 1982; R.A. Pedersen, K. Wu, and H. Balakier, unpublished results) and thus show some predisposition toward trophectoderm production (see Chapter 1). We do not know when these outside cells lose the potential to form inside cells, because outside cells cannot be readily isolated between the morula and blastocyst stage. As for primitive endoderm, there is evidence from both in vitro aggregation experiments and blastocyst-injection experiments (J. Rossant and M. Rosenstraus, unpublished observations) to suggest that outer cells of the ICM may also be committed to primitive endoderm prior to any obvious morphological differentiation.

Other experimental studies have indicated that the pluripotent cells of the ICM and primitive ectoderm may retain the potential to produce trophectoderm and primitive endoderm, respectively, for some time after differentiation of these cell types in the intact embryo. Several studies have shown that ICM cells from the early blastocyst retain the potential to form trophectoderm (Handyside 1978; Spindle 1978; Rossant and Lis 1979), but that property is lost by the expanded blastocyst stage (Rossant 1975a, 1975b). The situation is less clear cut with regard to the primitive ectoderm of the 4.5-day ICM. As mentioned previously, such cells never contribute to primitive endoderm after blastocyst injection (Gardner and Rossant 1979), but again one cannot be sure that the cells are exposed to an environment suitable for expression of this potential. It is clear that there are phenotypic differences between primitive ectoderm and primitive endoderm cells (Gardner and Rossant 1979), and it is possible that the properties of primitive ectoderm cells allowed them to be readily enclosed by the host ICM and therefore never exposed to an environment conducive to endoderm formation. There are clear examples of differential adhesiveness affecting cell behavior in earlier stages of development (Kimber, Surani, and Barton 1982). In vitro experiments in which the outer layer of primitive endoderm was removed from the ICM by immunosurgery suggested that regeneration of at least one cycle of primitive endoderm cells can occur (Pedersen et al. 1977; Dziadek 1979). Because of doubts about the efficacy of removing the endoderm with immunosurgery (Gardner 1982) in recent experiments Gardner (1985) compared immunosurgically and microsurgically isolated 4.5-day primitive ectoderms and showed that pure populations of these cells cannot regenerate primitive endoderm cells (Gardner 1985). By 5.5 days of development, it is clear that isolated primitive ectoderm cannot regenerate primitive endoderm in vitro (Rossant and Ofer 1977).

The exact relationship between time of cell commitment and specific cell cycles cannot be determined because of the asynchrony of cell divisions in the early embryo. This means that estimates of the time of cell commitment will always be somewhat broad, as exemplified by recent studies showing that some cells in the early ICM may already be committed to the ICM pathway, whereas other cells can still regenerate trophectoderm (Nichols and Gardner 1984; Chisholm et al. 1985). The important point is that all the cell types identified in the early embryo do eventually undergo commitment to distinct cell lineages.

VI. Special properties of extraembryonic lineages

A. Precocious differentiation

Although the foregoing sections have demonstrated that the first two specialized cell types to develop in the mouse embryo resemble committed lineage precursors in other embryonic systems, it must be noted that these two lineages do shown some unusual properties. The first of these is their apparent precocious differentiation, as exemplified by all the properties outlined in the first section of this chapter. Not only do the cell types appear to possess specialized properties at their time of formation, but also they rapidly differentiate further and by the early postimplantation stages produce several specific differentiation markers, such as alpha-fetoprotein in visceral endoderm (Dziadek and Adamson 1978) (see Chapter 11) and progesterone in trophoblast giant cells (Sherman 1983). During this time, the rest of the embryo is undergoing considerable developmental changes, but these are not characterized by such dramatic outward signs of differentiation. The reason for all this rapid activity within the two extraembryonic lineages surely rests in the requirement for immediate function of these cells after their formation to allow survival of the conceptus in the uterine environment (Rossant 1977). In the mouse, the initial role of fetal nutrition is probably carried out by the visceral and parietal yolk sacs. This role is then subsumed by the functional chorioallantoic placenta. Major fetal components of all these organs are derived from the trophectoderm and the primitive endoderm. Precocious differentiation should therefore not be considered as a property that makes these lineages abnormal and unlikely to provide information about general processes of cell commitment. Rather, the early differentiation markers shown by the lineages should be considered as convenient tools for analysis of the process of cell lineage development.

B. Limited developmental potential

The second unusual property of trophectoderm and primitive endoderm is that although they make up a large proportion of the cells of the preimplantation embryo, they do not give rise to any part of the resulting mouse. They are cell types whose role is complete once pregnancy is over. The mechanisms of their establishment and commitment might therefore differ from commitment events within the embryonic lineage itself. However, there is no real evidence that this is the case. Although there does appear to be some developmental lability within each of the two extraembryonic lineages, there is no clear evidence that they ever revert back to embryonic cell types, as would be predicted if commitment were not stable. The only observation that would support such an idea is the production of teratomas consisting of different cell types usually considered to be derived from embryonic lineages when rat yolk sacs are extruded from the uterus (Sobis and Vandeputte 1975, 1976). The exact origin of the cells producing these tumors is still unclear. Thus, the general conclusion seems to be that commitment is as stable in trophectoderm and primitive endoderm as in the embryo itself. This is perhaps not so surprising, because interference with the development of the various extraembryonic structures might well affect the well-being of the fetus indirectly and thus be equally as damaging as direct disturbance of the orderly development of the fetus itself.

C. Unusual molecular properties

In addition to showing precocious differentiation, trophecto-derm and primitive endoderm share some unusual features possibly related to control of gene expression. The first of these is preferential inactivation of the paternally derived X chromosome in all derivatives of these lineages in female embryos, as compared with the random inactiva-tion within the embryonic lineage. This subject is discussed further in Chapter 12. The second unusual feature is the extensive undermethyla-tion of DNA sequences within the two lineages. In most adult mammalian tissues there is little variation in total levels of methylation of cytosine residues, but tissue-specific patterns of methylation of specific sequences have been detected (reviewed by Razin and Riggs 1980). These variations often have been shown to be associated with differences in expression of the relevant gene sequences (Doerfler 1983). The exact relationship between DNA methylation and gene expression is complex and is outside the scope of this chapter. However, there is a general association between high levels of DNA methylation and nonexpression of genes. Further-more, the heritability of patterns of methylation suggested that patterns of DNA methylation might be involved in maintenance of the distinct cell lineages in the early mouse embryo. In particular, it seemed possible that the limited developmental potential of trophectoderm and primitive en-doderm might be reflected in permanent inactivation of genes required for fetal development and that this inactivation might be brought about by extensive DNA methylation. To this end we began to examine the levels of methylation at specific DNA sequences in derivatives of trophec-toderm and primitive endoderm.

Earlier studies on total methylation of cytosine in trophoblast and ICM from rabbit blastocysts (Manes and Manzel 1981) did not support the predicted difference between the two lineages. These studies indicated that trophoblast contained significantly less 5-methylcytosine than did ICM. Levels of methylation in trophoblast were similar to those in adult tissues, but the ICM was much more methylated than adult tissues. In the mouse, we examined methylation at MspI sites in both satellite and dispersed repetitive sequences and found that all the purified derivatives of the trophectoderm and primitive endoderm shared the property of extensive undermethylation of these sequences (Chapman et al. 1984). This undermethylation was particularly striking because the same sequences were completely methylated in adult tissues and in the primitive ectoderm lineage as early as 7.5 days of development. Further study revealed that this undermethylation was not confined to repetitive sequences but was also detected in several structural gene sequences probed (Rossant et al. in press). Studies by other workers on methylation of gene sequences in total visceral yolk sac and placenta (Razin et al. 1984; Young and Tilghman 1984) showed similar but less dramatic evidence for undermethylation in the extraembryonic lineages. Thus, it is apparent that DNA undermethylation is typical of a wide range of gene sequences in the first two specialized lineages to develop in the mouse embryo. The DNA is not completely devoid of methylation, but total methylcytosine levels are roughly 50 percent of those in the average adult tissue (R. Balling and J. Rossant, unpublished data). This residual methylation is not confined to genes that are inactive in the extraembryonic tissues, as might have been predicted from our original hypothesis. For example, the MspI sites in the genes for the major urinary-protein gene family were apparently completely unmethylated in all derivatives of the trophectoderm and primitive endoderm (Rossant et al. in press), although this gene family is not expressed at all in these lineages (Meehan et al. 1984).

Undermethylation of the DNA is clearly a heritable, lineage-restricted event in mouse embryogenesis, but it is not obvious that it has any direct role in controlling commitment to different lineages, especially because the patterns of methylation observed do not seem to show any correlation with the activity of genes. It seems possible that undermethylation of the extraembryonic lineages is another consequence of their early differentiation. We have examined the methylation of repetitive sequences in 8-cell embryos and early blastocysts and have shown that the preimplantation embryo shows extensive undermethylation of these sequences, similar to the later extraembryonic lineages (J. P. Sanford, J. Rossant, and V. M. Chapman, unpublished observations). This suggests that undermethylation may be typical of all cells in the early embryo and that this state is retained in the first two lineages to differentiate; the de novo methylation that is known to occur in the embryo (Jahner et al. 1982) would then be confined to the primitive ectoderm lineage. In this case, undermethylation of the DNA in extraembryonic lineages may not play any causal role in their differentiation but may be simply carried over from the early embryo. This essentially passive nature of undermethylation would imply that trophectoderm or primitive endoderm could differentiate even if heavily methylated. There is no proof for this contention, but the extensive loss of methylation from heavily methylated F9 cells on differentiation into primitive endoderm-like cells (Bestor, Hellewell, and Ingram 1984; Razin et al. 1984, Young and Tilghman 1984) might indicate a more important role for undermethylation. At this point, therefore, it is unclear whether or not the property of extensive undermethylation of DNA, which is unique to these mammalian extraembryonic lineages, has any direct role in differentiation of these cell types.

VII. Extraembryonic lineages in mouse as useful models for cell commitment

It is clear from the evidence in the previous sections that only the establishment of the two extraembryonic lineages in preimplantation development is understood to the extent that we understand lineage development in various invertebrate species. In the establishment of these extraembryonic lineages, we are dealing with commitment of a group of cells to a set of developmental options, not with commitment to a distinct cell lineage for every separate cell. Within each lineage, cell fate is not invariant; there is considerable regulative capacity. Cells can be added or taken away from a lineage experimentally without any effect on development, and, indeed, cell death seems to be a normal part of development (Copp 1979). Commitment to a given lineage is set up by cell-cell interactions during previous stages of development, but once differentiation has occurred, commitment is irreversible and presumably heritable. All of these properties resemble the establishment of polyclones in insect development (Crick and Lawrence 1975) or equivalence groups in nematodes (Sulston and White 1980). However, in many ways the establishment of the extraembryonic lineages in the mouse provides a system more amenable to experimental manipulation than the invertebrate systems described. The process of establishment of the different lineages can be studied directly in the readily manipulated preimplantation embryo; the progenitors of the different lineages can be isolated, and the later derivatives of each lineage can be isolated for molecular analysis. Is it valid, however, to use the development of these peculiarly mammalian lineages as models for embryonic developmental processes in general? We have seen that the two extraembryonic lineages do show some unusual properties, but I have argued that these properties are probably consequences of the precocious differentiation of these lineages and are not responsible for the actual process of commitment itself. All of the cellular events leading up to commitment have their counterparts in other embryonic systems.

How, then, are we to use all this accumulated information on the development of the extraembryonic lineages to gain insight into the underlying molecular mechanisms of development? Unfortunately, although the mouse embryo probably provides the most information at the cellular level on early developmental decisions of any embryonic system, it has not yet proved amenable to detailed genetic and molecular analyses of lineage development that are possible in some other systems. There are, for example, no mutations known that have been proved to affect developmental decisions in the early mouse embryo in the same way that homeotic mutations affect compartment development in *Drosophila* (Gehring 1985) (see Chapter 14). It is not immediately obvious how to search for such controlling genes in the early embryo nor, indeed, that they necessarily exist. It seems more likely that the clues to early lineage development in the mouse will come not from classic genetics but from molecular genetics. It is now possible to contemplate analyzing lineage-specific patterns of gene expression, searching for those genes whose patterns of expression suggest that they may play roles in cell commit-ment, and then investigating the roles of those genes by manipulating their expression in the embryo. This kind of approach, although pain-stakingly tedious, promises to provide some molecular basis for the cellu-lar events of early embryogenesis. Despite the unusual nature of the two lineages under discussion, the vastly superior knowledge we possess about their development, as opposed to development of cell lineages in the embryo itself, makes it likely that much research in early mammalian development will remain extraembryological for some time to come.

Acknowledgments

The author's work described here was supported by the Natural Sciences and Engineering Research Council, the Medical Research Council, and the Na-tional Cancer Institute of Canada and by an E. W. R. Steacie Memorial Fellowship from the NSERC.

References

Adamson, E. D., and Ayers, S. E. (1979) The localization and synthesis of some collagen types in developing mouse embryos. *Cell* 16, 953–965.

Bestor, T. H., Hellewell, S. B., and Ingram, V. M. (1984) Differentiation of two mouse cell lines is associated with hypomethylation of their genomes. *Mol. Cell. Biol.* 4, 1800–6.

Borland, R. M. (1977) Transport processes in the mammalian blastocyst. In *Development in Mammals* Vol. 1, ed. M. H. Johnson, pp. 31–67. Amsterdam: Elsevier/North Holland.

Brulet, P., Babinet, C., Kemler, R., and Jacob, F. (1980) Monoclonal antibod-ies against trophectoderm specific markers during mouse blastocyst forma-tion. *Proc. Natl Acad. Sci., U.S.A.* 77, 4113–17.

Brulet, P., Condamine, H., and Jacob, F. (1985) Spatial distribution of transcripts of the long repeated ETn sequence during early mouse embryogenesis. *Proc. Natl Acad. Sci. U.S.A.* 82, 2054–8.

Brulet, P., and Jacob, F. (1982) Molecular cloning of a DNA sequence encoding a trophectoderm specific marker during mouse blastocyst formation. *Proc. Natl Acad. Sci. U.S.A.* 79, 2328–32.

Calarco, P. G., and Brown, E. A. (1969) An ultrastructural and cytological study of the preimplantation development in the mouse. *J. Exp. Zool.* 171, 253–84.

Chapman, V., Forrester, L., Sanford, J., Hastie, N., and Rossant, J. (1984) Cell lineage-specific undermethylation of mouse repetitive DNA. *Nature* 307, 284–6.

Chisholm, J. C., Johnson, M. H., Warren, P. D., Fleming, T. P., and Pickering, S. J. (1985) Developmental variability within and between mouse expanding blastocysts and their ICMs. *J. Embryol. Exp. Morphol.* 86, 311–36.

Copp, A. J. (1979) Interaction between inner cell mass and trophectoderm of the mouse blastocyst. II. The fate of the polar trophectoderm. *J. Embryol. Exp. Morphol.* 51, 109–20.

Crick, F. H. C., and Lawrence, P. A. (1975) Compartments and polyclones in insect development. *Science* 189, 340–7.

Damjanov, I., Damjanov, A., and Andrews, P. A. (1985) Trophectodermal carcinoma: mouse teratocarcinoma-derived tumour stem cells differentiating into trophoblastic and yolk sac elements. *J. Embryol. Exp. Morphol.* 86, 125–141.

Doerfler, W. (1983) DNA methylation and gene activity. *Annu. Rev. Biochem.* 52, 93–124.

Ducibella, T., Albertini, D. F., Anderson, E., and Biggers, J. D. (1975) The preimplantation mammalian embryo; characterization of intercellular junctions and their appearance during development. *Dev. Biol.* 45, 231–50.

Dziadek, M. (1979) Cell differentiation in isolated inner cell masses of mouse blastocysts *in vitro*: onset of specific gene expression. *J. Embryol. Exp. Morphol.* 53, 367–79.

Dziadek, M., and Adamson, E. D. (1978) Localisation and synthesis of alpha foetoprotein in post-implantation mouse embryos. *J. Embryol. Exp. Morphol.* 43, 289–313.

Edirisinghe, W. R., Wales, R. G., and Pike, I. L. (1984) Studies of the distribution of glycogen between the inner cell mass and trophoblast cells of mouse embryos. *J. Reprod. Fertil.* 71, 533-8.

Enders, A. C., and Schalfke, S. J. (1965) The fine structures of the blastocyst; some comparative studies. In *Preimplantation Stages of Pregnancy*, eds. G. E. W. Wolstenholme and M. O'Connor, pp. 29–34. London: Churchill.

Enders, A. C., Given, R. L., and Schalfke, S. J. (1978) Differentiation and migration of the endoderm in the rat and mouse at implantation. *Anat. Rec.* 190, 65–77.

Fleming, T. P., Warren, P. D., Chisholm, J. C., and Johnson, M. H. (1984) Trophectodermal processes regulate the expression of totipotency within the inner cell mass of the mouse expanding blastocyst. *J. Embryol. Exp. Morphol.* 84, 63–90.

Gachelin, G., Kemler, R., Kelly, F., and Jacob, F. (1977) PCC4, a new cell surface antigen common to multipotential embryonal carcinoma cells, spermatozoa, and early mouse embryos. *Dev. Biol.* 57, 199–209.

Garcia-Bellido, A., Ripoll, P., and Morata, G. (1973) Developmental compartmentalisation of the wing disk of *Drosophila*. *Nature [New Biol.]* 245, 251–3.

Gardner, R. L. (1972) An investigation of inner cell mass and trophoblast tissues following their isolation from the mouse blastocyst. *J. Embryol. Exp. Morphol.* 28, 279–312.

–(1982) Investigation of cell lineage and differentiation in the extraembryonic endoderm of the mouse embryo. *J. Embryol. Exp. Morhpol.* 68, 175–98.

–(1984) An *in situ* cell marker for clonal analysis of development of the extraembryonic endoderm in the mouse. *J. Embryol. Exp. Morphol.* 80, 251–88.

–(1985) Regeneration of endoderm from primitive ectoderm in the mouse embryo: fact or artifact? *J. Embryol. Exp. Morphol.* 88, 303–26.

Gardner, R. L., and Papaioannou, V. E. (1975) Differentiation in the trophectoderm and inner cell mass. In *The Early Development of Mammals*, eds. M. Balls and A. E. Wild, pp. 107–132. Cambridge University Press.

Gardner, R. L., Papaioannou, V. E., and Barton, S. C. (1973) Origin of the ectoplacental cone and secondary giant cells in mouse blastocysts reconstituted from isolated trophoblast and inner cell mass. *J. Embryol. Exp. Morphol.* 30, 561–72.

Gardner, R. L., and Rossant, J. (1976) Determination during embryogenesis. In *Embryogenesis in Mammals*, Ciba Foundation Symposium, pp. 5–25. Amsterdam: Associated Scientific Publishers.

–(1979) Investigation of the fate of 4.5 day post-coitum mouse inner cell mass cells by blastocyst injection. *J. Embryol. Exp. Morphol.* 52, 141–52.

Gehring, W. J. (1985) The homeo box: a key to the understanding of development? *Cell* 40, 3–5.

Gooding, L. R., Hsu, Y.-C., and Edidin, M. (1976) Expression of teratoma-associated antigens on murine ova and early embryos. Identification of two early differentiation markers. *Dev. Biol.* 49, 479–86.

Gudas, L. J., and Wang, S.-Y. (1983) Isolation of cDNA clones specific for collagen IV and laminin from mouse teratocarcinoma cells. *Proc. Natl. Acad. Sci. U.S.A.* 80, 5880–4.

Handyside, A. H. (1978) Time of commitment of inside cells isolated from preimplantation mouse embryos. *J. Embryol. Exp. Morphol.* 45, 37–53.

Hogan, B. L. M., Barlow, D. P., and Tilly, R. (1983) F9 teratocarcinoma cells as a model for the differentiation of parietal and visceral endoderm in the mouse embryo. *Cancer Surveys* 2, 115–40.

Hogan, B. M. L., and Newman, R. (1984) A scanning electron microscope study of the extraembryonic endoderm of the 8th day mouse embryo. *Differentiation* 26, 138–43.

Hogan, B. L. M., and Tilly, R. (1981) Cell interactions and endoderm differentiation in cultured mouse embryos. *J. Embryol. Exp. Morphol.* 62, 379–94.

Jackson, B. W., Grund, C., Schmidt, E., Burki, K., Franke, W. W., and Illmensee, K. (1980) Formation of cytoskeletal elements during mouse embryogenesis. I. Intermediate filaments of cytokeratin type and desmosomes in preimplantation embryos. *Differentiation* 17, 161–79.

Jackson, B. W., Grund, C., Winter, S., Franke, W. W., and Illmensee, K. (1981) Formation of cytoskeletal elements during mouse embryogenesis. II. Epithelial differentiation and intermediate-sized filaments in early post-implantation embryos. *Differentiation* 20, 203–16.

Jahner, D., Stuhlmann, H., Stewart, C. L., Harbers, K., Lohler, J., Simon, J., and Jaenisch, R. (1982) De novo methylation and expression of retroviral genomes during mouse embryogenesis. *Nature* 298, 623–8.

Johnson, L. V., Calarco, P. G., and Siebert, M. L. (1977a) Alkaline phosphatase activity in the preimplantation mouse embryo. *J. Embryol. Exp. Morphol.* 40, 83–9.

Johnson, M. H. (1981) The molecular and cellular basis of preimplantation mouse development. *Biol. Rev.* 56, 463–98.

Johnson, M. H., Handyside, A. H., and Braude, P. R. (1977b) Control mechanisms in early mammalian development. In *Development in Mammals*, Vol. 2, ed. M. H. Johnson, pp. 67–97. Amsterdam: Elsevier.

Johnson, M. H., and Rossant, J. (1981) Molecular studies on cells of the trophectodermal lineage of the postimplantation mouse embryo. *J. Embryol. Exp. Morphol.* 61, 103–16.

Kemler, R., Brulet, P., Schnebelen, M. T., Gaillard, J., and Jacob, F. (1981) Reactivity of monoclonal antibodies against intermediate filament proteins during embryonic development. *J. Embryol. Exp. Morphol.* 64, 45–60.

Kimber, S. J., Surani, M. A. H., and Barton, S. C. (1982) Interactions of blastomeres suggest changes in cell surface adhesiveness during the formation of inner cell mass and trophectoderm in the preimplantation mouse embryo. *J. Embryol. Exp. Morphol.* 70, 133–52.

Kimble, J. E. (1981) Alterations in cell lineage following laser ablation of cells in the somatic gonad of *Caenorhabditis elegans*. *Dev. Biol.* 87, 286–300.

Lawson, K. A., Meneses, J. J., and Pedersen, R. A. (1984) Fate mapping of the endoderm in presomite mouse embryos by intracellular microinjection of horseradish peroxidase. *J. Embryol. Exp. Morphol.* Suppl., 82, 67.

Leivo, I., Vaheri, A., Timpl, R., and Wartiovaara, J. (1980) Appearance and distribution of collagens and laminin in the early mouse embryo. *Dev. Biol.* 76, 100–14.

Levine, R. A., La Rosa, G. J., and Gudas, L. J. (1984) Isolation of cDNA clones for genes exhibiting reduced expression after differentiation of murine teratocarcinoma stem cells. *Mol. Cell. Biol.* 4, 2142–50.

Levinson, J., Goodfellow, P., Vadenboncoeur, M., and McDevitt, H. (1978) Identification of stage-specific polypeptides synthesized during murine preimplantation development. *Proc. Natl. Acad. Sci. U.S.A.* 75, 3332–6.

Manes, C., and Menzel, P. (1981) Demethylation of CpG sites in DNA of early rabbit trophoblast. *Nature* 293, 589–90.

Martin, G. R., (1980) Teratocarcinomas and mammalian embryogenesis. *Science* 209, 768–76.

Meehan, R. R., Barlow, D. P., Hill, R. E., Hogan, B. L. M., and Hastie, N. (1984) Pattern of serum protein gene expression in mouse visceral yolk sac and foetal liver. *EMBO Journal* 3, 1881–5.

Morata, G., and Lawrence, P. (1976) Homeotic genes, compartments and cell determination in *Drosophila*. *Nature* 265, 211–16.

Nicolas, J. F., Avner, P., Gaillard, J., Guenet, J. L., Jakob, H., and Jacob, F. (1976) Cell lines derived from teratocarcinomas. *Cancer Res.* 36, 4224–31.

Nicols, J., and Gardner, R. L. (1984) Heterogeneous differentiation of external cells in individual isolated early mouse inner cell masses in culture. *J. Embryol. Exp. Morphol.* 80, 225–40.

Oshima, R. G. (1981) Identification and immunoprecipitation of cytoskeletal proteins from murine extra-embryonic endodermal cells. *J. Biol. Chem.* 256, 8124–33.

Oshima, R. G., Howe, W. E., Tabor, J. M., and Trevor, K. (1983) Cytoskeletal proteins as markers of embryonal carcinoma differentiation. In *Teratocarcinoma Stem Cells*, Cold Spring Harbor Conference on Cell Proliferation. Vol. 10, eds. L. M. Silver, G. R. Martin, and S. Strickland, pp. 51–61. Cold Spring Harbor, N.Y.: Cold Spring Harbor Laboratory.

Papaioannou, V. E. (1982) Lineage analysis of inner cell mass and trophectoderm using microsurgically reconstituted mouse blastocysts. *J. Embryol. Exp. Morphol.* 68, 199–209.

Paulin, D., Babinet, C., Weber, K., and Osborn, M. (1980) Antibodies as probes of cellular differentiation and cytoskeletal organization in the mouse blastocyst. *Exp. Cell Res.* 130, 297–304.

Pedersen, R. A., Spindle, A. E., and Wiley, L. M. (1977) Regeneration of endoderm by ectoderm isolated from mouse blastocysts. *Nature* 270, 435–7.

Razin, A., and Riggs, A. D. (1980) DNA methylation and gene function. *Science* 210, 604–9.

Razin, A., Webb, C., Szyf, M., Yisraeli, J., Rosenthal, A., Naveh-Many, T., Sciaky-Gallili, N., and Cedar, H. (1984) Variations in DNA methylation during mouse cell differentiation *in vivo* and *in vitro*. *Proc. Natl. Acad. Sci. U.S.A.* 81, 2275–9.

Rossant, J. (1975a) Investigation of the determinative stage of the mouse inner cell mass. I. Aggregation of inner cell masses with morulae. *J. Embryol. Exp. Morphol.* 33, 979–90.

–(1975b) Investigation of the determinative state of the mouse inner cell mass. II. The fate of isolated inner cell masses transfered to the oviduct. *J. Embryol. Exp. Morphol.* 33, 991–1001.

–(1977) Cell commitment in early rodent development. In *Development in Mammals*, Vol. 2, ed. M. H. Johnson, pp. 119–50. Amsterdam: North Holland.

–(1984) Somatic cell lineages in mammalian chimeras. In *Chimeras in Developmental Biology*, eds. N. LeDouarin and A. McLaren, pp. 89–109. New York: Academic Press.

Rossant, J.. and Croy, B. A. (1985) Genetic identification of tissue of origin of cellular populations within the mouse placenta. *J. Embryol. Exp. Morphol.* 86, 177–89.

Rossant, J., and Lis, W. T. (1979) Potential of isolated mouse inner cell masses to form trophectoderm derivatives *in vivo*. *Dev. Biol.* 70, 255–61.

–(1981) Effect of culture conditions on diploid to giant cell transformation in postimplantation mouse trophoblast. *J. Embryol. Exp. Morphol.* 62, 217–27.

Rossant, J., and Ofer, L. (1977) Properties of extraembryonic ectoderm isolated from postimplantation mouse embryos. *J. Embryol. Exp. Morphol.* 39, 183–94.

Rossant, J., Sanford, J. P., Chapman, V. M., and Andrews, G. K. (in press) Undermethylation of structural gene sequences in extraembryonic lineages of the mouse. *Dev. Biol.*

Rossant, J., and Vijh, K. M. (1980) Ability of outside cells from preimplantation mouse embryos to form inner cell mass derivatives *in vivo. Dev. Biol.* 76, 475–82.

Rossant, J., Vijh, K. M., Siracusa, L. D., and Chapman, V. M. (1983) Identification of embryonic cell lineages in histological sections of *M. musculus – M. caroli* chimaeras. *J. Embryol. Exp. Morphol.* 73, 179–91.

Sherman, M. I. (1983) Endocrinology of rodent trophoblast cells. In *Biology of Trophoblast*, eds. Y. W. Loke and A. Whyte, pp. 402–67. Amsterdam: North Holland.

Siracusa, L. D., Chapman, V. M., Bennett, K. L., Hastie, N. D., Pietras, D. F., and Rossant, J. (1983) Use of repetitive DNA sequences to distinguish *Mus musculus* and *Mus caroli* cells by *in situ* hybridization. *J. Embryol. Exp. Morphol.* 73, 163–78.

Slack, J. M. L. (1983) *From Egg to Embryo.* Cambridge University Press.

Sobis, H., and Vandeputte, M. (1975) Sequential morphological study of teratomas derived from displaced yolk sac. *Dev. Biol.* 45, 276–90.

–(1976) Yolk-sac derived rat teratomas are not of germ cell origin. *Dev. Biol.* 51, 320–3.

Solter, D., and Knowles, B. B. (1975) Immunosurgery of mouse blastocyst. *Proc. Natl. Acad. Sci. U.S.A.* 72, 5099–102.

–(1978) Monoclonal antibody defining a stage-specific mouse embryonic antigen (SSEA-1). *Proc. Natl. Acad. Sci. U.S.A.* 75, 5565–9.

Spindle, A. I. (1978) Trophoblast regeneration by inner cell masses isolated from cultured mouse embryos. *J. Exp. Zool.* 203, 483–9.

Stacey, A. J., and Evans, M. J. (1984) A gene sequence expressed only in undifferentiated EC, EK cells and testes. *EMBO Journal* 3, 2279–85.

Strickland, S., and Mahdavi, V. (1978) The induction of differentiation in teratocarcinoma stem cells by retinoic acid. *Cell* 15, 393–403.

Sulston, J. E., Schierenberg, E., White, J. G., and Thomson, J. N. (1983) The embryonic cell lineage of the nematode *Caenorhabditis elegans. Dev. Biol.* 100, 64–119.

Sulston, J., and White, J. G. (1980) Regulation and cell autonomy during postembryonic development of *C. elegans. Dev. Biol.* 78, 577–97.

Van Blerkom, J., Barton, S. C., and Johnson, M. H. (1976) Molecular differentiation in the preimplantation mouse embryo. *Nature* 259, 319–21.

Vasseur, M., Duprey, P., Brulet, P., and Jacob, F. (1985) One gene and one pseudogene for the cytokeratin endo A. *Proc. Natl. Acad. Sci. U.S.A.* 82, 1155–9.

Wang, S.-Y., La Rosa, G. J., and Gudas, L. J. (1985) Molecular cloning of gene sequences transcriptionally regulated by retinoic acid and dibutyryl cyclic AMP in cultured mouse teratocarcinoma cells. *Dev. Biol.* 107, 75–86.

Wartiovaara, J., Leivo, I., and Vaheri, A. (1979) Expression of the cell surface associated glycoprotein, fibronectin, in the early mouse embryo. *Dev. Biol.* 69, 247–57.

Weisblat, D. A., and Blair, S. S. (1984) Developmental indeterminacy in embryos of the leech *Helobdella triserialis*. *Dev. Biol.* 101, 326–35.

Weisblat, D. A., Kim, S. Y., and Stent, G. S. (1984) Embryonic origins of cells in the leech *Helobdella triserialis*. *Dev. Biol.* 104, 65–85.

Wiley, L. M. (1984) Cavitation in the mouse preimplantation embryo: Na/K-ATPase and the origin of nascent blastocoele fluid. *Dev. Biol.* 105, 330–42.

Wu, T.-C., Wan, Y.-J., Chung, A. E., and Damjanov, I. (1983) Immunohistochemical localisation of entactin and laminin in mouse embryo and fetuses. *Dev. Biol.* 100, 496–505.

Young, P. R., and Tilghman, S. M. (1984) Induction of α-fetoprotein synthesis in differentiating F9 teratocarcinoma cells is accompanied by a genome-wide loss of DNA methylation. *Mol. Cell. Biol.* 4, 898–907.

Zackson, S. L. (1984) Cell lineage, cell-cell interaction and segment formation in the ectoderm of a glossiphoniid leech embryo., *Dev. Biol.* 104, 143–60.

Ziomek, C. A., and Johnson, M. H. (1982) The roles of phenotype and position in guiding the fate of 16-cell mouse blastomeres. *Dev. Biol.* 91, 440–7.

Ziomek, C. A., Johnson, M. H., and Handyside, A. H. (1982) The developmental potential of mouse 16-cell blastomeres. *J. Exp. Zool.* 221, 345–55.

5 Analysis of tissue fate and prospective potency in the egg cylinder

ROSA BEDDINGTON

CONTENTS

I. Introduction

The construction of fate maps and the assessment of cell or tissue potency have preoccupied developmental biologists since the inception of experimental embryology. However, despite its nineteenth-century origins, the question of the relationship among cell ancestry, cell location, and cell commitment is not outdated. It remains a central issue in contemporary embryology.

Fate maps depict what cells in particular locations will give rise to in the undisturbed embryo. That is, they provide an indispensable description of the regular and predictable deployment of cells to different tissues. Prospective potency, on the other hand, constitutes an inventory of the variety of tissues that cells are capable of forming under different circumstances at a given time. Evidence from a number of organisms indicates that there is a progressive and heritable restriction in the prospective potency of cells during development. Thus, comparisons between prospective fate and potency can produce a more sophisticated picture of

121

embryogenesis that charts the divergence of cell or tissue lineages into different, and usually mutually exclusive, developmental pathways. An appreciation of when and where the progenitors of a particular tissue or pattern element originate is obviously a prerequisite for any causal explanation of their segregation. This has been exemplified by the recent investigations of gene expression related to segmentation and segment identity in *Drosophila*. Here, prior knowledge of segment lineage provided an immediate and necessary context for the molecular findings (Gehring 1985).

At present, very little is known about the diversification of tissues in the early postimplantation mammalian embryo, and we remain largely ignorant of how the dramatic and complex morphogenetic events of gastrulation generate not only many of the different fetal tissue primordia but also the definitive body plan of the animal. If the cell lineage relationships among the different founder tissues of the fetus could be resolved, the problem of their orderly segregation might become less intractable. Furthermore, recent evidence from studies in the mouse on X-chromosome inactivation and reactivation (West et al. 1977; Kratzer et al. 1983) and on levels of DNA methylation (Chapman et al. 1984) have suggested that the mechanisms underlying stable differentiation in the fetus may differ from those in certain extraembryonic tissues. Therefore, the binary stepwise divergence describing the origin of the trophectoderm and primitive endoderm tissue lineages in the preimplantation mouse embryo (Gardner 1983) may not necessarily be a prototype for the diversification of fetal primordia (see Chapter 4).

The difficulties of studying cell lineages during gastrulation in the mouse stem from the inaccessibility of the embryo once it has implanted in the uterus. After removal from its implantation site, normal development of the conceptus in vitro can be sustained only for relatively short periods during gastrulation (Tam and Snow 1980; Beddington 1981). As a result, tracing the fate or potency of single cells or tissue in a normal embryonic environment, which should provide the most realistic measure of lineage relationships and cell determination, is severely restricted. Consequently, many of the strategies adopted to evaluate prospective fate and potency in the mouse egg cylinder remain relatively crude, and we still cannot pinpoint the origin of a single fetal lineage during mammalian development.

II. Anatomy of gastrulation

The crux of early postimplantation development is gastrulation, for it is this dramatic morphogenetic upheaval that really heralds the onset of fetal organization and differentiation. Essentially, gastrulation entails both production of new tissues and radical rearrangement of cells within the egg cylinder. The net result of gastrulation is twofold. First, the basic body plan of the fetus is established. Second, the various novel juxtapositions of the different tissues set the stage for a subsequent cascade of tissue interactions.

The mechanics of gastrulation remain far from clear, and even the exact sequence of cell production and distribution is not fully understood. However, for the purpose of understanding studies on prospective potency and fate, only a very superficial grasp of the events leading up to organogenesis is required. This should be sufficient to appreciate the extraordinarily rapid and orderly diversification of the epiblast-derived fetal and extraembryonic tissues. More detailed descriptions and reviews of gastrulation and early organogenesis, and of the tissues involved, can be found elsewhere (Jolly and Ferester-Tadie 1936; Snell and Stevens 1966; Batten and Haar 1979; Beddington 1983b).

A diagram of mouse development from the time of implantation to the earliest stages of organogenesis is shown in Figure 5.1. The appearance of the primitive streak (Figure 5.1c) marks the beginning of gastrulation and defines the anteroposterior axis of the embryo. The epiblast in the region of the primitive streak loses its epithelial conformation and invaginates to produce mesoderm. This mesoderm moves anteriorly within the embryonic region, effectively converting a bilaminar cylinder into a trilaminar one (Figure 5.1d). At the anterior end of the primitive streak, the invaginated epiblast emerges as a quasi epithelium, the head process, which extends anteriorly along the midline (Figure 5.1d) and is thought to give rise to the notochord and possibly the gut (Jolly and Ferester-Tadie 1936). Mesoderm also moves into the extraembryonic region, contributing as it does so to the evagination of the amniotic folds (Figure 5.1c). These folds fuse, and as a result of the formation and coalescence of multiple lacunae within the extraembryonic mesoderm, the conceptus acquires three distinct chambers separated from one another by two membranes, the chorion and the amnion (Figure 5.1d). The distal chamber gives rise to the fetus, the intermediate chamber gives rise to the visceral yolk sac (VYS), and the proximal, or mesometrial, chamber produces constituents of the chorioallantoic placenta. Over the next 24 h, the relatively simple cylindrical structure of the eight-day conceptus is transformed into a much more complex entity that embodies many of the distinctive features of the fetus (Figure 5.1e). In the embryo itself, definitive organ primordia can be identified, and extraembryonically the allantois, one of the major components of the placenta, can be seen traversing the exocoelom on its way to fusing with the chorion.

In Figure 5.1, in order to give a better idea of the time elapsing between different developmental stages, embryonic age is measured in days from the time of *fertilization*. Elsewhere in this chapter, the age of embryos is

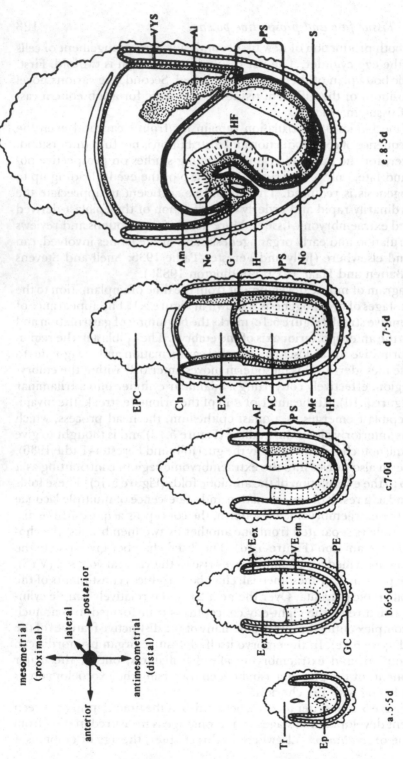

Figure 5.1. Development of the mouse embryo from the sixth to the ninth day of gestation. Ep, epiblast; Tr, trophectoderm; E$_{ex}$, extra-embryonic ectoderm; VE$_{em}$, visceral embryonic ectoderm; VE$_{ex}$, visceral embryonic endoderm; PE, parietal endoderm; AF, amniotic fold; PS, primitive streak; EPC, ectoplacental cone; Ch, chorion; AC, amniotic cavity; Me, mesoderm; HP, head process; HF, head fold; Am, amnion; Ne, neural epithelium; G, gut; H, heart; No, notochord; VYS, visceral yolk sac; Al, allantois; S, somite.

denoted by the day of *gestation,* the day on which a vaginal plug is detected being considered the first day of gestation. This convention has been adopted to avoid giving the false impression of very precise staging of embryos. In only a very few studies have embryos been carefully matched according to their developmental status. In most cases embryos have been classified by age, despite this being a poor criterion for ensuring developmental homogeneity. Even within a given litter, early postimplantation mouse embryos show a remarkable range of developmental stages, and, therefore, more assiduous classing of embryos according to stage probably would produce more precise and readily comparable results.

III. Estimates of prospective potency

Little has changed in the methods of prospective potency analyses since Weiss (1939) summarized the three basic types of experiments that test the range of cell or tissue differentiation. The first of these he called *isolation experiments,* in which the cells or tissues under consideration are removed from the embryo and placed in a supposedly neutral environment conducive to their continued growth and differentiation. This may be an in vitro milieu or a favorable site within an adult animal. The second type, *recombination experiments,* test the differentiation of isolated fragments either after they have been returned to their original locations in a host embryo or after they have been replaced in a different position or transferred to an embryo of a different developmental stage. In these experiments the environment is not neutral, because the relocated fragments have the opportunity of interacting with surrounding embryonic cells. Finally, *deletion experiments* involve removal or destruction of part of an embryo and assessment of the ability of the remaining tissues to compensate for this loss.

All three of these basic strategies have been applied to postimplantation mouse embryos. However, all experiments to date have dealt with *tissue* potency. Clonal studies either have failed or have not been attempted. Consequently, the developmental status of individual cells during gastrulation is unknown. Thus, heterogeneity within tissues may account for any diversity in their differentiation. Conversely, an observed restriction in tissue potency may not be a feature of its individual constituent cells but instead may reflect the limiting influences of tissue microenvironment or homotypic and heterotypic cell interactions.

As far as possible, representative experiments in the rodent egg cylinder will be classified according to the scheme of Weiss. However, inevitably in some studies more than one experimental principle may be employed.

A. Isolation experiments

i. In vitro isolation. The differentiation of defined fragments of the embryonic portion of the mouse egg cylinder has been examined at early, middle, and late primitive-streak stages (Snow 1981). In all isolates, the relationship between the germ layers remained intact, and therefore their development may be governed by sustained tissue interactions rather than by inherent differences in potency of individual tissues or regions. The consistent finding in these experiments was that each fragment, from whichever stage, exhibited a high degree of autonomy and an apparent lack of lability. A particular fragment after 24 h in culture formed only a limited but reproducible array of headfold-stage structures, and there was little or no overlap or duplication of the primordia formed by different fragments from the same-stage embryo. Instead, different fragments appeared to complement each other as if already determined to form specific elements of the fetus. However, unequivocal recognition of primordia on purely morphological grounds, particularly in necessarily deformed embryos, cannot be wholly reliable. The most compelling and interesting results in this study concern the origin of the primordial germ cells (PGCs). Here a histochemical marker, alkaline phosphatase, was used to supplement morphology in PGC identification. At both the early and middle primitive-streak stages, a region toward the posterior end of the primitive streak was found to be the only one capable of generating PGCs in isolation. Furthermore, those embryos from which this region had been removed (deletion) formed few if any PGCs. From these results, Snow concluded that "the behavior of the pieces removed from the primitive streak, at several developmental ages and in various combinations shows no evidence for lability in developmental potential."

Consideration of rather different isolation experiments on similar-stage embryos suggests that developmental lability may exist but simply not be expressed unless tissue interactions are disrupted or altered. For example, whereas visceral extraembryonic endoderm does not synthesize α-fetoprotein (AFP) or transcribe this gene in the intact embryo (Dziadek and Adamson 1978; Dziadek and Andrews 1983), once emancipated from its underlying extraembryonic ectoderm, either in isolation or associated with epiblast, it commences production of AFP (Dziadek 1978). When the same endoderm is isolated together with its adjacent extraembryonic ectoderm and cultured in suspension, it again changes phenotype, but under these conditions acquires the characteristics of parietal endoderm (Hogan and Tilly 1981). Likewise, visceral embryonic endoderm assumes parietal features when in contact with extraembryonic ectoderm (Hogan and Tilly 1981). Extraembryonic ectoderm in suspension culture commences transformation into trophoblast giant cells (Rossant and Ofer 1977; Ilgren 1981; Johnson and Rossant 1981; Rossant and

Tamura-Lis 1981), and this may be one of the positive cues for parietal differentiation, because visceral endoderm does not assume a parietal phenotype when isolated on its own. Interestingly, the giant cell transformation of extraembryonic ectoderm is itself subject to regulation by tissue interactions. Isolation of extraembryonic ectoderm or ectoplacental cone under various conditions in vitro indicates that both homotypic (Ilgren 1981) and heterotypic interactions with embryonic tissues (Rossant and Tamura-Lis 1981) serve to maintain, at least to some extent, mitotic activity and diploidy in the trophoblast.

Clearly, tissue interactions can profoundly affect the disposition of cells, even at the level of gene transcription, either by enhancing or suppressing certain intrinsic capabilities. One might imagine that in vitro isolation of particular germ layers or individual cells could expose these intrinsic capabilities. However, culture of germ layers leads to rather limited and unpredictable differentiation (Skreb and Svajger 1975; Skreb and Crnek 1977), and no cells from the egg cylinder have proved amenable to clonal expansion in vitro. This may, of course, be due to the inadequacy of tissue culture conditions, but it could also stem from a critical interdependence of the germ layers for their optimal growth and differentiation. Certainly, the survival and differentiation of isolated ICMs or ICM cores (presumptive epiblast) are augmented by addition of medium conditioned by blastocyst-derived or primitive-endoderm-like (PYS-2) cell lines (Atienza-Samols and Sherman 1978, 1979). Similarly, culture of preimplantation epiblast is improved if a fibroblast feeder layer is provided (Hogan and Tilly 1977; Rossant and Ofer 1977). Here, then, lies something of a dilemma. Culture conditions may never be able to sustain maximal growth and development of isolated egg cylinder tissues or cells, but if these isolates are cultured with their normal tissue neighbors to optimize their survival, then expression of potency may be influenced by tissue interactions. Therefore, the potency of individual tissues may better be tested in other environments.

ii. In vivo isolation. For many years the chorioallantoic membrane of the chick embryo has been a favorite site for testing the differentiation of avian tissues (Waddington 1952). Mammalian embryonic explants grow relatively poorly in this xenogeneic site (Nicholas and Rudnick 1933) and survive only for a limited period (Skreb and Svajger 1975). Analogous sites in rodents have been found that will support prolonged growth and differentiation of isolated embryonic tissues. The anterior chamber of the eye was one of the first of these to be used for early postimplantation rat or mouse embryos (Grobstein 1951, 1952). In this site, the developmental stage at which grafts are made affects the degree of differentiation. Once mesoderm formation is underway, grafts of the intact embryonic region (Grobstein 1951; Levak-Svajger and Skreb

1965) or of epiblast alone (Grobstein 1952) develop into a wide range of mature tissues representing derivatives of all three germ layers. Pre-primitive-streak-stage grafts produce less elaborate differentiation and consistently fail to form many mesodermal derivatives, presumably because the conditions are not suitable for initiation or maintenance of normal gastrulation.

Grobstein (1952) saw the pluripotency of primitive-streak-stage epiblast as incompatible with the then current classification of mammalian germ layers derived from histological description. His somewhat radical interpretation that epiblast could no longer be considered simply as the definitive ectoderm germ layer was, however, endorsed and reiterated only years later. Ectopic transfers to more permissive sites, such as beneath the kidney or testis capsule, where vascularization of the graft is improved (Skreb and Svajger 1975), confirmed that rat and mouse epiblast, before and during the first 24 h of gastrulation, develops into teratomas consisting of classic ectodermal, mesodermal, and endodermal derivatives (Levak-Svajger and Svajger 1971; Diwan and Stevens 1976; Beddington 1983a). Complementary transfers of visceral embryonic endoderm from the same stages were invariably resorbed in the rat (Levak-Svajger and Svajger 1971, 1974), and in the mouse, when they survived, they produced only parietal endoderm and Reichert's membranelike matrix (Diwan and Stevens 1976). This transformation from visceral to parietal phenotype is reminiscent of the behavior of visceral extraembryonic endoderm ectopic grafts (Solter and Damjanov 1973) strongly suggesting that embryonic endoderm at the onset of gastrulation is homologous to exraembryonic endoderm and is not the precursor of fetal gut.

Certainly, the prospective potency of primitive-streak-stage epiblast is consistent with its being the sole founder tissue of the fetus. Furthermore, unlike in vitro isolation (Snow 1981), grafts of different regions of the epiblast at this stage, either alone or with its enveloping mesoderm and endoderm, give little indication that different potencies reside in different locations (Skreb, Svajger, and Levak-Svajger 1976; Svajger et al. 1981; Beddington 1983a; Tam 1984). This discrepancy between in vivo and in vitro experiments probably reflects the rapid disorganization and disruption of normal tissue interactions that occur following ectopic transfer but not in short-term culture (Svajger et al. 1981; Tam 1984). Restriction in potential is seen at the headfold stage when "epiblast/ectoderm" no longer produces gut epithelium but does retain the capacity to generate mesodermal tissues (Levak-Svajger and Svajger 1974). It is not possible, however, to distinguish between a neural crest or a conventional mesoderm origin for these tissues (Svajger and Levak-Svajger 1974; Svajger et al. 1981), a problem that highlights the difficulty in extrapolating from the events occurring in teratomas to the developmental sequence in the embryo itself. Even at the headfold stage, isolated endoderm does not

survive beneath the kidney capsule, and mesoderm mysteriously differentiates exclusively into brown adipose tissue. Only if these two tissues are
grafted together do gut derivatives appear, as well as the normal repertoire of mesodermal tissues (Levak-Svajger and Svajger 1974). Thus, in
vivo isolation cannot resolve exactly when and where definitive endoderm
arises.

There is some evidence for axial regionalization in histogenetic potential by the headfold stage. Anterior and posterior fractions of the rat
embryo, in the absence of ectoderm, produce different but complementary gut derivatives (Svajger and Levak-Svajger 1974). Comparable results were obtained in the mouse, although ectoderm was always included
in the grafts, but there is some disagreement as to whether the qualitative
difference in gut derivatives is absolute (Fujimoto and Yanagisawa 1979)
or relative (Bennett et al. 1977; Tam 1984). Comparison of axial ectoderm, in which the neural groove was present, as opposed to lateral
ectoderm from headfold rat embryos, did not reveal any differences in
histogenic potential (Svajger and Levak-Svajger 1976). Thus, some morphogenetic changes associated with neural differentiation may precede
any corresponding restriction in prospective potency.

A peculiarity of early mouse embryos grafted to ectopic sites is the
production of teratocarcinomas, transplantable malignant tumors that
contain, in addition to the normal range of mature tissues, a population of
undifferentiated, dividing cells called embryonal carcinoma cells (EC
cells) (Solter, Skreb, and Damjanov 1970; Stevens 1970). It has been
shown that only the epiblast component of the egg cylinder on either the
seventh (Diwan and Stevens 1976) or eighth day of gestation (Beddington
1983a) can give rise to EC cells, but this neoplastic potential is lost by the
headfold stage (Damjanov, Solter, and Skreb 1971). Like epiblast, EC cells
can form derivatives of all three germ layers, not only in vitro and in vivo
but also following incorporation into the preimplantation embryo, and in
a very few of these chimaeras they have also given rise to viable germ cells
(Papaioannou and Rossant 1983). Germ cells, too, are precursors of
teratocarcinomas (Stevens 1983), and it is well documented that EC cells,
germ cells, and epiblast share a number of morphological and molecular
characteristics (Graham 1977). Thus, it has been argued that if grafts of
epiblast can generate EC cells, this tissue must also be the source of PGCs.
What is more difficult to determine is whether PGCs arise de novo from
the epiblast or constitute a discrete but as yet indistinguishable subpopulation within this epithelium. PGCs are first recognized by histochemical means late on the eighth day of gestation, when they are found
at the posterior aspect of the primitive streak (Ozdenski 1967). However,
epiblast from this region develops exceptionally poorly, if at all, in ectopic
grafts, whereas small fragments of epiblast from the anterior end of the
primitive streak or from the most anterior aspect of the embryo quite

frequently give rise to EC cells in an ectopic site (Beddington 1983a). Therefore, unless PGCs have a more widespread distribution, the neoplastic propensity of epiblast may be independent of its germ-line potential. This has been suggested before, because teratocarcinomas derived from early embryos are not subject to the same sex and strain effects as germ-cell-derived teratocarcinomas (Stevens 1970). Furthermore, grafts of mutant embryos (W/W and Sl^J/Sl^J) produce EC cells despite a supposed deficiency in PGCs (Mintz, Cronmiller, and Custer 1978). However, it is questionable that PGCs are actually absent, or even reduced, in these embryos at the time of grafting (Mintz and Russell 1957). Whatever the arguments regarding the origin of EC cells, the inescapable conclusion seems to be that ectopic transfer, in which tissues of unknown heterogeneity are being grafted, is too crude a tool for elucidating the status of the germ line during gastrulation.

B. Recombination experiments

i. In vitro recombination. Advances in whole-embryo culture (New 1978) have made possible the production and maintenance of early postimplantation chimeras (Beddington 1981). Unfortunately, culture regimens that support the development of pre-primitive-streak-stage rat embryos for several days (Buckley, Steele, and New 1978) have not proved so efficient with mouse embryos. Consistently normal growth and development of seven- or eight-day mouse embryos seldom can be sustained for more than about 36 h (Tam and Snow 1980; Beddington 1981), although slightly older embryos, explanted at the headfold or early somite stage, fare rather better (Sadler 1979; Sadler and New 1981). So far, only the mouse has been used to make early postimplantation chimeras in vitro (Beddington 1981, 1982), chiefly because most of the preimplantation lineage work has been done on this species. However, the limited duration of normal development precludes informative clonal analysis, because the size of clones after one or two days will be small, and consequently their distribution may be restricted by cell number rather than prospective potency. For this reason there may be an argument for using rat embryos, with which longer-term experiments would be possible.

The differentiation of three defined regions of the eight-day epiblast has been examined following orthotopic and heterotopic transplantation into synchronous embryos (Beddington 1981, 1982). The colonization pattern seen in orthotopic grafts approximates to the prospective fate of that region, whereas heterotopic grafts serve to test prospective potency. Because these experiments are necessarily short (36 h), [³H] thymidine, which appears not to be toxic to primitive-streak-stage embryos (Poelmann 1980; Beddington 1981), can be used to mark donor grafts, with the advantage that individual progeny can be recognized by autoradiography in histological sections. Analysis of chimeras formed by orthotopic grafts demonstrates, not surprisingly, that different regions of the epiblast have different prospective fates (Table 5.1). Anterior epiblast differentiates predominantly into neurectoderm and surface ectoderm, whereas tissue at the posterior aspect of the primitive streak colonizes mesodermal tissues, both extraembryonic and embryonic. Epiblast at the distal tip of the cylinder, which corresponds to the anterior end of the primitive streak, contributes to the gut and notochord as well as a variety of mesodermal derivatives.

Heterotopic grafts (Table 5.1) show that posterior epiblast and distal epiblast are quite capable of giving rise to definitive ectoderm derivatives when relocated in the anterior region. Similarly, distal epiblast forms only mesodermal tissues when placed in the posterior site. However, anterior epiblast shows a marked propensity to differentiate into

Table 5.1. *Differentiation of anterior, distal, and posterior epiblast in in vitro chimeras and ectopic grafts*

	IN VITRO CHIMERAS									ECTOPIC GRAFTS		
Site of injection	ANTERIOR			DISTAL			POSTERIOR			TESTIS		
Source of donor cells ⊗	A	D	P	A	D	P	A	D	P	A	D	P
Definitive ectoderm	■	□	□	■	░	□	▨	□	□	■	□	□
Fetal mesoderm	░	▨	░	□	■	□	▨	□	□	■	□	□
Gut endoderm	□	□	□	□	▨	□	□	□	□	■	□	□
Extraembryonic mesoderm	□	□	□	□	□	□	░	■	▨	□	□	□

⊗ : A = Anterior; D = Distal; P = Posterior

■ : Formation of tissue in more than 75% of chimeras or ectopic grafts

▨ : Formation of tissue in 25 – 75% of chimeras or ectopic grafts

░ : Formation of tissue in less than 25% of chimeras or ectopic grafts

□ : No detectable contribution

neurectoderm or surface ectoderm wherever it is located, and colonization of gut and notochord appears to be a property peculiar to distal epiblast. This could be seen as evidence for some regional differences in developmental potential, but this is not corroborated by evidence from ectopic grafts, in which experimental teratomas derived from anterior epiblast (Beddington 1983a) or from the proximal fraction of primitive-streak-stage rat epiblast (Skreb et al. 1976) invariably contain derivatives of all three germ layers, including gut derivatives (Table 5.1). In contrast, posterior epiblast readily differentiates after relocation in the embryo, but it fails to develop in an ectopic site (Beddington 1983a). Once again it appears that interactions within the embryo exert subtle effects on the expression of prospective potency.

Recently, in vitro chimeras have been used to investigate the origin of PGCs (A. J. Copp, personal communication). Donor epiblast and mesoderm tissue, consisting of the caudal end of the primitive streak, the allantoic bud and the root of the amnion, was taken from eight-day embryos that had been labeled for 4 h with [^{3}H]thymidine. This was grafted orthotopically, and the majority of resulting chimeras (analyzed 36–40 h later) contained, in addition to labeled somatic cells, alkaline-phosphatase-positive [^{3}H]thymidine-labeled cells, presumed to be donor-derived PGCs. Grafts of labeled extraembryonic endoderm, recovered from the same region, failed to produce labeled PGCs, as did heterotopic grafts of unseparated epiblast and mesoderm from the lateral region. However, both these grafts produced chimerism in somatic tissue, although most extraembryonic endoderm grafts gave rise to unincorporated endoderm vesicles. These experiments endorse the notion that PGCs are derived from the epiblast (Gardner et al. 1985) rather than the extraembryonic endoderm (Chiquoine 1954) and also suggest that by this stage, only certain regions of the epiblast can contribute to the germ line.

ii. In vivo recombination. So far, only retroviruses have been successfully microinjected into primitive-streak-stage mouse embryos in vivo (Jaenisch 1983). It is possible that cells could also be introduced into the conceptus at this stage, although it is difficult to see how very precise manipulations could be conducted on a cylinder less than 1 mm long that is buried within a mass of decidual tissue. Recent attempts to reimplant six- or seven-day mouse embryos, in order that the embryos might be manipulated precisely in vitro without forfeiting the option of letting them develop to later stages or even to term, have produced some encouraging results (Beddington 1985). Pre- or early-primitive-streak-stage embryos can survive after transfer to another decidua, and a few continue normal development, complete with placentation, for at least eight days (60-somite stage). However, at present, the frequency of successful reimplantation is too low for it to be viewed as an immediately viable procedure.

Although synchronous grafts of early postimplantation tissues in vivo are not yet feasible, certain tissues will tolerate transplantation into the blastocyst. For example, six- or seven-day extraembryonic ectoderm survives to colonize the ectoplacental cone and trophoblast giant cells after injection into four-day blastocysts (Rossant, Gardner, and Alexandre 1978). Likewise, six-day extraembryonic or embryonic visceral endoderm produces chimerism only in VYS endoderm (Rossant et al. 1978) and parietal endoderm (Gardner 1982). These results further substantiate the view that neither trophectoderm nor primitive endoderm lineages contribute tissue to the fetus. Unfortunately, postimplantation epiblast does not colonize the blastocyst (Rossant 1977), although retinal-pigmentation chimerism has been reported following injection of unspecified cells recovered from disaggregated eight-day embryos (Moustafa and Brinster 1972). Similarly prepared cells from six-day embryos, derived from blastocysts grown for 24 h in vitro, produced a low frequency of chimerism in ocular pigmentation, coat color, and the germ line (Moustafa and Brinster 1972). This suggests that there are cells in the postimplantation embryo whose potential to form fetal tissues can be tested by blastocyst injection. However, these cells have yet to be identified, and perhaps in the light of recent experiments in which single amphibian PGCs were seen to colonize somatic tissues after injection into the blastula (Wylie et al. 1985) the presumptive PGC population should be investigated more closely.

Chimeras have been produced in vivo later in development. Mononuclear blood island cells isolated from the VYS of 9- to 11-day conceptuses were injected into the VYS cavity in 9- to 12-day embryos in vivo (Weissman, Papaioannou, and Gardner 1978). Low levels of chimerism were detected by immunological assays in the bone marrow, thymus, and lymphoid tissue of newborn mice. However, neither glucose phosphate isomerase (GPI) electrophoresis nor a chromosomal marker could confirm this chimerism. High levels of erythroid and lymphoid colonization were achieved when normal fetal liver cells were injected into placental blood vessels of 12-day conceptuses homozygous for mutant genes at the *W* locus (Fleischman and Mintz 1979). *W/W* animals suffer from macrocytic anemia, and therefore the wild-type donor cells were at a selective advantage. This undoubtedly enhanced their colonization potential, because no chimerism was detected in injected heterozygotes. The choice of mutants, or even teratogen-treated embryos, as hosts in similar experiments might increase the liklihood of producing chimeras in utero, and perhaps this practice of donor cell selection could be successfully exploited at earlier developmental stages.

Recently, extensive chimerism has been obtained, in the absence of any selective pressure, following injection of neural-crest cells into 9- and 10-day conceptuses in vivo (Jaenisch 1985). Up to 300 presumptive neural-crest cells, isolated from cultures of explanted early somite-stage C57BL/6J neural tubes grown in vitro for several days (Ito and Takeuchi 1984), were injected into albino BALB/c embryos in utero. Approximately 15 percent of embryos injected late on the ninth day or early on the tenth day of gestation exhibited ocular and/or coat pigment chimerism after birth. Embryos injected 12 h earlier or later in gestation were not overtly chimeric, although it is conceivable that nonmelanocytic tissue may have been colonized. In all the chimeras, pigmentation was restricted to the head and posterior trunk, which would be expected if entry of donor cells into the fetus was limited to regions where the neural tube remained open at the time of injection. The distribution of donor pigment cells in the coat, choroid, and iris was reminiscent of that seen in aggregation chimeras, indicating that once incorporated into the embryo, injected cells and their progeny follow the normal melanocyte migration pathways. These experiments, therefore, introduce a new and potentially powerful method, particularly if used in conjunction with a cell marker that, unlike melanin, is both ubiquitous and expressed early in development, for analyzing the migration and differentiation of mammalian neural-crest cells in their normal embryonic environment. Furthermore, it is possible that this technique may also be applicable to other migratory cell types such as PGCs.

C. Deletion experiments

i. In vivo deletion.

All studies on in vivo deletion of tissues in the postimplantation mouse embryo employ teratogens. Often the precise nature and time course of the deletion are not known, although most agents used probably cause random and nonspecific cell death. In some cases, particularly where damage is caused by physical agents, such as radiation, the effect on the embryo may be direct and immediate, but in others, especially those involving treatment with drugs, the conceptus may also be affected via the mother and may suffer a more prolonged and less controllable insult (Saxen 1976). A further complicating factor is the variation in response found within litters, which may be due to variability in maternal blood supply to implantation sites or to some asynchrony in developmental stage between embryos or even to differences in genetic predisposition (Nebert and Bigelow 1982). Methodological problems apart, interpretation of teratogenic lesions is seldom straightforward. Defects in specific organs that occur as a result of an earlier insult do not necessarily stem from damage to the organ primordia themselves. Aberrant tissue interactions will also produce abnormalities. Consequently, teratology is usually a poor instrument for resolving cell lineage relationships.

A consistent finding in mouse and rat embryos is that prior to gastrulation they are refractory to most teratogens, in the sense that they either die or else regulate so that no abnormal survivors are seen (Austin 1973). This is interpreted as evidence of pluripotent cells in the embryo that, if the damage is not too extensive, can completely compensate for any loss before primitive-streak formation. However, once mesoderm formation is under way, the embryo's response changes, and specific defects start to arise. These occur predominantly in the head and vertebral column, although this bias may partly be a reflection of those systems most convenient for screening (Russell and Russell 1954; Skreb and Bijelic 1962; Skreb and Frank 1963; Gregg and Snow 1983). In general, these abnormalities are thought to result from perturbations in morphogenesis and the relative growth rates of different tissues during organogenesis and are not seen as evidence for determination of specific primordia during gastrulation (Gregg and Snow ·1983).

One *possible* progenitor target for teratogens during gastrulation is the PGC. Injection of mitomycin C (MMC) at the early primitive-streak stage drastically reduces cell number in the embryonic region, with over 85 percent of cells being lost (Snow and Tam 1979). Nonetheless, the embryos make a remarkable recovery, and approximately 90 percent survive to term, although postnatal death is common. Those that survive suffer a high incidence of sterility correlated with an obvious paucity of germ cells. Using alkaline phosphatase as a marker for PGCs, it is clear that by the beginning of somitogenesis (ninth day) their number in all MMC-treated embryos is depleted by almost two-thirds, and it is never restored to much more than half control values (Tam and Snow 1981). This would be expected if the germ line was segregated before or during gastrulation and was a self-contained lineage. It would be likely to suffer permanent damage after primitive-streak-stage cell loss if recovery was entirely dependent on the proliferative capacity of remaining PGCs.

ii. In vitro deletion. Apart from the deletion experiments undertaken by Snow (1981), which complement his isolation studies, the regulative capacity of primitive-streak-stage embryos has not been specifically tested in vitro, because until recently whole-embryo culture of early postimplantation mouse embryos was not sufficiently reliable to produce consistent results. A few teratological analyses have been performed on embryos at this stage, but on the whole, these experiments were designed to investigate the nature of drug actions rather than to examine embryonic regulation (Kochar 1975).

iii. Genetic deletion. In theory, genetic deletion, in the sense of mutations causing absence, loss, or deficiency of tissues, could be exploited in a number of ways to aid in the study of prospective potency. The use of mutants to give normal cells a selective advantage in chimeras was discussed earlier (also see Chapters 14 and 15). Genetically determined cell death, either ubiquitous or tissue-specific, can be used as a cell marker in chimeras, and an example of this will be given later in connection with retrospective analyses. *Conditional* cell autonomous lethality could also be informative. For example, the generality of restriction into anterior and posterior compartments in the organs of *Drosophila* has been examined in nuclear transplantation chimeras in which donor nuclei were homozygous for a lethal allele at the *engrailed* locus (Lawrence and Struhl 1982; Lawrence and Johnston 1984). The progeny of those cells that had incorporated *engrailed*-lethal nuclei died only if they were located in the posterior region of a structure in which posterior-compartment restriction had occurred. Lack of appropriate developmental mutants in the mouse precludes, at present, such an ingenious use of cell death to chart restriction in developmental potential.

Comprehensive comparisons of mutant and normal embryos, ideally encompassing all molecular, morphological, and behavioral characteristics, can produce compelling correlations between defects in certain embryonic cells and contingent lineage or pattern anomalies. Furthermore, in some circumstances it may be possible to confirm such correlations by effective rescue of mutants. Undoubtedly, this has proved a very profitable strategy in *Drosophila* (Gehring 1985). Once again, analogous studies on the postimplantation mouse embryo are severely restricted by the shortage of clearly defined mutants affecting pattern or tissue differentiation. Somewhat questionable conclusions have been drawn from parallel studies on t^{w18}/t^{w18} and wild-type embryos. This particular T-locus mutant is a recessive embryonic lethal, with homozygous embryos showing a deficiency in mesoderm and abnormalities of the primitive streak (Bennett and Dunn 1960; Moser and Gluecksohn-Waelsh 1967). The experimental teratomas produced from ectopic transfer of nine-day t^{w18}/t^{w18} embryos are deficient in mesoderm derivatives but contain the normal repertoire of definitive ectoderm tissues, although no very elaborate structures are seen, presumably because of the absence of mesodermal interactions (Artzt and Bennett 1972; Bennett et al. 1977). At the primitive-streak stage, the epiblast in these mutants exhibits an abnormally high mitotic index, and the plane of cleavage in dividing cells is different from that in wild-type embryos (Snow and Bennett 1978). Only one region appears normal, and that coincides with the proliferative zone (PZ), a small area described in normal embryos that is found at the anterior end of the primitive streak and has an elevated mitotic index (Snow 1977). The coincidence of respectable definitive ectoderm differentiation and a normal PZ in t^{w18}/t^{w18} embryos is seen as evidence for the PZ being the source of neurectoderm and surface ectoderm in the fetus. There is little foundation for this interpretation without evidence that the observed mitotic anomalies of the remaining epiblast prevent its ectodermal differentiation in an ectopic site. Even if this were so, the PZ might simply be acting as a supply of cells, and failure of mesoderm differentiation could be due to genetic or morphogenetic constraints quite unrelated to the function of the PZ. Certainly, in vivo recombination experiments (Beddington 1981, 1982) do not support the notion that the PZ is the source of definitive ectoderm.

So far, developmental mutants have contributed nothing concrete to our understanding of events during gastrulation in the mouse. Nonetheless, potentially they remain an invaluable tool. Indeed, they may even be an essential tool, for appropriate mutants may be the only decisive assay system for putative causal mechanisms in embryogenesis. However, it remains to be seen whether or not mutants already exist or can be produced experimentally (Jaenisch 1983) (see Chapter 14) that will affect the specification of pattern or lineage during gastrulation.

IV. Estimates of tissue fate

Strictly speaking, very few studies of tissue fate have been conducted on the mouse embryo, because most methods for introducing marked cells, such as the production of chimeras, involve some perturbation to the embryo and therefore constitute estimations of prospective potency. In the early postimplantation mouse embryo, two attempts have been made to label cells in situ and follow their development in the undisturbed embryo. By necessity, both these studies were carried out on embryos grown in vitro, and it might be argued that such embryos are not truly undisturbed, because they are grown in the absence of Reichert's membrane and under conditions that are unlikely to be optimal. However, there is no reason to believe that their development deviates from normal during the initial stages of culture.

The fate of embryonic endoderm in seven- and eight-day embryos has been monitored following iontophoretic injection of horseradish peroxidase (HRP) into individual endoderm cells located in specific regions of the egg cylinder (Lawson, Meneses, and Pedersen 1984; K. Lawson, personal communication). The distribution of labeled cells after culture (24−48 h) presumably approximates to their normal fate, because HRP does not seem to affect their development. As might be expected from in vitro chimeras (Beddington 1981), eight-day endoderm from the distal tip of the cylinder contributes to the gut and notochord along much of the embryonic axis. Endoderm cells or perhaps head-process cells anterior to this region are later found in the foregut, whereas endoderm from farther forward ends up in the VYS, with the endoderm overlying the

heart and a small part of the foregut. These results are compatible with
the cells in or near the head process being the precursors of the gut.
However, if cells in the posterior half of the axial endoderm are labeled on
the seventh day, before the head process has emerged, they, too, contribute to the gut, while the anterior endoderm colonizes the VYS. On the
face of it, this suggests that primitive endoderm may also contribute to
gut, contrary to its behavior in chimeras (Gardner 1982). It is conceivable
that primitive endoderm might be transiently incorporated into gut endoderm, but subsequently selected against, so that no contribution is detected in chimeras, which are invariably analyzed later in gestation. A
more plausible explanation is that epiblast-derived cells insert into the
superficial embryonic endoderm layer before or coincident with the
initial delamination of mesoderm in the primitive streak. The finding that
some HRP-injected cells on the surface of the posterior half of the seven-
day embryo can also colonize ectoderm and mesoderm tissues supports
this notion, because the potential to form these tissues is otherwise peculiar to epiblast at this stage (K. A. Lawson, unpublished observations).
Furthermore, in situ labeling of early streak-stage embryos has demonstrated that there is a rostral movement of endoderm during gastrulation,
and hence epiblast-derived, non-head-process endoderm could become
widely distributed along the axis and represent a second major precursor
of gut epithelium (K. A. Lawson and R. A. Pedersen, unpublished observations). Although not exhaustive, these studies have already provided
new and valuable information; in particular, they suggest that it may be
invalid to consider embryonic endoderm as a homogeneous tissue.

The second in situ technique was designed to follow neural-crest cell
migration in the early somite rat embryo (Smits-Van Prooije, Poelmann,
and Vermeij-Keers 1984). Wheat germ agglutinin (WGA) coupled to colloidal gold was injected into the amniotic cavity of embryos in vitro. The
presence of appropriate receptors leads to uptake of the WGA complex
by the ectoderm tissues on the surface of the embryo and, therefore, by
potential neural-crest cells. By virtue of the colloidal gold, the subsequent
invagination and destiny of neural-crest cells can be traced. No detailed
fate map can be constructed from such an experiment, because the initial
labeling is indiscriminate. However, if the colloidal gold is partitioned
more or less symmetrically at cell division and is not transferred between
cells, this could prove a useful method not only for tracing the neural crest
but also for examining the sequence of mesoderm production and development during gastrulation.

V. Tissue allocation

The relative or absolute spatial and numerical contributions of
the genetically distinct cell populations in adult aggregation chimeras

or X-inactivation mosaics have been used to deduce both the number of founder cells for particular tissues and their approximate time of segregation or allocation (West 1978). The use and abuse of retrospective analyses have been reviewed comprehensively elsewhere (West 1978), and only some recent examples will be discussed here. In addition to the problems of cell death, cell selection, and cell mixing that undermine any naive interpretation of adult pattern as a direct reflection of earlier embryological events (Lewis, Summerbell, and Wolpert 1972; McLaren 1972), the implicit assumption that development in aggregation chimeras is an exact replica of normal embryogenesis may be somewhat wishful thinking. Certainly, in terms of the initial allocation of cells to the ICM and trophectoderm, this is not the case – aggregation chimeras having a significantly higher proportion of ICM cells than single embryos (Rands 1985). Moreover, "unbalanced" chimerism in certain strain combinations is not uncommon (Mullen and Whitten 1971). Consequently, retrospective analyses are seldom straightforward.

An acceptable tenet is that particular tissues in a chimera or mosaic may share a common origin if the proportions of the two cell populations show a positive correlation in those tissues but not in others. Recently, proportional contributions were measured by quantitative phosphoglycerate kinase electrophoresis in extraembryonic and fetal tissues recovered from midgestation chimeras and X-inactivation mosaics (West et al. 1984). A positive correlation was found in each individual between the VYS mesoderm, the amnion, and the fetus, consistent with these all being derived from the epiblast. There was also a significant positive correlation in the relative contributions of the two populations to the parietal endoderm and visceral endoderm, but this was unrelated to that seen in epiblast derivatives. Therefore, this retrospective analysis has clearly identified and confirmed the separate primitive endoderm and epiblast tissue lineages.

Other retrospective analyses, where appropriate single-cell markers are available, have examined the distribution of the two populations in histological sections. The Purkinje cells (PCs) of the cerebellum are a suitable subject for this kind of analysis because effectively they constitute a monolayer. Calculations of the average PC clone size in adult chimeras, derived from the number of cells in spatially contiguous patches, have produced estimates of one cell (Mullen 1977) and two to four cells (Oster-Granite and Gearhart 1981), and only slight regional differences in the relative contributions of the two genotypes were observed (Mullen 1978). In other words, there is little evidence for sustained coherent clonal growth, and unlike the situation for the retina (Mintz and Sanyal 1970; Sanyal and Zeilmaker 1977), spatial patch size cannot be used as an indicator of progenitor cell number. The "numerical patch size," however, is more informative (Wetts and Herrup 1982a). In lurcher (*Lc*) ↔ wild-type aggregation chimeras, all PCs carrying the autosomal-dominant *Lc* gene die during early postnatal life. That lethality is cell-autonomous has been confirmed using genetically determined differences in β-glucuronidase activity as an independent cell marker (Wetts and Herrup 1982b). Counts of the total numbers of wild-type PCs remaining in these chimeras invariably produced figures that were almost exact integral multiples of the smallest population of surviving PCs (10,200 cells) ever seen in one half of the cerebellum. This was also the case for the total numbers of PCs in normal mice, in which integrals of 8 and 11 were calculated for two different inbred strains. The probability of this quantal effect occurring by chance in the seven animals examined is extremely remote, and therefore the results indicate that PCs are derived from a small number of progenitor cells (8–11 cells for each side of the cerebellum).

The facial-nerve nucleus has been analyzed in aggregation chimeras in a somewhat similar manner (Herrup, Diglio, and Letsou 1984). The total numbers of large motor neurons in 10 facial-nerve nuclei were counted, and each cell was ascribed to its embryo of origin using the β-glucuronidase marker. No constant numerical patch size was found, but the relative contributions of the two populations were not random. Instead, they were present in quantal fractions, with the smallest percentage contribution of either genotype being 8.3 percent, and all other contributions being approximate integral multiples of this. By simple arithmetic, one can calculate, therefore, that there are 12 progenitor cells for each facial-nerve nucleus. This could, of course, be a trivial finding in the sense that it might simply reflect that the whole central nervous system (CNS) originated from 12 cells. However, this is improbable, because the relative contributions, but not the quanta, differ between contralateral facial nuclei of a given individual and are independent of the genotype ratios found in neighboring retrofacial or fifth-cranial-nerve motor nuclei. Moreover, the presumed numbers of progenitors for the facial-nerve nucleus do not coincide with the numbers estimated for PCs.

Consideration of the average cell cycle time in the mouse neural tube and of the time when cell division ceases in the cerebellar PC layer or the facial-nerve nucleus suggests that the small numbers of progenitors for both these CNS components must arise on the eighth or ninth day of gestation (Wetts and Herrup 1982a; Herrup, Diglio, and Letson 1984). This is really surprisingly early, corresponding to the period between midgastrulation and the early somite stage. Of course, if these progenitors give rise to other neural tissue as well, which has not been altogether excluded, one would have to propose an even earlier time. Obviously, such calculations are based on many assumptions regarding cell division in neural primordia. Nonetheless, it does appear that cell allocation to different CNS structures not only is remarkably consistent and numerically precise but also occurs very early in the differentiation and morphogenesis of the neural plate.

VI. Conclusions and prospects

It is all too apparent that current methods for studying prospective potency and fate in the egg cylinder are far from ideal. The only point on which there seems to be a true consensus, regardless of experimental strategy, is that the fetus is derived from the epiblast, and undoubtedly the most persuasive evidence for this comes from blastocyst-injection chimeras (Gardner 1982). However, the exact nature of the epiblast at the onset of gastrulation remains unclear. The indirect methods used have failed to reveal any heterogeneity within epiblast tissue, except possibly with respect to the PGCs, but the absence of clonal studies makes it impossible to determine whether pluripotency is a feature of individual cells or only a property of cell populations.

The information derived from preimplantation chimeras regarding extraembryonic tissue lineages (see Chapter 4) provides a powerful demonstration of the value of long-term experiments in the embryo. Indeed, it is reasonable to suppose that the cell or tissue lineages of epiblast derivatives will be resolved only when similar experiments are possible in the postimplantation embryo. Only then can the full development of particular tissues or single cells be examined and the influences of position and tissue interactions be properly assessed. However, it is no mean task to devise a suitable experimental system. Ostensibly, there are three options: (1) to improve and extend the duration of normal development in vitro, (2) to produce a method for accurate microinjection in utero, (3) to achieve reliable reimplantation of manipulated postimplantation embryos. In the last few years, advances have been made in all three of these systems. Therefore, before too long, our understanding of the orderly diversification and segregation of the fetal tissue primordia, which currently lies largely in the realm of conjecture and extrapolation, may have a sounder basis, founded on experimental study of cells in their normal embryonic environment.

Acknowledgments

I would like to thank Professor R. L. Gardner for his discussion of the manuscript and Mrs. J. Williamson for typing it. I am also indebted to Dr. A. J. Copp, Prof. R. Jaenisch, and Dr. K. Lawson for allowing me to review their data prior to publication. The author holds a Lister Institute for Preventive Medicine Research Fellowship.

References

Artzt, K., and Bennett, D. (1972) A genetically caused embryonal ectodermal tumour in the mouse. *J. Natl. Cancer Inst.* 48, 141–58.

Atienza-Samols, S. B., and Sherman, M. I. (1978) Outgrowth promoting factor for the inner cell mass of the mouse blastocyst. *Dev. Biol.* 66, 220–31.

–(1979) *In vitro* development of core cells of the inner cell mass of the mouse blastocyst: effects of conditioned medium. *J. Exp. Zool.* 208, 67–72.

Austin, C. R. (1973) Embryo transfer and sensitivity to teratogenesis. *Nature (London)* 244, 333–4.

Batten, B. E., and Haar, J. L. (1979) Fine structural differentiation of germ layers in the mouse at the time of mesoderm formation. *Anat. Rec.* 194, 125–42.

Beddington, R. S. P. (1981) An autoradiographic analysis of the potency of embryonic ectoderm in the 8th day postimplantation mouse embryo. *J. Embryol. Exp. Morphol.* 64, 87–104.

–(1982) An autoradiographic analysis of tissue potency in different regions of the embryonic ectoderm during gastrulation in the mouse. *J. Embryol. Exp. Morphol.* 69, 265–85.

–(1983a) Histogenic and neoplastic potential of different regions of the mouse embryonic egg cylinder. *J. Embryol. Exp. Morphol.* 75, 189–204.

–(1983b) The origin of the foetal tissues during gastrulation in the rodent. In *Development in Mammals*, Vol. 5, ed. M. H. Johnson, pp. 1–32. Amsterdam: Elsevier.

–(1985) The development of 12th to 14th day foetuses following reimplantation of pre- and early primitive streak stage mouse embryos. *J. Embryol. Exp. Morphol.* 88, 281–91.

Bennett, D., Artzt, K., Magnuson, T., and Spiegelman, M. (1977) Developmental interactions studied with experimental teratomas derived from mutants at the T/t locus in the mouse. In *Cell Interactions in Differentiation*, eds. M. Karkinen-Jaaskelainen, L. Saxen, and L. Weiss, pp. 389–98. New York: Academic Press.

Bennett, D., and Dunn, L. C. (1960) A lethal mutant (t^{w18}) in the house mouse showing partial duplications. *J. Exp. Zool.* 143, 203–19.

Buckley, S. K. L., Steele, C. E., and New, D. A. T. (1978) In vitro development of early postimplantation rat embryos. *Dev. Biol.* 65, 396–403.

Chapman, V., Forrester, L., Sanford, J., Hastie, N., and Rossant, J. (1984) Cell lineage-specific undermethylation of mouse repetitive DNA. *Nature (London)* 307, 284–6.

Chiquoine, A. D. (1954) The identification, origin and migration of the primordial germ cells in the mouse embryo. *Anat. Rec.* 118, 135–46.

Damjanov, I., Solter, D., and Skreb, N. (1971) Teratocarcinogenesis as related to the age of embryos grafted under the kidney capsule. *Wilhelm Roux's Arch. Entwickl. Org.* 173, 228–84.

Diwan, S. B., and Stevens, L. C. (1976) Development of teratomas from ectoderm of mouse egg cylinders. *J. Natl. Cancer Inst.* 57, 937–42.

Dziadek, M. (1978) Modulation of alphafoetoprotein synthesis in the early postimplantation mouse embryo. *J. Embryol. Exp. Morphol.* 46, 135–46.

Dziadek, M., and Adamson, E. D. (1978) Localisation and synthesis of alpha-foetoprotein in postimplantation mouse embryos. *J. Embryol. Exp. Morphol.* 43, 289–313.

Dziadek, M., and Andrews, G. K. (1983) Tissue specificity of alphafetoprotein messenger RNA expression during mouse embryogenesis. *EMBO Journal* 2, 549–54.

Fleischman, R. A., and Mintz, B. (1979) Prevention of genetic anaemias in mice by microinjection of normal haematopoeitic stem cells into the fetal placenta. *Proc. Natl. Acad. Sci. U.S.A.* 76, 5736–40.

Fujimoto, H., and Yanagisawa, K. O. (1979) Effects of the T mutation on histogenesis of the mouse embryo under the testis capsule. *J. Embryol. Exp. Morphol.* 50, 21–30.

Gardner, R. L. (1982) Investigation of cell lineage and differentiation in the extraembryonic endoderm of the mouse embryo. *J. Embryol. Exp. Morphol.* 68, 175–98.

–(1983) Origin and differentiation of extraembryonic tissues in the mouse. *Int. Rev. Exp. Pathol.* 24, 63–133.

Gardner, R. L., Lyon, M. F., Evans, E. P., and Burtenshaw, M. D. (1985) Clonal analysis of X-chromosome inactivation and the origin of the germ line in the mouse embryo. *J. Embryol. Exp. Morphol.* 88, 349–63.

Gehring, W. J. (1985) The homeo box: a key to the understanding of development? *Cell* 40, 3–5.

Graham, C. F. (1977) Teratocarcinoma cells and normal mouse embryogenesis. In *Concepts in Mammalian Embryogenesis*, ed. M. I. Sherman, pp. 315–94. Cambridge: M.I.T. Press.

Gregg, B. C., and Snow, M. H. L. (1983) Axial abnormalities following disturbed growth in mitomycin C-treated mouse embryos. *J. Embryol. Exp. Morphol.* 73, 135–49.

Grobstein, C. (1951) Intraocular growth and differentiation of the mouse embryonic shield implanted directly and following in vitro cultivation. *J. Exp. Zool.* 116, 501–25.

–(1952) Intraocular growth and differentiation of clusters of mouse embryonic shields cultured with and without primitive endoderm and in the presence of possible inductors. *J. Exp. Zool.* 119, 355–80.

Herrup, K., Diglio, T. J., and Letsou, A. (1984) Cell lineage relationships in the development of the mammalian CNS. I. The facial nerve nucleus. *Dev. Biol.* 103, 329–36.

Hogan, B. L. M., and Tilly, R. (1977) In vitro culture and differentiation of normal mouse blastocysts. *Nature (London)* 265, 626–9.

–(1981) Cell interactions and endoderm differentiation in cultured mouse embryos. *J. Embryol. Exp. Morphol.* 62, 379–94.

Ilgren, E. B. (1981) On the control of the trophoblastic giant-cell transformation in the mouse: homotypic cellular interactions and polyploidy. *J. Embryol. Exp. Morphol.* 62, 183–202.

Ito, K., and Takeuchi, T. (1984) The differentiation *in vitro* of the neural crest cells of the mouse embryo. *J. Embryol. Exp. Morphol.* 84, 49–62.

Jaenisch, R. (1983) Endogenous retroviruses. *Cell* 32, 5–6.

–(1985) Mammalian neural crest cells participate in normal development when microinjected into postimplantation mouse embryos. *Nature* 318, 181–3.

Johnson, M. H., and Rossant, J. (1981) Molecular studies on cells of the tro-phectoderm lineage of the postimplantation mouse embryo. *J. Embryol. Exp. Morphol.* 61, 103–16.

Jolly, J., and Ferester-Tadie, M. (1936) Recherches sur l'oeuf du rat et de la souris. *Arch. Anat. Microsc. Morphol. Exp.* 32, 323–90.

Kochar, D. M. (1975) The use of *in vitro* procedures in teratology. *Teratology* 2, 273–88.

Kratzer, P. G., Chapman, V. M., Lambert, H., Evans, R. E., and Liskay, R. M. (1983) Differences in the DNA of the inactive X chromosome of fetal and extraembryonic tissues of mice. *Cell* 33, 37–42.

Lawrence, P. A., and Johnston, P. (1984) On the role of the *engrailed*[+] gene in the internal organs of *Drosophila*. *EMBO Journal* 3, 2839–44.

Lawrence, P. A., and Struhl, G. (1982) Further studies of the engrailed phe-notype in *Drosophila*. *EMBO Journal* 1, 827–34.

Lawson, K. A., Meneses, J. J., and Pedersen, R. A. (1984) Fate mapping of the endoderm in pre-somite mouse embryos by intracellular microinjection of horseradish peroxidase. *J. Embryol. Exp. Morphol.* Congress Suppl. 1, 67.

Levak-Svajger, B., and Skreb, N. (1965) Intraocular differentiation of rat egg cylinders. *J. Embryol. Exp. Morphol.* 13, 243–53.

Levak-Svajger, B., and Svajger, A. (1971) Differentiation of endodermal tis-sues in homografts of primitive ectoderm from two-layered rat embryonic shields. *Experientia* 27, 683–4.

–(1974) Investigation of the origin of definitive endoderm in the rat embryo. *J. Embryol. Exp. Morphol.* 32, 445–59.

Lewis, J. H., Summerbell, D., and Wolpert, L. (1972) Chimaeras and cell lin-eage in development. *Nature (London)* 239, 276–8.

McLaren, A. (1972) Numerology of development. *Nature (London)* 239, 274–6.

Mintz, B., Cronmiller, C., and Custer, R. C. (1978) Somatic cell origin of ter-atocarcinomas. *Proc. Natl. Acad. Sci. U.S.A.* 75, 2834–8.

Mintz, B., and Russell, E. S. (1957) Gene induced embryological modifications of primordial germ cells in the mouse. *J. Exp. Zool.* 134, 207–39.

Mintz, B., and Sanyal, S. (1970) Clonal origin of the mouse visual retina mapped from genetically mosaic eyes. *Genetics* Supp. 64, 43–4.

Moser, G. C., and Gluecksohn-Waelsh, S. (1967) Developmental genetics of a recessive allele at the complex T-locus in the mouse. *Dev. Biol.* 16, 564–76.

Moustafa, L. A., and Brinster, R. L. (1972) Induced chimaerism by trans-planting embryonic cells into mouse blastocysts. *J. Exp. Zool.* 181, 193–202.

Mullen, R. J. (1977) Site of *pcd* gene action and Purkinje cell mosaicism in the cerebella of chimaeric mice. *Nature (London)* 270, 245–7.

–(1978) Mosaicism in the central nervous system of mouse chimaeras. In *The Clonal Basis of Development*, eds. S. Subtelny and I. M. Suffex, pp. 82–101. New York: Academic Press.

Mullen, R. J., and Whitten, W. K. (1971) Relationship of genotype and degree of chimaerism in coat colour to sex ratios and gametogenesis in chimaeric mice. *J. Exp. Zool.* 178, 165–76.

Nebert, D. W., and Bigelow, S. W. (1982) Genetic control of drug metabolism: relationship to birth defects. *Semin. Perinatol.* 6, 105–15.

New, D. A. T. (1978) Whole embryo culture and the study of mammalian embryos during organogenesis. *Biol. Rev.* 53, 81–122.

Nicholas, J. S., and Rudnick, D. (1933) The development of embryonic rat tissues upon the chick chorioallantoic membrane. *J. Exp. Zool.* 66, 193–261.

Oster-Granite, M. L., and Gearhart, J. (1981) Cell lineage analysis of cerebellar Purkinje cells in mouse chimaeras. *Dev. Biol.* 85, 199–208.

Ozdenski, W. (1967) Observations on the origin of the primordial germ cells in the mouse. *Zool. Pol.* 17, 65–78.

Papaioannou, V. E., and Rossant, J. (1983) Effects of the embryonic environment on proliferation and differentiation of embryonal carcinoma cells. *Cancer Surveys* 2, 165–83.

Poelmann, R. E. (1980) Differential mitosis and degeneration patterns in relation to the alterations in shape of the embryonic ectoderm of early postimplantation mouse embryos. *J. Embryol. Exp. Morphol.* 55, 33–51.

Rands, G. F. (1985) Cell allocation in half and quadruple-sized preimplantation mouse embryos. *J. Exp. Zool.* 236, 67–70.

Rossant, J. (1977) Cell commitment in early rodent development. In *Development in Mammals*, Vol. 2, ed. M. H. Johnson, pp. 119–50, Amsterdam: Elsevier.

Rossant, J., Gardner, R. L., and Alexandre, H. (1978) Investigation of the potency of cells from the postimplantation mouse embryo by blastocyst injection: a preliminary report. *J. Embryol. Exp. Morphol.* 48, 239–47.

Rossant, J., and Ofer, L. (1977) Properties of extraembryonic ectoderm isolated from postimplantation mouse embryos. *J. Embryol. Exp. Morphol.* 39, 183–94.

Rossant, J., and Tamura-Lis, W. (1981) Effect of culture conditions on diploid to giant-cell transformation in postimplantation mouse trophoblast. *J. Embryol. Exp. Morphol.* 62, 217–27.

Russell, L. B., and Russell, W. L. (1954) Analysis of the changing radiation response of the developing mouse embryo. *J. Cell. Comp. Physiol.* Suppl. 1, 43, 103–47.

Sadler, T. W. (1979) Culture of early somite embryos during organogenesis. *J. Embryol. Exp. Morphol.* 49, 17–25.

Sadler, T. W., and New, D. A. T. (1981) Culture of mouse embryos during neurulation. *J. Embryol. Exp. Morphol.* 66, 109–16.

Sanyal, S., and Zeilmaker, G. H. (1977) Cell lineage in retinal development of mice studied in experimental chimaeras. *Nature (London)* 265, 731–3.

Saxen, L. (1976) Mechanisms of teratogenesis. *J. Embryol. Exp. Morphol.* 36, 1–12.

Skreb, N., and Bijelic, N. (1962) Effects of X-rays on rat embryos during mesoderm formation. *Nature (London)* 193, 292–3.

Skreb, N., and Crnek, U. (1977) Tissue differentiation in ectopic grafts after cultivation of rat embryonic shields in vitro. *J. Embryol. Exp. Morphol.* 42, 127–34.

Skreb, N., and Frank, Z. (1963) Developmental abnormalities in the rat induced by heat shock. *J. Embryol. Exp. Morphol.* 11, 445–57.

Skreb, N., and Svajger, A. (1975) Experimental teratomas in rats. In *Teratomas and Differentiation*, ed. D. Solter, pp. 83–97. London: Academic Press.

Skreb, N., Svajger, A., and Levak-Svajger, A. (1976) Developmental potentialities of the germ layers. In *Embryogenesis in Mammals*, CIBA Foundation Symposium 40 (new series), pp. 27–39. Amsterdam: Elsevier.

Smits-Van Prooije, A. E., Poelmann, R. E., and Vermeij-Keers, C. (1984) Mesoderm formation in the rat embryo, cultured in vitro, analysed with a lectin-coated colloidal gold marker. *J. Embryol. Exp. Morphol.* Congress Suppl. 1, 76.

Snell, G. D., and Stevens, L. C. (1966) Early embryology. In *Biology of the Laboratory Mouse*, ed. E. L. Green, pp. 205–45. New York: McGraw-Hill.

Snow, M. H. L. (1977) Gastrulation in the mouse: growth and regionalisation of the epiblast. *J. Embryol. Exp. Morphol.* 42, 293–303.

–(1981) Autonomous development of parts isolated from primitive-streak-stage mouse embryos. Is development clonal? *J. Embryol. Exp. Morphol.* Suppl. 65, 269–87.

Snow, M. H. L., and Bennett, D. (1978) Gastrulation in the mouse: the establishment of cell populations in the epiblast of t^{w18}/t^{w18} embryos. *J. Embryol. Exp. Morphol.* 47, 39–52.

Snow, M. H. L., and Tam, P. P. L. (1979) Is compensatory growth a complicating factor in mouse teratology? *Nature (London)* 279, 554–7.

Solter, D., and Damjanov, I. (1973) Explantation of extraembryonic parts of 7d mouse egg-cylinders. *Experientia* 29, 701–3.

Solter, D., Skreb, N., and Damjanov, I. (1970) Extrauterine growth of mouse egg-cylinders results in malignant teratomas. *Nature (London)* 227, 503–4.

Stevens, L. C. (1970) Experimental production of testicular teratomas in mice of strains 129, A/He and their F_1 hybrids. *J. Natl. Cancer Inst.* 44, 923–9.

–(1983) Testicular, ovarian and embryo-derived teratomas. *Cancer Surveys* 2, 75–91.

Svajger, A., and Levak-Svajger, B. (1974) Regional developmental capacities of the rat embryonic ectoderm at the headfold stage. *J. Embryol. Exp. Morphol.* 32, 461–7.

–(1976) Differentiation in renal homografts of isolated parts of rat embryonic ectoderm. *Experientia* 32, 378–9.

Svajger, A., Levak-Svajger, B., Kostovic-Knezevic, L., and Bradamante, Z. (1981) Morphogenetic behaviour of the rat embryonic ectoderm as a renal homograft. *J. Embryol. Exp. Morphol.* Suppl. 65, 243–67.

Tam, P. P. L. (1984) The histogenetic capacity of tissues in the caudal end of the embryonic axis of the mouse. *J. Embryol. Exp. Morphol.* 82, 253–66.

Tam, P. P. L., and Snow, M. H. L. (1980) The in vitro culture of primitive streak stage mouse embryos. *J. Embryol. Exp. Morphol.* 59, 131–43.

–(1981) Proliferation and migration of primordial germ cells during compensatory growth in mouse embryos. *J. Embryol. Exp. Morphol.* 64, 133–47.

Waddington, C. H. (1952) *The Epigenetics of Birds*. Cambridge University Press.

Weiss, P. (1939) *Principles of Development*. New York: Holt.

Weissman, I. L., Papaioannou, V. E., and Gardner, R. L. (1978) Fetal hematopoietic origins of the hematolymphoid system. In *Differentiation of Normal* and *Neoplastic Hematopoietic Cells,* eds. B. Clarkson, T. Markx, and J. Till, pp. 33–47. Cold Spring Harbor, N.Y.: Cold Spring Harbor Laboratory.

West, J. D. (1978) Analysis of clonal growth using chimaeras and mosaics. In *Development in Mammals,* Vol. 3, ed. M. H. Johnson, pp. 413–60. Amsterdam: Elsevier.

West, J. D., Bucher, T., Linke, I. M., and Dunnwald, M. (1984) Investigation of variability among mouse aggregation chimaeras and X chromosome inactivation mosaics. *J. Embryol. Exp. Morphol.* 84, 309–29.

West, J. D., Frels, I., Chapman, V. M., and Papaioannou, V. E. (1977) Preferential expression of the maternally derived X chromosome in the mouse yolk sac. *Cell* 12, 873–82.

Wetts, R., and Herrup, K. (1982a) Cerebellar Purkinje cells are descended from a small number of progenitors committed during early development: quantitative analysis of lurcher chimaeric mice. *J. Neurosci.* 2, 1494–8.

–(1982b) Interaction of granule, Purkinje, and inferior olivary neurons in lurcher chimaeric mice. I. Qualitative studies. *J. Embryol. Exp. Morphol.* 68, 87–98.

Wylie, C. C., Heasman, J., Snape, A., O'Driscoll, M., and Holwill, S. (1985) Primordial germ cells of *Xenopus laevis* are not irreversibly determined early in development. *Dev. Biol.* 112, 66–72.

Molecular and biochemical aspects

6 Regulation of gene expression during mammalian spermatogenesis

NORMAN B. HECHT

CONTENTS

I. Introduction

Spermatogenesis is the process of cellular differentiation in which diploid progenitor cells in the testis differentiate into haploid spermatozoa. During meiosis—one of the central events of spermatogenesis—chromosomes pair, and genetic recombination occurs in the functionally tetraploid pachytene spermatocyte. Numerous cytological studies have described in detail the dramatic morphological changes that the differentiating cells undergo and the unique structures that appear during the

151

development and maturation of the male gamete. These structural data provide an excellent framework on which to investigate at a molecular level the changes in testicular gene expression that underlie the developmental events. Moreover, the testis provides a superior system to study certain eukaryotic control mechanisms. The ordered differentiation of the germ cells allows analysis of the temporal regulation of gene expression for a series of defined proteins in a terminally differentiating system. Furthermore, because transcription terminates early during spermiogenesis, translational regulation of stored mRNA is essential to synthesize many of the proteins appearing during this haploid phase of spermatogenesis. A more complete understanding of the biochemistry of spermatogenesis and the roles played by the plethora of distinct testicular proteins will also disclose many essential regulatory mechanisms unique to the development of the male gamete.

In this chapter, the literature on gene expression during spermatogenesis in mammals will be examined. Although it will be directed toward mammals, pertinent information from nonmammalian species will be included. Because several excellent and detailed reviews of the biochemistry of spermatogenesis (Bellvé 1979; Bellvé and O'Brien 1983) and aspects of hormonal regulation (Steinberger 1971; Fritz 1978; Parvinen 1982) are available, I shall restrict most of this presentation to recent application of recombinant DNA technology to spermatogenesis. Special attention will be given to evidence demonstrating that unique genes are expressed from both the diploid and the haploid genomes during mammalian spermatogenesis.

II. Scheme of spermatogenesis

Spermatogenesis can be divided into three distinct intervals: (1) stem cell differentiation and renewal, (2) meiosis, and (3) spermiogenesis. The sequence of spermatogenesis in the mouse is presented in Figures 6.1 and 6.2 and will be discussed later.

A. Stem cell differentiation and renewal

During fetal development, primordial germ cells migrate to the genital ridge, where they continue to proliferate (Eddy 1970; Jeon and Kennedy 1973; Spiegelman and Bennett 1973; Zamboni and Merchant 1973; Eddy and Clark 1975; Fujimoto, Miyayama, and Fuyuta 1977). After a period of substantial cell growth and proliferation (Chiquoine 1954; Mintz and Russell 1955; Bennett 1956; Clark and Eddy 1975), the germ cells, now called gonocytes or prespermatogonia, enter a nonproliferative growth phase and move to a position adjacent to the basal lamina of the seminiferous cords, where they differentiate to a transitional cell

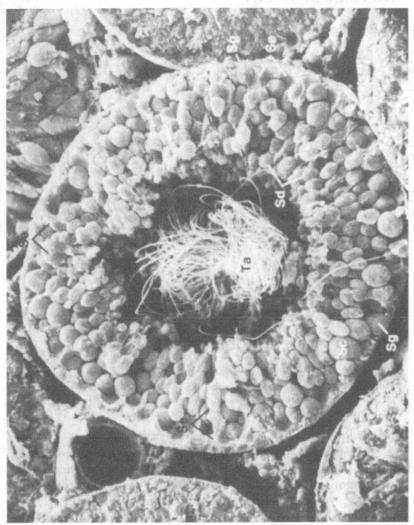

Figure 6.1. Scanning electron micrograph of a transverse section of a seminiferous tubule (×580): spermatogonia (Sg), primary spermatocyte (Sc), spermatid (Sd), sperm tails (Ta), Sertoli cells (Se). [Reproduced from Kessel and Kardon (1979) by permission of Freeman and Company.]

Spermatogenic Cells

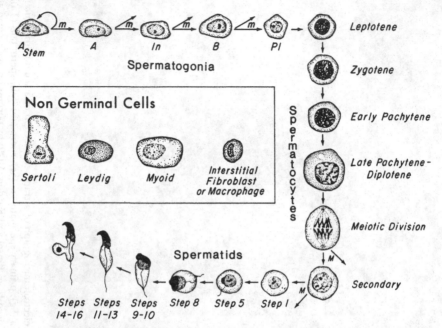

Figure 6.2. Schematic diagram of mouse testis cells. Mitotic and meiotic divisions are indicated by *m* and *M*, respectively. Other abbreviations: *In*, intermediate spermatogonium; *Pl*, proleptotene spermatogonium. [Reproduced from Meistrich (1977) by permission of Academic Press.]

type, primitive spermatogonia (Baillie 1964; Franchi and Mandl 1964; Hilscher and Makoski 1968; Huckins and Clermont 1968). These cells, after several mitotic divisions, serve as progenitors of the definitive spermatogonia that produce primitive type-A spermatogonia and, later, the presumptive type-A spermatogonia. In the mouse, the predominant experimental animal discussed in this chapter, type-A spermatogonia first appear in the seminiferous epithelium seven days postnatally, and intermediate and type-B spermatogonia are present in the testis by day nine (Nebel, Amarose, and Hackett 1961; Clermont 1972; Bellvé et al. 1977). The constant proliferation of germ cells necessary for the animal to maintain fertility is provided by a process of stem cell renewal and differentiation. Little is known at the molecular level of the regulatory processes in spermatogonial differentiation, a time interval that is experimentally difficult to study.

B. Meiosis

During the second major phase in the approximately 34-day cycle producing spermatozoa in mice, type-B spermatogonia divide to form

preleptotene primary spermatocytes, the cell type initiating meitoic prophase (Monesi 1962). Whereas testicular cell development up to this point
occurs in the peripheral compartment of the seminiferous epithelium,
the leptotene and zygotene-stage spermatocytes translocate into the ad-
luminal (central) compartment of the seminiferous epithelium (Dym and
Fawcett 1970; Fawcett, Leak, and Heidger 1970; Dym 1973; Ross 1977).
The final semiconservative DNA replication of the differentiating germ
cell in the preleptotene spermatocyte is notable because of an unusually
protracted S phase compared with the S phase of somatic cells (Lima de
Faria and Borum 1962; Monesi 1962). Although the reason preleptotene
spermatocytes undergo such a lengthy DNA replication interval is not
known, evidence from studies in *Triturus* (Callan 1972, 1973) suggests
that a marked reduction in number of DNA replication initiation sites is
the likely cause for the increased length of this terminal replicative DNA
synthesis. The high levels of the DNA-dependent DNA polymerases in
preleptotene spermatocytes (Hecht, Farrell, and Davidson 1976; Daentl,
Erickson, and Betlach 1977; Grippo et al. 1978; Hecht, Farrell, and
Williams 1978; Hecht and Parvinen 1981) suggest that the lengthening of
this last S phase cannot be explained by a limiting amount of the replicative DNA polymerase, but likely is due to the state of chromatin conformation or other essential regulatory replicative factors.

In mammals, meiosis is lengthy, requiring 11 to 12 days in the mouse
(Kofman-Alfaro and Chandley 1970) and 20 to 40 days in the Chinese
hamster (Utakoji 1966) and human (Heller and Clermont 1963). During
this time period, the chromosomes condense, and axial elements appear
between the two sister chromatids of each homologous chromosome (von
Wettstein 1977). By a mechanism yet to be defined, the homologous
chromosomes align, and the synaptonemal complex, a structure containing a central element, is formed (Moses 1956). By pachynema, the longest
stage of meiotic prophase, complete synaptonemal complexes are normally found between all homologues. This structure presumably facili-
tates genetic recombination in the pachytene spermatocyte by allowing
precise alignment and effective synapsis of the homologues. Although
much has been learned about the biochemical events associated with
synapsis and recombination (Stern and Hotta 1983), our knowledge of
the molecular events responsible for recombination in eukaryotes remains rudimentary.

Two meiotic divisions without DNA synthesis follow meiotic prophase,
thus reducing the nuclear DNA content in the primary spermatocyte
from a functional 4N level to N level in round spermatids.

C. Spermiogenesis
The phase of spermiogenesis, an interval of about 14 days in the
mouse (Nebel et al. 1961), is the period when the haploid round spermatid differentiates to the mature spermatozoon. Many unique morpholog-

ical transformations occur, including remodeling and condensation of the nucleus into species-specific shapes, assembly of a tail, reduction in number and localization of the remaining mitochondria of the cell to the midpiece region of the developing sperm, formation of the acrosome in the nucleus, and topographical changes in the cell surface (Bellvé 1979; Bellvé and O'Brien 1983). Toward the end of spermiogenesis, most of the cytoplasm of the maturing male gamete is eliminated in a structure called the residual body. Many of these differentiative changes in the developing germ cell occur concurrently with or after nuclear DNA condensation and termination of RNA transcription.

III. Cell separation methods

A major problem that has delayed and complicated interpretation of many biochemical studies of the testis has been the diverse number of cell types present in the testis. In attempts to surmount this problem, many investigators have utilized prepubertal animals. In the mouse and many other mammals, the timing of the stages of the spermatogenic cycle has been well characterized (Oakberg 1956a, 1956b), and the temporal appearance of successive cell types during the first wave of spermatogenesis following birth is precisely known. Thus, analysis of testes of 6-day-old mice provides a means to monitor events occurring in a population of Sertoli cells and primitive type-A spermatogonia, whereas the testes of 17-day-old mice allow one to assay events up to and including meiosis, and the testes of mice 24 days of age include early haploid cell types (Nebel et al. 1961; Bellvé et al. 1977).

Although prepubertal animals provide a means to roughly correlate the initial temporal appearance of a molecule or event with the presence of a newly appearing germ cell, data obtained from prepubertal studies *must* be confirmed with studies using isolated populations of cell types, for several reasons. First, the cell type of interest at the "leading edge" of the spermatogenic wave of the prepubertal testis represents only a small percentage (often 5–10%) of the total poulation of testicular cells in the prepubertal testis. Second, the often forgotten nongerminal Sertoli, Leydig, myoid, and interstitial fibroblast cells in total testicular extracts should be removed (Figure 6.2). A third, and perhaps most important, consideration in using animals that have not yet completed one wave of spermatogenesis is the question whether or not the cells present in the first spermatogenic wave are identical with those present in the testis of the sexually mature animal. By most morphological criteria the testicular cell types appear generally identical. Whether or not one can extrapolate such morphological similarities to assume that all biochemical events occurring in germ cells from prepubertal or sexually mature testes are identical is uncertain and unlikely. The higher incidence of cell death in the prepubertal testis compared with the sexually mature testis argues for caution (Allen and Altland 1952; Roosen-Runge 1973).

The ability to isolate "large" quantities of highly purified populations of mammalian testicular cell types has been a major breakthrough in the field of spermatogenesis, allowing biochemical analysis of specific "windows" of spermatogenesis. Pioneering work in testicular cell dissociation, preparation, and separation has come from several groups of investigators (Lam, Furrer, and Bruce 1970; Meistrich 1972, 1977, 1978; Loir and Lanneau 1975; Romrell, Bellvé, and Fawcett 1976; Meistrich et al. 1981). In general, testicular cells are dissociated into single-cell suspensions either by physical or enzymatic means, and the cells are separated into discrete fractions by the criteria of sedimentation rate or density. Highly purified populations of premeiotic, meiotic, and postmeiotic testicular cells can be obtained by the procedures to be described later.

Based on the criterion of sedimentation rate, testicular cells of different sizes can be isolated by (1) the Staput technique, a unit-gravity-sedimentation procedure in which testicular cells sediment through either a bovine-serum-albumin gradient or Ficoll gradient or (2) centrifugal elutriation, a velocity-sedimentation system in which a combination of adjusted flow rates in a specially designed centrifuge rotor allows separation of cells of different sizes. Both of these procedures have merit and have been used successfully in many laboratories. Advocates of the Staput procedure argue for its ability to separate several different cell types including spermatogonia and early meiotic prophase cells (when prepubertal testes are used) in a device not requiring a centrifuge (Romrell et al. 1976). Users of centrifugal elutriation are enthusiastic about its high cell capacity (about 10^9 cells) and speed of isolation (1 h once the system is operating) (Meistrich 1977, 1978). Both procedures are strongly dependent on methods of preparing single-cell suspensions that minimize damage and aggregation of cells and create the minimal number of multinucleate cells (multinucleate spermatids cosediment with pachytene spermatocytes, a serious problem for studies comparing meiotic and postmeiotic events). Recently, a modified unit-gravity-sedimentation procedure that maximizes the unit area over which sedimentation occurs and minimizes the volume of the gradient has also been used to isolate populations of testicular cells (Wolgemuth et al. 1985).

Equilibrium density-gradient centrifugation offers a second and different criterion by which to separate testicular cells. The gradient-forming materials, Renografin, Metrizamide (for elongated spermatid nuclei), and Percoll (Loir and Lanneau, 1975, 1977; Meistrich and Trostle 1975; Meistrich et al. 1981) have all been used to separate testicular cell types. Because of costs and limitations of gradient size, many laboratories use density gradients as a second step to increase the purity of enriched populations of testicular cells isolated by Staput sedimentation (Stern, Gold, and Hecht 1983a; Stern et al. 1983b) or elutriation (Meistrich et al.

1981; Fujimoto and Erickson 1982). The combination of sedimentation rate and density techniques provides a means to routinely isolate populations of meiotic and postmeiotic testicular cell types at levels of purity ranging up to 98 percent. Highly enriched populations of stem cells and early meiotic prophase cells can also be obtained when prepubertal animals are used (Bellvé et al. 1977).

IV. Differential gene expression in testis

A. Protein synthesis

Autoradiographic evidence has demonstrated that protein synthesis occurs throughout spermatogenesis, with maximal incorporation of radiolabeled amino acids in spermatogonia and pachytene spermatocytes (Monesi 1964b, 1965, 1967, 1971, 1978). Round spermatids are active in protein synthesis to a lesser extent (Monesi 1978). These conclusions have been extended by the work of Dadoune and associates (1981), who have described an incorporation gradient for several radiolabeled amino acids of decreasing incorporation intensity from the periphery to the lumen of the seminiferous tubule. They reported similar levels of protein synthesis in type-B spermatogonia and spermatocytes, with a decrease to about one-third in step 4−5 spermatids, suggesting a decrease in gene expression in postmeiotic cells. Assuming relatively constant amino acid pool sizes and no large differences in permeability in testicular cells, these autoradiographic studies demonstrate quantitative differences in levels of protein synthesis during spermatogenesis.

Evidence for differential gene expression of specific proteins has been accumulated from analyses of a wide variety of proteins, many of which are present in unique isozymic forms in the testis. Enzymes such as lactate dehydrogenase X, phosphoglycerate kinase 2, acrosin, hyaluronidase, and DNA polymerase have been demonstrated to be temporally regulated during spermatogenesis (Goldberg 1977). Because most of these studies have been dependent on the sensitivity and specificity of enzyme assays, it is difficult to demonstrate definitively "de novo" appearance of an enzyme activity rather than the investigator's ability to first detect it. Despite this limitation in assay sensitivity, definite temporal alterations in gene expression have been observed for many testicular enzymes. The initial temporal expressions during meiosis of other proteins, such as the testicular form of cytochrome c (cytochrome C_t) have also been well established by immunofluorescence studies (Goldberg et al. 1977). The use of antisera to the many sperm-specific proteins, some of which are first detected during spermiogenesis, makes this immunocytological approach a powerful one to localize specific proteins to cell types (Millette and Bellvé 1977, 1980; O'Rand and Romrell 1977; Millette and Moulding 1981). Appropriate controls can lessen a primary concern for this technique – the possibility of antigenic masking at specific stages of spermatogenesis.

The ability to analyze polypeptides from isolated populations of testicular cell types by two-dimensional polyacrylamide-gel electrophoresis has provided compelling evidence that large numbers of polypeptides are temporally regulated in the mammalian testis. Boitani, Palombi, and Stefanini (1980) found differences in the polypeptides present in meiotic and postmeiotic cells by examining radiolabeled testicular polypeptides that had been synthesized when seminiferous tubules or quasi-homogeneous populations (description of Boitani et al.) of germ cells were incubated with [^{3}H] leucine. By using cells obtained by centrifugal elutriation, Kramer and Erickson (1982) analyzed the in vivo [^{35}S] methionine-labeled proteins from pachytene spermatocytes, early spermatids, and late spermatids and found stage-specific synthesis for approximately 15 percent of the soluble proteins (100,000 × *g* supernatant) and 20 percent of the particulate proteins that were detectable on two-dimensional gels. Using a combination of Staput sedimentation and Percoll density gradients to separate cell types, Stern and associates (1983a) analyzed the polypeptides synthesized in pachytene spermatocytes, round spermatids, elongating spermatids, and residual bodies following intra-testicular injection of radiolabeled methionine or leucine. They concluded that for a constant number of cells, pachytene spermatocytes incorporated 5.4 times more isotope than round spermatids, which in turn incorporated 2.4 times more isotope than elongating spermatids – results in close agreement with the 10 : 2 : 1 incorporation ratio reported by Kramer and Erickson (1982) and supporting the gradient of protein synthesis in seminiferous tubules detected by autoradiography (Dadoune et al. 1981). At the level of detection provided by two-dimensional-gel electrophoretic methods, Stern and associates (1983a) demonstrated that pachytene spermatocytes and round spermatids synthesized approximately equivalent numbers of polypeptides, whereas the numbers of polypeptides synthesized in elongating spermatids and residual bodies were decreased. For each cell type examined, a minimum of 5 percent of the polypeptides appears to be either unique or greatly enriched. The similarity of the incorporation ratios for isolated cell types and those reported from autoradiography indicates that isolated populations of testicular cells can be used for biochemical analyses. The two-dimensional-gel analyses provide definitive evidence that each different testicular cell type contains a population of distinct polypeptides in addition to the majority of polypeptides common to all cell types studied.

Although the presence of a labeled polypeptide in a cell often means that it was synthesized in that cell type, the possibility of protein transfer between testicular cells must be considered. To establish the cellular site of synthesis of cell-type-specific polypeptides, mRNAs were purified from individual cell types, and the polypeptides they encoded were examined by cell-free translation (Wieben 1981; Fujimoto and Erickson 1982; Gold, Stern, and Hecht 1983b; Stern et al. 1983b). Wieben (1981), studying the regulation of lactate dehydrogenase X (LDH-X) in the mouse, demonstrated that LDH-X mRNA represented 0.17−0.18 percent of total functional mRNA in pachytene spermatocytes, but only 0.09−0.10 percent of the translation products of elongated spermatids. Translation of total mRNA from pachytene spermatocytes, round spermatids, and primitive type-A spermatogonia and Sertoli cells (the latter two isolated from the testes of six-day-old mice) demonstrated that for each cell type, a minimum of 5−10 percent of the mRNAs for polypeptides synthesized in intact cells either was specific to or was greatly enriched within that particular cell type (Gold et al. 1983b). A significant reduction in the number of polypeptides encoded by the total mRNA in elongating spermatids was observed. A comparison of the proteins encoded by poly(A)$^+$ RNA from spermatocytes and spermatids revealed a number of qualitatively new proteins appearing after meiosis (Fujimoto and Erickson 1982; Stern et al. 1983b). These experiments demonstrated that different cell types, including haploid cells, did indeed synthesize cell-type-specific proteins, confirming the in vivo labeling studies. However, movement of mRNA between testicular cells cannot be excluded until more is known about intracellular movement of RNA in eukaryotic cells.

B. RNA synthesis

The foundation for much of our current knowledge of RNA metabolism during spermatogenesis has been provided by a series of autoradiographic studies (Pelc 1956; Monesi 1964a, 1965, 1971, 1978; Utakoji 1966; Loir 1972; Kierszenbaum and Tres 1974a, 1974b, 1978; Tres and Kierszenbaum 1975). In the mouse, Monesi observed that following in vivo incorporation of [^{3}H]uridine, spermatogonia, midpachytene spermatocytes, and spermatids up to step 8 were most active in RNA synthesis. During the early stages of meiotic prophase, low levels of RNA synthesis were observed, whereas in midpachytene cells, a large increase in the amount of RNA synthesis was seen, followed by a decrease toward the end of meiotic prophase. The amount of [^{3}H]uridine incorporation increased again in round spermatids, which synthesized RNA until spermatid nuclear elongation occurred. Moore (1971), assaying endogenous RNA polymerase activity in fixed testicular cells, reached a similar conclusion.

The use of high-resolution autoradiography allowed Kierszenbaum and Tres (1974a, 1974b, 1975, 1978) to analyze both ribosomal RNA and hnRNA synthesis in the testis. Using [^{3}H]uridine, they detected incorporation into meiotic nucleoli but did not detect functional nucleoli in spermatids, suggesting that ribosomal RNA is synthesized during meiosis but not in postmeiotic cells. The presence of newly synthesized RNA in the perichromosomal regions of the autosomes of meiotic and spermatid nuclei argued that hnRNA, in contrast, was synthesized in the pachytene spermatocyte and also during early spermiogenesis.

i. During meiosis. Early biochemical studies of meiotic RNA synthesis in mammals indicated that pachytene spermatocyte nuclei synthesize heterogeneous high-molecular-weight RNA species (Muramatsu, Utakoji, and Sugano 1968). Grootegoed and associates (1977) demonstrated that isolated pachytene spermatocytes incorporate [^{3}H]uridine mainly into a heterogeneous population of RNAs with low electrophoretic motilities. Similar conclusions were reached in experiments in which [^{3}H]uridine-labeled RNA was isolated from 1−2-mm segments of seminiferous tubules after cell identification by means of the transillumination technique (Söderstrom 1976; Söderstrom and Parvinen 1976). A low rate of incorporation of radiolabeled precursor into ribosomal RNA (rRNA) and a retarded rate of cleavage of the 32S precursor of rRNA to 28S rRNA were also observed in RNA synthesized in tubule segments (Söderstrom 1976). These investigations suggested that meiotic cells contained high-molecular-weight RNA species that, by inference, were assumed to be precursor mRNAs. Although the tubules contain mixed cell populations, and the possibility of RNA aggregation cannot be excluded, because nondenaturing gels were not used in these early studies, the slow transport of [^{3}H]uridine-labeled RNA in meiotic cells from the nucleus to the cytoplasm (Monesi 1964a) suggests that the RNA metabolism of the primary spermatocyte differs substantially from that for most somatic cells. The accumulation of high-molecular-weight RNA molecules in the pachytene spermatocyte may be due to either slow processing of meiotic nuclear transcripts or a nontraditional role for RNA such as an involvement in chromosome pairing and/or recombination (for a subset of the nuclear meiotic transcripts). The recent discovery of novel chromosomal RNA species in meiotic nuclei (Hotta and Stern 1981; Hotta et al. 1985) suggests that a detailed examination of nuclear RNA metabolism in the primary spermatocyte using current techniques should be undertaken.

The fact that some of the RNA synthesized in mouse primary spermatocytes was polyadenylated and appeared on polysomes demonstrates that meiotic transcripts are used to synthesize protein (D'Agostino, Geremia, and Monesi 1978). Furthermore, the presence of labeled poly (A)$^+$ mRNA in intermediate and late spermatids up to 14 days after intratesticular injection of [^{3}H]uridine (Geremia et al. 1977) suggests that some of the mRNA synthesized in pachytene spermatocytes is long-lived. Considering the sizable uridine pools in haploid cells (Geremia et al. 1977), reutilization of the [^{3}H]uridine during spermiogenesis is unlikely. The synthesis of large numbers of polypeptides in elongating spermatids after the termination of transcription further argues for the need for storage and utilization of mRNAs from earlier cell types.

Definitive evidence for the storage and utilization of long-lived mRNAs in haploid cells comes from studies analyzing the regulation of protamine and phosphoglycerate kinase 2 (PGK-2). Dixon and associates (Dixon et al. 1977; Iatrou and Dixon 1978) demonstrated that trout protamine mRNAs are synthesized in the pachytene spermatocyte, stored in the cytoplasm as ribonucleoprotein particles, and later activated and translated in condensing spermatids. Kleene, Distel, and Hecht (1984) have demonstrated that in mice, protamine mRNA is first detectable in round spermatids, although the protamines are first synthesized much later in stages 12–15 of spermiogenesis. Gold and associates (1983a), studying the regulation of PGK-2 synthesis, have discovered that PGK-2 mRNA is first present in the cytoplasm of pachytene spermatocytes, although the protein is synthesized only during spermiogenesis (Vandeberg, Cooper, and Close 1976; Kramer and Erickson 1981a, 1981b). The continued increase of PGK-2 mRNA levels in postmeiotic cells prevents a definitive determination of the actual utilization of meiotic PGK-2 mRNA transcripts in spermatids, because no measurements of the half-life of the PGK-2 mRNAs have been made.

ii. During spermiogenesis. In the previous section it was demonstrated that meiotic mRNAs are likely to play a prominent role in the complex sequence of events occurring during differentiation of the haploid cell. The large numbers of proteins synthesized during spermiogenesis, coupled with the high level of RNA synthesis during meiosis and evidence for long-lived mRNA, have been interpreted as evidence that the diploid nucleus transcribes all the mRNAs necessary for male germ cell differentiation. As is often the case for biological systems, this hypothesis has proved to be an oversimplification.

Proof that mRNA is synthesized during spermiogenesis was obtained by isolation of radiolabeled heterogeneous poly(A)$^+$ RNA from postmeiotic cells following a 1–3-h labeling interval with [^{3}H]uridine (Geremia, D'Agostino, and Monesi 1978; Erickson et al. 1980a). Accumulation of increased levels of poly(A)$^+$RNA in haploid cells is supported by quantitative studies of RNA levels in isolated populations of mouse testicular cells (Table 6.1). When normalized to DNA content, the ratio of cytoplasmic RNA to DNA is constant in pachytene spermatocytes and round spermatids. However, the concentration of cytoplasmic poly(A)$^+$ RNA is twofold *higher* in round spermatids than in pachytene spermatocytes (Kleene et al. 1983). Although this increase in poly(A)$^+$ RNA could be ascribed to selective stabilization of mRNA or increased polyadenylation in postmeiotic cells, there is strong evidence for synthesis and accumula-

Table 6.1. *Amounts of cytoplasmic RNA and poly(A) in various spermatogenic and nonspermatogenic cells*

Type of RNA	Picograms RNA/cell[a] (mean ± SD)	Nanograms poly(A)/μg RNA[a] (mean ± SD)
Seminiferous tubules, cytoplasmic[b] (phenol : chloroform extraction)	2.2 ± 0.4 (6)	N.M.
Seminiferous tubules, cytoplasmic[b,c] (Orcinol)	1.9 ± 0.3 (4)	N.M.
Pachytene spermatocyte, cytoplasmic (phenol : chloroform extraction)	9.1 ± 2.0 (9)	6.3 ± 1.5 (9)
Round spermatid, cytoplasmic (phenol : chloroform extraction)	2.2 ± 0.2 (8)	10.8 ± 2.2 (7)
Elongating spermatid, cytoplasmic[d] (phenol : chloroform extraction)	0.86 ± 0.3 (8)	3.1 ± 1.6 (6)
Residual body (phenol : chloroform extraction)[e]	0.91 ± 0.82 (2)	4.1, 3.7 (2)
Pachytene spermatocyte, total cellular (guanidine extraction)	6.0 ± 2.2 (4)	N.M.
Round spermatid, total cellular (guanidine extraction)	1.5 ± 0.4 (4)	N.M.
Elongating spermatid, total cellular (guanidine extraction)	0.55 ± 0.11 (4)	N.M.
16- and 17-day mouse testis, cytoplasmic	N.M.	1.5 ± 0.2 (4)
Liver, cytoplasmic	N.M.	1.8 ± 0.2 (4)
Mouse Sertoli cell, total cellular[f]	N.M.	1.2
Rat Sertoli cell, total cellular[f]	N.M.	1.1

Note:

[a] Cell counts were performed with a hemocytometer. Picograms of RNA per cell were calculated from the absorbance at 260 nm assuming that 1 O.D. unit equals 40 μg RNA. The 260/230 and 260/280 ratios were consistently greater than 2.1. The amount of poly(A) was measured by hybridization with [^{3}H]poly(U), and cpm hybridized were converted to ng using a standard curve of poly(A). The data are expressed as the mean and standard deviation with the number of independent preparations given in parentheses. N.M., not measured.

[b] The seminiferous tubules were prepared from sexually mature animals by dissociation with collagenase and trypsin and filtration through Nitex. Only nucleated cells were counted.

[c] The absorbance at 660 nm with Orcinol reagent was converted to amount of RNA using a standard curve of AMP.

[d] The amount of RNA in elongating spermatids was calculated relative to cell counts of nucleated cells.

[e] The amounts of RNA and poly(A) in the residual body fraction were corrected for round spermatid contaminants by subtracting the amount of RNA due to these contaminants using the values given in this table.

[f] Sertoli cell RNA was obtained from the testicular cell line TMA (mouse) and TRST-R (rat) by the guanidine thiocyanate procedure.

Source: Reproduced from Kleene and associates (1983) by permission of *Developmental Biology.*

tion of "haploid" mRNAs, including some that encode gene products unique to spermiogenesis (see Section V). In an early effort to study the regulation of proteins known to be synthesized during spermiogenesis, Erickson and associates (1980b) used a rabbit reticulocyte cell-free system to detect mouse PGK-2 and protamine mRNAs. Assaying mRNAs isolated from populations of several germ cells, they reported that mRNA levels for both protamine and PGK-2 increased after meiosis, with the amount of PGK-2 mRNA increasing sixfold in late spermatids. However, because of the low levels of [^{3}H]arginine incorporated by their cell-free translations and the difficulty in resolving the mouse protamines from testis-specific protein (TP) by the acetic-acid/urea-gel electrophoresis system they used, definitive proof that the [^{3}H]arginine was incorporated into protamine was lacking. Thus, these studies suggest, but do not prove, that some of the increased poly(A)$^+$ RNA found in round spermatids codes for unique gene products of the differentiating haploid cell. Whether such mRNAs are transcribed during meiosis or are exclusively transcribed from the haploid genome requires the availability of appropriate cDNA probes to measure mRNA levels by hybridization.

iii. Translational regulation of mRNAs. Concurrent with the morphological changes seen as the spermatid differentiates to a spermatozoon, RNA transcription decreases and terminates (Monesi 1964a). Thus, translational regulation of stored mRNAs of either meiotic or early postmeiotic origin is thought to play a crucial role in the orderly synthesis of essential sperm components during spermiogenesis.

Cell fractionation experiments demonstrate that translational regulation occurs in meiotic and postmeiotic testicular cells. When the mRNAs from the polysomal and nonpolysomal (ribonucleoprotein) fractions of a mouse testis extract are analyzed by a cell-free translation assay, approximately two-thirds of the total mRNA or poly(A)$^+$ RNA is present in the nonpolysomal compartment (Gold and Hecht 1981). In contrast, over 80 percent of the total liver mRNAs is found in the polysome fraction. Approximately two-thirds of the total or poly(A)$^+$ mRNA from pachytene spermatocytes, round spermatids, and elongating spermatids is also found in the nonpolysomal fraction, suggesting that in the testis some form of selective mRNA activation (or inactivation) or cytoplasmic sequestration regulates the loading of specific mRNAs onto polysomes to allow their temporal expression (Gold et al. 1983b; Stern et al. 1983b). Analysis of the polypeptides encoded by the nonpolysomal and polysomal mRNAs from meiotic and postmeiotic cells provides further evidence for translational regulation during spermiogenesis, because specific mRNAs are seen to move between the ribonucleoprotein and polysomal compartments in different cell types as spermatogenesis proceeds (Gold et al. 1983b; Stern et al. 1983b). Although the majority of testicular mRNAs are in the nonpolysomal compartment, the distribution of individual mRNAs during spermatogenesis varies greatly. For instance, 82 percent of PGK-2 mRNA is present in the polysomal fraction in total testis extracts; whereas 100 percent of the PGK-2 mRNA is present in the nonpolysomal fraction in pachytene spermaocytes, and more than half of the PGK-2 mRNA is found on polysomes in round spermatids (Gold et al. 1983a). The translational regulation of the protamines, which are the best studied of the proteins synthesized in postmeiotic cells, will be described in the next section.

V. Application of recombinant DNA techniques to spermatogenesis

A. Differential screening of cDNA libraries

Recombinant DNA techniques have been used by several laboratories to isolate probes to study haploid gene expression during mammalian spermatogenesis (Kleene et al. 1983; Dudley et al. 1984; Fujimoto et al. 1984). The rationale for this approach is based on the premise that cell-type-specific cDNA probes for mRNAs present in haploid cells, but absent from other testicular cell types, can be identified by a differential screening of a testicular cDNA library. To identify such cDNAs in a cDNA library prepared from total cytoplasmic testicular poly(A)$^+$ RNA, we hybridized ^{32}P-labeled cDNA prepared from cytoplasmic poly(A)$^+$ RNA from either pachytene spermatocytes or round spermatids to duplicate filters containing bacterial colonies and looked for colonies enriched or solely present on the filter hybridized with the radiolabeled round spermatid cDNAs (Kleene et al. 1983). Seventeen out of 750 cDNA clones, which did not hybridize with cytoplasmic poly(A)$^+$ from 16–17-day-old testis, adult liver or brain, total RNA from cultured Sertoli cells, poly(A)$^-$RNA from adult testis, or the mouse mitochondrial genome, reproducibly hybridized much more strongly with round spermatid cDNA. Two of the cDNA clones have been sequenced and identified as encoding the two protamine-sequence variants present in the mouse (Kleene et al. 1985; Yelick et al. 1985).

Dudley and associates (1984) also used a differential hybridization screening procedure to identify testicular cDNAs that correspond to RNAs that are expressed in mouse testis and at lower levels in liver and spleen. Focusing on 10 cDNAs that identified RNAs that were larger than 1.6 kB, they cleverly used mRNA isolated from mice carrying the sex-reversed and testicular-feminization mutations to confirm that specific mRNAs were expressed in the haploid phase of spermatogenesis. Their analysis to date has identified eight clones corresponding to RNAs expressed uniquely or at much higher levels in meiotic or postmeiotic cells.

Fujimoto and associates (1984) prepared cDNAs from round spermatid poly(A)$^+$ RNA and identified four cDNAs that detected mRNAs that increase in abundance in postmeiotic germ cells. One clone, pPM459, encodes an RNA that increases 15-fold in amount in the testes of 21-day-old mice compared with those from 14-day-old mice. Hybridization of pPM459 to radiolabeled RNA isolated from round spermatids following a 3-h incubation with [^{3}H]uridine suggests that RNA homologous to the pPM549 probe is synthesized in round spermatids.

Distel, Kleene, and Hecht (1984) have taken a different approach to identify tubulin cDNAs in postmeiotic testicular cells. On the basis of evidence from two-dimensional-gel electrophoresis that testis-specific isoforms of the tubulins are present in postmeiotic cell types (Hecht et al. 1984), we screened a mouse testis cDNA library with a cDNA probe for rat brain α tubulin. Sequence analysis of one isolated clone revealed that about 20 percent of the predicted amino acids from this cDNA differed from the highly conserved sequence reported for somatic α tubulins.

These cloning efforts have produced a sizable number of cDNA probes for testicular proteins. Although the postmeiotic appearance of a large number of enzymes, proteins, and sperm antigens is well established, definitive proof that certain haploid gene products are solely encoded by transcripts from the haploid genome has awaited the availability of cDNA probes for proteins such as the protamines and the testicular α tubulin. Similar approaches can be and should be used to identify cDNAs that code for "marker" proteins for spermatogonia, meiotic prophase cell types, or Sertoli cells. In cases where antibodies have been prepared to a protein of interest, expression vector libraries can be prepared and screened with the antibody to obtain the desired cDNA probes.

B. Utilization of cDNAs encoding haploid-expressed proteins

i. Protamines. Protamines are small, arginine-rich nuclear proteins of 27–65 amino acids that are synthesized and deposited on DNA during spermatid differentiation. In mammals, histones are not directly replaced by protamines but by a heterogeneous group of basic proteins, called testis proteins (TP) (Kistler, Geroch, and Williams-Ashman 1975; Kistler, Keims, and Heinrikson 1975; Grimes et al. 1977). TP are transiently associated with the spermatid nucleus during the transition of its chromatin from a nucleosome-like structure to a smooth branching fibril (Kierszenbaum and Tres 1978). In mice, the nucleoprotamine complex forms in step-12–15 spermatids. Although the protamines are often considered to be responsible for the termination of transcription and for the condensation of sperm chromatin, in mammals both processes initiate before protamine is first synthesized. Once the DNA : protamine complex becomes established, disulfide bridges are formed between the cysteine residues (present in mammalian protamine), providing a means to maximally stabilize the sperm chromatin (Bedford and Calvin 1974).

Protamines have been isolated from a large number of different species, including several fish and domestic fowl, bull, boar, ram, stallion, guinea pig, mouse, rabbit, rat, and human (Coelingh, Rozijn, and Monfoort 1969; Dixon 1972; Monfoort et al. 1973; Bellvé, Anderson, and Hanley-Bowdoin 1975; Nakano, Tobita, and Ando 1975; Calvin 1976; Kistler et al. 1976; Pongswasdi and Svasti 1976; Balhorn, Gledhill, and Wyrobek 1977; Bellvé and Carraway 1978; Gaastra, Lukkes-Hofstra, and Holk 1978; Tanphaichitr et al. 1978; Rodman et al. 1979; Tobita et al. 1983; Sautiere et al. 1984; McKay, Renaux, and Dixon 1985; Mazrimas et al. 1986). Most mammals, with the exception of the mouse and human, which have two and three amino acid variants, respectively, are believed to contain one form of protamine. The significance of protamine variants is unknown but noteworthy, especially considering the relatively recent divergence of rat and mouse. The complete amino acid sequences of protamine from the bull (Coelingh et al. 1972; Mazrimas et al. 1986), boar (Tobita et al. 1983), ram (Sautiere et al. 1984), and human protamine P1 (McKay et al. 1985) have been established, and the predicted amino acid sequence has been derived from the nucleotide sequence of the cDNA for mouse protamine 1 (Kleene et al. 1985) (Table 6.2). Partial amino acid sequences have also been reported for the rat (Kistler et al. 1975) and for one of the human protamines (Gaastra et al. 1978) (Table 6.2). The difference in the reported sequences for the bull protamine suggests that the sequence cys-arg-arg between Coelingh's Cys 38 and Tyr 39 was missed. The differences reported for the human protamine remain to be resolved but may be clarified when the sequence of all the human protamines is known. Mammalian protamines are organized into three domains, a central basic core with clusters of arginine, and two less basic regions at the amino and carboxy termini (Balhorn 1982). Although the protamines from trout (Aiken et al. 1983) and mammals differ significantly in length and sequence, the central, highly basic region containing four arginine clusters has been conserved except in mouse protamine 2, to be discussed later (Kleene et al. 1985).

Several groups of investigators, using radiolabeled amino acids, have demonstrated that protamines are synthesized in mice and rats late during spermiogenesis (Lam and Bruce 1971; Bellvé et al. 1975; Geremia, Goldberg, and Bruce 1976; Mayer and Zirkin 1979; Mayer, Chang, and Zirkin 1981). Balhorn and associates (1984) have shown that the two mouse protamines are synthesized slightly asynchronously during stages 12–15 of spermiogenesis. The recent identification of a cDNA encoding

Table 6.2. *Amino acid sequence homologies of type I protamines*[a]

```
                          10                20              30            40          50
Mouse 1   A R Y R C C R S K S R S R C R R R R R   R C R R R R R R   C C R R R R R   Y T   I R C K K Y
Rat       - - - - - - - - - - - P - - - - - - -   - - - - - - - -   - - - - - - -   - -   A V - - - R R C
Boar      - - - - - L T H - G - - - - - - - - -   - R C - - - - -   R F G - - - -   - -   R V - - - T R Q
Bull*     - - - - - L T H - G - - - - - - - - -   - R C - - - - -   R F G - - - -   - -   R V - - - T R Q
Bull**    - - - - - L T H - R - - - - - - - - -   - R C - - - - -   R F G - - - -   - -   R V - - - T R Q
Ram       - - - - - L T H - R - - - - - - - - -   - R C - - - - -   - - - - - - -   - -   V V - - - T R Q
Human     - - - ? A - - - Q - - - Y Y ? Q ? Q - Q  S - - - - - - -   S - Q T - A M R - - P   - R   P - - R R H
Human I   - - - Q - - Y Y - Q - Q - - - - - - -    S - - - - - - -   S - - - - - -         M Y H R R R C
Trout     M P - - S*** P V - - - - - - * R  V S - - - - - - - G G - - - - -   - R   P - - R R H
```

Note:

[a]The amino acids of the mouse protamine 1 have been numbered, and gaps have been introduced to maximize the homologies between sequences. Amino acids that are the same as in mouse 1 are denoted by a dash. Regions of nucleotide heterogeneity in six different trout protamine genes are represented by asterisks. Sequence sources are mouse 1 (Kleene et al. 1985), rat (Kistler et al. 1976), boar (Tobita et al. 1983), bull* (incomplete sequence) (Coelingh et al. 1972), bull** (complete sequence) (Mazrimas et al. in press), ram (Sautiere et al. 1984), human 1 (partial sequence) (Gaastra et al. 1978), human 1 (complete sequence) (McKay et al. 1985), and trout (Aiken et al. 1983). A, alanine; C, cysteine; F, phenylalanine; G, glycine; H, histidine; I, isoleucine; K, lysine; L, leucine; M, methionine; P, proline; Q, glutamine, R, arginine; S, serine; T, threonine; V, valine; Y, tyrosine.

mouse protamine 1 has allowed additional regulatory studies of this postmeiotically expressed protein to be made (Kleene et al. 1985). The cDNA for mouse protamine 1 was obtained by Kleene and associates (1983) following differential screening of a mouse testis cDNA library with radiolabeled round spermatid cDNAs. Sequence analysis of one clone encoding a low-molecular-weight abundant poly(A)$^+$ mRNA that was first detectable during spermatogenesis in round spermatids revealed a predicted amino acid composition virtually identical with that reported for the cysteine-rich tyrosine-containing mouse protamine (MP1) (Bellvé and Carraway 1978) and identical with that reported for the first 15 amino acids of rat protamine (Kistler et al. 1975). MP1 contains 50 amino acids, of which 28 are arginine, 9 are cysteine, 3 are tyrosine, and none is histidine (Table 6.2).

Using the MP1 cDNA to probe Northern blots, Kleene and associates (1984) demonstrated that the vast majority of MP1 mRNA was present in the nonpolysomal fraction of the cytoplasm of the testis. Analyzing the distribution of MP1 mRNA in isolated testicular cells, they found that MP1 mRNA is first present in the nonpolysomal fraction of round spermatids and is later present on polysomes in elongating spermatids. In addition, MP1 mRNA can be detected in two distinct size classes of approximately 580 and 450 nucleotides. Before translation, the MP1 mRNA is 580 nucleotides in length, whereas the polysomal form of MP1 mRNA is 450 nucleotides. This difference has been shown to be due to a poly(A) tract of about 160 nucleotides present on untranslated MP1 mRNA that is reduced to 30 nucleotides on polysomal MP1 mRNA. No MP1 mRNA is detectable in pachytene spermatocytes or in the testis of 16–17 day CD-1 mice, a prepubertal stage in which the first wave of spermatogenesis has advanced only into midmeiosis.

Additional studies using prepubertal animals have been conducted to ascertain when protamine mRNA is first present in mammalian testicular cells. Northern blots first detect protamine mRNA around day 22 in CD-1 mice, a time when approximately 10 percent of the testicular cells have differentiated to round spermatids. These results suggest that, in contrast to the trout, where protamine mRNA is synthesized during meiosis, mammalian protamine mRNAs are first synthesized during early stages of spermiogenesis, stored as an untranslated ribonucleoprotein particle in round spermatids, and translated in elongating spermatids.

The reason for the loss of a substantial part of the 3' end of the protamine mRNA concurrent with translation is not understood but may be characteristic of protamine mRNA regulation, because most polysomal trout protamine mRNA also has a short poly(A) tail (Iatrou and Dixon 1978). Efforts to translate the mouse protamines in a cell-free system have been unsuccessful, leading to the suggestion that protamine may be synthesized as precursors and later processed to the species found in sperm (Elsevier 1982). The sequence determination of the cDNA for MP1 (Kleene et al. 1985) and the successful translation of a bovine protamine in a cell-free system (Krawetz and Dixon 1984) rule out this explanation for all the protamines. However, DNA sequence analysis of two independent cDNA clones that encode a protein with the amino acid composition of mouse protamine 2 (MP2) (Bellvé and Carraway 1978) suggests that MP2 is synthesized in a larger form than found in sperm (Yelick et al. 1985). MP2 mRNA (previously called HSAR 900) is first detectable by Northern blots in round spermatids (Kleene et al. 1983) and is also translationally regulated.

Both of the mouse protamine genes map on chromosome 16, as shown by hybridizing Southern blots of a series of mouse-hamster somatic cell hybrids containing subsets of mouse chromosomes and a complete set of hamster chromosomes with radiolabeled protamine cDNAs (Hecht et al. 1986). Surprisingly, the cDNAs of the two mouse protamines do not cross-hybridize, raising the possibility that they have evolved from different sequences or that the sequence of MP2 has rapidly evolved from one primordial mammalian protamine sequence. It is of interest that both bull protamine and MP1, which share significant sequence homology (Table 6.2), are not synthesized as precursors, whereas the distinct MP2 appears to be.

In addition to providing a useful probe to monitor protamine mRNA utilization, MP1 cDNA has been employed as a molecular marker to evaluate the extent of spermiogenesis in wild-type, mutant, and experimentally induced sterile mammals (Hecht et al. 1985). The assay is predicated on the temporal appearance and changes in the poly(A) length of MP1 mRNA during spermatogenesis: No MP1 mRNA is detectable in testicular RNA preparations of up to 20-day-old animals, the 580-nucleotide form of MP1 is present in the testes of 24-day-old mice, and a heterogeneous population of the 450- and 580-nucleotide forms of MP1 mRNA is found in sexually mature animals (Figure 6.3). To determine the predictive value of this molecular marker for testicular maturation, we have examined three male sterile mouse mutants (Lyon and Hawkes 1970; Bennett et al. 1971; Varnum 1983).

Blind sterile (bs) is an autosomal mutation causing cataracts in both sexes and sterility in male mice (Sotomayor and Handel 1986; Varnum 1983). Homozygous males are vitually azoospermic. Analysis of testicular extracts of bs animals with MP1 cDNA revealed drastic diminutions in amounts of both size classes of MP1 mRNA, a result in agreement with the cytological reports of markedly reduced numbers of haploid spermatogenic cells in these animals.

Quaking (qk) is an autosomal mutation causing deficient myelinization of the central nervous system and virtual azoospermia in homozygous males (Bennett et al. 1971). However, in contrast to bs, male sterility in

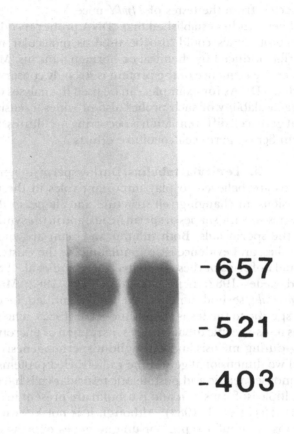

Figure 6.3. Measurement of the sizes of MP1 mRNAs in testicular extracts of prepubertal and sexually mature mice. Following purification, 10-ug aliquots of testicular RNA from CD-1 mice were denatured with glyoxal, electrophoresed on a 2% agarose gel, transferred to nitrocellulose, and hybridized to nick-translated pMP1. Lane 1, RNA from a 16-day-old animal; lane 2, RNA from a 24-day-old animal; lane 3, RNA from a sexually mature animal. [Reproduced from Hecht et al. (1985a) by permission of *Differentiation*.]

qk is due to defects in spermatid differentiation involving sperm head shaping and tail morphology. Analysis of testicular RNA from *qk/qk* males reveals the presence of both the 450- and 580-nucleotide MP1 mRNAs, demonstrating that a substantial proportion of the spermiogenic cells have developed to at least stage-12 spermatids, a result in agreement with cytological analysis (Bennett et al. 1971).

Testicular feminization (*Tfm*) is an X-linked mutation in which chromosomally XY male mice fail to respond to testosterone and have vestigial testes, with no differentiation of germ cells (Lyon and Hawkes 1970; Lyon, Cattanach, and Charlton 1981). As expected, no MP1 mRNA is detected from the testes of *Tfm*/Y mice.

These studies established that cDNA probes encoding proteins such as the protamines could also be used as molecular markers to monitor sterility induced by chemical or hormonal means. Moreover, in cases in which the sequence of the protein is strongly conserved between species, rodent cDNAs for example, can be used in analyses of human infertility. The availability of such probes also provides a sensitive means to prove that germ cell differentiation is occurring in cultures of spermiogenic cells or in Sertoli/germ cell coculture efforts.

ii. Testicular tubulins. During spermatogenesis the tubulins and actins are believed to play important roles in the mitotic and meiotic divisions, in changing cell structure and shape, in the formation of the species-specific shapes of sperm heads, and in the synthesis of the axoneme of the sperm tails. Both tubulin and actin are encoded by multigene families, and evidence is accumulating for the existence of distinct functional isoforms of these proteins (Cleveland et al. 1980; Engel, Gunning, and Kedes 1982; Lemischka and Sharp 1982; Minty et al. 1982). In *Drosophila*, several tubulins have been identified, including one shown to be specific to the testis (Kemphues et al. 1982). The genetic alteration of this testis-specific tubulin causes disruption of microtubule function starting during meiosis and throughout spermiogenesis.

Two-dimensional gels of the radiolabeled proteins synthesized in vivo in mouse meiotic and postmeiotic testicular cells have demonstrated that multiple isoforms of α and β tubulin are present in haploid cells (Figure 6.4) (Hecht et al. 1984). Although it is not known whether these new forms of tubulin represent unique genes expressed postmeiotically or posttranslational modifications of mitotic or meiotic forms of tubulin, the recent identification of a unique testicular α-tubulin cDNA (Distel et al. 1984) demonstrates that at least one novel tubulin gene is expressed in the mammalian testis. This cDNA, containing an approximately 1,000-base-pair insert, was isolated by Distel and associates (1984) after screening a mouse testicular cDNA library with a rat brain α-tubulin cDNA. Although the coding region of this mouse testicular cDNA shares sequence homology with both mouse and rat somatic α-tubulin mRNAs, a subclone containing about 360 bases from the 3' untranslated region of the testis cDNA hybridizes to several species of testicular poly(A)$^+$ RNAs, including two mRNAs that are present in postmeiotic testicular cells but are not found in RNA preparations from mouse brain, kidney, spleen, ovary, lung, or liver. DNA sequence analysis of this cDNA has confirmed its uniqueness, with about 20 percent of the predicted amino acid sequence differing from the highly conserved sequences of chicken, rat, and human α tubulins.

The cellular location(s) and function(s) of the testicular tubulin are unknown. Recently, Cherry and Hsu (1984) have used antitubulin immunofluorescent staining to identify microtubule constructs throughout mouse spermatogenesis. It is possible that one or more unique tubulins are involved in spermatid nuclear shaping as a component of the manchette, involved in sperm axoneme function, or are present as a nontraditional nonmicrotubular protein during spermiogenesis. Whatever function(s) postmeiotic tubulin(s) may play during spermatogenesis, the isolation of a cDNA for a haploid α tubulin provides additional evidence for testicular haploid expression of a defined cytoplasmic protein.

iii. Actins. At least six distinct but closely related forms of actin have been isolated from vertebrates. Sequence analysis has indicated that these actin isotypes have evolved from two major classes, the cytoplasmic and muscle actins. In addition to the general functions of somatic cytoplasmic actins, such as cell motility and division, secretion, organelle transport, and maintenance of cellular cytoarchitecture, the actins in the testis are involved in additional germ-cell-specific functions. Immunofluorescence studies have shown that actin is present at sites where chromosomes bind to the nuclear inner membrane in the meiotic nucleus in mice, suggesting a possible involvement with chromosome pairing (Martino et al. 1980). In the rabbit, Welch and O'Rand (1985) have shown that filamentous actin is present in spermatogenic cells, but in sperm they conclude that actin is present in a nonfilamentous form. They speculate that actin may function during spermatogenesis to shape the acrosome to the nucleus and to anchor inner acrosomal membrane proteins. Actin has also been detected in the sperm tail, connecting piece, and postacrosomal region of the sperm in many different mammals (Clarke and Yanagimachi 1978; Talbot and Kleve 1978; Clarke, Clarke, and Wilson 1982).

To date, two proteins that comigrate with β and γ actin have been identified in the mammalian testis by two-dimensional-gel electrophoresis of [^{35}S]methionine-labeled proteins (Figure 6.4) (Hecht et al. 1984). The relative proportions of synthesis of β and γ actins change during spermatogenesis, with the amount of radiolabeled β actin increasing between meiosis and the termination of spermiogenesis, and the labeling of γ actin decreasing during this interval.

Using actin-isotype-specific cDNA probes, actin mRNA of two sizes, 2.1 and 1.5 kB, has been detected in the total and poly(A)$^+$ RNA extracts of the mouse testis (Waters, Distel, and Hecht 1985). The 2.1-kB size class, containing mRNAs that encode β and γ actins, is present in both meiotic

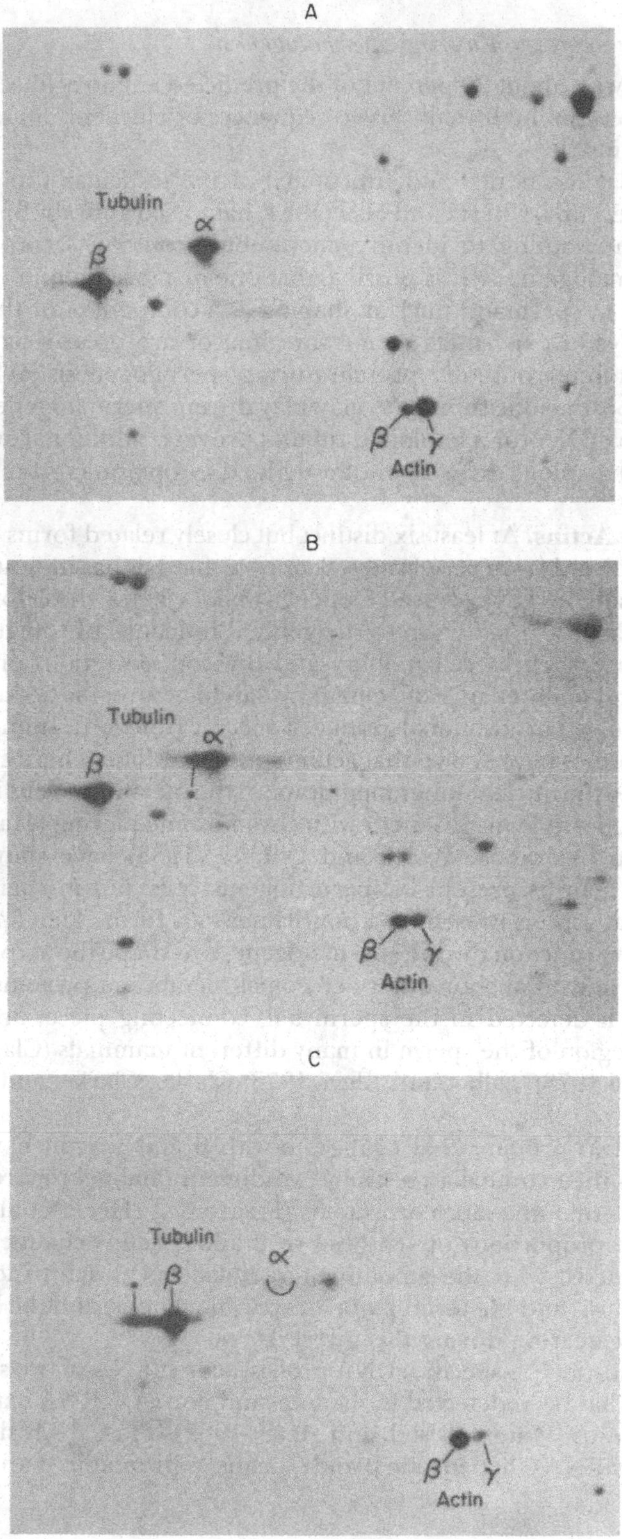
A
Tubulin
α
β
β
γ
Actin
B
Tubulin
α
β
β
γ
Actin
C
Tubulin
α
β
β
γ
Actin

and postmeiotic testicular cells. However, the 1.5-kB actin mRNA is not detectable in pachytene spermatocytes, liver, or kidney, but is present in round and elongating spermatids and residual bodies. Although both the 2.1- and 1.5-kB mRNAs hybridize to the coding region of actin cDNAs, the 1.5-kB mRNA does not hybridize to a series of actin cDNAs containing the 3′ untranslated regions of cardiac and skeletal muscle α actin and the cytoplasmic β and γ actins. This suggests that the 1.5-kB testicular actin mRNA may represent transcriptional or posttranscriptional processing of a longer actin mRNA to the 1.5-kB form present in haploid testicular cells, thereby eliminating much of the 3′ untranslated region and making it undetectable by the probes used. Alternatively, the 1.5-kB actin sequence may be the product of a unique actin gene that is expressed only during spermiogenesis in the testis. Either possibility argues for haploid-specific regulation of actin biosynthesis in the mouse testis.

iv. Proto-oncogenes. Cellular oncogenes are believed to play a fundamental role in controlling normal cell growth and differentiation. In the mouse during normal embryogenesis, tissue-specific expression of several cellular oncogenes has been reported. Ponzetto and Wolgemuth (1985) have analyzed the temporal expression of the c-*abl* transcripts during spermatogenesis in the mouse and have found that in addition to the 6.2- and 8.0-kB transcripts characteristic of the c-*abl* RNA in embryonic and adult somatic tissues, the testis contains an additional 4.7-kB c-*abl* transcript. Using RNA isolated from the testes of prepubertal mice and from populations of pachytene spermatocytes, early spermatids, and residual bodies, Ponzetto and Wolgemuth (1985) have demonstrated that the 4.7-kB transcript appears coincident with spermiogenesis. They postulate that this regulated trancription of the multiple c-*abl* RNAs during spermatogenesis suggests an important function for the gene product in the development of male germ cells. A similar analysis of *myc* gene transcripts in mouse testis by Stewart, Bellvé, and Leder (1984) has revealed that, in contrast to testicular somatic cells, germ cells have very few, if any, *myc* transcripts. The differences in temporal expression of the c-*abl* and *myc* gene transcripts make it unlikely they share a common mode of action in the testis. Hopefully, in the near future, investigators will study the expression of other cellular on-

Figure 6.4. Fluorographs of portions of two-dimensional polyacrylamide gels analyzing the [^{35}S] methionine incorporated into protein in vivo. A: Pachytene spermatocytes. B: Round spermatids. C: Elongating spermatids. The β and γ actins are identified by their identical migrations with rabbit, rat, and mouse actins. A possible new isoform of α tubulin is identified by an asterisk in B, and a possible new isoform of β tubulin is identified by an asterisk in C. [Reproduced from Hecht et al. (1984) by permission of *Experimental Cell Research*.]

cogenes during spermatogenesis to see if proto-oncogene expression falls into two distinct patterns for germ and somatic cells in the testis. The testis could provide an excellent system to clarify the function(s) of oncogene products in normal cell growth and differentiation.

The appearance of shorter transcripts in haploid cells for both the c-*abl* oncogene and actin is striking. Such findings suggest that these changes in mRNA length may represent a general mechanism of mRNA regulation operating during spermiogenesis. The testicular form of pro-opiomelanocortin (POMC) mRNAs, transcripts believed to be present primarily in Leydig cells, is also considerably shorter than the POMC mRNAs in rat and mouse pituitary (Pinter et al. 1984; Gizang-Ginsberg and Wolgemuth 1985). The sizable changes in lengths of the transcripts (2.1 to 1.5 kB for actin, 1.15 to 0.6−0.8 kB for POMC mRNA, and 8.0−6.2 to 4.7 kB for c-*abl*) suggest that differential transcription or differential posttranscriptional processing, not deadenylation, is the likely basis for such differences.

C. Postsegregational effects of haploid gene expression

Gametic selection (i.e., an unequal competition between gametes producing non-Mendelian ratios at the time of fertilization) is most easily explained by differential haploid gene expression in the male gamete. As discussed earlier, the use of recombinant DNA technology has provided clearcut evidence for the existence of postmeiotically expressed genes. From the numerous cell-type-specific proteins detected by two-dimensional-gel electrophoresis and the large number of cDNAs encoding poly(A)$^+$ RNA enriched or specific to haploid testicular cells, it is conceivable that hundreds of other "haploid" genes exist.

The observation that different mRNAs are transcribed from the meiotic nucleus and the early haploid nucleus in mammals raises intriguing regulatory and physiological considerations for gene products common to all spermatozoa versus unique phenotypic allelic variation of spermatozoa. For haploid gene products to produce a functional haploid phenotypic expression in the spermatozoon, several criteria must be met. First, structural proteins must remain in the spermatozoon, rather than serve a transient function during spermiogenesis and later disappear from the maturing male gamete. Second, because the spermatids are differentiating in a syncytium, the proteins encoded by "haploid-specific" genes must not freely exchange with other developing spermatids (Dym and Fawcett 1970, 1971; Erickson 1973). Clearly, a major contribution to answering this question would come from a critical test of the oft-stated, but never experimentally tested, assumption that proteins freely move throughout the syncytium of maturing sperm cells. To date, no proof exists that either soluble small proteins or large insoluble proteins move or do not move between spermatids. A third criterion that must be satisfied in any evaluation of the postsegregational effects of haploid gene expression is an ability to recognize the result of haploid phenotypic expression. Without appropriate markers, the results of haploid gene expression would remain undetected. Thus, improved means to examine and monitor sperm-specific antigens and proteins are essential to extend our knowledge of the effects of haploid gene expression. Currently, the only effects readily attributable to haploid expression are distorted transmission ratios of t alleles and unbalanced chromosomal states in males (Gluecksohn-Waelsch and Erickson 1970; Bennett 1975).

VI. Areas for future research

In the past 10 years, significant advances have been made in defining cell structures, components, and macromolecules required to produce a spermatozoon. Combining our substantial biochemical and morphological knowledge of germ cell development with techniques of cell separation and molecular biology, investigators can now concentrate more on mechanisms that regulate spermatogenesis. In this section, I shall discuss future applications of recombinant DNA techniques to the study of the regulation of gene expression during spermatogenesis. Because of space limitations, other powerful approaches to study gene regulation in the testis, such as the use of immunological techniques, will not be presented.

One area especially promising for study is gene regulation at the level of chromatin structure. Many studies have demonstrated that in eukaryotes, changes in chromatin structure are associated with transcriptional activation of genes. The mammalian testis provides a special system to evaluate the effects of chromatin changes on transcription, because protamines replace histones in the maturing gamete, and on fertilization the protamines are in turn replaced with histones. Seeking to determine how information is stored and maintained in the paternal genome, Groudine and Conklin (1985) have elegantly demonstrated that although the somatic chromatin of constitutively expressed genes, tissue-specific genes, and inactive genes is absent in the sperm nucleus, point sites of undermethylation corresponding to hypersensitive sites of somatic and spermatogonial cells remain in the sperm chromatin for constitutively expressed genes. They have also found that a specific de novo DNA methylation process occurs between the spermatogonial and primary spermatocyte stages of spermatogenesis, leading to the speculation that methylation operant during spermatogenesis may play a role in creating the "template" of information in sperm DNA. A continuation of this work examining the chromatin state of germ-line-specific genes and genes expressed during spermatogenesis and early embryogenesis will be an important contribution to the field.

A detailed analysis of DNA methylation patterns during spermatogenesis for temporally regulated testicular proteins would enhance the chromatin structure studies. Kaput and Sneider (1979) have reported methylation differences in somatic and sperm cells, as detected by restriction enzyme digestion. In the rooster, total DNA from meiotic and postmeiotic testicular cells contains approximately 30 percent less methylcytosine than DNA isolated from premeiotic and somatic cells (Rocamora and Mezquita 1984). Jagiello and associates (1982) have used an immunoperoxidase method with antibody to 5-methylcytidine to demonstrate increasing methylation of total DNA during pachytene in the mouse. Studies concentrating on specific classes of mouse genomic sequences revealed that repetitive (satellite) DNA sequences were undermethylated throughout spermatogenesis (Ponzetto-Zimmerman and Wolgemuth 1984; Sanford, Forrester, and Chapman 1984), whereas the genes encoding a major β-globin, a pancreatic amylase, a type-I histocompatibility-2 antigen, and an unidentified spermatid cDNA were hypermethylated early in testicular development and remained so throughout spermatogenesis (Rahe, Erickson, and Quinto 1983). The growing availability of cDNA and genomic probes for temporally regulated testicular proteins will allow future experiments with specific subclones of the 5' and 3' regions of the probes to critically evaluate the role of methylation, if any, in regulating testicular genes.

Evidence from viral and somatic eukaryotic systems has demonstrated that cis-DNA sequences and trans-acting factors play major roles in regulating gene transcription (Dean et al. 1984; Ephrussi et al. 1985; Garabedian, Hung, and Wensink 1985; Kandala, Kistler, and Kistler 1985). The extended process of spermatogenesis provides an excellent system to define the roles of such factors during the temporal expression of meiotically activated genes such as lactate dehydrogenase X (Goldberg and Hawtrey 1967; Meistrich et al. 1977; Wieben 1981) and genes expressed during spermiogenesis. For instance, will we find similar/identical cis-DNA sequences in groups of coordinately regulated genes? Sequence analysis of upstream and downstream regions of the genomic DNA of coordinately regulated premeiotic, meiotic, and postmeiotic genes should help answer this question.

An explanation for the many organ-specific isozyme variants of the testis is central to our understanding of the metabolism and regulation of the testis. Many testis-specific proteins are assumed to be essential for metabolic intricacies peculiar to spermatogenesis, thus offering the reproductive biologist novel means to perturb the testicular differentiative process and provide specific procedures for fertility control. The elegant immunocontraceptive studies in progress with lactate dehydrogenase X demonstrate the feasibility of such an approach (Goldberg et al. 1981; Wheat and Goldberg 1981; Gonzales-Prevatt, Wheat, and Goldberg 1982). Our growing ability to alter gene expression with techniques of targeted mutagenesis to specific testicular isozymes offers a promising means to define how essential such proteins are to spermatogenesis. For instance, what would happen to spermatogenesis if the testicular form of cytochrome c could not be expressed? The many sperm-specific proteins in the acrosome and sperm cell (O'Brien and Bellvé 1980a, 1980b) and testis-specific organelles, such as the helical sperm mitochondria (Pallini, Baccetti, and Burrini 1979; Hecht and Bradley 1981) and the chromatoid body (Eddy 1970; Fawcett, Eddy, and Phillips 1970a; Comings and Okada 1972), provide other targets for perturbation.

Of the approaches that promise to impact strongly in future efforts to understand the regulation of gene expression during spermatogenesis, four that stand out are in situ hybridization, culture systems allowing spermatogenesis to proceed in vitro, mutagenesis of genes by transgenic or retrovirus methods, and inhibition of gene expression by anti-sense RNA. A discussion of each follows.

Building on a variety of applications of "in situ" hybridization studies with viral and eukaryotic probes (Hudson et al. 1981; Gee and Roberts 1983; Jeffery 1984; Southern, Blount, and Oldstone 1984; Lawrence and Singer 1985), Gizang-Ginsberg and Wolgemuth have successfully established fixation and hybridization procedures for RNA : cDNA hybridization "in situ" for mouse testis. Using cDNA probes for α tubulin and POMC mRNAs, they found α-tubulin mRNA in virtually all testis cells, whereas they localized POMC mRNAs primarily to Leydig cells, although lower levels of hybridization of the POM probe to spermatogonia and spermatocytes within the seminiferous epithelium were detected. These pioneering efforts provide a means to quantify relative levels of mRNA in a continuum of differentiating substages within individual cell populations that have not been disturbed by loss of cell-to-cell contacts, as well as a means to confirm the cell type identity of cDNA probes isolated from cell types that are difficult to isolate. Moreover, when used with electron microscopy, in situ hybridization can monitor movement of macromolecules into and out of specific regions of cells and organelles, such as has been speculated to occur in the chromatoid body (Parvinen and Parvinen 1979).

A second technique that will greatly facilitate advances in our understanding of spermatogenesis is a culture system in which germ cell differentiation proceeds in vitro. Although individual purified testicular cell types are often able to survive for short periods of time in culture (Jutte et al. 1981; Gerton and Millette 1984), for the most part germ cells do not continue their development (Steinberger 1975). The substantial progress that has been made in isolating and maintaining Sertoli cell cultures that carry out many of the metabolic reactions characteristic of Sertoli cells suggests that germ cell culture may also be possible, if not autonomously, then in coculture with Sertoli cells (Boitani et al. 1983; Tres and Kierszenbaum 1983). The ability to study differentiating germ cells in culture would allow investigators to define the importance of cell-to-cell contact and analyze the transport of essential macromolecules and factors between cells (Skinner and Griswold 1982). It would also provide an opportunity to perturb germ cell differentiation with chemical inhibitors or monoclonal antibodies and to utilize mixed cultures of cell types with different genotypes.

Insertion of foreign DNA by microinjection into fertilized eggs or by retrovirus infection offers a third promising method to discover genes essential for events for which a screening procedure is available. For example, the retrovirus sytem has produced embryonic lethal mutations (Schnieke, Harbers, and Jaenisch 1983). Ruddle and associates have produced completely sterile male mice following embryo gene-injection experiments in which an interferon gene has been randomly inserted into the genome (Friedman 1983). Palmiter and associates (1984) have produced a mosaic transgenic mouse carrying transmission distortion, suggesting that the foreign DNA in this case has disrupted a gene that is expressed during spermiogenesis. The isolation, identification, and DNA sequence analysis of the junction fragments of this *Myk* insert and other inserts expressing transmission distortion offer an elegant means to identify haploid gene products of the testis. Similarly, mutations of genes active in other stages of spermatogenesis can be identified by recessive mutations that specifically inhibit spermatogenesis in homozygous males. The production of male sterile mutants by random insertion of foreign DNA provides a valuable addition to the currently available group of male sterile mouse mutants that, unfortunately, exhibit pleiotropic phenotypes because the defects are in loci that affect far more than spermatogenesis.

Gene expression in prokaryotic and eukaryotic cells can be selectively prevented with anti-sense RNAs (i.e., oligonucleotides complementary to specific RNA sequences that form RNA duplexes in vivo and prevent the translation of that sequence) (Coleman, Green, and Inouye 1984; Izant and Weintraub 1984, 1985; Melton 1985). The application of this method to prevent the expression of particular genes during spermatogenesis offers an alternative approach to classic genetic analysis or to the introduction of foreign DNA by microinjection or viral vectors for the study of spermatogenesis. Once it is possible to overcome such technical problems as introducing anti-sense plasmids into testicular cells, this method of specific inactivation of a particular protein can help define the function of many testis-specific proteins, including protein isozyme variants unique to spermatogenesis.

In addition, several more global questions merit thought. For instance, what is the trigger that routes stem cells into the differentiative pathway producing germ cells? What are the molecular mechanisms, operative during meiosis but suppressed in mitotic cells, whereby homologous chromosomes pair and genetic recombination occurs? What is the role of the Y chromosome in sex determination and spermatogenesis? Lastly, consideration should be given to the conditions, *if any*, in which society would consider it appropriate to transfect functional genes into the germ cell line of individuals carrying hereditary genetic defects. Clearly, application of the fast-moving advances in molecular biology to spermatogenesis will yield many exciting insights into eukaryotic cell differentiation and lead to novel methods of fertility control and aid for the infertile.

Acknowledgments

The preparation of this chapter was funded in part by research grants GM-29224 and HD-11878. The author thanks P. Bower, R. Distel, S. Ernst, L. Hake, P. Johnson, R. Schatz, S. Waters, and P. Yelick for comments and helpful suggestions on this manuscript and Mrs. Louise Tricca and Ms. Kimberly Hegdahl for their excellent typing and editorial suggestions.

References

Aiken, J. M., McKenzie, D., Zhao, H. Z., States, H. C., and Dixon, G. H. (1983) Sequence homologies in the protamine gene family of rainbow trout. *Nucleic Acids Res.* 11, 4907–22.

Allen, E., and Altland, P. D. (1952) Studies on degenerating sex cells in immature mammals. II. Mode of degeneration in the normal differentiation of the definitive germ cells in the male albino rat from age 12 days to maturity. *J. Morphol.* 91, 515–32.

Baillie, A. H. (1964) The histochemistry and ultrastructure of the gonocyte. *J. Anat.* 98, 641–5.

Balhorn, R. (1982) A model for the structure of chromatin in mammalian sperm. *J. Cell Biol.* 93, 298–305.

Balhorn, R., Gledhill, B. L., and Wyrobeck, A. J. (1977) Mouse sperm chromatin proteins; quantitative isolation and partial characterization. *Biochemistry* 16, 4074–80.

Balhorn, R., Weston, S., Thomas, C., and Wyrobek, A. (1984) DNA packaging in mouse spermatids. Synthesis of protamine variants and four transition proteins. *Exp. Cell Res.* 150, 298–308.

Bedford, J. M., and Calvin, H. I. (1974) The occurrence and possible functional significance of -S-S- crosslinks in sperm heads, with particular reference to eutherian mammals. *J. Exp. Zool.* 188, 137–56.

Bellvé, A. R. (1979) The molecular biology of mammalian spermatogenesis. In *Oxford Reviews of Reproductive Biology*, ed. C. A. Finn, pp. 159–261. Oxford University Press.

Bellvé, A. R., Anderson, E., and Hanley-Bowdoin, L. (1975) Synthesis and amino acid composition of basic proteins in mammalian sperm nuclei. *Dev. Biol.* 47, 349–65.

Bellvé, A. R., and Carraway, R. (1978) Characterization of two basic chromosomal proteins isolated from mouse spermatozoa. *J. Cell Biol.* 79, 177.

Bellvé, A. R., Cavicchia, J. C., Millette, C. F., O'Brien, D. A., Bhatnagar, Y. M., and Dym, M. (1977) Spermatogenic cells of the prepubertal mouse. Isolation and morphological characterization. *J. Cell Biol.* 74, 68–85.

Bellvé, A. R., and O'Brien, D. A. (1983) The mammalian spermatozoon: structure and temporal assembly. In *Mechanisms and Control of Fertilization*, ed. J. F. Hartman, pp. 55–137. New York: Academic Press.

Bennett, D. (1956) Developmental analysis of a mutation with pleiotropic effects in the mouse. *J. Morphol.* 98, 199–234.

–(1975) The T-locus of the mouse. *Cell* 6, 441–54.

Bennett, W. I., Gall, A. M., Southard, J. L., and Sidman, R. L. (1971) Abnormal spermiogenesis in Quaking, a myelin-deficient mutant mouse. *Biol. Reprod.* 5, 30–58.

Boitani, C., Geremia, R., Rossi, R., and Monesi, V. (1980) Electrophoretic pattern of polypeptide synthesis in spermatocytes and spermatids of the mouse. *Cell Differ.* 9, 41–9.

Boitani, C., Palombi, F., and Stefanini, M. (1983) Influence of Sertoli cell products upon the *in vitro* survival of isolated spermatocytes and spermatids. *Cell Biol. Intern. Reports* 7, 383–93.

Callan, H. G. (1972) Replication of DNA in the chromosomes of eukaryotes. *Proc. R. Soc. Lond. [Biol.]* 181, 19–41.

–(1973). DNA replication in chromosomes of eukaryotes. *Cold Spring Harbor Symp. Quant. Biol.* 38, 195–204.

Calvin, H. I. (1976) Comparative analysis of the nuclear basic proteins in rat, human, guinea pig, mouse and rabbit spermatozoa. *Biochim. Biophys. Acta* 434, 377–89.

Cherry, L. M., and Hsu, T. C. (1984) Antitubulin immunofluorescence studies of spermatogenesis in the mouse. *Chromosoma* 90, 265–74.

Chiquoine, A. D. (1954) The identification, origin, and migration of the primordial germ cells in the mouse embryo. *Anat. Rec.* 118, 135–46.

Clark, J. M., and Eddy, E. M. (1975) Fine structural observations on the origin and associations of primordial germ cells of the mouse. *Dev. Biol.* 47, 136–55.

Clarke, G. N., Clarke, F. M., and Wilson, S., (1982) Actin in human spermatozoa. *Biol. Reprod.* 26, 319–27.

Clarke, G. N., and Yanagimachi, R. (1978) Actin in mammalian sperm heads. *J. Exp. Zool.* 205, 125–32.

Clermont, Y. (1972) Kinetics of spermatogenesis in mammals: seminiferous epithelium cycle and spermatogonial renewal. *Physiol. Rev.* 52, 198–236.

Cleveland, D. W., Lopata, M. A., McDonald, R. J., Cowan, N. J., Rutter, W. J., and Kirschner, M. W. (1980) Number and evolutionary conservation of α and β tubulin and cytoplasmic β and γ actin genes using specific cloned cDNA probes. *Cell* 20, 95–105.

Coelingh, J. P., Monfoort, C. H., Rozijn, T. H., Gevers Leuven, J. A., Schiphof, R., Steyn-Parve, E. P., Braunitzer, G., Schrank, B., and Ruhfus, A. (1972) The complete amino acid sequence of the basic nuclear protein of bull spermatozoa. *Biochim. Biophys. Acta* 285, 1–14.

Coelingh, J. P., Rozijn, T. H., and Monfoort, C. H. (1969) Isolation and partial characterization of a basic protein from bovine sperm head. *Biochim. Biophys. Acta* 188, 353–6.

Coleman, J., Green, P. J., and Inouye, M. (1984) The use of RNAs complementary to specific mRNAs to regulate the expression of individual bacterial genes. *Cell* 37, 429–36.

Comings, D. E., and Okada, T. A. (1972) The chromatoid body in mouse spermatogenesis: evidence that it may be formed by the extrusion of nucleolar components. *J. Ultrastruct. Res.* 39, 15–23.

Dadoune, J. P., Fain-Maurel, M. A., Alfonsi, M. F., and Katsanis, G. (1981) *In vivo* and *in vitro* radioautographic investigation of amino acid incorporation into male germ cells. *Biol. Reprod.* 24, 153–62.

Daentl, D., Erickson, R. P., and Betlach, C. J. (1977) DNA synthetic capabilities of differentiating sperm cells. *Differentiation* 8, 159–66.

D'Agostino, A., Geremia, R., and Monesi, V. (1978) Post-meiotic gene activity in spermatogenesis of the mouse. *Cell Differ.* 7, 175–83.

Dean, D. C., Gope, R., Knoll, B. J., Riser, M. E., and O'Malley, B. W. (1984) A similar 5' flanking region is required for estrogen and progesterone induction of ovalbumin gene expression. *J. Biol. Chem.* 259, 9967–70.

Distel, R. J., Kleene, K. C., and Hecht, N. B. (1984) Haploid expression of a mouse testis α tubulin gene. *Science* 224, 68–70.

Dixon, G. H. (1972) The basic proteins of trout testis chromatin: aspects of their synthesis, post-synthetic modifications and binding to DNA. In *Gene Transcription in Reproductive Tissue*, Karolinska Symposium No. 5, ed. E. Diczfalusy, pp. 130–54. Stockholm: Karolinska Institutet.

Dixon, G. H., Davies, P. L., Ferrier, L. N., Gedamu, L., and Iatrou, K. (1977) The expression of protamine genes in developing trout sperm cells. In *The Molecular Biology of the Genetic Apparatus*, ed. P. Ts'o, pp. 355–79. Amsterdam: Elsevier/North Holland.

Dudley, K., Potter, J., Lyon, M. F., and Willison, K. R. (1984) Analysis of male sterile mutations in the mouse using haploid-stage expressed cDNA probes. *Nucleic Acids Res.* 12, 4281–93.

Dym, M. (1973) The fine structure of the monkey (*Macaca*) Sertoli cell and its role in maintaining the blood-testis barrier. *Anat. Rec.* 175, 639–56.

Dym, M., and Fawcett, D. W. (1970) The blood-testis barrier in the rat and the physiological compartmentation of the seminiferous epithelium. *Biol. Reprod.* 3, 308–26.

–(1971) Further observations on the numbers of spermatogonia, spermatocytes, and spermatids connected by intercellular bridges in the mammalian testis. *Biol. Reprod.* 4, 195–215.

Eddy, E. M. (1970) Cytochemical observations on the chromatoid body of the male germ cells. *Biol. Reprod.* 2, 114–28.

Eddy, E. M., and Clark, J. M. (1975) Electron microscopic study of migrating primordial germ cells in the rat. In *Electron Microscopic Concepts of Secretion. The Ultrastructure of Endocrine and Reproductive Organs*, ed. M. Hess, pp. 151–68. New York: Wiley.

Elsevier, S. M. (1982) Messenger RNA encoding basic chromosomal proteins of mouse testis. *Dev. Biol.* 90, 1–12.

Engel, J. P., Gunning, P., and Kedes, L. (1982) Human cytoplasmic actin proteins are encoded by a multigene family. *Mol. Cell Biol.* 2, 674–84.

Ephrussi, A., Church, G. M., Tonegawa, S., and Gilbert, W. (1985) B cell lineage-specific interactions of an immunoglobin enhancer with cellular factors "in vivo." *Science* 227, 134–40.

Erickson, R. P. (1973) Haploid gene expression versus meiotic drive: the relevance of intercellular bridges during spermatogenesis. *Nature [New Biol.]* 243, 210–13.

Erickson, R. P., Erickson, J. M., Betlach, C. J., and Meistrich, M. L. (1980a) Further evidence for haploid gene expression during spermatogenesis: heterogenous, poly[A]-containing RNA is synthesized postmeiotically. *J. Exp. Zool.* 215, 13–19.

Erickson, R. P., Kramer, J. M., Rittenhouse, J., and Salkeld, A. (1980b) Quantitation of mRNAs during mouse spermatogenesis: protamine-like histone and phosphoglycerate kinase 2 mRNAs increase after meiosis. *Proc. Natl. Acad. Sci. U.S.A.* 77, 6086–90.

Fawcett, D. W., Eddy, E. M., and Phillips, D. M. (1970a) Observations on the fine structure and relationships of the chromatoid body in mammalian spermatogenesis. *Biol. Reprod.* 2, 129–53.

Fawcett, D. W., Leak, L. V., and Heidger, P. M. (1970b) Electron microscope observations on the structural components of the blood-testis barrier. *J. Reprod. Fertil.* (Suppl.) 10, 105–22.

Franchi, L. L., and Mandl, A. M. (1964) The ultrastructure of germ cells in foetal and neonatal male rats. *J. Embryol. Exp. Morphol.* 12, 289–308.

Friedman, T. (1983) *Gene Therapy. Fact and Fiction.* Cold Spring Harbor, N.Y.: Cold Spring Harbor Laboratory.

Fritz, I. B. (1978) Sites of action of androgens and follicle stimulating hormone on cells of the seminiferous tubule. In *Biochemical Actions of Hormones*, Vol. 5, ed. G. Litwack, p. 249. New York: Academic Press.

Fujimoto, H., and Erickson, R. P. (1982) Functional assays for mRNA detect many new messages after male meiosis in mice. *Biochem. Biophys. Res. Commun.* 108, 1369–75.

Fujimoto, H., Erickson, R. P., Quinto, M., and Rosenberg, M. P. (1984) Post-meiotic transcription in mouse testes detected with spermatid cDNA clones. *Biosc. Reports* 4, 1037–44.

Fujimoto, T., Miyayama, Y., and Fuyuta, M. (1977) The origin, migration and fine morphology of human primordial germ cells. *Anat. Rec.* 188, 315–30.

Gaastra, W., Lukkes-Hofstra, J., and Holk, A. H. J. (1978) Partial covalent structure of two basic chromosomal proteins from human spermatozoa. *Biochem. Genet.* 5, 525–9.

Garabedian, M. J., Hung, M. C., and Wensink, P. C. (1985) Independent control elements that determine yolk protein gene expression in alternative *Drosophila* tissues. *Proc. Natl. Acad. Sci. U.S.A.* 82, 1396–400.

Gee, C. E., and Roberts, J. L. (1983) *In situ* hybridization histochemistry: a technique for the study of gene expression in single cells. *DNA* 2, 157–63.

Geremia, R., Boitani, C., Conti, M., and Monesi, V. (1977) RNA synthesis in spermatogenesis and spermatids and preservation of meiotic RNA during spermiogenesis in the mouse. *Cell Differ.* 5, 343−55.

Geremia, R., D'Agostino, A., and Monesi, V. (1978) Biochemical evidence of haploid gene activity in spermatogenesis of the mouse. *Exp. Cell Res.* 111, 23−30.

Geremia, R., Goldberg, R. B., and Bruce, W. R. (1976) Kinetics of histone and protamine synthesis during meiosis and spermiogenesis in the mouse. *Andrologia* 8, 147−56.

Gerton, G. L., and Millette, C. F. (1984) Generation of flagella by cultured mouse spermatids. *J. Cell Biol.* 98, 619−28.

Gizang-Ginsberg, E., and Wolgemuth, D. J. (1985) Localization of mRNAs in mouse testes by "in situ" hybridization: distribution of β-tubulin and developmental stage specificity of pro-opiomelanocortin transcripts. *Dev. Biol.* 111, 293−305.

Gluecksohn-Waelsch, S., and Erickson, R. P. (1970) The T-locus of the mouse: implications for mechanisms of development. *Curr. Top. Dev. Biol.* 5, 281−316.

Gold, B., Fujimoto, H., Kramer, J. M., Erickson, R. P., and Hecht, N. B. (1983a) Haploid accumulation and translational control of phosphoglycerate kinase-2 messenger RNA during mouse spermatogenesis. *Dev. Biol.* 98, 392−9.

Gold, B., and Hecht, N. B. (1981) Differential compartmentalization of messenger ribonucleic acid in murine testis. *Biochemistry* 20, 4871−7.

Gold, B., Stern, L., and Hecht, N. B. (1983b) Gene expression during mammalian spermatogenesis. II. Evidence for stage-specific differences in mRNA populations. *J. Exp. Zool.* 225, 123−34.

Goldberg, E. (1977) Isozymes in testes and spermatozoa. In *Isozymes: Current Topics in Biological Research*, Vol. 1, eds. M. C. Rattazzi, J. C. Scandalios, and G. S. Litt, pp. 79−124. New York: Alan R. Liss.

Goldberg, E., and Hawtrey, C. (1967) The ontogeny of sperm specific lactate dehydrogenase in mice. *J. Exp. Zool.* 164, 309−16.

Goldberg, E., Sberna, D., Wheat, T. E., Urbanski, G. J., and Margoliash, E. (1977) Cytochrome c: immunofluorescent localization of the testis-specific form. *Science* 196, 1010−12.

Goldberg, E., Wheat, T. E., Powell, J. E., and Stevens, V. C. (1981) Reduction of fertility in female baboons immunized with lactate dehydrogenase C_4. *Fert. Steril.* 35, 214−17.

Gonzales-Prevatt, V., Wheat, T. E., and Goldberg, E. (1982) Identification of an antigenic determinant of mouse lacate dehydrogenase C_4 *Molec. Immunol.* 19, 1579−85.

Grimes, S. J., Jr., Meistrich, M. L., Platz, R. D., and Hnilica, L. S. (1977) Nuclear protein transitions in rat testis spermatids. *Exp. Cell Res.* 110, 31−9.

Grippo, P., Geremia, R., Locorotondo, G., and Monesi, V. (1978) DNA-dependent DNA polymerase species in male germ cells of the mouse. *Cell Diff.* 7, 237−48.

Grootegoed, J. A., Grolle-Hey, A. H., Rommerts, F. F. G., and Van der

Molen, H. J. (1977) Ribonucleic acid synthesis in vitro in primary spermato-cytes isolated from rat testis. *Biochem. J.* 168, 23–31.

Groudine, M., and Conklin, K. F. (1985) Chromatin structure and "de novo" methylation of sperm DNA: implications for the developmental acti-vation of the paternal genome. *Science* 228, 1061–8.

Hecht, N. B., and Bradley, F. M. (1981) Changes in mitochondrial protein composition during testicular differentiation in mouse and bull. *Gam. Res.* 4, 433–49.

Hecht, N. B., Bower, P. A., Kleene, K. C., and Distel, R. J. (1985) Size changes of protamine 1 mRNA provide a molecular marker to monitor spermatogenesis in wild type and mutant mice. *Differentiation* 29, 189–93.

Hecht, N. B., Farrell, D., and Davidson, D. (1976) Changing DNA polymer-ase activities during the development of the mammalian testis. *Dev. Biol.* 48, 56–66.

Hecht, N. B., Farrell, D., and Williams, J. L. (1978) DNA polymerases in mouse spermatogenic cells separated by sedimentation velocity. *Biochim. Biophys. Acta* 561, 358–68.

Hecht, N. B., Kleene, K. C., Distel, R. J., and Silver, L. M. (1984) The dif-ferential expression of the actins and tubulins during spermatogenesis in the mouse. *Exp. Cell Res.* 153, 275–80.

Hecht, N. B., Kleene, K. C., Yelick, P. C., Johnson, P. A., Pravtcheva, D. D., and Ruddle, F. H. (1986) Haploid gene mapping: the genes for both mouse protamines are located on chromosome 16. *Som. Cell. Mol. Gen.* 12, 191–6.

Hecht, N. B., and Parvinen, M. (1981) DNA synthesis catalyzed by endoge-nous DNA-dependent DNA polymerases and template in spermatogenic cells from mouse and rat. *Exp. Cell Res.* 135, 103–14.

Heller, C. G., and Clermont, Y. (1963) Spermatogenesis in man: an esti-mate of its duration. *Science* 140, 184–6.

Hilscher, W., and Makoski, H. B. (1968) Histologische und autoradiograph-ische Untersuchungen zur Praspermatogenese und Spermatogenese der Ratte. *Z. Zellforsch. Mikrosk. Anat.* 86, 327–50.

Hotta, Y., and Stern, H. (1981) Small nuclear RNA molecules that regulate nuclease accessibility in specific chromatin regions of meiotic cells. *Cell* 27, 309–19.

Hotta, Y., Tabata, S., Stubbs, L., and Stern, H. (1985) Meiosis-specific transcripts of a DNA component replicated during chromosome pairing: homology across the phylogenetic spectrum. *Cell* 40, 785–93.

Huckins, C., and Clermont, Y. (1968) Evolution of gonocytes in the rat tes-tis during late embryonic and post-natal life. *Extrait Arch. Anat. Hist. Embryol. Norm. Exp.* 51, 343–54.

Hudson, P., Penschow, J., Shine, J., Ryan, G., Niall, H., and Coghlan, J. (1981) Hybridization histochemistry: use of recombinant DNA as a "homing probe" for tissue localization of specific mRNA populations. *Endrocrinology* 108, 353–6.

Iatrou, K., and Dixon, G. H. (1978) Protamine messenger RNA: its life his-tory during spermatogenesis in rainbow trout. *Fed. Proc.* 37, 2626–33.

Izant, J. G., and Weintraub, H. (1984) Inhibition of thymidine kinase gene expression by anti-sense RNA: a molecular approach to genetic analysis. *Cell* 36, 1007–15.

–(1985) Constitutive and conditional suppression of exogenous and endogenous genes by anti-sense RNA. *Science* 229, 345–52.

Jagiello, G., Tantravahi, U., Fang, J. S., and Erlanger, B. F. (1982) DNA methylation of human pachytene spermatocytes. *Exp. Cell Res.* 141, 253–9.

Jeffrey, W. R. (1984) Spatial distribution of messenger RNA in the cytoskeletal framework of ascidian eggs. *Dev. Biol.* 103, 482–92.

Jeon, K. W., and Kennedy, J. R. (1973) The primordial germ cells in early mouse embryos: light and electron microscope studies. *Dev. Biol.* 31, 275–84.

Jutte, N. H. P. M., Grottegoed, J. A., Rommerts, F. F. G., and Van der Molen, H. J. (1981) Exogenous lactate is essential for metabolic activities in isolated rat spermatocytes and spermatids. *J. Reprod. Fertil.* 62, 399–405.

Kandala, J. C., Kistler, M. K., and Kistler, W. S. (1985) Androgen-regulated genes from prostate and seminal vesicle share upstream sequence homologies. *Biochem. Biophys. Res. Commun.* 126, 948–52.

Kaput, J., and Sneider, T. W. (1979) Methylation of somatic vs. germ cell DNAs analyzed by restriction endonuclease digestions. *Nucleic Acids Res.* 7, 2303–21.

Kemphues, K. J., Kaufman, T. C., Raff, R. A., and Raff, E. C. (1982) The testis-specific β-tubulin subunit in *Drosophila melanogaster* has multiple functions in spermatogenesis. *Cell* 31, 655–70.

Kessel, R. G., and Kardon, R. H. (1979) *Tissues and Organs: A Text-Atlas of Scanning Electron Microscopy.* San Francisco: Freeman.

Kierszenbaum, A. L., and Tres, L. L. (1974a) Nucleolar and perichromosomal RNA synthesis during meiotic prophase in the mouse testis. *J. Cell Biol.* 60, 39–53.

–(1974b) Transcription sites in spread meiotic prophase chromosomes from mouse spermatocytes. *J. Cell Biol.* 63, 923–35.

–(1975) Structural and transcriptional features of the mouse spermatid genome. *J. Cell Biol.* 65, 258–70.

–(1978) RNA transcription and chromatin structure during meiotic and postmeiotic stages of spermatogenesis. *Fed. Proc.* 37, 2512–16.

Kistler, W. S., Geroch, M. E., and Williams-Ashman, H. G. (1975) The amino acid sequence of a testis-specific basic protein that is associated with spermatogenesis. *J. Biol. Chem.* 250, 1847–53.

Kistler, W. S., Keim, P. S., and Heinrikson, R. L. (1976) Partial amino acid sequence of the basic chromosomal protein of rat spermatozoa. *Biochim. Biophys. Acta* 427, 752–7.

Kleene, K. C., Distel, R. J., and Hecht, N. B. (1983) cDNA clones encoding cytoplasmic poly(A)$^+$ RNAs which first appear at detectable levels in haploid phases of spermatogenesis in the mouse. *Dev. Biol.* 98, 455–64.

–(1984) Translational regulation and coordinate deadenylation of a haploid mRNA during spermiogenesis in the mouse. *Dev. Biol.* 105, 71–9.

–(1985) The nucleotide sequence of a cDNA clone encoding mouse protamine I. *Biochemistry* 24, 719–22.

Kofman-Alfaro, S., and Chandley, A. C. (1970) Meiosis in the male mouse. An autoradiographic investigation. *Chromosoma* 31, 404–20.

Kramer, J. M., and Erickson, R. P. (1981a) Immunofluorescent localization of PGK-1 and PGK-2 isozymes within specific cells of the mouse testis. *Dev. Biol.* 87, 30–6.

–(1981b) Developmental programs of PGK-1 and PGK-2 isozymes in spermatogenic cells of the mouse: specific activities and rates of synthesis. *Dev. Biol.* 87, 37–45.

–(1982) Analysis of stage-specific protein synthesis during spermatogenesis of the mouse by two-dimensional gel electrophoresis. *J. Reprod. Fertil.* 64, 139–44.

Krawetz, S. A., and Dixon, G. H. (1984) Isolation and "in vitro" translation of a mammalian protamine mRNA. *Biosci. Rep.* 4, 593–603.

Lam, D. M. K., and Bruce, W. R. (1971) The biosynthesis of protamine during spermatogenesis of the mouse: extraction, partial characterization, and site of synthesis. *J. Cell. Physiol.* 78, 13–24.

Lam, D. M. K., Furrer, R., and Bruce, W. R. (1970) The separation, physical characterization, and differentiation kinetics of spermatogonial cells of the mouse. *Proc. Natl. Acad. Sci. U.S.A.* 65, 192–9.

Lawrence, J. B., and Singer, R. H. (1985) Quantitative analysis of "in situ" hybridization methods for the detection of actin gene expression. *Nucleic Acids Res.* 13, 1777–99.

Lemischka, I., and Sharp, P. A. (1982) The sequences of an expressed rat α-tubulin gene and a pseudogene with an inserted repetitive element. *Nature* 300, 330–5.

Lima de Faria, A., and Borum, K. A. (1962) The period of DNA synthesis prior to meiosis in the mouse. *J. Cell Biol.* 14, 381–8.

Loir, M. (1972) Protein and ribonucleic acid metabolism in spermatocytes and spermatids of the ram (*Ovis aries*). I. Incorporation and fate of [^{3}H]uridine. *Ann. Biol. Anim. Biochim. Biophys.* 12, 203–19.

Loir, M., and Lanneau, M. (1975) Isopycnic separation of ram spermatids in colloidal silica gradients. *Exp. Cell Res.* 92, 499–505.

–(1977) Separation of mammalian spermatids. In *Methods in Cell Biology*, Vol. 15, ed. D. M. Prescott, pp. 55–7. New York: Academic Press.

Lyon, M. F., Cattanach, B. B., and Charlton, H. M. (1981) Genes affecting sex differentiation in mammals. In *Mechanisms of Sex Differentiation in Animals and Man*, eds. C. R. Austin, R. G. Edwards, pp. 329–86. New York: Academic Press.

Lyon, M. F., and Hawkes, S. G. (1970) X-linked gene for testicular feminization in the mouse. *Nature* 227, 1217–19.

McKay, D. J., Renaux, B. S., and Dixon, G. H. (1985) The amino acid sequence of human sperm protamine P1. *Biol. Reprod.* 5, 383–91.

Martino, C. D., Capanna, E., Nicotra, M. R., and Natali, P. G. (1980) Immunochemical localization of contractile proteins in mammalian meiotic chromosomes. *Cell Tissue Res.* 213, 159–78.

Mayer, J. F., Chang, T. S. K., and Zirkin, B. R. (1981) Spermatogenesis in the mouse. 2. Amino acid incorporation into basic nucleoproteins of mouse spermatids and spermatozoa. *Biol. Reprod.* 25, 1041–51.

Mayer, J. F., and Zirkin, B. R. (1979) Spermatogenesis in the mouse. 1. Autoradiographic studies of nuclear incorporation and loss of [^{3}H]-aminoacids. *J. Cell Biol.* 81, 403–10.

Mazrimas, J. A., Corzett, M., Campos, C., and Balhorn, R. (in press) A corrected primary sequence for bull protamine. *Biochim. Biophys. Acta.*

Meistrich, M. L. (1972) Separation of mouse spermatogenic cells by velocity sedimentation. *J. Cell Physiol.* 80, 299–312.

–(1977) Separation of spermatogenic cells and nuclei from rodent testes. In *Methods in Cell Biology*, Vol. 15, ed. D. M. Prescott, pp. 15–24. New York: Academic Press.

–(1978) Cell separation by centrifugal elutriation. *Pulse-Cytophotometry* 3, 161–8.

Meistrich, M. L., Bruce, W. R., and Clermont, Y. (1973) Cellular composition of fractions of mouse testis cells following velocity sedimentation separation. *Exp. Cell Res.* 79, 213–27.

Meistrich, M. L., Longtin, J., Brock, W. A., Grimes, S. J., and Mace, M. L. (1981) Purification of rat spermatogenic cells and preliminary biochemical analysis of these cells. *Biol. Reprod.* 25, 1065–77.

Meistrich, M. L., and Trostle, P. K. (1975) Separation of mouse testis cells by equilibrium density centrifugation in Renografin gradients. *Exp. Cell Res.* 92, 231–44.

Meistrich, M. L., Trostle, P. K., Fraport, M., and Erickson, R. P. (1977) Biosynthesis and localization of lactate dehydrogenase X in pachytene spermatocytes and spermatids of mouse testes. *Dev. Biol.* 60, 428–41.

Melton, D. A. (1985) Injected anti-sense RNAs specifically block messenger RNA translation "in vivo." *Proc. Natl. Acad. Sci. U.S.A.* 82, 144–8.

Millette, C. F., and Bellvé, A. R. (1977) Temporal expression of membrane antigens during mouse spermatogenesis. *J. Cell Biol.* 74, 86–97.

–(1980) Selective partitioning of plasma membrane antigens during mouse spermatogenesis. *Dev. Biol.* 79, 309–24.

Millette, C. F., and Moulding, C. T. (1981) Cell surface marker proteins during mouse spermatogenesis: two-dimensional gel electrophoretic analysis. *J. Cell Sci.* 48, 367–82.

Minty, A. J., Alonso, S., Caravatti, M., and Buckingham, M. E. (1982) A fetal skeletal muscle actin mRNA in the mouse and its identity with cardiac actin mRNA. *Cell* 30, 185–92.

Mintz, B., and Russell, E. S. (1955) Developmental modifications of primordial germ cells, induced by the W-series gene in the mouse embryo. *Anat. Rec.* 122, 443.

Monesi, V. (1962) Autoradiographic study of DNA synthesis and the cell cycle in spermatogonia and spermatocytes of the mouse testis using tritiated thymidine. *J. Cell. Biol.* 14, 1–18.

–(1964a) Ribonucleic acid synthesis during mitosis and meiosis in the mouse testis. *J. Cell Biol.* 22, 521–32.

–(1964b) Autoradiographic evidence of a nuclear histone synthesis during mouse spermiogenesis in the absence of detectable quantities of nuclear ribonucleic acid. *Expl. Cell Res.* 36, 683–8.

−(1965) Synthetic activities during spermatogenesis in the mouse: RNA and protein. *Expl. Cell Res.* 39, 197−224.

−(1967) Ribonucleic acid and protein synthesis during differentiation of male germ cells in the mouse. *Arch. Anat. Microsc. Morphol. Exp.* 56, 61−74.

−(1971) Chromosome activities during meiosis and spermiogenesis. *J. Reprod. Fertil. Suppl.* 13, 1−14.

−(1978) Biochemistry of male germ cell differentiation in mammals: RNA synthesis in meiotic and post-meiotic cells. *Curr. Top. Dev. Biol.* 12, 11−36.

Monfoort, C. H., Schiphof, R., Rozijn, T. H., and Steyn-Parvé, E. P. (1973) Amino acid composition and carboxy-terminal structure of some basic chromosomal proteins of mammalian spermatozoa. *Biochim. Biophys. Acta* 322, 173−7.

Moore, G. P. M. (1971) DNA-dependent RNA synthesis in fixed cells during spermatogenesis in mouse. *Exp. Cell Res.* 68, 462−6.

Moses, M. J., (1956) Chromosomal structures in crayfish spermatocytes. *J. Biophys. Biochem. Cytol.* 2, 215−18.

Muramatsu, M., Utakoji, T., and Sugano, H. (1968) Rapidly-labeled nuclear RNA in Chinese hamster testis. *Exp. Cell Res.* 53, 278−83.

Nakano, M., Tobita, T., and Ando, T. (1975) Studies on a protamine (galline) from fowl sperm. *Int. J. Pept. Protein Res.* 7, 31−46.

Nebel, B. R., Amarose, A.P., and Hackett, E. M. (1961) Calendar of gametogenic development in the prepubertal male mouse. *Science* 134, 832−3.

Oakberg, E. F. (1956a) A description of spermatogenesis in the mouse and its use in analysis of the cycle of the seminiferous epithelium and germ cell renewal. *Am. J. Anat.* 99, 391−413.

−(1956b) Duration of spermatogenesis in the mouse and timing of stages of the cycle of the seminiferous epithelium. *Am. J. Anat.* 99, 507−16.

O'Brien, D. A., and Bellvé, A. R. (1980a) Protein constituents of the mouse spermatozoon. I. An electrophoretic characterization. *Dev. Biol.* 75, 386−404.

−(1980b) Protein constituents of the mouse spermatozoon. II. Temporal synthesis during spermatogenesis. *Dev. Biol.* 75, 405−418.

O'Rand, M. G., and Romrell, L. J. (1977) Appearance of cell surface auto- and isoantigens during spermatogenesis in the rabbit. *Dev. Biol.* 55, 347−58.

Pallini, V., Baccetti, B., and Burrini, A. G. (1979) A peculiar cysteine-rich polypeptide related to some unusual properties of mammalian sperm mitochondria. In *The Spermatozoon*, eds. D. W. Fawcett and J. M. Bedford, pp. 141−51. Baltimore: Urban & Schwarzenberg.

Palmiter, R. D., Wilkie, T. M., Chen, H. Y., and Brinster, R. L. (1984) Transmission distortion and mosaicism in an unusual transgenic mouse pedigree. *Cell* 36, 869−77.

Parvinen, M. (1982) Regulation of the seminiferous epithelium. *Endocrinol. Rev.* 3, 404−17.

Parvinen, M., and Parvinen, L. M. (1979) Active movements of the chromatoid body. A possible transport mechanism for haploid gene products. *J. Cell Biol.* 80, 621−8.

Pelc, S. R. (1956) On the connection between the synthesis of RNA and DNA in the testis of the mouse. *Exp. Cell Res.* 12, 320−4.

Pintar, J. E., Schachter, B. S., Herman, A. B., Durgerian, S., and Krieger, D. T., (1984) Characterization and localization of pro-opiomelanocortin messenger RNA in the adult rat testis. *Science* 225, 632−4.

Pongswasdi, P., and Svasti, J. (1976) The heterogeneity of the protamines from human spermatozoa. *Biochim. Biophys. Acta* 434, 462−73.

Ponzetto-Zimmerman, C., and Wolgemuth, D. J. (1984) Methylation of satellite sequences in mouse spermatogenic and somatic DNAs. *Nucleic Acids Res.* 12, 2807−22.

Ponzetto, C., and Wolgemuth, D. J. (1985) Haploid expression of a unique c-*abl* transcript in the mouse male germ line. *Mol. Cell Biol.* 5, 1791−4.

Rahe, B., Erickson, R. P., and Quinto, M. (1983) Methylation of unique sequence DNA during spermatogenesis in mice. *Nucleic Acids Res.* 11, 7947−59.

Rocamora, N., and Mezquita, C. (1984) Hypomethylation of DNA in meiotic and postmeiotic rooster testis cells. *FEBS Lett.* 177, 81−4.

Rodman, T. C., Litwin, S. D., Romani, M., and Vidali, G. (1979) Life history of mouse sperm protein. Intratesticular stages. *J. Cell Biol.* 80, 605−20.

Romrell, L. J., Bellvé, A. R., and Fawcett, D. W. (1976) Separation of mouse spermatogenic cells by sedimentation velocity. A morphological characterization. *Dev. Biol.* 49, 119−31.

Roosen-Runge, E. C. (1973) Germinal cell loss in normal metazoan spermatogenesis. *J. Reprod. Fertil.* 35, 339−48.

Ross, M. H. (1977) Sertoli-Sertoli junctions and Sertoli-spermatid junctions after ductule ligation and lanthanum treatment. *Am. J. Anat.* 148, 49−56.

Sanford, J., Forrester, L., and Chapman, V. (1984) Methylation patterns of repetitive DNA sequences in germ cells of *Mus musculus*. *Nucleic Acids Res.* 12, 2823−36.

Sautiere, P., Belaiche, D., Martinage, A., and Loir, M. (1984) Primary structure of the ram (*Ovis aries*) protamine 1. *Eur. J. Biochem.* 144, 121−5.

Schnieke, A., Harbers, K., and Jaenisch, R. (1983) Embryonic lethal mutation in mice induced by retrovirus in the α1(I) collagen gene. *Nature* 304, 315−20.

Skinner, M. K., and Griswold, M. D. (1982) Secretion of testicular transferrin by cultured Sertoli cells is regulated by hormones and retinoids. *Biol. Reprod.* 27, 211−21.

Söderstrom, K. O. (1976) Characterization of RNA synthesis in mid-pachytene spermatocytes of the rat. *Exp. Cell Res.* 102, 237−45.

Söderstrom, K. O., and Parvinen, M. (1976) RNA synthesis in different stages of rat seminiferous epithelial cycle. *Mol. Cell Endocrinol.* 5, 181−99.

Sotomayor, R. E., and Handel, M. A. (1986) Failure of acrosome assembly in a male sterile mouse mutant. *Biol. Reprod.* 34, 171−82.

Southern, P. J., Blount, P., and Oldstone, M. B. A. (1984) Analysis of persistent virus infections by "in situ" hybridization to whole mouse sections. *Nature* 312, 555−8.

Spiegelman, M., and Bennett, D. (1973) A light- and electron-microscopic study of primordial germ cells in the early mouse embryo. *J. Embryol. Exp. Morphol.* 30, 97−118.

Steinberger, A. (1975) *In vitro* techniques for the study of spermatogenesis. *Methods Enzymol.* 39, 283–96.

Steinberger, E. (1971) Hormonal control of mammalian spermatogenesis. *Physiol. Rev.* 51, 1.

Stern, H., and Hotta, Y. (1983) Meiotic aspects of chromosome organization. *Stadler Symposia* 15, 25–41.

Stern, L., Gold, B., and Hecht, N. B. (1983a) Gene expression during mammalian spermatogenesis. I. Evidence for stage-specific synthesis of polypeptides "in vivo." *Biol. Reprod.* 28, 483–96.

Stern, L., Kleene, K. C., Gold, B., and Hecht, N. B. (1983b) Gene expression during mammalian spermatogenesis. III. Changes in populations of mRNA during spermiogenesis. *Exp. Cell Res.* 143, 247–55.

Stewart, T. A., Bellvé, A. R., and Leder, P. (1984) Transcription and promoter usage of the *myc* gene in normal somatic and spermatogenic cells. *Science* 226, 707–10.

Subirana, J. A. (1975) On the biological role of basic proteins in spermatozoa and during spermiogenesis. In *The Biology of the Male Gamete*, eds. J. G. Duckett and P. A. Racey, pp. 239–44. New York: Academic Press.

Talbot, P., and Kleve, M. G. (1978) Hamster sperm cross react with anti-actin. *J. Exp. Zool.* 204, 131–6.

Tanphaichitr, N., Sobhon, P., Taluppeth, N., and Chalermisarachai, P. (1978) Basic nuclear proteins in testicular cells and ejaculated spermatozoa in man. *Exp. Cell Res.* 117, 347–56.

Tobita, T., Tstutsumi, H., Kato, A., Suzuki, H., Nomoto, M., Nakano, M., and Ando, T. (1983) Complete amino acid sequence of boar protamine. *Biochim. Biophys. Acta* 744, 141–6.

Tres, L. L., and Kierszenbaum, A. L. (1975) Transcription during mammalian spermatogenesis with special reference to Sertoli cells. In *Hormonal Regulation of Spermatogenesis*, eds. F. S. French, V. Hansson, E. M. Ritzen, and S. N. Nayfeh, pp. 455–78. New York: Plenum.

–(1983) Viability of rat spermatogenic cells *in vitro* is facilitated by their co-culture with Sertoli cells in serum-free hormone-supplemented medium. *Proc. Natl. Acad. Sci. U.S.A.* 80, 3377–81.

Utakoji, T. (1966) Chronology of nucleic acid synthesis in meiosis of the male Chinese hamster. *Exp. Cell Res.* 42, 585–96.

Vandeberg, J. L., Cooper, D. W., and Close, P. J. (1976) Testis specific phosphoglycerate kinase B in mouse. *J. Exp. Zool.* 198, 231–40.

Varnum, D. S. (1983) Blind-sterile: a new mutation in chromosome 2 of the house mouse. *J. Hered.* 74, 206–7.

von Wettstein, D. (1977) The assembly of the synaptonemal complex. *Philos. Trans. R. Soc. Lond. [Biol.]* 277, 235–43.

Waters, S. H., Distel, R. J., and Hecht, N. B. (1985) Identification and distribution of actin in spermatogenic cells and spermatozoa of the rabbit. *Mol. Cell Biol.* 5, 1649–54.

Welch, J. E., and O'Rand, M. G. (1985) Identification and distribution of actin in spermatogenic cells and spermatozoa of the rabbit. *Dev. Biol.* 109, 411–17.

Wheat, T. E., and Goldberg, E. (1981) Immunologically active peptide fragments of the sperm-specific lactate dehydrogenase C_4 isozyme. In *Peptides: Synthesis-Structure-Function,* eds. D. H. Rich and E. Gross, pp. 557−60. Rockford, Ill.: Pierce.

Wieben, E. D. (1981) Regulation of the synthesis of lactate dehydrogenase X during spermatogenesis in the mouse. *J. Cell Biol.* 88, 492−8.

Wolgemuth, D. J., Gizang-Ginsberg, E., Engelmyer, E., Gavin, B. J., and Ponzetto, C. (1985) Separation of mouse testis cells on a Celsep apparatus and their usefulness as a source of high molecular weight DNA or RNA. *Gam. Res.* 12, 1−10.

Yelick, P. C., Johnson, P. A., Kleene, K. C., and Hecht, N. B. (1985) Sequence analysis of cDNAs encoding mouse protamine 1 and 2 (MP1 and MP2) suggests that MP2 is synthesized as a precursor whereas MP1 is not. *J. Cell Biol.* 101, 366a.

Zamboni, L., and Merchant, H. (1973) The fine morphology of mouse primordial germ cells in extragonadal locations. *Am. J. Anat.* 137, 299−336.

7   Molecular aspects of mammalian oocyte growth and maturation

RICHARD M. SCHULTZ

CONTENTS

I. Introduction

    The process of oogenesis generates the egg whose central role in biology is exemplified by the statement "Omne vivum ex ovo"—"All living things come from eggs," which is attributed to William Harvey. In the mouse, oogenesis begins with the formation of the primordial germ cells in the 8-day-old embryo. These cells are the sole source of germ cells and are readily identifiable by a variety of histochemical and ultrastructural criteria. By day 14 post fertilization, some of the primordial germ cells, which are initially found in the region of the allantois, have migrated to and colonized the genital ridge of the presumptive gonad, which is situated near the kidney. The oogonia then undergo a last round of DNA synthesis and are transformed into oocytes that enter meiotic prophase by day 14 post fertilization. This prophase is characterized by a series of changes in chromosome morphology. By day 5 post partum the primary oocytes have entered the dictyate stage in which the chromosomes are highly diffuse and presumably transcriptionally active. The ovary is now populated with thousands of small oocytes about 12–20 μm in diameter that are arrested in the dictyate stage of the first meiotic prophase. They remain at this stage until just prior to ovulation, a period extending from several weeks to the length of the reproductive life span of the animal; this feature is common to all mammalian species.

Following a period of oocyte growth (oocyte diameter increases to about 80 μm during a period of about 14 days), ovulation and resumption of meiosis are initiated by a hormonal stimulus. The fully grown dictyate oocyte then enters the period of meiotic maturation, which is characterized by dissolution of the nuclear membrane, condensation of diffuse chromatine into distinct bivalents, separation of homologous chromosomes, and emission of the first polar body, with arrest at metaphase II. The ability of oocytes to mature properly is critical for development, because only oocytes that have successfully completed meiotic maturation are capable of being fertilized and giving rise to normal development. A schematic of mouse oogenesis and the ultrastructural changes that occur during growth are shown in Figures 7.1 and 7.2, respectively.

There are two aspects of oogenesis. Oocyte growth is concerned with accumulation of materials that constitute the maternal contribution to early development, and meiosis generates the haploid gamete. This review will focus on how follicle cells regulate oocyte growth, the molecular differentiation of the oocyte during the growth phase, and current models for regulation of meiotic maturation.

II. Oocyte growth

The growth phase of the oocytes is different from that of somatic cells, because there is growth in the absence of cell division. Mammalian oocytes are transcriptionally active during this period and accumulate macromolecules and organelles. The oocyte also differentiates dramatically, at both the morphological and biochemical levels, during the growth phase, eventually acquiring meiotic competence (i.e., the ability to resume meiosis when removed from the follicle). The following discussion exam-

OOGENESIS IN THE MOUSE

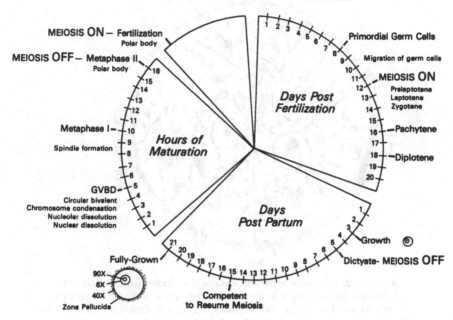

Figure 7.1. Schematic representation of developmental aspects of oogenesis in the mouse. The onset of meiosis (meiosis ON) occurs around day 12 post fertilization and involves entry of the proliferating oogonia into a nonproliferative meiotic cell cycle. Around day 5 post partum, the oocytes arrest in the first meiotic prophase (meiosis OFF) and enter the growth phase. *In vivo*, a hormonal stimulus initiates resumption of meiosis. Germinal vesicle breakdown (GVDB) is one of the first easily observable manifestations of resumption meiosis. The oocytes progress through metaphase I, emit a polar body, and arrest at metaphase II (meiosis OFF). Fertilization triggers resumption of meiosis and emission of the second polar body (meiosis ON). [Reproduced from Schultz and Wassarman (1977a) by permission of the Company of Biologists, Ltd.]

ines how follicle cells influence oocyte growth and what factors may be involved in acquisition of meiotic competence.

A. Role of intercellular communication in oocyte growth

Follicle cells invest the oocyte and send out processes that traverse the zona pellucida and terminate on the oolemma. At points of contact between these processes and the oocyte, gap junctions can occur (Albertini and Anderson 1974; Anderson and Albertini 1976; Gilula, Epstein, and Beers, 1978). Recent experimental evidence indicates that follicle cells furnish nutrients to the oocyte via gap junctions and more-

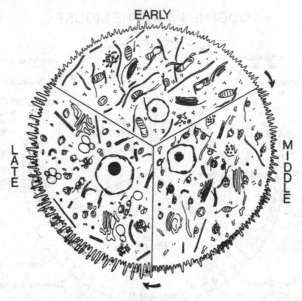

Figure 7.2. Schematic representation of the ultrastructural changes that accompany oocyte growth in the mouse. Early growth (3–5 days post partum) is characterized by a thin, diffuse *zona pellucida*, a few relatively short microvilli, an extensive network of smooth endoplasmic reticulum, elongated and dumbell-shaped mitochondria that are associated with the endoplasmic reticulum and have cristae in an "orthodox" configuration, numerous clusters of free ribosomes, a Golgi apparatus consisting solely of flattened stacks of lamellae arranged in a parallel manner, a few small multivesicular bodies, and a rapidly enlarging fibrillogranular nucleolus. During the middle stages of growth (8–14 days post partum) the *zona pellucida* thickens and becomes more dense, more microvilli are present per unit surface area, and the smooth endoplasmic reticulum becomes more vesicular and less abundant than before. Mitochondria are round or oval, are smaller but more abundant than before, and possess columnar-shaped cristae. Ordered lattices begin to appear, while free ribosomes decrease in number. The Golgi apparatus consists of parallel lamellae, vacuoles, and granules, and the nucleolus becomes larger, fibrillar, and dense. During the late stages of growth (beyond 14 days post partum) the microvilli are longer than before, and the *zona pellucida* is thicker and denser. Lipid droplets appear, sometimes in association with the Golgi complex, which is now highly vacuolated and granular. The mitochondria are round or oval and possess arched or concentrically arranged cristae. Small vesicles of smooth endoplasmic reticulum are found free in the cytoplasm and in association with mitochondria. Ordered lattices are abundant, large multivesicular bodies are plentiful, and the nucleolus is very dense. [Reproduced from Wassarman and Josefowicz (1978) by permission of The Company of Biologists, Ltd.]

over that this heterologous intercellular communication is essential for oocyte growth.

The development of two culture systems that support oocyte growth in vitro was critical for analyzing the role of follicle cells in oocyte growth. In the system developed by Eppig (1977), some of the granulosa cells migrate away from the follicle, but oocytes remain invested by at least one layer of granulosa cells. The oocytes grow at a rate approaching the in vivo rate and undergo many of the structural and functional changes associated with growth in vivo, including the ability to resume meiosis and to undergo fertilization and preimplantation development (J.J. Eppig, personal communication).

In the system developed by Bachvarova (Bachvarova, Baran, and Tejblum 1980), ovarian cells attach to the culture dish and leave denuded oocytes that rest on a monolayer of ovarian cells. Although the growth rate in this system is lower than that in the Epigg system, some of the oocytes acquire the ability to resume meiosis.

In either in vitro culture system, oocyte growth occurs only when the oocyte is in direct contact with the follicle cells (Eppig 1979a; Bachvarova et al. 1980). Evidence from results of uptake experiments is consistent with a nutritional role for intercellular communication between follicle cells and the oocyte in oocyte growth. After a brief incubation, uptake of radiolabeled leucine or uridine by cumulus-cell-enclosed dictyate oocytes is about 60 percent greater than that of denuded oocytes (Cross and Brinster 1974; Wassarman and Letourneau 1976). This difference is not found in ovulated oocytes, which are arrested at metaphase II, because ovulation results in mucification and cumulus expansion that physically disrupts intercellular communication between follicle cells and the oocyte (Gilula et al. 1978; Heller and Schultz 1980; Moor, Smith, and Dawson 1980; Eppig 1982).

The follicle cells provide the major route of entry into the oocyte for most metabolites, which then reach the oocyte via gap junctions. For compounds that serve as energy sources, and precursors for protein, RNA, and phospholipid biosynthesis, this coupling usually accounts for greater than 90 percent of the influx into the oocyte; the remainder is the direct uptake of the metabolite by the oocyte (Heller and Schultz 1980; Moor et al. 1980; Heller, Cahill, and Schultz 1981). In addition to containing more radiolabeled metabolites, follicle-cell-enclosed oocytes have different distributions of phosphorylated metabolites of ribonucleosides and of a glucose analog when compared with denuded oocytes (Table 7.1). As expected, the distribution of phosphorylated metabolites in follicle-cell-enclosed oocytes is similar to that in follicle cells (Heller et al. 1981; Brower and Schultz 1982a). In addition, oocytes do not contain the energy-dependent A transport system for amino acid uptake, whereas

Table 7.1. *Distribution of metabolites among denuded oocytes, follicle-enclosed oocytes, and follicle cells*

Compound	Percentage of total[a]		
	Denuded oocyte	Follicle-enclosed oocyte	Follicle cells
Guanosine	10	3	2
GMP	42	3	5
GDP	22	8	8
GTP	25	86	85
	100	100	100
2-DG	67	8	8
2-DG-6-P	33	92	92
	100	100	100

Note:
[a]2-DG, 2-deoxyglucose; 2-DG-6-P, 2-deoxyglucose-6-phosphate.
Source: Data from Heller et al. (1981) and Brower and Schultz (1982a).

follicle cells do (Colonna et al. 1983). In this case, the coupling pathway provides the route of entry to the oocyte for amino acids that use this transport system.

Pulse-chase experiments are also consistent with metabolic cooperativity between follicle cells and the oocyte, because the amount of radioactivity in oocytes increases during the chase (Heller et al. 1981). Furthermore, treatments known to disrupt gap junctions in other cell types result in reversible uncoupling of intercellular communication between follicle cells and the oocyte (Heller et al. 1981), and, lastly, the extent of intercellular communication between follicle cells and the oocyte is a linear function of the number of attached follicle cells (Brower and Schultz 1982a).

Although the results of these experiments indicate that the heterologous gap junctions are functional throughout the period of oocyte growth and that for most cases examined the follicle cells furnish the oocyte most of the nutrients presumably used in oocyte growth, they do not bear directly on the physiological role for such intercellular communication. If a nutritional role is provided by intercellular communication, higher levels of communication should promote greater rates of oocyte growth. This was addressed using the in vitro culture systems, where the rate of growth is proportional to the extent of intercellular communication (Brower and Schultz 1982a) (Figure 7.3).

A nutritional role for intercellular communication between follicle cells and oocytes in oocyte growth is buttressed by the observation that reestablishment of intercellular communication between follicle cells and oocytes

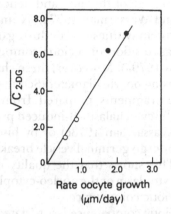

Figure 7.3. Relationship between extent of intercellular communication between follicle cells and oocytes and rate of growth *in vitro*. The experiments were performed as described in Brower and Schultz (1982a); 2-deoxyglucose (2-DG) was used as a marker for intercellular communication. Because the extent of communication is a function of the number of attached follicle cells, and this in turn is a function of the surface area (Brower and Schultz 1982a), the data were plotted as the square root of the coupling index to obtain the linear relationship. Filled circles: Oocytes grown using the Eppig culture system. Open circles: Oocytes grown using the Bachvarova culture system. [Reproduced from Schultz (1985) by permission of Academic Press, Inc.]

results in oocyte growth (Herlands and Schultz 1984). Reestablishment of communication was determined by a pulse-chase assay similar to that originally used to demonstrate the existence of metabolic coupling (Pitts and Simms 1977). On the other hand, oocytes incubated on a monolayer of communication-incompetent mouse L cells neither establish communication nor grow. Thus, it is very likely that this type of heterologous intercellular communication provides a nutritional role in oocyte growth. It would be interesting to determine if other cell types can establish intercellular communication with the oocyte and, if so, promote oocyte growth or if this is a unique property of follicle cells.

B. Acquisition of meiotic competence

In all mammals examined to date, oocyte growth is associated with the acquisition of meiotic competence at a specific stage of oocyte growth. Oocytes isolated from mice less than 15 days of age are about 60 μm in diameter and fail to resume meiosis when placed in a suitable culture medium (Sorensen and Wassarman 1976), whereas oocytes isolated from mice 15 days of age or older are greater than 60 μm in diameter and can resume meiosis. In addition, the frequency of resumption of meiosis increases with increasing age of the mice and hence with increasing oocyte diameter (Schultz and Wassarman 1977a). Numerous changes in the qualitative pattern of protein synthesis occur during oocyte growth, and these changes precede acquisition of meiotic competence (Schultz, Letourneau, and Wassarman 1979a). Moreover, these changes become more pronounced with increasing oocyte diameter (Schultz et al. 1979a). Experiments using nucleate fragments prepared from fully grown, meiotically competent oocytes by cytochalasin-B-induced pseudo-cleavage (Schultz, Letourneau, and Wassarman 1978c) or by bisection (Balakier and Czolowska 1977) still undergo germinal vesicle breakdown (GVBD) and polar-body emission. This shows that the quality of the cytoplasm, rather than the amount of cytoplasm or the nucleo-cytoplasmic ratio, is involved in acquisition of meiotic competence.

Although in vivo, acquisition of meiotic competence is associated with oocyte growth, results of recent experiments suggest that acquisition of meiotic competence can be dissociated from oocyte growth (Canipari et al. 1984). After 8 days of culture in a conditioned medium that does not support oocyte growth, but does support oocyte viability, oocytes obtained from 7-day-old juvenile mice under go GVBD and chromosome condensation. Similarly, oocytes obtained from 10-day-old juvenile mice acquire the ability to undergo GVBD after 5 days of in vitro culture. Results of these experiments indicate that a constant amount of time, which totals 15 days of in vivo growth or in vitro culture, is required before oocytes can resume meiosis. Thus, initiation of oocyte growth may trigger an internal clock, which is independent of or can be dissociated from oocyte growth in vitro, that sets in motion the program for acquisition of meiotic competence. It would be interesting to determine if the changes in pattern of protein synthesis observed during oocyte growth also occur under these conditions of in vitro culture that do not foster oocyte growth, but allow resumption of meiosis. Results of such experiments may suggest that the synthesis of these polypeptides is responsible, at least in part, for acquisition of meiotic competence.

III. Macromolecular synthesis during oocyte growth and maturation

The ability to obtain oocytes at discrete stages of growth (Mangia and Epstein 1975; Eppig 1977) has permitted analysis of various biochemical and morphological aspects of oocyte differentiation during the growth phase. The following discussion will focus on RNA and protein synthesis and the accumulation of specific molecular species where information is available.

A. RNA synthesis and accumulation during oocyte growth

The fully grown mouse oocyte contains about 0.4−0.6 ng total RNA, as determined by independent chemical methods (Olds, Stern, and Biggers 1973; Bachvarova 1974; Sternlicht and Schultz 1981). The kinetics of accumulation of RNA are biphasic with respect to oocyte volume; oocytes accumulate essentially all of their RNA by the time they have reached about 65−70 percent of their final volume (Sternlicht and Schultz 1981). Similar kinetics are also observed for accumulation of rRNA and poly(A)-containing RNA (i.e., presumptive mRNA) (Sternlicht and Schultz 1981; Kaplan, Abreu, and Bachvarova 1982). For further details of RNA synthesis and accumulation during oogenesis, consult the review by Bachvarova (1985).

RNA synthesized during in vivo growth is very stable, because about 80 percent of the acid-insoluble radioactive material two days after labeling is retained as acid-insoluble radioactive material in oocytes ovulated 10−20 days later (Bachvarova 1974; Jahn, Baron, and Bachvarova 1976). In these ovulated oocytes, 65−70 percent of the RNA is rRNA, 18 percent tRNA, and 15 percent heterogeneous RNA (hnRNA)(Bachvarova 1974; Jahn et al. 1976; Brower et al. 1981). About 10 percent of the RNA in ovulated oocytes behaves as poly(A)-containing RNA after chromatography on poly(U)-Sepharose. This fraction is essentially constant during the period of oocyte growth 7 to 19 days prior to ovulation (Bachvarova and DeLeon 1980; Brower et al. 1981). The proportion of polyadenylated RNA in the mammalian oocyte is quite high when compared with oocytes of lower species, in which 1−2 percent of the total RNA is polyadenylated (Davidson 1976).

The synthesis and turnover of hnRNA classes in growing mouse oocytes have also been examined using the in vitro culture sytems (Bachvarova 1981). The results of these studies indicate that the steady-state amount and half-life of hnRNA >36S and <36S are 12 pg and 20 min, respectively, and 1.5 pg and 20 min, respectively. Pulse-chase experiments, using the in vitro culture systems, indicate that poly(A)$^-$ RNA, which is predominantly rRNA and tRNA, is essentially stable during the growth phase (t½ about 28 days) and that poly (A)$^+$ RNA is also quite stable (t½ about 10−14 days) (Bachvarova 1981; Brower et al. 1981). Under conditions that do not support oocyte growth, however, the RNA is rapidly degraded (Brower et al. 1981). Furthermore, the subpolysomal polyadenylated RNA, which is apparently not translated, is completely stable, whereas translated polyadenylated RNA turns over with a half-life of about six days (DeLeon, Johnson, and Bachvarova 1983). The marked stability of many types of RNA species during the growth period is a common property of oocytes of various species, including *Xenopus laevis* (Ford, Maethieson, and Rosbash 1977). This stability probably reflects a strategy to optimize accumulation of substances that will constitute the maternal contribution to early development. The synthesis of 28S, 18S, 5S, and 4S RNA is coordinate during oocyte growth (Brower et al. 1981; Boreen, Gizang, and Schultz 1983), which is in contrast to the situation in *X. laevis* oocytes (Davidson 1976). The concentration of U1 RNA, which is thought to be involved in RNA splicing, is similar to that in somatic cells (Kaplan, Jelinek, and Bachvarova 1985).

Although the polyadenylated RNA present in mouse oocytes at various stages of growth is quite heterodisperse, it sediments predominantly between 18S and 28S (Bachvarova and DeLeon 1980; Brower et al. 1981). The majority of this mRNA present in ovulated oocytes is capped (Schultz, Clough, and Johnson 1980).

During growth of rodent oocytes, fibrillar or lamellar structures accumulate and become quite abundant in the cytoplasm (Weakley 1967, 1968; Calarco and Brown 1969; Zamboni 1970; Burkholder, Comings, and Okada 1971; Kang and Anderson 1975; King and Tibbits 1977; Wassarman and Josefowicz 1978). Mouse, but not rat or hamster, oocytes examined by whole-mount electron microscopy show lattice-like structures that contain particles whose diameter is similar to that of ribosomes. Treatment of these preparations with either RNase or trypsin, but not DNase, destroys the integrity of the lattices, as do acid or urea treatments, leading to the proposal that the lattices are a storage form of inactive ribosomes embedded in a proteinaceous matrix (Burkholder et al. 1971).

Results of a number of experiments are consistent with this proposal. Ribosomes obtained from ovulated mouse oocytes appear inactive, because only 13 percent of the mouse oocyte ribosomes can form high-salt-stable complexes with poly(U) (Bachvarova and DeLeon 1977). In addition, 40–70 percent of radiolabeled oocyte RNA can be sedimented at low centrifugal force (9,000 × g), and examination of the material in the pellet by electron microscopy reveals the presence of structures that could be disordered lattices containing ribosomal particles (Bachvarova, DeLeon, and Spiegelman 1981; Brower and Schultz 1982b). In addition, treatments known to disrupt protein structure reduce the amount of radioactive material sedimenting at low centrifugal forces (Bacharova et al. 1981; Brower and Schultz 1982b). Analysis of this pellet reveals the presence of rRNA, tRNA, and hnRNA and shows that the RNA is associated with a large mass of protein, because the buoyant density is significantly lower than that of ribosomes (Bachvarova et al. 1981; Brower and Schultz 1982b). The stability of this sedimentable RNA appears similar to that of the RNA that remains in the supernatant, and in vitro translation of RNA in either the pellet or the supernatant reveals the same size distribution of synthesized polypeptides (Brower and Schultz 1982b). Although a number of these properties are consistent with the idea that the lattices are composed of inactive ribosomes complexed with mRNA, comparisons of the numbers of ribosomes by electron microscopy, morphometry, the amount of RNA by biochemical methods, staining properties, and sensitivity to alkalai treatment do not support this conclusion (Piko and Clegg 1982). Because the lattice-like structures have not been purified and characterized, the question of their composition and functions remains open.

B. Protein synthesis during oocyte growth

i. Absolute rates of protein synthesis during oocyte growth. Similar results have been obtained using three different methods to determine the absolute rate of protein synthesis during oocyte growth; two approaches use [^{35}S]methionine and one uses [^{3}H]leucine. The rate of incorporation of [^{35}S]methionine into radiolabeled acid-insoluble material was converted to absolute rates of protein synthesis by determining the specific activity of the oocyte methionine pool (Schultz, La Marca, and Wassarman 1978b; Schultz et al. 1979a). In one instance, the specific activity was determined by using [^{3}H]fluorodinitrobenzene (FDNB), which yields the specific activity of the total methionine pool and also permits calculation of the size of the methionine pool. In the other instance, the endogenous methionine pool was differentially expanded, and the specific activity and the size of the kinetic pool (i.e., the pool that participates in protein synthesis) were determined by solving a pair of simultaneous equations. Results of these experiments indicated that the total and kinetic methionine pools were equivalent; thus, there is no compartmentalization, and the entire intracellular free methionine pool participates in protein synthesis. The third method to determine absolute rates of protein synthesis is based on expanding the endogenous leucine pool by increasing the concentration of [^{3}H]leucine in the medium, so that the specific activity of the endogenous leucine pool is essentially equal to that in the medium (Canipari, Pietrolucci, and Mangia 1979). When this is achieved, the rate of incorporation of [^{3}H]leucine into acid-insoluble material becomes constant even though the external leucine concentration is increasing.

The rate of protein synthesis increases from 1.1 pg/h in nongrowing oocytes to about 45 pg/h in the fully grown oocyte (Figure 7.4), which corresponds to a 40-fold increase (compared with the 350-fold increase in oocyte volume). In contrast, the size of the methionine pool increases in direct proportion to oocyte volume, and thus the concentration of methionine remains constant during oocyte growth.

Can oocytes synthesize all of their protein during the growth phase? Calculations using the measured absolute rates of protein synthesis and assuming no protein turnover suggest that denuded oocytes could syn-

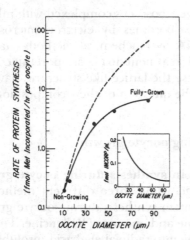

Figure 7.4. Relationship between diameters of growing mouse oocytes and absolute rates of protein synthesis. [Reproduced from Schultz et al. (1979a) by permission of Academic Press, Inc.]

thesize about half of their protein (Schultz et al. 1979a). The actual fraction is probably smaller, because in mid-growth-phase oocytes, about 40 percent of the newly synthesized protein turns over with a half-life of about 11 h (Kaplan et al. 1982). Absolute rates of oocyte protein synthesis in intact three-layered follicles, which contain oocytes about 45–50 μm in diameter, are about 30 percent greater than that of denuded oocytes (Salustri and Martinozzi 1983). Even taking this into consideration, calculations indicate that the oocyte cannot synthesize all of its protein. Although Bachvarova (1985) has proposed that the growing oocyte could synthesize all of its protein, based on the number of ribosomes in polysomes and allowing for 40 percent turnover, the range of rates of peptide elongation is so wide that such calculations are not reliable. Thus, part of the oocyte protein may be derived by uptake of serum proteins, which has been demonstrated in vivo (Glass 1971).

ii. Absolute rate of protein synthesis during maturation. The absolute rates of protein synthesis measured by the methods described decrease about 30 percent during meiotic maturation (Schultz et al. 1978b) (Table 7.2). This contrasts with the *Xenopus* oocyte, in which a twofold increase in the rate of protein synthesis is observed (Wasserman, Richter, and Smith 1982).

The decrease in the rate of protein synthesis is consistent with turnover of oocyte RNA that is initiated during the resumption of meiosis. A 50 percent decline in the amount of polyadenylated RNA occurs during oocyte maturation (Brower et al. 1981; Bachvarova et al. 1985). In addi-

Table 7.2. *Absolute rates of total protein synthesis during maturation and early embryogenesis*

Stage of maturation/ embryogenesis	Method	Rate (pg/h)
GV	FDNB	42.9
GV	Kinetic (simultaneous equations)	41.8
GV	Kinetic (pool expansion)	48.0
GVBD−metaphase I	FDNB	36.3
Met I−Met II	FDNB	31.1
Met II	FDNB	33.0
1-cell	FDNB	45.1
8-cell	FDNB	51.2

Source: Data from Schultz et al. (1978b, 1979b) and Canipari et al. (1979).

tion, the total amount of RNA decreases about 20 percent during meiotic maturation, and thus it is likely that ribosomes and tRNA are degraded (Bachvarova et al. 1985). Consistent with this is the observation that the fraction of ribosomes in polysomes in ovulated oocytes decreases about 25 percent when compared with unovulated oocytes (De Leon et al. 1983), which is similar to the decrease in the rate of protein synthesis that occurs during maturation.

iii. Changes in patterns of protein synthesis during oocyte growth and meiotic maturation. Mouse oocytes accumulate total protein as a linear function of oocyte volume (Schultz and Wassarman 1977a). The fully grown oocyte contains about 30 ng of protein (Schultz and Wassarman 1977a), of which the zona pellucida accounts for about 6 ng (Schultz, Bleil, and Wassarman 1978a). Differential gene expression occurs in the differentiating and growing oocyte, as evidenced by the spectrum of polypeptides synthesized. Two-dimensional-gel electrophoresis of [^{35}S]methionine-labeled protein obtained from oocytes at discrete stages of growth reveals that although the overall patterns of protein synthesis are quite similar, changes in the relative rates of synthesis of specific polypeptides do occur, and, moreover, these changes become more pronounced with increasing oocyte diameter (Schultz et al. 1979a). In addition, changes in the pattern of protein synthesis in porcine oocytes obtained from antral follicles at different stages of development correlate with the ability to resume meiosis (McGaughey, Montgomery, and Richter 1979), although a causal correlation has not been established.

Dramatic changes in the pattern of protein synthesis occur during or after GVBD in oocytes of different species, including pig (McGaughey and Van Blerkom 1977), rabbit (Van Blerkom and McGaughey 1978), and mouse (Schultz and Wassarman 1977b). These changes are not dependent on nuclear progression, because drugs that inhibit nuclear progression at specific stages without inhibiting GVBD do not inhibit the changes in the pattern of protein synthesis in mouse oocytes (Wassarman, Josefowicz, and Letourneau 1976; Schultz and Wassarman 1977b).

Although the physiological function of these changes is not known, some of them appear required for nuclear progression through polar-body emission. Oocytes incubated in the continuous presence of protein synthesis inhibitors undergo GVBD, but arrest at the circular bivalent stage. When protein synthesis is allowed to occur for a brief period subsequent to GVBD (during a time when the changes in the patterns of protein synthesis occur) and then halted with a protein synthesis inhibitor, these oocytes emit polar bodies (Wassarman et al. 1979). Thus, protein synthesis during specific periods of maturation is both necessary and sufficient for polar-body formation. In addition, the changes in the pattern of protein synthesis that occur during maturation persist in the one-cell embryo, suggesting that the "program" for early embryogenesis is initiated during meiotic maturation.

iv. Synthesis of specific proteins during oogenesis. Data concerning the synthesis and accumulation of specific proteins during oocyte growth are quite limited and have, for technical reasons, focused for the most part on either abundant proteins or enzymes that are readily assayed in mouse oocytes.

Zona pellucida proteins. The synthesis and biological function of the zona proteins in sperm binding and inducing the acrosome reaction have been best studied in the mouse and reviewed recently by Wassarman and associates (Wassarman, Florman, and Greve 1984). Three glycoproteins, ZP1, ZP2, and ZP3, with molecular masses of 200,000, 120,000, and 80,000 daltons, respectively, compose the zona pellucida (Bleil and Wassarman 1980a). Synthesis of the zona proteins constitutes a major portion of protein synthesis in the growing oocyte, and the change in the structure of the oocyte Golgi apparatus is consistent with it becoming increasingly active in secretory processes. The rate of synthesis of the zona proteins in fully grown oocytes is essentially undetectable.

Competitive inhibition studies indicate that ZP3 possesses all of the sperm receptor activity present in the zona pellucida (Bleil and Wassarman 1980b), and furthermore, ZP3 can induce the acrosome reaction (Bleil and Wassarman 1983). Subsequent studies employing small glycopeptides obtained from ZP3 indicated the sperm receptor and acrosome induction functions are likely to depend on different parts of the glycoprotein, because ZP3 glycopeptides could inhibit sperm binding but not induce the acrosome reaction (Florman, Becktol, and Wassarman 1984).

The carbohydrate moiety presumably is involved in sperm receptor activity. Mild alkali treatment of ZP3, which selectively removes O-linked oligosaccharides, inactivates the sperm receptor activity, whereas sperm binding is unaltered by treatments that remove N-linked oligosaccharides (Florman and Wassarman 1985). These results are consistent with the observation that galactosyl transferase associated with the sperm surface mediates sperm binding to the zona pellucida (Shur and Hall 1982).

ZP2 is the major zona protein, and gel-electrophoretic analysis indicates that it is modified in embryos (Bleil and Wassarman 1981); under reducing conditions, ZP2 isolated from embryos has an apparent molecular mass of 90,000 daltons, whereas ZP2 obtained from oocytes or unfertilized eggs has an apparent molecular mass of 120,000 daltons. This modification probably results from proteolysis, the fragments being held together by disulfide bonds. The modification is also observed in eggs activated by the calcium ionophore A23187 and may play a role in the block to polyspermy.

The zona proteins are synthesized and secreted by the oocyte, not to any significant extent by the follicle cells (Bleil and Wassarman 1980c). The zona pellucida in other species is also composed of three glycoproteins. There is a report, however, that follicle cells also can synthesize zona proteins in the rabbit (Wolgemuth et al. 1984); ovarian sections treated with antibodies to zona proteins, which were then localized by indirect immunocytochemical procedures, revealed antigenic material present in follicle cells of growing follicles. It was not determined, however, if follicle cells synthesize the zona proteins.

Electron-microscopic examination of zonae pellucidae partially dissolved by either low pH or protease treatment, suggests that the three zona proteins are organized in long filaments composed of "beads on a string"; the beads are about 9.5 nm in diameter and are located about every 17 nm (Greve and Wassarman 1985). ZP1 may connect the filaments to produce the three-dimensional structure, because partial digestion of zonae with chymotrypsin causes proteolysis of ZP1, but not of ZP2 or ZP3, and disrupts the filamentous structure.

Tubulin. The absolute rates of tubulin synthesis were determined from the rate of incorporation of $[^{35}S]$methionine into tubulin subunits resolved on two-dimensional gels and from the specific activity of the endogenous radiolabeled methionine pool. Although the absolute rates of synthesis of α and β tubulin increase during oocyte growth, the percentage of total protein synthesis devoted to tubulin synthesis decreases (Schultz et al. 1979a, 1979b) (Table 7.3). The rate of tubulin synthesis decreases about 39 percent during meiotic maturation and increases

Table 7.3. *Absolute rates of synthesis of specific proteins during oocyte growth, maturation, and embryogenesis*

Protein	Oocyte size/ stage of maturation/embryogenesis	Rate (pg/h)	Percentage total protein synthesis
Tubulin[a]	45−50 μm diameter	0.40	2.0
	50−60 μm diameter	0.45	1.8
	Fully grown	0.61	1.5
	GVBD/Met I	0.45	1.2
	Met II	0.36	1.1
	1-cell	0.60	1.3
	8-cell	0.66	1.3
Histone H4[b]	60 μm diameter	0.038	0.07
	Fully grown	0.043	0.05
	Met II	0.026	0.04
Ribosomal[c]	55 μm diameter	0.42	1.5
	Fully grown	0.62	1.5
	Met II	0.37	1.1
	8-cell	4.17	8.1
Lactate dehydrogenase[d]	30−40 μm diameter	0.1	0.7
	40−50 μm diameter	0.3	1.5
	50−60 μm diameter	0.5	1.9
	Fully grown	0.7	1.8
	Met II	0.1	0.3
	1-cell	0.04	<0.1
	8-cell	0.04	<0.1

Note:
[a]Data from Schultz et al. (1979a, 1979b).
[b]Data from Wassarman and Mrozak (1981).
[c]Data from La Marca and Wassarman (1979).
[d]Data from Cascio and Wassarman (1982).

about 80 percent by the eight-cell embryo stage of development (Abreu and Brinster 1978; Schultz et al. 1979b). These changes parallel changes in the rate of total protein synthesis during these developmental stages (Schultz et al. 1979a, 1979b) (Table 7.3).

Actin. Although the absolute rates of actin synthesis have been determined during embryogenesis (Abreu and Brinster 1978), the rate of actin synthesis has not been determined during oocyte growth. The relative rate of synthesis of actin, however, is quite significant, because actin is a major spot on fluorograms of radiolabeled oocyte proteins subjected to two-dimensional-gel electrophoresis (Kaplan et al. 1982). The relative rate of actin synthesis decreases during maturation, remains low during the 1-cell embryo, but then becomes a major spot by the 2- to 4-cell embryo stage (Abreu and Brinster 1978; Howe and Solter 1979; Cullen, Emigholz, and Monahan 1980).

Although the amount of actin mRNA, as determined by Northern analysis, remains constant during maturation, the actin mRNA is deadenylated (Bachvarova et al. 1985). Northern analysis of actin mRNA obtained from oocytes and ovulated eggs reveals that the mRNA from the egg has a lower molecular weight. Hydridization of mRNA from oocytes and eggs with oligo(dT), followed by RNase H digestion, which cleaves poly(dT/A) sequences, reduces the molecular weight of actin mRNA obtained from oocytes to a size similar to that of actin mRNA obtained from eggs, which itself is unaltered by RNase H treatment.

Histone. The absolute rate of synthesis of H4 decreases about 40 percent during maturation (Table 7.3) (Wassarman and Mrozak 1981). About 2 percent of total newly synthesized protein is associated with the germinal vessicle (GV), whose volume is about 2 percent of the total oocyte volume (Wassarman and Mrozak 1981). Pulse-chase experiments demonstrated that greater than 80 percent of the radioactivity initially associated with isolated GVs is retained. Of this protein, H4 is highly concentrated in the GV; about 50 percent of the radiolabeled H4 is present in the GV, whereas only 1 percent of radiolabeled tubulin becomes associated with GV. Similar results were obtained in *Xenopus* oocytes (Adamson and Woodland 1977).

Results of these and other experiments in mouse oocytes also suggest that the total amount of histones (i.e., H2a, H2b, H3, and H4) could compose about 10 percent of the total protein present in the GV. If the amount of protein in the GV is proportional to its fractional oocyte volume, then the GV will contain about 500 pg of protein, or about 50 pg of histone. The somatic mouse nucleus contains about 6 pg of DNA, and because chromatin contains equal masses of histone and DNA, the amount of histone stored in the oocyte GV could support two to three cell divisions. This contrasts with the *Xenopus* oocyte, which contains enough histone to support development until the late blastula stage (20,000 cells).

Ribosomal proteins. Absolute synthesis rates have been measured in mouse oocytes and early embryos for 11 proteins associated with the large ribosomal subunit and one protein associated with the small subunit (La Marca and Wassarman 1979). Assuming that the rates of synthesis of these proteins are representative of the 70 ribosomal proteins, ribosomal protein synthesis increases during oocyte growth and constitutes about 1.5 percent of total protein synthesis (Table 7.3). This rate decreases about 40 percent during maturation, so that ribosomal protein synthesis composes about 1.1 percent of total protein synthesis in the unfertilized egg, and then increases during embryogenesis so that in the 8-cell embryo, ribosomal protein synthesis has increased over 11-fold and constitutes about 8 percent of total protein synthesis.

Like histones, ribosomal proteins accumulate rapidly in the germinal vesicle and may constitute 25 percent of the total and newly synthesized proteins present in the GV (La Marca and Wassarman 1984). In addition, synthesis and accumulation of ribosomal proteins in the GV occur in the presence of transcriptional inhibitors, which indicates that synthesis of ribosomal proteins is not coupled to rRNA synthesis.

Although ribosomal proteins are present in ribosomes in equimolar amounts, ribosomal proteins are synthesized throughout oogenesis in nonequimolar amounts (La Marca and Wassarman 1979). The oocyte appears to compensate for this difference by concentrating to a greater extent in the GV ribosomal proteins that are synthesized at lower rates. This results in essentially equimolar concentrations of the various ribosomal proteins in the GV, where ribosomes are assembled. During embryogenesis, ribosomal proteins are synthesized in more nearly equimolar amounts (La Marca and Wassarman 1979).

Mitochondrial proteins. Although much is known about changes in mitochondrial morphology during oocyte growth, very little is known about mitochondrial protein synthesis during this period. The mitochondria undergo a dramatic change in morphology and increase in number during the growth phase – the fully grown mouse oocyte contains about 100,000 mitochondria (Piko and Matsumoto 1976). Early in the growth phase, mitochondria are elongated and possess transverse cristae, whereas at later stages of oocyte growth the mitochondria become round, and the cristae become concentrically arranged. A gradual transition to the normal morphology between the 4-cell and 8-cell stages correlates with the ability to metabolize glucose.

Synthesis of mitochondrial proteins was examined by incubating oocytes in the presence of [^{35}S]methionine and emetine, which inhibits cytoplasmic protein synthesis but not mitochondrial protein synthesis (Cascio and Wassarman 1981). Mitochondrial protein synthesis accounted for about 1−2 percent that of total oocyte protein synthesis. Following isolation of oocyte mitochondria, the radiolabeled proteins were displayed on polyacrylamide gels, and 5−10 polypeptides could be visualized. Chloramphenicol, which inhibits mitochondrial protein synthesis, inhibited synthesis of these polypeptides. Polypeptides of molecular masses 40 kd, 33 kd, and 22 kd could correspond to the three largest subunits of cytochrome c oxidase, the band at 31 kd with cytochrome c reductase, and the 30 kd, 20 kd, and 8 kd polypeptides with subunits of the oligomycin-sensitive ATPase. Because similar results were obtained with 8-cell embryos, it is unlikely that these polypeptides initiate the changes in mitochondrial morphology that occur during development from the oocyte to the 8-cell embryo.

RNA polymerases. The activities of RNA polymerases I and II during oocyte growth have been measured in transcriptionally active complexes by the enzymes' nuclear or nucleolar location and by insensitivity or sensitivity to α-amanitin, respectively (Moore 1978). The results of these studies indicate that both activities increase with oocyte growth; oocytes 45−85 percent of the volume of fully-grown oocytes and present in stage-4 follicles possess nearly maximal levels of RNA polymerase activity (Moore, Lintern-Moore, and Peters 1974; Moore and Lintern-Moore 1978). Both RNA polymerase I and II activities decrease markedly as the oocyte reaches its final volume, the lowest level of activity being present in fully grown oocytes present in antral follicles (Moore and Lintern-Moore 1978). The changes in incorporation of [^{3}H]uridine into oocytes in vivo correlate very well with the changes in RNA polymerase activity detected by these assays (Moore et al. 1974; Rodman and Bachvarova 1976). In addition, the kinetics of rRNA and poly(A)-containing RNA accumulation, which are biphasic during oocyte growth, are consistent with the levels of endogenous RNA polymerase I and II observed during oocyte growth.

Glucose-6-phosphate dehydrogenase. Mouse oocytes contain high levels of glucose-6-phosphate dehydrogenase activity that increases during oocyte growth; the specific activity, expressed as units of activity per picoliter of oocyte volume, however, remains essentially constant during growth (Mangia and Epstein 1975). The high level of activity may reflect a need for the oocyte to generate reducing equivalents in the form of NADPH that are required for oocyte anabolic pathways.

Lactate dehydrogenase. Mouse oocytes contain isozyme 1 of lactate dehydrogenase (LDH) whose activity increases about 10-fold during growth (Mangia and Epstein 1975). LDH synthesis constitutes about 2−5 percent that of total protein synthesis in the fully grown oocyte, and the absolute rate of synthesis increases about sevenfold during oocyte growth (Table 7.3) (Mangia, Erickson, and Epstein 1976; Cascio and Wassarman 1982). The high level of LDH activity is probably involved in oocyte energy metabolism, because oocytes cannot utilize glucose as an energy source but can use pyruvate (Biggers, Whittingham, and Donahue 1967; Donahue and Stern 1968; Biggers 1971). Lactate, formed by glucose catabolism in the follicle cells, and either (1) secreted into the follicular milieu and then taken up by the oocyte or (2) transferred directly to the oocyte via the coupling pathway, could then be oxidized by the oocyte. A likely example of this is that whereas the rate of protein synthesis in denuded oocytes is stimulated when pyruvate is added to the medium, as compared with glucose, the rate of protein synthesis in follicle-enclosed oocytes, which is greater than that of denuded oocytes, is about the same using either glucose or pyruvate as an energy source (Salustri and Martinozzi 1982).

The level of LDH synthesis decreases about sevenfold during oocyte maturation and decreases further during preimplantation development

(Table 7.3) (Mangia et al. 1976; Cascio and Wassarman 1982). The decrease in LDH activity follows this decrease in LDH synthesis by about two or three days (Brinster 1965; Mangia et al. 1976). In vitro translation of RNA isolated from oocytes shows high levels of LDH synthesis, whereas mRNA obtained from one-cell embryos supports much lower levels. It would be interesting to know if this decrease is due to decreased levels and/or deadenylation of LDH mRNA.

Creatine kinase. Mouse oocytes contain the brain isozyme form of creatine kinase, whose activity increases linearly in relation to oocyte volume (Iyengar et al. 1983). The oocyte is one of the richest known sources of this enzyme; creatine kinase probably constitutes about 1–3 percent of the total oocyte protein. The activity remains constant during maturation and then increases during cleaveage to the 8-cell stage. The activity declines markedly during the 8-cell-to-blastocyst transition and correlates negatively with the ability of the embryo to metabolize glucose (Biggers 1971). The deficiency in glucose metabolism at earlier stages is not in the ability of either the oocyte or embryo to transport or phosphorylate glucose, but rather appears to be an inhibition of phosphofructokinase activity (Barbehenn, Wales, and Lowry 1974) (see Chapter 9).

A possible physiological role for creatine kinase is to maintain the high ATP/ADP ratio that is essential for biosynthetic activities of the oocyte and embryo during a period when it cannot utilize glucose as an energy source. The brain isozyme form is more suited than the muscle form to catalyze the production of ATP from creatine phosphate and ADP, because the brain isozyme has a lower K_m for both ADP and creatine phosphate. The decline in creatine kinase activity, correlated with the embryo's ability to utilize glucose, may reflect a lesser requirement for this shuttle pathway to serve as an ATP buffer.

IV. Regulation of oocyte maturation

The first part of this discussion will focus on events occurring in the oocyte that are involved in resumption of meiosis. The latter part of this section will discuss the various models that have been proposed for gonadotropin-induced resumption of meiosis. The discusson will focus mainly on results obtained with mouse and rat oocytes, because these species have been the most frequently studied.

It has been known for over 50 years that meiotically competent oocytes removed from antral follicles will spontaneously resume maturation in vitro (Pincus and Enzmann 1935). Oocyte maturation in vitro is physiologically relevant, because mouse oocytes matured and fertilized in vitro and then tranferred to foster mothers give rise to viable offspring at a similar frequency as oocytes matured in vivo and fertilized in vitro (Schroeder and Eppig 1984). In addition, both sheep and rabbit oocytes that are matured and fertilized in vitro undergo normal preimplantation development at a high frequency (Moor and Trounson 1977; Van Blerkom and McGaughey 1978). Thus, oocyte maturation in vitro provides a suitable and valid system to study certain aspects of regulation of resumption of meiosis.

A. Roles of cAMP and protein phosphorylation in maintenance of meiotic arrest and resumption of meiosis

It should be pointed out that although resumption of meiosis is unique to oocytes and is requisite for fertilization and normal development, the results of the studies to be described later also provide information regarding regulation of the G_2-to-M transition of the mitotic cell cycle. Oocytes are naturally arrested in a G_2-like state, and resumption of meiosis entails a G_2-to-M transition. Cyclic AMP is implicated in promoting arrest of somatic cells in G_2, and a decrease in cAMP is correlated with entry into and transit of M (Friedman 1982). As will be discussed later, oocyte cAMP is involved in maintenance of meiotic arrest, and a drop in oocyte cAMP occurs prior to GVBD. This suggests a common role for cAMP in both meiotic and mitotic cells in regulating this aspect of the cell cycle.

It has been proposed that cAMP is involved in maintenance of meiotic arrest, because membrane-permeable cAMP analogs, such as dibutyryl cAMP (db cAMP) and 8-bromo cAMP, but not the corresponding cGMP analogs, and phosphodiesterase inhibitors, such as theophylline and 3-isobutyl-1-methylxanthine (IBMX), reversibly inhibit maturation in vitro of denuded oocytes derived from either mice, rats, or pigs (Cho, Stern, and Biggers 1974; Wassarman et al. 1976; Magnusson and Hillensjo 1977; Dekel and Beers 1978; Rice and McGaughey 1981). These results have led to a model for regulation of oocyte maturation by the concentration of oocyte cAMP in the oocyte (Schultz and Wassarman 1977a; Dekel and Beers 1978).

Results of several additional lines of experimentation strengthen the conclusion drawn from the cAMP analog experiments that a drop in oocyte cAMP concentration is required for resumption of meiosis. Forskolin, a reversible activator of adenylate cyclase in every mammalian system tested to date (Seamon, Padgett, and Daly 1981), induces dose-dependent increases in both mouse and hamster oocyte cAMP levels and inhibits GVBD (Schultz, Montgomery, and Belanoff 1983a; Urner et al. 1983; Sato and Koide 1984; Racowsky 1985a). Forskolin also stimulates cAMP synthesis and inhibits GVBD in rat oocytes (Olsiewski and Beers 1983).* In addition, GVBD is transiently inhibited in mouse oocytes

---

* However, it was reported that forskolin (0.2–100 μM) did not inhibit GVBD when the oocytes were examined after 4 h of culture or increased cAMP levels (Racowsky 1984). Dekel, Aberdam, and Sherizly (1984) did not observe any inhibition of maturation by 300 μM forskolin, whereas Eckholm and associates (1984) reported a very transient inhibition induced by 100 μM forskolin. The basis for these differences is unresolved.

microinjected with cAMP, but not in oocytes microinjected with 2'-deoxy cAMP, which does not activate cAMP-dependent protein kinase (Bornslaeger, Mattei, and Schultz 1986). These results also suggest that a cAMP-dependent protein kinase mediates the inhibitory effect of cAMP in maintenance of meiotic arrest, as will be described later. The transient nature of inhibition of GVBD is not surprising, because oocytes contain a high level of phosphodiesterase activity (Bornslaeger, Wilde, and Schultz 1984).

Although the forskolin-induced increase in cAMP could be due to forskolin activation of adenylate cyclase in cumulus cell remnants that remain associated with the oocyte following denudation, the observation that the forskolin-induced elevation of oocyte cAMP in zona-free oocytes is similar to that in zona-intact oocytes strongly suggests that oocytes possess adenylate cyclase (Bornslaeger and Schultz 1985a). The existence and function of regulatory subunits of oocyte adenylate cyclase, however, are still unclear. The regulation of adenylate cyclase by either cholera toxin or pertussis toxin is thought to be mediated by the toxin's ability to catalyze ADP ribosylation of either the G_s or G_i subunit of the cyclase, respectively. IBMX and cholera toxin have been reported to increase oocyte cAMP and inhibit GVBD (Vivarelli et al. 1983). On the other hand, other investigators have not been able to ascertain either an inhibitory effect of cholera toxin on GVBD in rat (Dekel and Beers 1980) or mouse (E. A. Bornslaeger and R. M. Schultz, unpublished results) or an increase in mouse oocyte cAMP level. Results of recent experiments indicate that there is a synergistic effect of forskolin and cholera toxin in elevating mouse oocyte cAMP levels, which suggests the presence of some functional G_s (E. A. Bornslaeger and R. M. Schultz, unpublished results).

The model for cAMP regulation of meiotic maturation predicts that a drop in oocyte cAMP is involved in resumption of meiosis and that this drop should occur prior to GVBD. A decrease in mouse oocyte cAMP is associated with oocyte maturation both in vitro (Schultz et al. 1983a; Vivarelli et al. 1983) and in vivo (Schultz et al. 1983a). Moreover, this decrease occurs during a period of time in which oocytes become committed to resume meiosis, as defined by the observation that these oocytes undergo GVBD when returned to medium containing either IBMX or db cAMP (Schultz et al. 1983a). A similar result has also been obtained regarding the time that rat oocytes become committed to resume meiosis (Dekel and Beers 1980). It is likely that the decrease in oocyte cAMP is causally related to resumption of meiosis, because treatments that block this decrease, such as IBMX, also inhibit maturation (Schultz et al. 1983a), and microinjection of purified calmodulin-modulated sheep brain phosphodiesterase (PDE) into oocytes incubated in IBMX results in resumption of meiosis (Bornslaeger et al. 1986). These and other observations are summarized in the model shown in Figure 7.5.

Maintenance of Meiotic Arrest

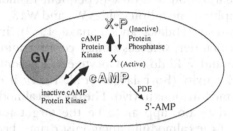

Resumption of Meiotic Maturation

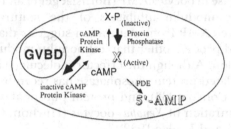

Figure 7.5. Model for maintenance of meiotic arrest and resumption of meiosis. The bold symbols depict the predominating form of a hypothetical maturation-regulating protein X, as well as cAMP or 5'-AMP. The model postulates that X promotes GVBD. A cAMP-dependent protein kinase converts X to a phosphorylated form X-P, which is inactive, and a protein phosphatase catalyzes the dephosphorylation of X-P. Agents that elevate oocyte cAMP levels would activate a cAMP-dependent protein kinase to generate more X-P and thus inhibit GVBD. A decrease in oocyte cAMP would initiate maturation, because this would reduce the level of cAMP-dependent protein kinase activity. This would in turn shift the equilibrium between X-P and X to X, which would ultimately lead to GVBD.

A role for PDE in oocyte maturation is suggested by the observation that IBMX inhibits both the maturation-associated decrease in oocyte cAMP and GVBD (Schultz et al. 1983a). This conclusion is buttressed by the observation that a dose-dependent inhibition of PDE in oocyte extracts by three different PDE inhibitors correlates well with their ability to inhibit GVBD (Bornslaeger et al. 1984). It is unlikely that the inhibition of GVBD by methylxanthines is due to perturbation of calcium fluxes or interaction with adenosine receptors (Wells and Kramer 1981), because Ro 1724/1, which is structurally dissimilar to the methylxanthines IBMX and theophylline and does not exert the calcium flux perturbations or interact with adenosine receptors, also inhibits GVBD.

A likely modulator of PDE activity is calmodulin, which accounts for 0.3 percent of the total oocyte protein (Bornslaeger et al. 1984). The calmodulin antagonists trifluoperazine and calmidizolium reversibly, but transiently, inhibit oocyte maturation in a dose-dependent fashion. In addition, the halogenated naphthalene sulfonamides W7 and W13, which are also calmodulin antagonists (Hidaka and Tanaka 1983), inhibit GVBD in a dose-dependent manner, whereas similar concentrations of the less active congeners W5 and W12 do not. Oocyte extracts contain a calmodulin-modulated PDE activity (Bornslaeger et al. 1984); it is not known, however, if this enzyme functions in vivo. The oocyte calmodulin-modulated PDE, however, does not appear to be the target for the inhibitory effect on GVBD of the calmodulin antagonist drugs, because W7 at concentrations that inhibit maturation does not inhibit the maturation-associated decrease in oocyte cAMP (Bornslaeger et al. 1984). The dissociation by the calmodulin inhibitors of the maturation-associated decrease in oocyte cAMP from maturation suggests that another calmodulin-modulated process, other than calmodulin-modulated PDE, is involved in maturation. Although the effector molecule(s) is not known, a calmodulin-modulated protein phosphatase has recently been described (Ingebritsen and Cohen 1983), and protein phosphatase appears to be involved in maturation of *Xenopus* oocytes (Huchon, Ozon, and Demaille 1981; Foulkes and Maller 1982).

Implicit in the model for mammalian oocyte maturation shown in Figure 7.5 is that the decrease in oocyte cAMP leads to a decrease in cAMP-dependent protein kinase activity, which in turn results in dephosphorylation of specific phosphoproteins prior to GVBD. Two-dimensional-gel electrophoresis of oocyte phosphoproteins reveals that specific proteins undergo dephosphorylation during, but not prior to, the commitment period (Schultz et al. 1983a; Bornslaeger et al. 1986). In addition, an increase occurs during this time in spot intensity of several phosphoproteins. Similar increases and decreases in the pattern of phosphoprotein synthesis have also been observed during maturation of sheep oocytes (Crosby, Osborn, and Moor 1984). Also implicit in this model is that inhibition of the catalytic subunit of cAMP-dependent protein kinase (C) by protein kinase inhibitor (PKI), which specifically interacts with free catalytic subunit but not with C in the holoenzyme, should induce maturation in the presence of high cAMP levels, because the inactive form, X-P, of the hypothetical maturation-regulating protein, X, can no longer be generated, and the protein phosphatase should shift the equilibrium to X. Microinjecting oocytes with increasing amounts of protein kinase inhibitor induces a dose-dependent increase in resumption of meiosis in oocytes incubated in a totally inhibiting concentration of db cAMP (Bornslaeger et al. 1986). This result strongly suggests that, as proposed previously, cAMP mediates inhibition of meiotic arrest through activation of a cAMP-dependent protein kinase. In addition, PKI-injected oocytes induced to resume GVBD manifest most of the changes in phosphoprotein metabolism that occur during the commitment period (Bornslaeger et al. 1986). Microinjection of mouse oocytes with purified catalytic subunit of cAMP-dependent protein kinase inhibits GVBD in a dose-dependent manner, as well as the changes in phosphoprotein metabolism, as predicted by the model. Thus, to date, these changes in phosphoprotein metabolism are tightly correlated with resumption of meiosis. These changes, however, have not been shown to be causally related to GVBD.

Proteins with short half-lives may also be important for resumption of meiosis. Oocytes incubated in medium containing inhibitors of protein synthesis, such as puromycin, undergo GVBD with kinetics similar to oocytes in medium not containing an inhibitor of protein synthesis; such treated oocytes, however, arrest at the circular bivalent stage. On the other hand, if the oocytes are incubated in medium containing puromycin and db cAMP (to prevent resumption of meiosis) and then transferred to medium containing puromycin but not db cAMP, inhibition of GVBD is observed (Eckholm and Magnusson 1979). A possible interpretation of these results is that proteins with short half-lives, which are essential for resumption of meiosis, are turned over during the incubation period in puromycin. Protein synthesis also appears to be required for maintenance of arrest at metaphase II, because incubation of oocytes arrested at metaphase II in medium containing inhibitors of protein synthesis undergo activation (Siracusa et al. 1978).

It should be noted that a similar mechanism for resumption of meiosis appears to occur in *X. laevis* oocytes (Maller and Krebs 1980). Unlike mammalian oocytes, which spontaneously resume maturation when removed from their follicles, amphibian oocytes require a steroid hormone to reinitiate maturation. The steroid, which interacts with a surface receptor (Sadler and Maller 1982), decreases the level of adenylate cyclase activity (Finidori-Lepicard et al. 1981; Jordana, Allende, and Allende 1981; Sadler and Maller 1981) by inhibiting the exchange of GTP for bound GDP on the regulatory subunit of adenylate cyclase (Sadler and Maller 1983; Jordana et al. 1984). The drop in oocyte cAMP that ensues appears causally related to resumption of meiosis, and a similar mechanism for resumption of meiosis leading to GVBD has been proposed in this system.

B. Role of oocyte calcium in meiotic maturation

Although calcium appears to be involved in mammalian oocyte maturation, because a calmodulin-dependent step appears to be involved in resumption of meiosis, its role is not known. The requirement for extracellular calcium is complicated by the observation that fully grown, denuded mouse oocytes become necrotic within 1.5 h in calcium-free medium (Paleos and Powers 1981; De Felici and Siracusa 1982), although cumulus-cell-enclosed rat or bovine oocytes do not degenerate in calcium-free medium (Tsafriri and Bar-Ami 1978; Liebfried and First 1979). In addition, the sensitivity of denuded oocytes to calcium-free medium coincides with the acquisition of meiotic competence (De Felici and Siracusa 1982). External calcium may not play a central role in resumption of meiosis, because either verapamil or tetracaine, both of which inhibit transmembrane calcium movement, only transiently inhibits GVBD, although they do inhibit polar-body formation (Paleos and Powers 1981). It should be noted, however, that the effects of these drugs on calcium transport were not examined in these studies and that there exist other types of calcium channels that are not inhibited by either verapamil or tetracaine.

The effects of verapamil and tetracaine on denuded mouse oocytes incubated in medium containing db cAMP suggest that there may be an interaction between cAMP levels and free intracellular calcium levels (Powers and Paleos 1982). Increasing the external calcium concentration from 1.7 mM to 20 mM decreases the inhibitory effect of db cAMP (concentrations of 150 μM or less) on GVBD when scored after 23 h in culture. Most striking is the observation that 2.5 to 20 μM A23187 can reduce the inhibitory effect of 0.2 mM db cAMP on oocyte maturation. Furthermore, concentrations of either verapamil or tetracaine that do not inhibit GVBD potentiate the inhibitory effect of GVBD of subinhibitory concentrations of db cAMP. Taken together, these results suggest that raising the intracellular concentration of free calcium may be involved in oocyte maturation. Clearly, spatial and temporal analysis of intracellular free calcium concentrations during oocyte maturation using fluorescent calcium chelator probes, such as fura-2 (Grynkiewicz et al. 1985), will provide useful information regarding the role of calcium in oocyte maturation.

C. Role of steroid hormones in meiotic maturation

Steroid hormones can induce maturation in amphibian oocytes, and the follicle fluid bathing the mammalian oocytes is rich in various steroid hormones. For these reasons, much attention has focused on the role of steroid hormones in mammalian oocyte maturation (McGaughey 1983). The results of these studies have suggested that nonphysiologically high concentrations of certain steroid hormones (e.g., pregnenolone, progesterone, androstenedione, and testosterone) can reversibly inhibit GVBD in mouse oocytes (Smith and Tenney 1980). Estrogens and the biologically inactive androgen dihydrotestosterone were reported not to inhibit GVBD. Conflicting data have been reported on the ability of estradiol-17β to inhibit GVBD in porcine oocytes (McGaughey 1977; Richter and McGaughey 1979; Racowksy and McGaughey 1982a).

Concentrations of steroid hormones, such as estradiol-17β, testosterone, dihydrotestosterone, or progesterone, but not estradiol-17α, that do not inhibit GVBD can potentiate the inhibitory effect on GVBD elicited by subinhibitory concentrations of db cAMP, 8-bromo cAMP, or forskolin (Rice and McGaughey 1981; Eppig et al. 1983; Racowsky 1985a). Given the role of cAMP-dependent protein kinase in maintenance of meiotic arrest, a possible explanation for this synergistic interaction may be steroid-hormone-induced phosphorylation of the regulatory subunit (R) of cAMP-dependent protein kinase, which has been reported (Liu, Walter, and Greengard 1981). Such phosphorylation of the regulatory subunit increases the activity of the enzyme, because phosphorylation of R inhibits the reassociation of R and C and its concomitant inhibition. Consistent with this hypothesis is the finding that the antiestrogen tamoxifen inhibits the synergistic inhibition of GVBD induced by forskolin and estradiol-17β (Racowsky 1985b). Thus, under basal conditions, steroid hormones may play a role in maintenance of meiotic arrest.

D. Maturation-promoting factor

Maturation-promoting factor (MPF) is a ubiquitous protein(s) that is involved in the G_2-to-M transition in both meiotic and mitotic cells (see Masui and Clark 1979 for review; Gerhart, Wu, and Kirchner 1984). MPF is thought to control many of the early events in this cell cycle transition, such as chromosome condensation and nuclear membrane breakdown, and its activity appears to oscillate during the cell cycle (Gerhart et al. 1984). Although most studies have focused on amphibian and starfish oocytes and somatic cells as sources of MPF, there is evidence that mammalian oocytes generate MPF during maturation. Inactivated Sendai-virus-induced fusion of fully grown, meiotically mature mouse oocytes with meiotically incompetent oocytes (i.e., <60 μm in diameter) resulted in GVBD of the nucleus of the incompetent oocyte when examined immediately after the fusion process (Balakier 1978). In addition, injection of cytoplasm obtained from oocytes that had undergone GVBD into *X. laevis* (Sorensen, Cyert, and Pedersen 1985) or *Asterina Pectinifera* oocytes (Kishimoto et al. 1984) induced GVBD in the absence of added progesterone or 1-methyladenine, respectively. Cytoplasm obtained from oocytes inhibited from resuming maturation did not elicit GVBD in these cases (Sorensen et al. 1985).

The mammalian MPF appears to be generated in the cytoplasm and has been found to be independent of nuclear control (Balakier and Czolowska 1977). Fully grown oocytes were bisected into nucleate and anucleate fragments. When the nucleate fragment had undergone GVBD, the anucleate fragment was fused with interphase blastomeres obtained from 2-cell embryos, and chromosome condensation was immediately observed in the interphase nuclei. What remains to be determined is whether or not the MPF activity is generated in the mammalian oocyte cytoplasm during the commitment period (i.e., prior to GVBD).

Meiotically incompetent mouse and pig oocytes contain a maturation-inhibiting activity, because fusion of a meiotically competent oocyte with an intact GV with a meiotically incompetent oocyte results in maintenance of both GVs (Fulka et al. 1985). Loss of this "anti-MPF" activity, which requires futher characterization, may be essential for acquisition of meiotic competence.

E. Mechanism of gonadotropin-induced resumption of meiosis

Meiotically competent oocytes in cultured antral follicles do not resume meiosis, whereas exposure of such follicles to gonadotropins or liberation of the oocytes from their follicles results in resumption of meiosis (Pinus and Enzmann 1935). This observation led to the proposal that the follicle exerts an inhibitory effect on oocyte maturation. Granulosa cells are implicated in this inhibition, because cumulus-cell-enclosed oocytes cocultured with granulosa cells (Tsafriri and Channing, 1975a) or grafted to granulosa cells in experimentally opened follicles do not resume meiosis (Foote and Thibault 1969; Tsafriri and Channing 1975a; Liebfried and First 1980a). Meiosis does ensue if graft formation is prevented or if complexes are grafted to hemisections in which the granulosa cells are removed. This inhibition appears to be physiological, because luteinizing hormone (LH) induces resumption of meiosis of cumulus-cell-enclosed oocytes grafted to follicle wall hemisections (Liebfried and First 1980a). Thus, it appears that (1) the follicle exerts an inhibitory influence on oocyte maturation, (2) the functional syncytium generated by the gap junctions between granulosa cells, cumulus cells, and the oocyte is required for maintenance of meiotic arrest, and (3) gonadotropins, which act on the follicle cells, can overcome this inhibitory effect.

During the past decade, two models have emerged to account for these observations. One is based on the observation reported over 30 years ago (Chang 1955) that follicular fluid can inhibit spontaneous meiotic maturation in vitro. This model proposes that follicular fluid contains a low-molecular-weight polypeptide, oocyte maturation inhibitor (OMI), that is synthesized by granulosa cells. OMI is proposed to inhibit oocyte maturation, and its amount is reduced by LH. The other model is based on the well-documented observation that cAMP is involved in maintenance of meiotic arrest. This model proposes that LH terminates intercellular communication such that the flux of follicle cell cAMP to the oocyte is no longer sufficient to maintain inhibitory levels of cAMP. This results in a maturation-associated decrease in cAMP and resumption of meiosis. The following discussion examines the current status of these two models.

i. Oocyte maturation inhibitor. The early work concerning OMI suggested that maturation of porcine cumulus-cell-enclosed oocytes was inhibited when they were cocultured with porcine granulosa cells (Tsafriri and Channing 1975a). A similar inhibitory effect was observed with porcine follicular fluid, granulosa-cell-conditioned medium, and granulosa cell extracts (Tsafriri and Channing 1975a, 1975b; Centola, Anderson, and Channing, 1981). None of these treatments inhibited maturation of denuded oocytes, and thus it was proposed that the inhibitory activity is mediated by the cumulus cells. In addition, porcine follicular fluid can inhibit maturation of mouse (Downs and Eppig 1984) and rat (Tsafriri et al. 1977) oocytes, and thus the inhibitor is not species-specific.

Inhibition of oocyte maturation by OMI appears to be specific, because maturation is not inhibited in cumulus-cell-enclosed oocytes cocultured with fibroblasts under the same conditions in which granulosa cells inhibit maturation (Tsafriri and Channing 1975a). In addition, porcine serum does not inhibit maturation, whereas corresponding concentrations of follicular fluid do. Inhibition by follicular fluid is reversible, and LH can overcome inhibition of maturation exerted by follicular fluid (Tsafriri, Pomerantz, and Channing 1976; Stone et al. 1978). Follicular fluid, granulosa-cell-conditioned medium, or granulosa cell extracts isolated or prepared from small follicles possess a greater inhibitory activity than those obtained from large follicles (Tsafriri and Channing 1975a, 1975b; Centola et al. 1981). The inhibitory activity in follicular fluid is not removed by charcoal treatment and is heat-stable. Trypsin, however, removes the inhibitory activity (Tsafriri et al. 1976).

The bioassay for OMI, which is tedious and has a large inherent error, has seriously impeded its purification from follicular fluid. The activity, however, has been partially purified using the filtrate following Amicon PM-10 filtration, gel-permeation chromatography, and paper electrophoresis. The results of these studies suggest that OMI is a polypeptide of molecular mass less than 2,000 daltons (Tsafriri et al. 1976; Stone et al. 1978). It should be noted that during purification of OMI, more than one peak of inhibitory activity was observed, and it is unfortunate that the sensitivity of the partially purified material to charcoal and trypsin treatment has not been examined.

Granulosa cells are a likely source of OMI, because granulosa cell extracts and granulosa-cell-conditioned medium possess a maturation inhibitory activity. It must be borne in mind, however, that because complexes grafted to follicle wall hemisections are inhibited from resuming meiosis in the absence of follicular fluid, the inhibitory substance present in follicular fluid may be unrelated to or a metabolite of that in the granulosa cells.

The general lack of acceptance of OMI as a true physiological regulator of gonadotropin-induced resumption of meiosis comes from the inability of other investigators to observe the inhibitory effect of granulosa cell cocultures or follicular fluid on oocyte maturation (Liebfried and First 1980a, 1980b; Racowsky and McGaughey 1982b). Racowsky and

McGaughey (1982b) proposed that previous investigators may have incorrectly scored the extent of maturation, because porcine oocytes that have reached metaphase II during the first day of culture can undergo activation during the second day of culture and exhibit a diploid pronucleus. These oocytes could have been misclassified as GV oocytes. Thus, follicular fluid would not inhibit maturation, but rather would promote events that lead to a misclassification of meiotic stage.

The effect of porcine follicular fluid on rat oocyte maturation has yielded conflicting results. Tsafriri and associates (1977) reported that porcine follicular fluid exerted an inhibitory effect when GVBD was scored after 6 h in culture, whereas Fleming, Kahlil, and Armstrong (1983) reported no inhibition when GVBD was scored after 9–10 h of culture. The reason(s) for this discrepancy is not known, but it may be due to the transient nature of the inhibition, because Tsafriri and associates (1977) did not observe inhibition of maturation when GVBD was scored after 20 h of culture. Porcine follicular fluid does exert a transient inhibitory effect on mouse oocyte maturation (Downs and Eppig 1984).

Results of recent studies cast serious doubt that OMI is a low-molecular-weight polypeptide. Porcine follicular fluid transiently inhibits maturation of denuded mouse oocytes, and this inhibition is potentiated in complexes (Downs and Eppig 1984). In addition, porcine follicular fluid interacts synergistically with subinhibiting concentrations of either IMBX or db cAMP to inhibit mouse oocyte maturation. The inhibitory activity present in porcine follicular fluid has been unequivocally demonstrated to be hypoxanthine (Downs et al. 1985). The concentration of hypoxanthine in porcine follicular fluid is about 1.4 mM. This concentration of hypoxanthine added to medium totally mimics the synergistic inhibitory effects of porcine follicular fluid on oocyte maturation. In addition, serum, which does not inhibit maturation, contains very low levels of hypoxanthine.

The inhibitory effect of hypoxanthine on oocyte maturation is consistent with it being an inhibitor of PDE activity. As discussed previously, oocyte PDE is involved in the maturation-associated decrease in oocyte cAMP, and inhibiting this enzyme inhibits maturation. The concentration of hypoxanthine in bovine follicular fluid is extremely low, which probably accounts for its inability to inhibit oocyte maturation (J.J. Eppig, personal communication). The physiological role that hypoxanthine plays in maintenance of meiotic arrest is not known.

ii. Termination of intercellular communication. The original model of Dekel and Beers (1978, 1980) postulated that gap-junction-mediated transmission of follicle cell cAMP to the oocyte inhibits maturation and that LH terminates the flux of follicle cell cAMP to the oocyte. A decrease in oocyte cAMP below inhibitory levels occurs because oocytes lack an adenylate cyclase but contain a PDE. This model explains the paradoxical situation in which cAMP can act directly on the oocyte to inhibit maturation, but can act in the follicle to induce maturation. The following evidence is consistent with the model:

1.   The role for cAMP in maintenance of meiotic arrest and resumption of meiosis is fairly well established as discussed previously.

2.   Cholera toxin does not inhibit maturation of denuded oocytes, which is consistent with a lack of functional adenylate cyclase (Dekel and Beers 1978, 1980); cumulus cells are known to possess an active adenylate cyclase (Schultz et al. 1983b).

3.   Oocytes have an active PDE (Bornslaeger et al. 1984).

4.   The suggestion that cAMP can be transmitted between cumulus cells and the oocytes is based on the observation that cholera toxin inhibits maturation of cumulus-cell-enclosed oocytes (Dekel and Beers 1978, 1980). Previous experiments involving cocultures of cells responsive to hormones that activate adenylate cyclase suggested that cAMP can move through heterologous gap junctions (Lawrence, Beers, and Gilula 1978; Murray and Fletcher 1984).

  Cumulus cell cAMP apparently can be transferred to the oocyte. Exposure of cumulus-cell-oocyte complexes to forskolin, FSH, or cholera toxin elevates cumulus cell cAMP levels, as well as intraoocyte cAMP levels for oocytes derived from such treated complexes, relative to oocytes obtained from unstimulated complexes, or denuded oocytes exposed to these agents (Racowsky 1984, 1985a; Bornslaeger and Schultz 1985b). Given the direct inhibitory effect of cAMP on oocyte maturation, transfer of cumulus cell cAMP to the oocyte explains the observation that, in vitro, oocytes present in complexes stimulated by agents that elevate cumulus cell cAMP are inhibited from resuming meiosis, when compared with oocytes present in unstimulated complexes (Dekel and Beers 1978, 1980; Eppig et al. 1983; Schultz et al. 1983b; Racowsky 1984, 1985b).

5.   Examination of intercellular communication by electrical coupling, dye transfer, or metabolic coupling reveals that exposure of follicles to LH results in termination of intercellular communication between cumulus cells and the oocyte (Gilula et al. 1978; Moor et al. 1980; Eppig, 1982; Salustri and Siracusa 1983). In response to elevated cAMP levels, cumulus cells mucify, a cAMP-induced process that involves synthesis and secretion of hyaluronic acid (Eppig, 1979b, 1979c), which leads to physical disruption of contact of cumulus cells with the oocyte, and thus terminates intercellular communication between the two cell types.

  Certain aspects of this model have come into question on closer scrutiny. The original model proposed that oocyte cAMP is derived from an exogenous source, because the oocyte lacks an adenylate cyclase. Oocytes do, however, contain adenylate cyclase, as previously described, and thus are not obligatorily dependent on exogenous cAMP to maintain meiotic arrest.

Although cAMP appears to be transferred from cumulus cells to the oocyte under very nonphysiological conditions, as described earlier, there are no data regarding the basal level of transfer of follicle cell cAMP to the oocyte. If this movement does occur, the level may be quite low, because conditions that elevate cumulus cell cAMP by 50–100-fold (which is far greater than the increase that is observed in vivo) result in only a twofold increase in oocyte cAMP (Bornslaeger and Schultz 1985b). Although the follicle cell contribution of cAMP to the oocyte may be small under basal conditions, this contribution could be potentially significant in maintaining meiotic arrest. For example, spontaneous meiotic maturation could be triggered by termination of this basal level of transfer of follicle cell cAMP to the oocyte, because the oocyte adenylate cyclase activity may not be able to compensate for the PDE activity. The basal level of cAMP transfer could be the critical factor in maintaining oocyte cAMP levels necessary to inhibit maturation. Determining the relative contributions of oocyte adenylate cyclase and follicle cell cAMP to the pool of oocyte cAMP should help resolve the role of the follicle cell in maintaining meiotic arrest.

The major experimental finding that is apparently inconsistent with the model is the temporal relationship between resumption of meiosis and termination of intercellular communication between the cumulus cells and the oocyte. Results of experiments using either sheep (Moor et al. 1980) or mouse oocytes (Eppig 1982; Salustri and Siracusa 1983) indicate that resumption of meiosis, which was induced in vivo by administration of hCG, occurs prior to a reduction in the extent of intercellular communication. However, other evidence obtained with rat and hamster oocyte-cumulus complexes suggests that a reduction in intercellular communication precedes resumption of meiosis (Dekel and Sherizly 1985; Racowsky and Satterlie 1985).

A quantitative morphometric study of freeze-fractured rat cumulus cell complexes revealed a 10–20-fold reduction in the extent of gap-junctional surface area between cumulus cells during resumption of meiosis in vivo; this decrease correlates well with resumption of meiosis (Larsen, Wert, and Brunner 1986). As previously pointed out (Eppig 1982; Freter and Schultz 1984), intercellular communication between cumulus cells and the oocyte is probably mediated by the inner layers of cumulus cells, which are the last cells to undergo the mucification reaction and become uncoupled from one another and the oocyte. A reduction in the extent of intercellular communication among the outer cumulus cells of the complex, therefore, would not result in a noticeable reduction in the extent of communication between cumulus cells and the oocyte, but would effectively isolate the oocyte from communicating with the entire follicle cell mass. Thus, such a reduction in communication between cumulus cells could result in the maturation-associated decrease in oocyte cAMP at a time when overall follicle cell cAMP levels are increasing. It should be pointed out that even if a reduction in intercellular communication precedes or is concurrent with GVBD, uncoupling need not be the trigger event, because a commitment event (e.g., inhibition of oocyte adenylate cyclase or activation of oocyte PDE) could occur prior to the reduction in communication.

In conclusion, it is likely that cAMP is involved in maintaining meiotic arrest and that a gonadotropin-induced reduction of communication between cumulus cells is involved in resumption of meiosis. What remains to be demonstrated is whether this reduction in intercellular communication initiates the maturation-associated decrease in oocyte cAMP by reducing the influx of follicle cell cAMP to the oocyte or whether this decrease reduces the influx of another substance of follicle cell origin that regulates oocyte cAMP levels. Given the substantial increase in our understanding of mammalian oocyte maturation that has accrued during the past decade, the answer to the mechanism of gonadotropin-induced resumption of meiosis of follicle-enclosed oocytes may be forthcoming in the near future.

References

Abreu, S. L., and Brinster, R. L. (1978) Synthesis of tubulin and actin during the preimplantation development of the mouse. *Exp. Cell Res.* 114, 135–141.

Adamson, E. D., and Woodland, H. R. (1977) Changes in the rate of histone synthesis during oocyte maturation and very early development of *Xenopus laevis*. *Dev. Biol.* 57, 136–49.

Albertini, D. F., and Anderson, E. (1974) The appearance and structure of intercellular connections during the ontogeny of the rabbit ovarian follicle with particular reference to gap junctions. *J. Cell Biol.* 63, 234–50.

Anderson, E., and Albertini, D. F. (1976) Gap junctions between the oocyte and companion follicle cells in the mammalian ovary. *J. Cell Biol.* 71, 680–6.

Bachvarova, R. (1974) Incorporation of tritiated adenosine into mouse ovum RNA. *Dev. Biol.* 40, 52–8.

−(1981) Synthesis, turnover, and stability of heterogeneous RNA in growing mouse oocytes. *Dev. Biol.* 86, 384–92.

−(1985) Gene expression during oogenesis and oocyte development in mammals. In *Developmental Biology. A Comprehensive Synthesis, Vol. 1, Oogenesis,* ed. L. W. Browder, pp. 453–524. New York: Plenum.

Bachvarova, R., Baran, M. M., and Tejblum, A. (1980) Development of growing mouse oocytes *in vitro. J. Exp. Zool.* 211, 159–69.

Bachvarova, R., and DeLeon, V. (1977) Stored and polysomal ribosomes of mouse ova. *Dev. Biol.* 58, 248–54.

−(1980) Polyadenylated RNA of mouse ova and loss of maternal RNA in early development. *Dev. Biol.* 74, 1–8.

Bachvarova, R., De Leon, V., Johnson, A., Kaplan, G., and Paynton, B. V.

(1985) Changes in total RNA, polyadenylated RNA, and actin mRNA during meiotic maturation of mouse oocytes. *Dev. Biol.* 108, 325–31.

Bachvarova, R., De Leon, V., and Spiegelman, I. (1981) Mouse egg ribosomes: biochemical evidence for storage in lattices. *J. Embryol. Exp. Morphol.* 62, 153–64.

Balakier, H. (1978) Induction of maturation in small oocytes from sexually immature mice by fusion with meiotic or mitotic cells. *Exp. Cell Res.* 112, 137–41.

Balakier, H., and Czolowska, R. (1977) Cytoplasmic control of nuclear maturation in mouse oocytes. *Exp. Cell Res.* 110, 466–9.

Barbehenn, E. K., Wales, R. G., and Lowry, O. H. (1974) The explanation for the blockade of glycolysis in early mouse embryos. *Proc. Natl. Acad. Sci. U.S.A.* 71, 1056–60.

Biggers, J. D. (1971) Metabolism of mouse embryos. *J. Reprod. Fertil.* Suppl. 14, 41–54.

Biggers, J. D., Whittingham, D. G., and Donahue, R. P. (1967) The pattern of energy metabolism in the mouse oocyte and zygote. *Proc. Natl. Acad. Sci. U.S.A.* 58, 560–7.

Bleil, J. D., and Wassarman, P. M. (1980a) Structure and function of the zona pellucida: identification and characterization of the proteins of the mouse oocyte's zona pellucida. *Dev. Biol.* 76, 185–203.

–(1980b) Mammalian sperm-egg interactions: identification of a glycoprotein in mouse egg zonae pellucidae possessing sperm receptor activity for sperm. *Cell* 20, 873–82.

–(1980c) Synthesis of zona pellucida proteins by denuded and follicle-enclosed mouse oocytes during culture *in vitro*. *Proc. Natl. Acad. Sci. U.S.A.* 77, 1029–33.

–(1981) Mammalian sperm-egg interaction: fertilization of mouse eggs triggers modification of the major zona pellucida glycoprotein, ZP2. *Dev. Biol.* 86, 189–97.

–(1983) Sperm-egg interactions in the mouse: sequence of events and induction of the acrosome reaction by a zona pellucida glycoprotein. *Dev. Biol.* 95, 317–24.

Boreen, S. M., Gizang, E., and Schultz, R. M. (1983) Biochemical studies of mammalian oogenesis: synthesis of 5S and 4S RNA during growth of the mouse oocyte. *Gam. Res.* 8, 379–83.

Bornslaeger, E. A., Mattei, P. M., and Schultz, R. M. (1986) Involvement of cAMP-dependent protein kinase and protein phosphorylation in regulation of mouse oocyte maturation. *Dev. Biol.* 114, 453–62.

Bornslaeger, E. A., Schultz, R. M. (1985a) Adenylate cyclase activity in zona-free mouse oocytes. *Exp. Cell Res.* 156, 277–81.

–(1985b) Regulation of mouse oocytes maturation: effect of elevating cumulus cell cAMP on oocyte cAMP levels. *Biol. Reprod.* 33, 698–704.

Bornslaeger, E. A., Wilde, M. W., and Schultz, R. M. (1984) Regulation of mouse oocyte maturation: involvement of cyclic AMP phosphodiesterase and calmodulin. *Dev. Biol.* 105, 488–99.

Brinster, R. L. (1965) Lactate dehydrogenase activity in the preimplanted mouse embryo. *Biochem. Biophys. Acta* 110, 439–41.

Brower, P. B., Gizang, E., Boreen, S., and Schultz, R. M. (1981) Biochemical studies of mammalian oogenesis: synthesis and stability of various classes of RNA during growth of the mouse oocyte *in vitro. Dev. Biol.* 86, 373–83.

Brower, P. T., and Schultz, R. M. (1982a) Intercellular communication between granulosa cells and mouse oocytes: existence and possible nutritional role during oocyte growth. *Dev. Biol.* 90, 144–53.

–(1982b) Biochemical studies of mammalian oogenesis: possible existence of a ribosomal and poly(A)-containing RNA-protein supramolecular complex in mouse oocytes. *J. Exp. Zool.* 220, 251–60.

Burkholder, G. P., Comings, D. E., and Okada, T. (1971) A storage form of ribosomes in mouse oocytes. *Exp. Cell Res.* 69, 361–71.

Calarco, P. G., and Brown, E. H. (1969) An ultrastructural and cytological study of preimplantation development in the mouse. *J. Exp. Zool.* 171, 253–84.

Canipari, R., Palombi, F., Riminucci, M., and Mangia, F. (1984) Early programming of maturation competence in mouse oogenesis. *Dev. Biol.* 102, 519–24.

Canipari, R., Pietrolucci, A., and Mangia, F. (1979) Increase of total protein synthesis during mouse oocytes growth. *J. Reprod. Fertil.* 57, 405–13.

Cascio, S. M., and Wassarman, P. M. (1981) Program of early development in the mammal: synthesis of mitochondrial proteins during oogenesis and early embryogenesis in the mouse. *Dev. Biol.* 83, 166–72.

–(1982) Program of early development in the mammal: post-transcriptional control of a class of proteins synthesized by mouse oocytes and early embryos. *Dev. Biol.* 89, 397–408.

Centola, G. M., Anderson, L. D., and Channing, C. P. (1981) Oocyte maturation inhibitor (OMI) activity in porcine granulosa cells. *Gam. Res.* 4, 451–61.

Chang, M. C. (1955) The maturation of rabbit oocytes in culture and their maturation, activation, fertilization, and subsequent development in the fallopian tubes. *J. Exp. Zool.* 128, 378–405.

Cho, W. K., Stern, S., and Biggers, J. D. (1974) Inhibitory effect of dibutyryl cAMP on mouse oocyte maturation *in vitro. J. Exp. Zool.* 187, 383–6.

Colonna, R., Cecconi, S., Buccione, R., and Mangia, F. (1983) Amino acid transport systems in growing mouse oocytes. *Cell Biol. Int. Reports* 7, 1007–15.

Crosby, I. M., Osborn, J. C., and Moor, R. M. (1984) Changes in protein phosphorylation during the maturation of mammalian oocytes *in vitro. J. Exp. Zool.* 229, 459–66.

Cross, P. C., and Brinster, R. L. (1974) Leucine uptake and incorporation at three stages of mouse oocyte maturation. *Exp. Cell Res.* 86, 43–6.

Cullen, B. R., Emigholz, K., and Monahan, J. J. (1980) Protein patterns of early mouse embryos during development. *Differentiation* 17, 151–60.

Davidson, E. H. (1976) *Gene Activity in Early Development.* New York: Academic Press.

De Felici, M., and Siracusa, G. (1982) Survival of isolated, fully grown mouse ovarian oocytes is strictly dependent on external Ca^{2+}. *Dev. Biol.* 92, 539–43.

Dekel, N., Aberdam, E., and Sherizly, I. (1984) Spontaneous maturation *in vitro* of cumulus-enclosed rat oocytes is inhibited by forskolin. *Biol. Reprod.* 31, 244–50.

Dekel, N., and Beers, W. H. (1978) Rat oocyte maturation *in vitro*: relief of cyclic AMP inhibition by gonadotropins. *Proc. Natl. Acad. Sci. U.S.A.* 75, 4369–73.

–(1980) Development of the rat oocyte *in vitro*: inhibition and induction of maturation in the presence or absence of the cumulus oophorus. *Dev. Biol.* 75, 247–54.

Dekel, N., and Sherizly, I. (1985) Epidermal growth factor induces maturation of rat follicle-enclosed oocytes. *Endocrinology* 116, 406–9.

De Leon, V., Johnson, A., and Bachvarova, R. (1983) Half-lives and relative amounts of stored and polysomal ribosomes and poly(A)$^+$ RNA in mouse oocytes. *Dev. Biol.* 98, 400–8.

Donahue, R. P., and Stern S. (1968) Follicular cell support of oocyte maturation: production of pyruvate *in vitro*. *J. Reprod. Fertil.* 17, 395–8.

Downs, S. M., Coleman, D. L., Ward-Bailey, P. F., and Eppig, J. J. (1985) Hypoxanthine is the principal inhibitor of murine oocyte maturation in a low molecular weight fraction of porcine follicular fluid. *Proc. Natl. Acad. Sci. U.S.A.* 82, 454–8.

Downs, S. M., and Eppig, J. J. (1984) Cyclic adenosine monophosphate and ovarian follicular fluid act synergistically to inhibit mouse oocyte maturation. *Endocrinology* 114, 418–27.

Eckholm, C., Hillensjo, T., Magnusson, C., and Rosberg, S. (1984) Stimulation and inhibition of rat oocyte meiosis by forskolin. *Biol. Reprod.* 30, 537–43.

Eckholm, C., and Magnusson, C. (1979) Rat oocyte maturation: effects of protein synthesis inhibitors. *Biol. Reprod.* 21, 1287–93.

Eppig, J. J. (1977) Mouse oocyte development *in vitro* with various culture systems. *Dev. Biol.* 60, 371–88.

–(1979a) A comparison between oocyte growth in co-culture with granulosa cells and oocytes with granulosa cell-oocyte junctional contact maintained. *J. Exp. Zool.* 209, 345–53.

–(1979b) Gonadotropin stimulation of the expansion of cumulus oophori isolated from mice: general conditions for expansion in vitro. *J. Exp. Zool.* 208, 111–20.

–(1979c) FSH stimulates hyaluronic acid synthesis by oocyte-cumulus cell complexes from mouse preovulatory follicles. *Nature (London)* 281, 483–4.

–(1982) The relationship between cumulus cell–oocyte coupling, oocyte meiotic maturation, and cumulus expansion. *Dev. Biol.* 89, 268–72.

Eppig, J. J., Freter, R. R., Ward-Bailey, P. F., and Schultz, R. M. (1983) Inhibition of oocyte maturation in the mouse: participation of cAMP, steroid hormones, and a putative maturation-inhibitory factor. *Dev. Biol.* 100, 39–49.

Finidori-Lepicard, J., Schorderet-Slatkine, S., Hanoune, J., and Baulieu, E. E. (1981) Progesterone inhibits membrane-bound adenylate cyclase in *Xenopus laevis* oocytes. *Nature (London)* 272, 255–7.

Fleming, A. D., Khalil, W., and Armstrong, D. T. (1983) Porcine follicular fluid does not inhibit rat oocytes *in vitro*. *J. Reprod. Fertil.* 69, 665–70.

Florman, H. M., Bechtol, K. B., and Wassarman, P. M. (1984) Enzymatic dissection of the functions of the mouse egg's receptor for sperm. *Dev. Biol.* 106, 243–55.

Florman, H. M., and Wassarman, P. M. (1985) O-linked oligosaccharides of mouse egg ZP3 account for sperm receptor activity. *Cell* 41, 313–24.

Ford, P. J., Meathieson, T., and Rosbash, M. J. (1977) Very long-lived messenger RNA in ovaries of *Xenopus laevis*. *Dev. Biol.* 57, 417–26.

Foulkes, J. G., and Maller, J. L. (1982) *In vivo* actions of protein phosphatase inhibitor-2 in *Xenopus* oocytes. *FEBS Lett.* 150, 155–60.

Freter, R. R., Schultz, R. M. (1984) Regulation of murine oocyte meiosis: evidence for a gonadotropin-induced cAMP-dependent reduction in a maturation inhibitor. *J. Cell Biol.* 98, 1119–28.

Friedman, D. L. (1982) Regulation of the cell cycle and cellular proliferation by cyclic nucleotides. In *Handbook of Experimental Pharmacology*, Vol. 58, Part 2, eds. J. W. Kebabian and J. A. Nathanson, pp. 151–88. New York: Springer-Verlag.

Fulka, J., Jr., Motlik, J., Fulfa, J., and Crozet, N. (1985) Inhibition of nuclear maturation in fully grown porcine and mouse oocytes after their fusion with growing porcine oocytes. *J. Exp. Zool.* 235, 255–9.

Gerhart, J., Wu, M., and Kirschner, M. (1984) Cell cycle dynamics of an M-phase-specific cytoplasmic factor in *Xenopus laevis* oocytes and eggs. *J. Cell Biol.* 98, 1247–55.

Gilula, N. B., Epstein, M. L., and Beers, W. H. (1978) Cell-to-cell communication and ovulation. A study of the cumulus cell-oocyte complex. *J. Cell Biol.* 78, 58–75.

Glass, L. E. (1971) Transmission of maternal proteins into oocytes. *Adv. Biosci.* 6, 29–58.

Greve, J. M., and Wassarman, P. M. (1985) Mouse egg extracellular coat is a matrix of interconnected filaments possessing a structural repeat. *J. Mol. Biol.* 181, 253–64.

Grynkiewicz, G., Poenie, M., and Tsein, R. Y. (1985) A new generation of Ca^{2+} indicators with greatly improved fluorescence properties. *J. Biol. Chem.* 260, 3440–50.

Heller, D. H., Cahill, D. M., and Schultz, R. M. (1981) Biochemical studies of mammalian oogenesis: metabolic cooperativity between granulosa cells and growing mouse oocytes. *Dev. Biol.* 84, 455–64.

Heller, D. H., and Schultz, R. M. (1980) Ribonucleoside metabolism by mouse oocytes: metabolic cooperatively between the fully-grown oocyte and cumulus cells. *J. Exp. Zool.* 214, 355–64.

Herlands, R. L., and Schultz, R. M. (1984) Regulation of mouse oocyte growth: probable nutritional role for intercellular communication between follicle cells and oocytes in oocyte growth. *J. Exp. Zool.* 229, 317–25.

Hidaka, H., and Tanaska, T. (1983) Naphthalenesulfonamides as calmodulin inhibitors. In *Methods in Enzymology*, Vol. 102, eds. A. R. Means and B. W. O'Malley, pp. 185–94. New York: Academic Press.

Howe, C. C., and Solter, D. (1979) Cytoplasmic and nuclear protein synthesis in preimplantation mouse embryos. *J. Embryol. Exp. Morphol.* 52, 209–25.

Huchon, D., Ozon, R., and Demaille, J. G. (1981) Protein phosphatase-1 is involved in *Xenopus* oocyte maturation. *Nature (London)* 294, 358–9.

Ingebritsen, T. S., and Cohen, P. (1983) Protein phosphatases: properties and role in cellular regulation *Science* 221, 331–8.

Iyengar, M. R., Iyengar, C. W. L., Chen, H. Y., Brinster, R. L., Bornslaeger, E., and Schultz, R. M. (1983) Expression of creatine kinase isoenzyme during oogenesis and embryogenesis in the mouse. *Dev. Biol.* 96, 263–8.

Jahn, C. L., Baran, M. M., and Bachvarova, R. (1976) Stability of RNA synthesized by the mouse oocyte during its major growth phase. *J. Exp. Zool.* 197, 161–72.

Jordana, X., Allende, C. C., and Allende, J. E. (1981) Guanine nucleotides are required for progesterone inhibition of adenylate cyclase. *Biochem. Int.* 3, 527–32.

Jordana, X., Olate, J., Allende, C. C., and Allende, J. E. (1984) Studies on the mechanism of inhibition of amphibian oocyte adenylate cyclase by progesterone. *Arch. Biochem. Biophys.* 228, 379–87.

Kang. Y. H., and Anderson, W. A. (1975) Ultrastructure of oocytes of the Egyptian spiny mouse (*Acomys cahirinus*). *Anat. Rec.* 182, 175–200.

Kaplan, G., Abreu, S. L., and Bachvarova, R. (1982) rRNA accumulation and protein synthetic patterns in growing mouse oocytes. *J. Exp. Zool.* 220, 361–70.

Kaplan, G., Jelinek, W. R., and Bachvarova, R. (1985) Repetitive sequence transcripts and U1 RNA in mouse oocytes and eggs. *Dev. Biol.* 109, 15–24.

King, B. F., and Tibbitts, F. D. (1977) Ultrastructural observations on cytoplasmic lamellar inclusion in oocytes of the rodent *Thomomys. Anat. Rec.* 189, 263–72.

Kishimoto, T., Yamazaki, K., Kato, Y., Koide, S. S., and Kanatani, H. (1984) Induction of starfish oocyte maturation by maturation-promoting factor of mouse and surf clam oocytes. *J. Exp. Zool.* 231, 293–5.

La Marca, M. J., and Wassarman, P. M. (1979) Program of early development in the mammal: changes in absolute rates of synthesis of ribosomal proteins during oogenesis and early embryogenesis in the mouse. *Dev. Biol.* 73, 103–19.

—(1984) Relationship between rates of synthesis and intracellular distribution of ribosomal proteins during oogenesis in the mouse. *Dev. Biol.* 102, 525–30.

Larsen, W. J., Wert, S. E., and Brunner, G. D. (1986) A dramatic loss of cumulus cell gap junctions is correlated with germinal vessicle breakdown in rat oocytes. *Dev. Biol.* 13, 517–21.

Lawrence, T. S., Beers, W. H., and Gilula, N. B. (1978) Transmission of hormonal stimulation by cell-to-cell communication. *Nature (London)* 272, 501–6.

Liebfried, L., and First, N. L. (1979) Effects of divalent cations on *in vitro* maturation of bovine oocytes. *J. Exp. Zool.* 210, 575–80.

—(1980a) Follicular control of meiosis in the porcine oocyte. *Biol. Reprod.* 23, 705–9.

−(1980b) Effect of bovine and porcine follicular fluid and granulosa cells in maturation of oocyte *in vitro. Biol. Reprod.* 23, 699−704.

Liu, A. Y. -C., Walter, U., and Greengard, P. (1981) Steroid hormones may regulate autophosphorylation of adenosine-3′, 5′ monophosphate-dependent protein kinase in target tissues. *Eur. J. Biochem.* 114, 539−48.

McGaughey, R. W. (1977) The culture of pig oocytes in minimal medium, and the influence of progesterone and estradiol-17β on meiotic maturation. *Endocrinology*, 100, 39−45.

−(1983) Meiotic maturation of mammalian oocytes. *Oxford Rev. Reprod. Biol.* 5, 106−30.

McGaughey, R. W., Montgomery, D. H., and Richter, J. D. (1979) Germinal vesicle configurations and patterns of polypeptide synthesis of porcine oocytes from antral follicles of different size, as related to their competency for spontaneous maturation. *J. Exp. Zool.* 209, 239−54.

McGaughey, R. W., and Van Blerkom, J. (1977) Patterns of polypeptide synthesis of porcine oocytes during maturation. *Dev. Biol.* 56, 241−54.

Magnusson, C., and Hillensjo, T. (1977) Inhibition of maturation and metabolism in rat oocytes by cyclic AMP. *J. Exp. Zool.* 201, 139−47.

Maller, J. L., and Krebs, E. G. (1980) Regulation of oocyte maturation. *Curr. Top. Cell. Regul.* 16, 271−311.

Mangia, F., and Epstein, C. J. (1975) Biochemical studies of growing mouse oocytes: preparation of oocytes and analyses of glucose-6-phosphate dehydrogenase and lactate dehydrogenase activities. *Dev. Biol.* 45, 211−20.

Mangia, F., Erickson, R. P., and Epstein, C. J. (1976) Synthesis of LDH-1 during mammalian oogenesis and early development. *Dev. Biol.* 54, 146−50.

Masui, Y., and Clark, H. J. (1979) Regulation of oocyte maturation. *Int. Rev. Cytol.* 57, 185−282.

Moor, R. M., Smith, M. W., and Dawson, R. M. C. (1980) Measurement of intercellular coupling between oocytes and cumulus cells using intracellular markers. *Exp. Cell Res.* 126, 15−29.

Moor, R. M., and Trounson, A. O. (1977) Hormonal and follicular factors affecting maturation of sheep oocytes *in vitro* and their subsequent developmental capacity. *J. Reprod. Fertil.* 49, 101−9.

Moore, G. P. M. (1978) RNA synthesis in fixed cells by endogenous RNA polymerase. *Exp. Cell Res.* 111, 317−26.

Moore, G. P. M., and Lintern-Moore, S. (1978) Transcription of the mouse oocyte genome. *Biol. Reprod.* 17, 217−26.

Moore, G. P. M., Lintern-Moore, S., and Peters, H. (1974) RNA synthesis in the mouse oocyte. *J. Cell Biol.* 60, 416−22.

Murray, S. A., and Fletcher, W. H. (1984) Hormone-induced intercellular signal transfer dissociates cyclic AMP-dependent protein kinase. *J. Cell Biol.* 98, 1710−19.

Olds, P. J., Stern, S., and Biggers, J. D. (1973) Chemical estimates of the RNA and DNA contents of the early mouse embryo. *J. Exp. Zool.* 186, 39−46.

Olsiewski, P. A., and Beers, W. H. (1983) cAMP synthesis in the rat oocyte. *Dev. Biol.* 100, 287−93.

Paleos, G. A., and Powers, R. D. (1981) The effect of calcium on the first meiotic division of the mammalian oocyte. *J. Exp. Zool.* 217, 409−16.

Piko, L., and Clegg, K. B. (1982) Quantitative changes in total RNA, total poly(A), and ribosomes in early mouse embryos. *Dev. Biol.* 89, 362–78.

Piko, L., and Matsumoto, L. (1976) Number of mitochondria and some properties of mitochondrial DNA in the mouse egg. *Dev. Biol.* 49, 1–10.

Pincus, G., and Enzmann, E. V. (1935) The comparative behaviour of mammalian eggs *in vivo* and *in vitro*. *J. Exp. Med.* 62, 665–75.

Pitts, J. D., and Simms, J. W. (1977) Permeability of junctions between animal cells. Intercellular transfer of nucleotides but not of macromolecules. *Exp. Cell Res.* 104, 153–63.

Powers, R. D., and Paleos, G. A. (1982) Combined effects of calcium and dibutyryl cAMP on germinal vesicle breakdown in the mouse oocyte. *J. Reprod. Fertil.* 66, 1–8.

Racowsky, C. (1984) Effect of forskolin on the spontaneous maturation and cyclic AMP content of rat oocyte-cumulus complexes. *J. Reprod. Fertil.* 72, 107–16.

–(1985a) Effect of forskolin on the spontaneous maturation and cyclic AMP content of hamster oocyte-cumulus complexes. *J. Exp. Zool.* 234, 87–96.

–(1985b) Antagonistic actions of estradiol and tamoxifen upon forskolin-dependent meiotic arrest, intercellular coupling, and the cAMP content of hamster oocyte-cumulus complexes. *J. Exp. Zool.* 234, 251–60.

Racowsky, C., and McGaughey, R. W. (1982a) In the absence of protein, estradiol suppresses meiosis of porcine oocytes *in vitro*. *J. Exp. Zool.* 224, 103–10.

–(1982b) Further studies of the effects of follicular fluid and membrana granulosa cells on the spontaneous maturation of pig oocytes. *J. Reprod. Fertil.* 66, 505–12.

Racowsky, C., and Satterlie, R. A. (1985) Metabolic, fluorescent dye and electrical coupling between hamster oocytes and cumulus cells during meiotic maturation in vivo and in vitro. *Dev. Biol.* 108, 191–202.

Rice, C., and McGaughey, R. W. (1981) Effect of testosterone and dibutyryl cAMP on the spontaneous maturation of pig oocytes. *J. Reprod. Fertil.* 62, 245–56.

Richter, J. D., and McGaughey, R. W. (1979) Specificity of inhibition by steroids of porcine oocyte maturation *in vitro*. *J. Exp. Zool.* 209, 81–90.

Rodman, T. C., and Bachvarova, R. (1976) RNA synthesis in preovulatory mouse oocytes. *J. Cell Biol.* 70, 251–7.

Sadler, S. E., and Maller, J. L. (1981) Progesterone inhibits adenylate cyclase in *Xenopus* oocytes. Action on the guanine nucleotide regulatory protein. *J. Biol. Chem.* 256, 6368–73.

–(1982) Identification of a steroid receptor on the surface of *Xenopus* oocytes by photoaffinity labeling. *J. Biol. Chem.* 257, 355–61.

–(1983) Inhibition of *Xenopus* oocyte adenylate cyclase by progesterone and 2′, 5′ dideoxyadenosine is associated with slowing of guanine nucleotide exchange. *J. Biol. Chem.* 258, 7935–41.

Salustri, A., and Martinozzi, M. (1982) Glucose utilization by growing mouse oocytes mediated *in vitro* by follicle cells. *Cell Biol. Int. Reports* 6, 420.

–(1983) A comparison of protein synthesis activity in *in vitro* cultured denuded and follicle-enclosed oocytes. *Cell Biol. Int. Reports* 7, 1049–55.

Salustri, A., and Siracusa, G. (1983) Metabolic coupling, cumulus expansion, and meiotic resumption in mouse cumuli oophori cultured *in vitro* in the presence of FSH or dbcAMP, or stimulated *in vivo* by hCG. *J. Reprod. Fertil.* 68, 335–41.

Sato, E., and Koide, S. S. (1984) Forskolin and mouse oocyte maturation *in vitro. J. Exp. Zool.* 230, 125–9.

Schroeder, A. C., and Eppig, J. J. (1984) The developmental capacity of mouse oocytes that matured spontaneously *in vitro* is normal. *Dev. Biol.* 102, 493–7.

Schultz, G. A., Clough, J. R., and Johnson, M. H. (1980) Presence of cap structures in the messenger RNA of mouse eggs. *J. Embrol. Exp. Morphol.* 56, 139–56.

Schultz, R. M. (1985) Roles of cell-to-cell communication in development. *Biol. Reprod.* 32, 27–42.

Schultz, R. M., Bleil, J. D., and Wassarman, P. M. (1978a) Quantitation of nanogram amounts of protein using [^{3}H]dinitrofluorobenzene. *Anal. Biochem.* 91, 353–6.

Schultz, R. M., La Marca, M. J., and Wassarman, P. M. (1978b) Absolute rates of protein synthesis during meiotic maturation of mammalian oocytes *in vitro. Proc. Natl. Acad. Sci. U.S.A.* 75, 4160–4.

Schultz, R. M., Letourneau, G. E., and Wassarman, P. M. (1978c) Meiotic maturation of mouse oocytes *in vitro*: protein synthesis in nucleate and anucleate oocyte fragments. *J. Cell Sci.* 30, 251–264.

–(1979a) Program of early development in the mammal: changes in the patterns and absolute rates of tubulin and total protein synthesis during oocyte growth in the mouse. *Dev. Biol.* 73, 120–33.

–(1979b) Program of early development in the mammal: changes in the patterns and absolute rates of tubulin and total protein synthesis during oogenesis and early embryogenesis in the mouse. *Dev. Biol.* 68, 341–59.

Schultz, R. M., Montgomery, R. R., and Belanoff, J. R. (1983a) Regulation of mouse oocyte maturation: implication of a decrease in oocyte cAMP and protein dephosphorylation in commitment to resume meiosis. *Dev. Biol.* 97, 264–73.

Schultz, R. M., Montgomery, R. R., Ward-Bailey, P. F., and Eppig, J. J. (1983b) Regulation of oocyte maturation in the mouse: possible roles of intercellular communication, cAMP, and testosterone. *Dev. Biol.* 95, 294–304.

Schultz, R. M., and Wassarman, P. M. (1977a) Biochemical studies of mammalian oogenesis: protein synthesis during oocyte growth and meiotic maturation in the mouse. *J. Cell Sci.* 24, 167–94.

–(1977b) Specific changes in the pattern of protein synthesis during meiotic maturation of mammalian oocytes *in vitro. Proc. Natl. Acad. Sci. U.S.A.* 74, 538–41.

Seamon, K. B., Padgett, W., and Daly, J. W. (1981) Forskolin: a unique diterpene activator of adenylate cyclase in membrane and intact cells. *Proc. Natl. Acad. Sci. U.S.A.* 78, 3363–7.

Shur, B. D., and Hall, N. G. (1982) A role for mouse sperm surface galactosyltransferase in sperm binding to the egg zona pellucida. *J. Cell Biol.* 95, 574–9.

Siracusa, G., Whittingham, D. G., Molinaro, M., and Vivarelli, E. (1978)

Siracusa, G., Whittingham, D. G., Molinaro, M., and Vivarelli, E. (1978) Parthenogenic activation of mouse oocytes induced by inhibitors of protein synthesis. *J. Embryol. Exp. Morphol.* 43, 157–66.

Smith, D. M., and Tenney, D. Y. (1980) Effects of steroids on mouse oocyte maturation *in vitro. J. Reprod. Fertil.* 60, 331–8.

Sorensen, R. A., Cyert, M. S., and Pedersen, R. A. (1985) Active maturation-promoting factor is present in mature mouse oocytes. *J. Cell Biol.* 100, 1637–40.

Sorensen, R. A., and Wassarman, P. M. (1976) Relationship between growth and meiotic maturation of the mouse oocyte. *Dev. Biol.* 50, 531–6.

Sternlicht, A. L., and Schultz, R. M. (1981) Biochemical studies of mammalian oogenesis: kinetics of accumulation of total and poly(A)-containing RNA during growth of the mouse oocyte. *J. Exp. Zool.* 215, 191–200.

Stone, S. L., Pomerantz, S. H., Schwartz-Kripner, A., and Channing, C. P. (1978) Inhibitor of oocyte maturation from porcine follicular fluid: further purification and evidence for reversible action. *Biol. Reprod.* 19, 585–92.

Thibault, C. G. (1972) Final stages of mammalian oocyte maturation. In *Oogenesis*, eds. J. D. Biggers and A. W. Schuetz, pp. 397–411. Baltimore: University Park Press.

Tsafriri, A, and Bar-Ami, S. (1978) Role of divalent cations in the resumption of meiosis in rat oocytes. *J. Exp. Zool.* 205, 293–300.

Tsafriri, A., and Channing, C. P. (1975a) An inhibitory influence of granulosa cells and follicular fluid upon porcine oocyte meiosis *in vitro. Endocrinology* 96, 922–7.

–(1975b) Influence of follicular maturation and culture conditions on the meiosis of pig oocytes *in vitro. J. Reprod. Fertil.* 43, 149–52.

Tsafriri, A., Channing, C. P., Pomerantz, S. H., and Lindner, H. R. (1977) Inhibition of maturation of isolated rat oocytes by porcine follicular fluid. *J. Endocrinol.* 75, 285–91.

Tsafriri, A., Pomerantz, S. H., and Channing, C. P. (1976) Inhibition of oocyte maturation by porcine follicular fluid: partial characterization of the inhibitor. *Biol. Reprod.* 14, 511–16.

Urner, F., Herrmann, W. L., Baulieu, E. E., and Schorderet-Slatkine, S. (1983) Inhibition of denuded mouse oocyte meiotic maturation by forskolin, an activator of adenylate cyclase. *Endocrinology* 113, 1170–2.

Van Blerkom, J., and McGaughey, R. W. (1978) Molecular differentiation of the rabbit ovum. I. During oocyte maturation *in vivo* and *in vitro. Dev. Biol.* 63, 139–50.

Vivarelli, E., Conti, M., De Felici, M., and Siracusa, G. (1983) Meiotic resumption and intracellular cAMP levels in mouse oocytes treated with compounds which act on cAMP metabolism. *Cell Differ.* 12, 271–6.

Wassarman, P. M., Florman, H. M., and Greve, J. M. (1984) Receptor-mediated sperm-egg interactions in mammals. In *Biology of Fertilization*, Vol. 2, eds. C. B. Metz and A. Monroy, pp. 341–60. New York: Academic Press.

Wassarman, P. M., and Josefowicz, W. J. (1978) Oocyte development in the mouse: an ultrastructural comparison of oocyte isolated at various stages of growth and meiotic competence. *J. Morphol.* 156, 209–36.

Wassarman, P. M., Josefowicz, W. J., and Letourneau, G. E. (1976) Meiotic maturation of mouse oocytes *in vitro*: inhibition of maturation at specific stages of nuclear progression. *J. Cell Sci.* 22, 431–45.

Wassarman, P. M., and Letourneau, G. E. (1976) RNA synthesis in fully grown mouse oocytes. *Nature (London)* 361, 73–4.

Wassarman, P. M., and Mrozak, S. C. (1981) Program of early development in the mammal: synthesis and intracellular migration of histone H4 during oogenesis in the mouse. *Dev. Biol.* 84, 364–71.

Wassarman, P. M., Schultz, R. M., Letourneau, G. E., La Marca, M. J., Josefowicz, W. J., and Bleil, J. D. (1979) Meiotic maturation of mouse oocyte in vitro. In *Ovarian, Follicular, and Corpus Luteum Function*, eds. C. P. Channing, J. M. Marsh, and W. A. Sadler, pp. 251–7. New York: Plenum.

Wasserman, W. J., Richter, J. D., and Smith, L. D. (1982) Protein synthesis during maturation promoting factor and progesterone-induced maturation in *Xenopus* oocytes. *Dev. Biol.* 89, 152–8.

Weakley, B. S. (1967) Investigations into the structure and fixation properties of cytoplasmic lamellae of hamster oocytes. *Z. Zellforsch Mikrosk. Anat.* 81, 91–9.

–(1968) Comparison of cytoplasmic lamellae and membranous elements in oocytes of five mammalian species. *Z. Zellforsch. Mikrosk. Anat.* 85, 109–23.

Wells, J. N., and Kramer, G. L. (1981) Phosphodiesterase inhibitors as tools in cyclic nucleotide research: a precautionary comment. *Mol. Cell. Endocrinol.* 23, 1–9.

Wolgemuth, D. J., Celenza, J., Bundman, D. S., and Dunbar, B. S. (1984) Formation of the rabbit zona pellucida and its relationship to ovarian follicular development. *Dev. Biol.* 106, 1–14.

Zamboni, L. (1970) Ultrastructure of mammalian oocytes and ova. *Biol. Reprod. Suppl.* 2, 44–63.

8 Utilization of genetic information in the preimplantation mouse embryo

GILBERT A. SCHULTZ

CONTENTS

I. Introduction

During the preimplantation period, the early zygote is transported from the ampullary end of the oviduct through the utero-tubal junction into the uterus. Over this period, the fertilized egg increases in cell number through several cleavage divisions and differentiates into the blastocyst, which contains two morphologically and biochemically distinct cell types, the inner cell mass (ICM) and trophectoderm. The length of time the embryo spends in the preimplantation period and the cell number achieved prior to implantation vary considerably from one mammalian species to another (see Chapter 3). Nonetheless, the formation of the blastocyst is a common feature of early development in all mammals. Most of the information regarding the molecular events underlying blastocyst formation derives from studies on embryos of the mouse and rabbit, although for early postfertilization and cleavage events, by far the bulk of the information comes from studies on mouse embryos. The subject also depends heavily on recent advances made in understanding the control of gene expression in eukaryotic cells in general. In this chapter, our discussion will largely be confined to eggs and early stages of mouse development. Particular emphasis is placed on the qualitative and quantitative changes that occur within mRNA populations in mouse embryos as they proceed in development from the first cleavage, which is largely under the control of informational macromolecules accumulated during oogenesis (see Chapter 7), to blastocyst formation, when there is active transcription and accumulation of new mRNA from the zygote genome. Information on posttranscriptional modification of mRNA through processing, capping, and polyadenylation is included. The protein synthesis changes that accompany changes in mRNA during this early developmental interval are also reviewed. Because homologous recombinant DNA probes for genes coding for actin, histone, and heat-shock proteins are now available, data on the expression of these specific genes are highlighted. Special emphasis is given to histone gene utilization, because there are qualitative shifts in the accumulation of mRNAs from various members of the histone gene family in blastocysts, as compared with eggs.

The cellular phenotypes generated during blastocyst formation are, as noted earlier, accompanied by differential expression of genes. Complex molecular mechanisms exist in different cell types to control the number of specific proteins generated from a common pool of genetic information. The relative amounts of these proteins vary enormously in a typical mammalian cell, from less than 0.01 percent for certain enzymes to as much as 2 percent of the total for some structural proteins like histone, tubulin, and actin. Much of the attention of molecular embryologists has therefore focused on transcriptional processes and on measuring the rates of synthesis of given mRNA species that code for developmentally interesting proteins. Recently, posttranscriptional events and translational control factors have also gained some attention in terms of regulation.

A key feature in the analysis of gene expression in the early embryo is determining the number of genes actually involved. In the past decade, the sequence complexities of transcripts from egg to early embryonic stages of sea urchin and *Xenopus* embryos have been accurately quantitated; for current discussions, see Davidson, Hough-Evans, and Britten (1982) and Anderson and associates (1982). Because of restrictions imposed by the amounts of RNA available from limited numbers of preimplantation mammalian embryos, few equivalent data from nucleic acid hybridization experiments are available. However, the diversity of transcription products from unique DNA sequences has been measured in the late rabbit blastocyst, which contains almost 10^5 cells (Daniel 1964). Total RNA (primarily heterogeneous RNA in terms of complexity) is estimated to be homologous to 1.8 percent of the unique-sequence DNA (Schultz, Manes, and Hahn 1973). Using a refined phenol-emulsion reassociation technique, Manes, Byers, and Carver (1981) were able to determine that the cytoplasmic mRNA population in these embryos hybridized to 0.41 percent of the single-copy genomic DNA. This fraction can potentially specify the synthesis of some 6,000 diverse polypeptides on the basis of a rabbit single-copy DNA value of about 3×10^9 base pairs (Manes et al. 1981) and an average messenger RNA length of 2,000 nucleotides (Lewin 1975). Similar complexity might be expected in earlier stages of development for both rabbit and mouse embryos.

Hybridization reactions between excess mRNA and complementary DNA have been used to estimate the average numbers of molecules of various mRNAs within the total mass of RNA in a cell. For a variety of eukaryotic cells, a few mRNA species (perhaps up to 10) are highly abundant, with many thousands of copies per cell. A few hundred are of intermediate abundance and are present in about 1,000 copies per cell, whereas the vast majority (several thousand different mRNAs) are rare and are present in 10 copies or fewer per cell (Lewin 1980). The first two abundance classes typically account for half of the mass of the mRNA, and the rare species account for the other half. Two-dimensional electrophoresis has provided one method to resolve newly synthesized polypeptides during the early stages of development. It follows that the approximately 300 polypeptides resolved in such procedures reflect the products of the more abundant templates. Thus, although two-dimensional electrophoretic analyses of protein synthesis in early embryos are valuable in outlining changes in gene expression patterns, other assays will ultimately be required to analyze rare mRNAs that code for low-frequency proteins in the egg-to-blastocyst transition period. In this chapter, the utilization of genetic information in the early embryo will be emphasized, concentrating on current approaches and new data subsequent to previous reviews on this subject (Johnson 1981; Schultz et al. 1981).

II. Transition from maternal to embryonic control: quantitative changes in RNA populations

Studies conducted a number of years ago reported that the newly ovulated mouse egg contains about 23 ng of protein (Brinster 1967) and 0.5 ng of RNA (Olds, Stern, and Biggers 1973). Subsequently, estimates of the amounts of all major classes of RNA in eggs and early embryos were made in a number of laboratories (for review, see Johnson 1981). Most of these data have been reevaluated in a detailed and thorough set of experiments by Piko and Clegg (1982). The major features are summarized as follows:

1. The total RNA content of unfertilized eggs is observed to be 0.35 ng when alkaline hydrolysates of acid-precipitable extracts are made and RNA concentration is measured by absorbance at 260 nm (corrected for recovery of [^{14}C]labeled 28S marker RNA added to embryo lysates). Corresponding values for the late 2-cell embryo, 8–16-cell embryo, and 32-cell (early) blastocyst are 0.24, 0.69, and 1.47 ng per embryo, respectively (Piko and Clegg 1982) (Table 8.1).

2. By estimating the total number of ribosomes in mouse eggs and embryos using electron-microscopic morphometry (Piko and Clegg 1982), the amount of ribosomal RNA has been calculated to compose between 58 and 70 percent of the total RNA for all stages examined (Table 8.1). These values support the estimate of 0.2 to 0.4 ng per egg for RNA species comigrating with 28S and 18S rRNA on acrylamide gels (Young, Stull, and Brinster 1973; Bachvarova 1974).

3. Quantitative data on transfer RNA content are scarce. Using an aminoacyl synthetase system for mouse liver, Young and associates (1973) reported detection of 10.2 pg of leucyl tRNA per mouse egg. On the assumption that this tRNA made up 7.4 percent of the total tRNA population, Young and associates (1973) calculated the total tRNA content per egg to be 0.14 ng. Using electrophoretic analysis, Bachvarova (1974) measured the 5S plus tRNA component to be 0.16 ng per egg. These estimates suggest that in the egg, tRNA composes about 40 percent of the total RNA pool. These values appear to be slightly overestimated, considering that ribosomal RNA accounts for about 60 percent of the total, whereas putative mRNA, to be described later, composes 6 to 8 percent.

4. The amount of mRNA in mouse eggs has been estimated indirectly through hybridization of the poly(A) tail of putative RNA molecules with radiolabeled polyuridylic acid (Levy, Stull, and Brinster, 1978; Piko and Clegg 1982). The poly(A) content in mouse eggs is about 0.8 pg (Piko and Clegg 1982). It decreases to 0.26 pg in the late 2-cell embryo, before increasing progressively to 1.42 pg per early (32-cell) blastocyst. Based on the amount of 0.83 pg of poly(A) in the 1-cell fertilized egg, an average poly(A) length of 63 nucleotides (Clegg and Piko 1983a) and an average length of 1,700 nucleotides for poly(A)$^+$ RNA (Clegg and Piko 1983a), the fraction of putative mRNA can be calculated as 23 pg (6.6%). Similarly, the early blastocyst contains about 40 pg of mRNA (Table 8.1).

Observations of changes in abundance of total mRNA have recently been extended to include relative changes in histone and actin mRNAs as markers of specific nonpolyadenylated and polyadenylated mRNA classes, respectively (Giebelhaus, Heikkila, and Schultz 1983; Giebelhaus et al. 1985; Graves et al. 1985a). An example of a Northern blot of RNA isolated from various stages of preimplantation mouse embryos hybridized with a radiolabeled mouse histone H3.2 DNA probe (Sittman,

Table 8.1. *Content of total RNA, rRNA, poly(A)+ RNA, histone mRNA, and actin mRNA in early mouse embryos*

Stage	Total RNA (ng/embryo)[a]	rRNA (ng/embryo)[a]	poly(A)+ RNA (pg/embryo)[b]	Histone H3 mRNA (fg/embryo)[c]	Actin mRNA (fg/embryo)[d]
Ovulated egg	0.35	0.22	19	167	431
2-cell	0.24	0.17	8	23	35
8-cell	0.69	0.40	14	41	182
Early blastocyst (32-cell)	1.47	1.00	37	149	854

Note:
[a]Data from Piko and Clegg (1982).
[b]Calculated from Clegg and Piko (1983a) estimates of number of mRNA molecules on the basis of average length of 2,000 nucleotides.
[c]Data from Graves et al. (1985a).
[d]Data from Giebelhaus, Weitlauf, and Schultz (1985).

Graves, and Marzluff 1983; Graves et al. 1985b) is shown in panel A of Figure 8.1. The relative amount of histone H3 mRNA in 1,000 eggs is quite high. The amount decreases markedly through the transition from the 2-cell to 4-cell stage before histone H3 mRNA reaccumulates (Figure 8.1). The early blastocyst has about the same amount of histone H3 mRNA as the egg (in Figure 8.1., panel A, compare RNA from 600 blastocysts in lane 7 with RNA from 1,000 eggs in lane 2). Similar patterns for the mRNAs for two other nucleosomal histone proteins, histone H2a and H2b, have been observed, indicating that all three histone mRNAs are regulated coordinately during this period (Graves et al. 1985a). Northern blots of actin mRNA demonstrate a similar decrease at the 2-cell stage, then gradual accumulation (Giebelhaus et al. 1983) that parallels that of total poly(A)$^+$ mRNA (Piko and Clegg 1982; Clegg and Piko 1983a).

The amount of specific mRNA species in total RNA of a given stage of development can be determined by comparing the intensity of hybridization of a labeled complementary DNA probe for the RNA sample in question to that of purified mRNA or an RNA with known specific mRNA content. In panel B of Figure 8.1, a [^{32}P]labeled actin cDNA has been hybridized to total RNA from cultured mouse L cells. This RNA preparation was previously analyzed and shown to contain 200 pg of actin mRNA per microgram of total RNA. By comparison, the dots of RNA from 500 eggs are similar to that of 1.0 μg of L-cell RNA (panel B, Figure 8.1). RNA from 500 blastocysts hybridizes with an intensity equivalent to 400 to 500 pg of actin mRNA, whereas that from 500 2-cell embryos is barely detectable (panel B, Figure 8.1).

Quantitative estimates of histone H3 and actin mRNA are summarized in Table 8.1. Data for histone H2a and H2b mRNA are available (Graves et al. 1985a) but have not been included here because they are almost identical with the H3 mRNA values. The lowest histone and actin mRNA content (15 to 20 fg per embryo) was actually observed at the late 2-cell to early 4-cell (48 to 50 h post hCG) stage (Giebelhaus et al. 1985; Graves et al. 1985a). After the 4-cell stage, the embryo has about 15,000 to 20,000 molecules of mRNA per cell for both histone and actin proteins (Table 8.2). The total mRNA content per embryo increases with nuclear number as the embryo divides, but the amount of mRNA per cell remains relatively constant through late cleavage and blastocyst formation.

Actin and histone mRNAs are abundant, and each accounts for up to 2

Figure 8.1. Northern and dot blots of histone and actin mRNA in early mouse embryos. A: Hybridization of histone H3.2 DNA to RNA isolated from mouse eggs and early embryos. Conditions of electrophoresis, transfer to DBM paper, and hybridization were exactly as described by Giebelhaus et al. (1983). Lane 1, 1 μg of RNA from mouse L cells; lane 2, RNA from 1,000−

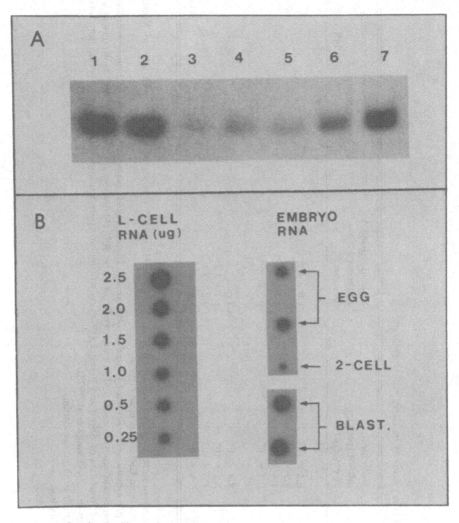

Caption to Figure 8.1 *(cont.)*

1,200 unfertilized eggs; lane 3, RNA from 1,000–1,200 42-h post hCG mid-2-cell-stage embryos; lane 4, RNA from 1,000–1,200 embryos (48 h post hCG) in the 2-cell-to-4-cell transition; lane 5, RNA from 500 66-h post-hCG 8-cell embryos; lane 6, RNA from 300 96-h post-hCG early (32-cell) blastocysts; lane 7, RNA from 600 early blastocysts. B: Autoradiographs of RNA dots from L cells and mouse embryos hybridized with mouse actin cDNA. Procedures for dots and hybridization were exactly as described by Giebelhaus et al. (1985). On the left side, increasing amounts of L-cell RNA from 0.25 μg to 2.5 μg were applied to the membrane; 1 μg of this L-cell RNA preparation contains approximately 200 pg of actin mRNA. On the right side, dots of RNA from 500 eggs (two samples), 500 2-cell embryos (one sample), and 500 blastocysts (two samples) hybridized at the same time as the L-cell RNA standards are shown. Autoradiographic exposure time was 48 h.

Table 8.2. *Rates of histone and actin synthesis in early mouse embryos*

Stage of development	Molecules mRNA (per egg or embryo)[a]	Molecules protein synthesized (per min)[b]	Theoretical translation efficiency (polypeptides/mRNA/min)[c]
Histone H3			
Egg	6.1×10^5	1.9×10^5	0.3
2-cell	8.2×10^4	1.5×10^5	1.9
8-cell	1.5×10^5	5.0×10^5	3.3
Early blastocyst (32–cell)	5.5×10^5	2.7×10^6	5.1
Actin			
Egg	3.8×10^5	Trace	n.c.[d]
2-cell	3.1×10^4	1.1×10^5	3.5
8-cell	1.6×10^5	9.2×10^5	5.8
Early blastocyst (32–cell)	7.5×10^5	8.3×10^6	11.1

Note:

[a] Molecules of histone H3 mRNA per embryo were calculated from the amounts of histone H3 mRNA as reported in Graves et al. (1985a). Numbers of actin mRNA molecules were obtained from Giebelhaus et al. (1985).

[b] Molecules of histone H3 synthesized per minute were calculated from data of Kaye and Church (1983) expressed as moles of histone synthesized per hour and corrected for histone extraction recovery estimated in rabbit embryos to be 30% (Matheson and Schultz 1980). Molecules of actin synthesized per minute were converted from data of Abreu and Brinster (1978) on leucine incorporation into actin on the basis that there are 26 leucine residues out of 374 amino acids in actin (Elzinga et al. 1973).

[c] All mRNAs are assumed to be translatable in these calculations.

[d] Not calculated, but value would be very low.

to 4 percent of the total protein synthesized in early blastocysts (Abreu and Brinster 1978; Kaye and Church 1983). To date, no equivalent analyses on other mRNA species have been published, except for RNA transcripts of viral origin (Piko, Hammons, and Taylor 1984). Nonetheless, techniques are becoming more sensitive, and it is not unreasonable, in the future, to expect data on rarer mRNAs present in as few as 100 copies per cell.

In summary, the mouse embryo inherits a large supply of "maternal" rRNA, tRNA, and mRNA that is stored in the egg. The marked drop in poly(A)$^+$ mRNA soon after the first cleavage division is noteworthy. This observation is compatible with the loss of prelabeled maternal mRNA reported by Bachvarova and De Leon (1980) over the same period. It is interesting that exogenous globin mRNA injected into newly fertilized mouse eggs is actively translated 15 to 17 h later (approximately 29–31 h post hCG) but also appears to be eliminated on a functional basis within 48 h (approximately 62 h post hCG) after injection (Brinster et al. 1980). The loss (decay) of maternal mRNA and reaccumulation due to new synthesis have been well documented in Northern and dot-blot experiments on histone and actin mRNAs described here. Finally, because the rate of protein synthesis is about the same in the 2-cell embryo as in the fertilized egg (Brinster, Wiebold, and Brunner 1976), it follows that most of the abundant maternal mRNA lacks access to functional translational machinery or is utilized at a very reduced efficiency.

III. Protein synthesis during early cleavage period

The newly ovulated mouse egg contains rRNA, mRNA, tRNA, and ribosomes and is fully equipped to synthesize proteins. An ordered set of changes in the pattern of synthesis and modification of proteins occurs during oocyte maturation, fertilization, and development to the early 2-cell stage (Van Blerkom and Brockway 1975; Levinson et al. 1978; Braude et al. 1979; Howe and Solter 1979; Cullen, Emigholz, and Monahan 1980). These changes appear to be intrinsically regulated, because they occur in the absence of concurrent transcription from the genome in physically enucleated eggs or in eggs treated with the transcriptional inhibitor α-amanitin (Braude et al. 1979; Petzoldt, Hoppe, and Illmensee 1980; Flach et al. 1982; Bolton, Oades, and Johnson 1984). Many of the changes in protein pattern or cytoplasmic structure (such as mitochondrial translocation) that occur up to the stage of pronuclear fusion appear to be controlled not so much by translation of preexisting mRNA as by posttranslational means (Van Blerkom 1981, 1985; Van Blerkom and Runner 1984). Other changes that take place during development to the 2-cell stage depend on sequential activation of selected mRNA subsets by some translational control mechanism (Braude et al.

1979; Cascio and Wassarman 1982). In recent studies of mouse eggs fertilized in vitro, fertilization-independent, fertilization-accelerated, and fertilization-dependent changes in polypeptide synthesis were observed in the first few hours of development (Howlett and Bolton 1985). As reported earlier, some of the changes in polypeptide synthetic profile are due to postranslational modifications, some are due to differential mRNA activation, and still others are due to differential polypeptide turnover (Howlett and Bolton 1985).

Two-dimensional electrophoretic analyses of polypeptides produced in vivo or in vitro (reticulocyte lysate) from mRNA in late 1-cell to 2-cell mouse embryos have also led to the identification of two periods of α-amanitin-sensitive protein synthesis activity (Flach et al. 1982). These occur just before and after DNA replication, which takes place 1 to 5.5 h after the first cleavage in synchronized embryos (Bolton et al. 1984). Treatment with α-amanitin during the first period prevents the appearance of a set of 67,000–70,000-dalton polypeptides on the two-dimensional fluorogram (Flach et al. 1982; Bolton et al. 1984). These polypeptides have been putatively identified as heat-shock proteins on the basis of one-dimensional polypeptide maps (Bensaude et al. 1983). Neither the DNA replication event nor the loss of maternal mRNA that takes place during the 2-cell stage is affected by the synthesis of the α-amanitin-sensitive polypeptides; both appear to be postranscriptionally regulated (Bolton et al. 1984). Finally, many of the changes in the polypeptide synthesis profile in the late 2-cell stage appear to be dependent on putative mRNA synthesis events (Bolton et al. 1984).

During the ensuing developmental period covering compaction, cavitation, and blastocyst formation, transcriptional and translational processes are more or less coupled. In the case of compaction, new transcription is necessary, although it may be completed in advance of this morphogenetic event by the 4-cell stage (Kidder and McLachlin 1985). New synthesis of protein and the rise in protein synthesis rate accompanying the morula-blastocyst transition are also dependent on ongoing transcriptional activity (Braude 1979a, 1979b). These events ultimately lead to the appearance of tissue-specific polypeptides in ICM and trophectoderm at the blastocyst stage (Van Blerkom, Barton, and Johnson 1976; Handyside and Johnson 1978; Brulet et al. 1980; Howe, Gmur, and Solter 1980).

IV. Activation of embryonic gene expression following fertilization

Many attempts have been made over the last 20 years to measure the rate of incorporation of radiolabeled precursors into RNA at various stages of postfertilization mouse development. Few studies, however, have paid attention to determining the absolute rates of synthesis based on the specific activity of the precursor pool. This is largely because of the difficulty of such measurements, especially where the amount of material is limited. Clegg and Piko (1977) measured the specific activities of UTP and ATP pools in early embryos by the method devised by Maxson and Wu (1976) for sea urchin eggs in which *Escherichia coli* RNA polymerase is used to synthesize poly(rA-rU) from a synthetic poly(dA-dT) template. Only trace incorporation of [^{3}H]uridine was observed in the 1-cell embryo, but absolute rates of synthesis were measurable by the 2-cell stage. The rate of uridine incorporation per embryo was observed to increase some 50-fold from the 2-cell stage to the blastocyst. On a cellular basis, however, the rate of synthesis was observed to increase only modestly from a value of 1.25 pg per cell per hour in the 2- to 4-cell embryo to 2.5 pg per cell per hour in the 8-cell embryo and to 5 pg per cell per hour in the blastocyst. These values are not very different from the rate of total RNA synthesis (5.7 pg/cell/h) in exponentially growing HeLa cells (Brandhorst and McConkey 1974) and are higher than rates of synthesis of heterogeneous RNA in cleavage- and blastula-stage sea urchin embryos, where the rate is about 0.5 per cell per hour (Brandhorst and Humphreys 1971; Wu and Wilt 1974).

Low levels of incorporation of [^{3}H]uridine and [^{3}H]guanosine into RNA in the 1-cell embryo have been observed (Clegg and Piko 1977; Young, Sweeney, and Bedford 1978). The validity of these observations was not clear at first, because failure to detect RNA polymerase activity in the pronuclei of 1-cell mouse embryos (Moore 1975) suggested that the embryonic genome was transcriptionally inactive until some time after the first cleavage. To reexamine RNA synthesis in the 1-cell embryo, Clegg and Piko (1983a, 1983b) used [^{3}H]adenosine as a precursor, because earlier studies had shown it to be about 1,000 times more readily taken up than uridine (Clegg and Piko 1977). Subsequent analyses have demonstrated the following: (1) The majority of labeled adenosine incorporated into tRNA is due to turnover of the 3'-terminal AMP, but some new synthesis of tRNA takes place in both 1-cell and 2-cell embryos at a rate of 0.2 pg per embryo per hour (Clegg and Piko 1983b). (2) Stable heterogeneous RNA devoid of poly(A) tracts (but nonribosomal) is synthesized by both 1-cell and 2-cell embryos at a rate of about 0.3 pg per embryo per hour. It is not clear if this is a putative mRNA fraction (Clegg and Piko 1983b). (3) Internally labeled poly(A)$^+$ RNA is synthesized in the 1-cell embryo, but at a very low rate of 0.045 pg per embryo per hour. At the 2-cell stage, the rate is five times higher (Clegg and Piko 1983b). (4) Ribosomal RNA synthesis is not detectable until the 2-cell stage, where the rate is 0.4 pg per embryo per hour (Clegg and Piko 1983b). (5) There is considerable incorporation of adenosine into poly(A) tails, due in large part to cytoplasmic adenylation of preexisting mRNA molecules, suggestive of turnover of poly(A) tails [and perhaps poly(A)$^+$ mRNA] in the 1-cell embryo (Young and Sweeney 1979; Clegg and Piko, 1983a, 1983b). In summary, a very low level of heterogenous RNA synthesis is occurring in 1-cell mouse embryos. Some of this nucleic acid metabolism is related to posttranscriptional modifications, as will be discussed subsequently. By the 2-cell stage, however, all major classes of RNA appear to be synthesized.

Because spermatozoa do not appear to contribute mRNA to the egg, detection of paternal gene products is another method to establish the expression of genes from the embryonic genome. Electrophoretic and structural variants of a number of enzymes have been used to advantage to identify paternal gene expression in embryos of crosses of appropriate mouse lines. The earliest of these, β-glucuronidase, has been detected as early as 30 to 57 h of development (late 2−4-cell stage) by Wudl and Chapman (1976). Similarly, the existence of an electrophoretic variant allowed Sawicki, Magnuson, and Epstein (1982) to detect paternally derived β$_2$-microglobulin synthesis (following immunoprecipitation, two-dimensional electrophoresis, and autoradiography) in 2-cell mouse embryos. This genetic evidence and the biochemical studies described earlier confirm that some newly derived transcripts from the embryonic genome are utilized at least by the 2-cell stage.

V. Posttranscriptional modification of mRNA

During the first 24 h of postfertilization development in the mouse, total protein content, net protein synthesis rate, and protein turnover do not change in any significant way (Brinster et al. 1976; Merz et al. 1981). Spare translational capacity in the mouse egg is extremely limited (Ebert and Brinster 1983). Hence, although there is an abundance of mRNA in the unfertilized egg, no postranscriptional or translational control mechanisms seem to be available to produce an increase in protein synthesis rate after fertilization.

In the sea urchin embryo, by contrast, there is a well-characterized mobilization of stored mRNA from free cytoplasmic messenger ribonucleoprotein particles (mRNP) to polysomes, although the in vivo mechanisms that control this event are not fully elucidated (see Davidson et al. 1982 for review). The majority of the mRNA that is stored in the sea urchin egg codes for the same set of proteins that are translated after fertilization (Brandhorst 1976). Thus, although there are dramatic quantitative changes subsequent to fertilization, there is little selection regarding the set of sequences being translated. An exception is the selected translation and resynthesis of cyclin proteins from maternal mRNA through each cell during early cleavage (Evans et al. 1983).

Qualitative rather than quantitative changes in protein synthesis pattern occur during the first two cleavages in early mouse development. Some of these seem postranslationally controlled, but others seem to rely on some selective utilization of stored mRNA (Howlett and Bolton 1985; Van Blerkom 1985). Although some of the mRNA in the newly ovulated egg is functional (Braude et al. 1979), little is known about the physical nature of stored mRNA as a whole. For example, limiting amounts of mouse embryo material have precluded the type of analyses performed in sea urchin and *Xenopus* eggs demonstrating that much of the cytoplasmic poly(A)$^+$ mRNA is unprocessed and contains interspersed repetitive-sequence elements (Costantini, Britten, and Davidson 1980; Anderson et al. 1982). Such elements are not present in polysomal poly(A)$^+$ mRNA of the sea urchin or *Xenopus* gastrulae. The *Xenopus* egg RNA that contains repetitive sequences is nonfunctional when added to in vitro translation systems (Richter et al. 1984). A similar situation does not appear to occur during mouse preimplantation development. A recent study has shown that full-grown oocytes and ovulated eggs do not contain any larger proportion of transcripts complementary to the Alu repetitive sequence family than differentiated cell types such as liver and brain (Kaplan, Jelinek, and Bachvarova 1985). In addition, the molecular-weight distributions of molecules containing complementary Alu sequences are similar in mouse egg, liver, and brain cytoplasmic RNA, a finding inconsistent with the notion of large unprocessed RNA molecules in mouse egg cytoplasm (Kaplan et al. 1985). Further evidence against the presence of unprocessed RNA is provided by studies on U1 small nuclear RNA, an essential component for splicing and ligation of messenger RNA precursors (Kramer et al. 1984). Growing oocytes, full-grown oocytes, and newly ovulated eggs of the mouse all contain about 0.7 pg of U1 RNA (Kaplan et al. 1985). This is equal to a concentration of about 250 pg/nl of nuclear (germinal vesicle) volume in the oocyte and is very similar to the concentration (350 pg/nl) in HeLa cell and liver nuclei (Kaplan et al. 1985). By contrast, in the *Xenopus* system, where unprocessed RNA accumulates in the egg cytoplasm, there is a relative depletion of U1 RNA in fully grown oocytes and early embryos as compared with growing oocytes and postgastrula stages of development (Forbes, Kronberg, and Kirschner 1983; Zeller, Nyffenegger, and DeRobertis 1983).

Considerable effort has been given to analyzing [^{3}H]adenosine incorporation into poly(A) tracts and poly(A)$^+$ mRNA in 1-cell and 2-cell mouse embryos (Young and Sweeney 1979; Clegg and Piko 1983a; 1983b). The major conclusion is that there is some turnover of the poly(A)$^+$ mRNA population in 1-cell embryos as a result of polyadenylation of new RNA species and degradation of some of the preexisting poly(A)$^+$ RNA. Similar poly(A) addition to maternal mRNA was observed in sea urchin embryos (Slater, Gillespie, and Slater 1973; Wilt

1973), but this seems to have no role in the postfertilization rise in protein synthesis rate in these embryos (Mescher and Humphreys 1974). Because protein synthesis rates do not rise in the 1-cell or 2-cell period in the mouse (Brinster et al. 1976), the polyadenylation events are not related to quantitative changes in protein synthesis activity. Moreover, nearly every polypeptide synthesized in reticulocyte lysate from unfertilized egg mRNA is already represented in the poly(A)$^+$ mRNA fraction (Johnson 1981). Although the in vivo function of cytoplasmic adenylation and accompanying poly(A)$^+$ mRNA degradation remains speculative, it seems integrally associated with a shift from maternal to embryo-derived mRNA templates during this period. In this regard, it is interesting to note that translation of actin, which stops in the unfertilized egg when follicle cells are removed (Osborn and Moor 1982), may be regulated by concomitant deadenylation of actin mRNA (Bachvarova et al. 1985).

In addition to their 3′ poly(A) tails, most eukaryotic mRNAs contain at the 5′ end a structure of the form . . . $m^7G^{5'}ppp^{5'}X^mpY$ . . . that is added rapidly after initiation of transcription and is called the cap structure. The cap structure contributes to the stability of mRNAs by protecting them against 5′ exonucleolytic degradation (Furuichi, LaFiandra, and Shatkin 1977). The cap also plays an important role in mammalian cells in formation of an initiation complex along with cap binding protein (Sonenberg et al. 1978) for mRNA translation (see Banerjee 1980 for review).

Young (1977) observed a low level of [^{3}H]guanosine incorporation into RNA obtained from 1-cell mouse embryos 1−3 h after fertilization. Although the limited amount of material did not allow direct demonstration of [^{3}H]guanosine addition to preexisting RNA, digestion with ribonucleases A, T_1, and T_2 followed by nucleotide pyrophosphatase led to the resolution of both pm^7G and GP, with more radioactivity in the former than the latter. The m^7G should have been released from the cap, but not other 5′ termini. In an alternative approach to assess whether or not capping might be a regulatory mechanism to activate the utilization of maternal mRNAs after fertilization, the degree of capping in mouse egg and embryo RNA was assessed by (1) translational inhibition with cap analogues in reticulocyte lysate and (2) labeling egg and embryo poly(A)$^+$ with γ-^{32}P-ATP and polynucleotide kinase after enzymatic removal of cap structures with tobacco acid phosphatase (Schultz, Clough, and Johnson 1980). Reductions in the synthesis of all types of polypeptides coded for by mRNA from all stages of development analyzed (from egg to blastocyst) were observed in the presence of the cap analogue. There was no evidence for poly(A)$^+$ mRNA molecules without cap structures in the end-labeling experiments. The degree of capping was observed to be the same in unfertilized and fertilized egg mRNA with both approaches (Schultz et al. 1980).

Renewed interest in capping of maternal mRNAs as a means of activating stored mRNA in sea urchin eggs has emerged since the discovery that, unlike mammalian cells, sea urchin eggs lack the machinery to initiate translation of uncapped mRNAs and have an absolute requirement for 5' cap structures (Winkler, Bruening, and Hershey 1983). In particular, postranscriptional addition of [^{3}H]methyl cap structures to maternally derived histone mRNAs occurs following fertilization of the sea urchin egg (Caldwell and Emerson 1985). Quantitatively, more than 90 percent of the cap methylation is recovered in small (less than 12S) RNA molecules lacking poly(A) segments, and the methylation process is temporally coordinated with DNA replication (Caldwell and Emerson 1985). It is not known if the remaining cap methylated mRNA molecules, which account for only a few percent of the maternal mRNA molecules in the sea urchin egg, are translationally activated by a similar mechanism. The significance of the foregoing findings for the mouse zygote is unclear. Because there is no marked quantitative increase in protein synthesis rate following fertilization, and because histone and DNA syntheses are uncoordinated in the early mouse embryo (Kaye and Church 1983), the systems are quite different. The low level of cap addition suggested by Young's (1977) data could be significant if it indeed leads to increased protein coding potential of a few selectively activated mRNAs. However, because uncapped mRNAs in mammalian cells are unstable (Banerjee 1980), and because cap structures are necessary on pre-mRNAs of mammalian cells for correct splicing and excision of intervening sequences (Konarska, Padgett, and Sharp 1984; Kramer et al. 1984), any proposed mechanism of postranscriptional regulation via capping should be viewed with caution.

VI. Changes in protein synthesis rate and theoretical translation efficiency

The rate of protein synthesis in the early mouse embryo remains constant through fertilization and development to the 2-cell stage. Between the 2-cell stage and blastocyst stage, the rate of protein synthesis increases sixfold to eightfold (Brinster et al. 1976). There is an actual loss of some ribosomes from the fertilized egg to the 2-cell stage (Piko and Clegg 1982). It appears that not all ribosomes in the newly ovulated egg are functional when assayed by ability to form initiation complexes with a synthetic messenger RNA (Bachvarova and De Leon 1977). Messenger RNA levels are high in the egg compared with later cleavage stages (Piko and Clegg 1982; Giebelhaus et al. 1985; Graves et al. 1985a). Hence, it is not surprising that synthesis of polypeptides from available mRNA templates in the egg is at a low efficiency.

On the basis of the number of mRNA molecules per embryo and the absolute rate of synthesis of a given protein per unit time, it is possible to calculate theoretical translation efficiencies. This mathematical treatment has been applied to data for both histone H3 and actin synthesis (Table 8.2). That the egg is rather inefficient in protein synthesis is clear for both histone H3 and, especially, actin production. From two-dimensional electrophoretic studies, actin synthesis occurs at very low levels in the egg (Howe and Solter 1979; Cullen et al. 1980). Yet mRNA levels are high at this time. Actin synthesis is clearly detectable by the 2-cell stage and becomes an increasingly major product of synthesis during further development to the blastocyst stage (Howe and Solter 1979). Quantitatively, actin synthesis represents 0.25 percent of the total protein synthesis activity at the 2-cell stage, 2 percent of the total at the 8-cell stage, and 5.7 percent of the total in the blastocyst (Abreu and Brinster 1978). This increase is accompanied by a progressive improvement in the number of polypeptides synthesized per mRNA per unit time over this developmental period (Table 8.2). The theoretical values calculated here for actin and histone proteins in the blastocyst are within the observed range of translational efficiencies of other protein products in eukaryotic cells maintained at 37°C (Kafatos 1972).

The increase in translational efficiency after the 2-cell stage might well be explained on the basis of an increase in ribosomal numbers. Initially it was hypothesized that the fibrillar lattices observed in mouse eggs consist of ribosomes embedded in a proteinaceous structure that were gradually dispersed during implantation development (Schlafke and Enders 1967; Burkholder, Comings, and Okada 1971) (see Chapter 7). This now appears incorrect, but ribosome numbers nonetheless increase because of active synthesis of rRNA (between 1.25 and 2.5 pg per cell per hour at the 8-cell and morula stages) (Piko and Clegg 1982) and active synthesis of ribosomal proteins (4.17 pg per embryo per hour at the 8-cell stage) (La Marca and Wassarman 1979). These rates of synthesis are sufficient for the production of about 2.5×10^6 ribosomes per embryo per hour and can account for the increase in ribosomal content observed in early mouse development (Piko and Clegg 1982). Certainly changes in initiation and elongation rates may also be involved in the predicted increase in translational efficiency, but the major reason for increased protein synthesis rate appears simply to be the increase in the amount of protein synthesis machinery. A final component may involve recruitment of mRNA from subribosomal ribonucleoprotein fractions to polysomes in cleavage and morula stages up to the point of blastocyst expansion (Kidder and Conlon 1985).

On the basis of a histone H3 mRNA content of 20,000 molecules per cell and a translational efficiency of five polypeptides per mRNA per minute (Table 8.2), it can be predicted that it would take approximately 8 to 10 h to produce the 1.5 pg of this nucleosomal core protein required for one cell doubling in the early blastocyst. This is possible, because the doubling time at this stage of development is about 12 h and histone and DNA synthesis are uncoordinated in mouse embryos during most of the preimplantation period (Kaye and Church 1983). It is not known if there is a carryover of histone protein from the egg stage through early cleavage stage. At least some histone protein must be present in the unfertilized egg to displace protamines after fertilization (Eklund and Levine 1975).

VII. Differential expression of genes during preimplantation development

A number of cell-specific polypeptides unique to ICM and trophectoderm have been identified by two-dimensional electrophoretic methods (Van Blerkom et al. 1976; Handyside and Johnson 1978). In the following section, discussion of changes in gene expression between the egg-to-cleavage-stage embryo and the blastocyst is confined to heat-shock genes and the histone gene family. These have been chosen because data are available at both the protein level and mRNA level (availability of homologous recombinant DNA probes), the products are relatively abundant, and, in the case of heat-shock gene expression, the induction of expression is simple in terms of experimental manipulation.

A. Response to heat shock

Exposure of virtually all prokaryotic and eukaryotic cells to heat shock or other environmental stresses results in the induction of synthesis of a small number of heat-shock proteins (hsps). Heat-shock proteins from a wide variety of organisms are immunologically cross-reactive and have a number of structural similarities (Kelly and Schlesinger 1982; Voellmy, Bromley, and Kocher 1983). Although the cellular mechanism remains to be elucidated, hsp induction and synthesis are correlated with an increased capacity for thermotolerance (McAlister and Finkelstein 1980; Li and Werb 1982).

Heat shock genes can be activated by a number of environmental stresses and are also developmentally regulated at specific stages of embryogenesis. For example, hsp 70 mRNA accumulates in *Xenopus* oocytes (Bienz and Gurdon 1982) but is rapidly lost after fertilization (Bienz 1984). A number of hsp mRNAs also accumulate in *Drosophila* oocytes (Zimmerman, Petri, and Meselson 1983) and late larval and early pupal stages (Cheney and Shearn 1983) in the absence of stress-inducing stimuli. During the early 2-cell stage in the mouse, hsp 70 appears to be synthesized transiently as an early product from the zygote genome (Bensaude et al. 1983).

A unique heat-shock-incompetent period, in which induction of hsps in response to environmental stresses does not occur, has been observed prior to the blastoderm stage of *Drosophila* embryos (Dura 1981), blastula stage in sea urchins (Roccheri, Bernardo, and Guidice 1981a), and mid-blastula stage in *Xenopus* embryos (Bienz 1984; Heikkila et al. 1985). The acquisition of thermotolerance correlates with the accumulation of hsp 68–70 mRNA and the synthesis of the respective proteins. Although hsp 68–70 synthesis appears to be developmentally regulated at the 2-cell stage of mouse development, the ability to respond to elevated temperature by the induction of hsp synthesis is not acquired until the blastocyst stage in the mouse (Wittig et al. 1983; Morange et al. 1984) and the rabbit (Heikkila and Schultz 1984). Again, the most heat-sensitive period in early mammalian development is between early cleavage and implantation (Ulbert and Sheean 1973), and it is noteworthy that in the mouse the most sensitive event seems to be that of activation of RNA synthesis at the 2-cell stage (Bellvé 1976).

The mechanism by which heat-shock competence and the ability to transcribe mRNAs from hsp genes in response to induction are acquired in development is unknown. Clearly there is a difference in the genetic apparatus or necessary transcriptional factors between the hsp-refractory cleavage stages and the hsp-inducible blastocyst stage. Even in the latter, heat-shock responsiveness may be restricted to specific regions of the embryo, because in the sea urchin embryo, heat-shock protein synthesis is predominantly associated with ectodermal cells (Roccheri et al. 1981b). The differential response to heat shock during preimplantation mammalian development awaits further study.

B. Histone gene expression

The histone genes of mammals are part of a small multigene family with 10 to 20 different genes for each histone protein. Nine histone genes (four H3, three H2b, and two H2a genes) that are expressed in cultured mouse cells and fetal mice have been isolated from three different mouse genomic fragments. All of these genes have been characterized (Sittman et al. 1893; Graves et al. 1985b). Two of the gene clusters, MM221 and MM229, are located on chromosome 13 of the mouse, and the other, MM614, is located on chromosome 3.

Because the coding regions of the various histone genes within a family are highly homologous (95%), but regions complementary to untranslated regions of the mRNA are divergent, it has been possible to develop an S1-nuclease mapping technique to quantitatively measure the expression of individual genes (Graves et al. 1985b). The two genes on chromosome 3 (one H3 and one H2a gene) are expressed in quite high amounts in cultured mouse cells, each contributing about 30 percent of the total transcripts in the pool. The remaining seven genes (on chromosome 13), on the other hand, are all expressed in small amounts in cultured cells and are homologous to less than 10 percent of the total transcript pool in each case (Graves et al. 1985b).

The S1-nuclease mapping techniques have recently been extended to analysis of maternal mRNA derived from newly ovulated mouse eggs and to zygote-genome-derived mRNA at the early blastocyst stage (Graves et al. 1985a). The H3 gene on chromosome 3 (H3.614) is expressed in very high amounts in the egg (40−50%), as in cultured mouse myeloma cells, but is complementary to only 14 percent of the histone H3 mRNA in the blastocyst. One of the genes from chromosome 13 (H3.221) is also represented in high amounts (40%) in the egg mRNA, whereas in this case the low level in blastocyst mRNA (7%) is similar to that of mouse myeloma cells (Graves et al. 1985a). The most striking variation between egg and blastocyst RNA was observed with the H2a.614 gene from chromosome 3. This gene accounts for nearly all of the H2a mRNA (>90%) in the egg, but only 30 percent of the H2a mRNA in blastocysts and myeloma cells. These S1-nuclease assays demonstrate that although the same sets of histone genes are expressed in eggs and early embryos, there are large quantitative differences in the amounts of certain histone mRNAs.

The amount of a particular mRNA is determined by both its rate of synthesis and rate of degradation. At the present time, it is not known if the changes in proportions of individual histone mRNAs are controlled at the level of mRNA stability. A further consequence of the variation in histone mRNA levels is a change in the histone protein variants (e.g., H2a.2 protein versus H2a.1 protein) (Franklin and Zweidler 1977) expressed in egg and blastocyst. The significance of these changes with respect to control of gene expression is not clear, but changes in histone gene sets expressed during sea urchin development are also well documented (Newrock et al. 1978; Maxson et al. 1983).

VIII. Future directions

The S1-nuclease mapping study of histone mRNAs (Graves et al. 1985a) demonstrated that it is possible to measure specific mRNAs in mouse eggs and early embryos. The rare histones detected in that study (where RNA from about 100 eggs or blastocysts was utilized) can be calculated to be present at about 600 copies per cell in the blastocyst. By increasing probe-specific activity and autoradiographic exposure time, further sensitivity to a level of detection of mRNAs present in only 100 copies per cell should be possible. Recently, in situ hybridization methods have been developed that can also potentially detect mRNAs in the concentration range of 50 to 100 copies per cell (Cox et al. 1984). Such methods can be applied to small numbers of embryos and have the power to localize the expression of particular genes to defined cell types or lineages (Lynn et al. 1983; Angerer and Davidson 1984). Indeed, the cytological application of molecular probes to the study of gene expression is now an active pursuit of many laboratories.

It is theoretically possible to clone any gene for which the protein product is available in nanomolar quantities by a combination of protein microsequencing and synthesis of minimally redundant oligodeoxynucleotide probes, followed by cloning of the cognate cDNA and finally the gene itself. At the protein level, as nucleic acid sequences become known, they can be converted to peptide sequences suitable for raising antibodies for immunocytochemical studies. These approaches do have promise for localization of the transcript of a gene or its protein product, or both, to particular cell types in early mammalian development.

The small amount of mRNA in eggs and early mouse embryos has hampered the process of identifying stage-specific gene utilization in the preimplantation period and the ultimate molecular characterization of the genes. It is now possible, however, to construct cDNA libraries from as little as 100 ng of poly(A)$^+$ mRNA. For the 2-cell embryo, where the poly(A)$^+$ mRNA content is lowest (8 pg, Table 8.1), this would require the collection of 10,000 to 15,000 embryos. This sounds formidable, but it is not an impossible task. Once a set of libraries for stages of interest is constructed, screening for stage-specific clones is performed at the library level without the use of embryonic material. Again, several laboratories have begun the cDNA approach, which could lead to rapid progress in analyzing stage-specific gene expression and transcriptional regulatory components.

A final approach to problems of gene expression in preimplantation development might well involve the use of anti-sense RNAs (Coleman, Green, and Inouye 1984; Izant and Weintraub 1984) to abolish or block endogenous gene function. This approach has been used elegantly in *Drosophila* to elucidate the action of the Kruppel gene (Preisa et al. 1985). Attempts to identify the function of early gene products in the mouse embryo may be hampered by the likelihood that microinjected anti-sense RNAs will be degraded along with maternal mRNA through the 2-cell stage. If anti-sense genes are used, limitations of expression may occur because of "de novo" methylation and random chromosomal integration of microinjected sequences. Nonetheless, considerable information regarding the integration and expression of exogenous genetic material introduced into mouse eggs has been acquired in recent years (see Chapter 16), providing a basis for novel experimental approaches to gene expression in the near future.

Acknowledgments

Original work from the laboratory of the author referred to in this review was supported by a grant from the Medical Research Council of Canada. I am grateful for the expert assistance provided by Ms. Marnie Cudmore in the preparation of this manuscript.

References

Abreu, S. L., and Brinster, R. L. (1978) Synthesis of tubulin and actin during the preimplantation development of the mouse. *Exp. Cell Res.* 114, 135–41.

Anderson, D. M., Richter, J. D., Chamberlin, M. E., Price, D. H., Britten, R. J., Smith, L. D., and Davidson, E. H. (1982) Sequence organization of the poly(A) RNA synthesized and accumulated in lampbrush chromosome stage *Xenopus laevis* oocytes. *J. Mol. Biol.* 155, 281–309.

Angerer, R. C., and Davidson, E. H. (1984) Molecular indices of cell lineage specification in sea urchin embryos. *Science* 226, 1153–60.

Bachvarova, R. (1974) Incorporation of tritiated adenosine into mouse ovum RNA. *Dev. Biol.* 40, 52–8.

Bachvarova, R., and De Leon, V. (1977) Stored and polysomal ribosomes of mouse ova. *Dev. Biol.* 58, 248–54.

– (1980) Polyadenylated RNA of mouse ova and loss of maternal RNA in early development. *Dev. Biol.* 74, 1–8.

Bachvarova, R., De Leon, V., Johnson, A., Kaplan, G., and Paynton, B. V. (1985) Changes in total RNA, polyadenylated RNA, and actin mRNA during meiotic maturation of mouse oocytes. *Dev. Biol.* 108, 325–31.

Banerjee, A. K. (1980) 5'-terminal cap structure in eukaryotic messenger ribonucleic acids. *Microbiol. Rev.* 44, 175–205.

Bellvé, A. R. (1976) Incorporation of [^{3}H]uridine by mouse embryos with abnormalities induced by parental hyperthermia. *Biol. Reprod.* 15, 632–46.

Bensaude, O., Babinet, C., Morange, M., and Jacob, F. (1983) Heat shock proteins, first major products of zygotic gene activity in the mouse embryo. *Nature* 305, 331–1.

Bienz, M. (1984) Developmental control of the heat shock response in *Xenopus. Proc. Natl. Acad. Sci. U.S.A.* 81, 3138–42.

Bienz, M., and Gurdon, J. B. (1982) The heat shock response in *Xenopus* oocytes is controlled at the translational level. *Cell* 27, 811–19.

Bolton, V. N., Oades, P. J., and Johnson, M. H. (1984) The relationship between cleavage, DNA replication, and gene expression in the mouse 2-cell embryo. *J. Embryol. Exp. Morphol.* 79, 139–63.

Brandhorst, B. P. (1976) Two dimensional gel patterns of protein synthesis before and after fertilization of sea urchin eggs. *Dev. Biol.* 52, 310–17.

Brandhorst, B. P., and Humphreys, T. (1971) Synthesis and decay rates of major classes of deoxyribonucleic acid-like ribonucleic acid in sea urchin embryos. *Biochemistry* 10, 877–81.

Brandhorst, B. P., and McConkey, E. H. (1974) Stability of nuclear RNA in mammalian cells. *J. Mol. Biol.* 85, 451–63.

Braude, P. R. (1979a) Control of protein synthesis during blastocyst formation in the mouse. *Dev. Biol.* 68, 440–52.

– (1979b) Time-dependent effects of α-amanitin on blastocyst formation in the mouse. *J. Embryol. Exp. Morphol.* 52, 193–202.

Braude, P. R., Pelham, H., Flach, G., and Lobatto, R. (1979) Post-transcriptional control in the early mouse embryo. *Nature* 282, 102–5.

Brinster, R. L. (1967) Protein content of the mouse embryo during the first five days of development. *J. Reprod. Fertil.* 13, 413–20.

Brinster, R. L., Chen, H. Y., Trumbauer, M. E., and Avarbock, M. R. (1980) Translation of rabbit globin messenger RNA by the mouse ovum. *Nature* 283, 499–501.

Brinster, R. L., Wiebold, J. L., and Brunner, S. (1976) Protein metabolism in preimplanted mouse ova. *Dev. Biol.* 51, 215–24.

Brulet, P., Babinet, C., Kemler, R., and Jacob, F. (1980) Monoclonal antibodies against trophectoderm specific markers during mouse blastocyst formation. *Proc. Natl. Acad. Sci. U.S.A.* 77, 4113–17.

Burkholder, G. D., Comings, D. E., and Okada, T. A. (1971) A storage form of ribosomes in mouse oocytes. *Exp. Cell Res.* 69, 361–71.

Caldwell, D. C., and Emerson, C. P., Jr. (1985) The role of cap methylation in the translational activation of stored maternal histone mRNA in sea urchin embryos. *Cell* 42, 691–700.

Cascio, S. M., and Wassarman, P. M. (1982) Program of early development in the mammal: post-transcriptional control of a class of proteins synthesized by mouse oocytes and early embryos. *Dev. Biol.* 89, 397–408.

Cheney, C. M., and Shearn, A. (1983) Developmental regulation of *Drosophila* imaginal disc proteins: synthesis of a heat shock protein under non-heat-shock conditions. *Dev. Biol.* 95, 325–30.

Clegg, K. B., and Piko, L. (1977) Size and specific activity of the UTP pool and overall rates of RNA synthesis in early mouse embryos. *Dev. Biol.* 58, 76–95.

– (1983a) Poly(A) length, cytoplasmic adenylation and synthesis of poly(A)$^+$ RNA in early mouse embryos. *Dev. Biol.* 95, 331–41.

– (1983b) Quantitative aspects of RNA synthesis and polyadenylation in 1-cell and 2-cell mouse embryos. *J. Embryol. Exp. Morphol.* 74, 169–82.

Coleman, J., Green, P. J., and Inouye, M. (1984) The use of RNAs complementary to specific mRNAs to regulate the expression of individual bacterial genes. *Cell* 37, 429–36.

Costantini, F. D., Britten, R. J., and Davidson, E. H. (1980) Message sequences and short repetitive sequences are interspersed in sea urchin egg poly(A)$^+$ RNAs. *Nature* 287, 111–17.

Cox, K. H., De Leon, D. V., Angerer, L. M., and Angerer, R. (1984) Detection of mRNAs in sea urchin embryos by *in situ* hybridization using asymmetric RNA probes. *Dev. Biol.* 101, 485–502.

Cullen, B., Emigholz, K., and Monahan, J. (1980) The transient appearance of specific proteins in one-cell mouse embryos. *Dev. Biol.* 76, 215–21.

Daniel, J. C., Jr. (1964) Early growth of rabbit trophoblast. *Am. Natur.* 98, 85–98.

Davidson, E. H., Hough-Evans, B. R., and Britten, R. J. (1982) Molecular biology of the sea urchin embryo. *Science* 217, 17–26.

Dura, J. M. (1981) Stage-dependent synthesis of heat shock induced proteins in early embryos of *Drosophila melanogaster*. *Mol. Gen. Genet.* 184, 381–5.

Ebert, K. M., and Brinster, R. L. (1983) Rabbit α-globin messenger RNA translation by the mouse ovum. *J. Embryol. Exp. Morphol.* 74, 159–68.

Eklund, P. F., and Levine, L. (1975) Mouse sperm basic nuclear proteins.

Electrophoretic characterization and fate after fertilization. *J. Cell Biol.* 66, 251–62.

Elzinga, M., Collins, J. H., Kuehl, W. M., and Adelstein, R. S. (1973) Complete amino acid sequence of actin from rabbit skeletal muscle. *Proc. Natl. Acad. Sci. U.S.A.* 70, 2687–91.

Evans, T., Rosenthal, E. T., Youngblom, J., Distel, D., and Hunt, T. (1983) Cyclin: a protein specified by maternal mRNA in sea urchin eggs that is destroyed at each cleavage division. *Cell* 33, 389–96.

Flach, G., Johnson, M. H., Braude, P. R., Taylor, R. A. S., and Bolton, V. N. (1982) The transition from maternal to embryonic control in the 2-cell mouse embryo. *EMBO Journal* 1, 681–6.

Forbes, D. J., Kornberg, T. B., and Kirschner, M. W. (1983) Small nuclear RNA transcription and ribonucleoprotein assembly in early Xenopus development. *J. Cell Biol.* 97, 62–72.

Franklin, S. A., and Zweidler, A. (1977) Non-allelic variants of histones 2a, 2b and 3 in mammals. *Nature* 266, 273–5.

Furuichi, Y., La Fiandra, A., and Shatkin, A. J. (1977) 5′-terminal structure and mRNA stability. *Nature* 266, 235–9.

Giebelhaus, D. H., Heikkila, J. J., and Schultz, G. A. (1983) Changes in the quantity of histone and actin messenger RNA during the development of preimplantation mouse embryos. *Dev. Biol.* 98, 148–54.

Giebelhaus, D. H., Weitlauf, H. M., and Schultz, G. A. (1985) Actin mRNA content in normal and delayed implanting mouse embryos. *Dev. Biol.* 107, 407–13.

Graves, R. A., Marzluff, W. F., Giebelhaus, D. H., and Schultz, G. A. (1985a) Quantitative and qualitative changes in histone gene expression during early mouse development. *Proc. Natl. Acad. Sci. U.S.A.* 82, 5685–9.

Graves, R. A., Wellman, S. E., Chiu, I. M., and Marzluff, W. F. (1985b) Differential expression of two clusters of mouse genes. *J. Mol. Biol.* 183, 179–94.

Handyside, A. H., and Johnson, M. H. (1978) Temporal and spatial patterns of the synthesis of tissue-specific polypeptides in the preimplantation mouse embryo. *J. Embryol. Exp. Morphol.* 44, 191–9.

Heikkila, J. J., Kloc, M., Bury, J., Schultz, G. A., and Browder, L. W. (1985) Acquisition of the heat shock response and thermo-tolerance during early development of *Xenopus laevis. Dev. Biol.* 107, 483–9.

Heikkila, J. J., and Schultz, G. A. (1984) Different environmental stresses can activate the expression of a heat shock gene in rabbit blastocyst. *Gam. Res.* 10, 45–56.

Howe, C. C., Gmur, R., and Solter, D. (1980) Cytoplasmic and nuclear protein synthesis during *in vitro* differentiation of murine ICM and embryonal carcinoma cells. *Dev. Biol.* 74, 351–63.

Howe, C. C., and Solter, D. (1979) Cytoplasmic and nuclear protein synthesis in preimplantation mouse embryos. *J. Embryol. Exp. Morphol.* 52, 209–25.

Howlett, S. K., and Bolton, V. N. (1985) Sequence and regulation of morphological and molecular events during the first cell cycle of mouse embryogenesis. *J. Embryol. Exp. Morphol.* 87, 175–206.

Izant, J. G., and Weintraub, H. (1984) Inhibition of thymidine kinase gene expression by anti-sense RNA: a molecular approach to genetic analysis. *Cell* 36, 1007–15.

Johnson, M. H. (1981) The molecular and cellular basis of preimplantation mouse development. *Biol. Rev.* 56, 463–98.

Kafatos, F. C. (1972) The cocoonase zymogen cells of silkmoths: a model of terminal differentiation for specific protein synthesis. In *Current Topics in Developmental Biology*, Vol. 7, eds. A. A. Moscona and A. Monroy, pp. 125–91. New York: Academic Press.

Kaplan, G., Jelinek, W. R., and Bachvarova, R. (1985) Repetitive sequence transcripts and U1 RNA in mouse oocytes and eggs. *Dev. Biol.* 109, 15–24.

Kaye, P. L., and Church, R. B. (1983) Uncoordinated synthesis of histone and DNA by mouse eggs and preimplantation embryos. *J. Exp. Zool.* 226, 231–7.

Kelley, P. M., and Schlesinger, M. J. (1982) Antibodies to two major heat shock proteins cross-react with similar proteins in widely divergent species. *Mol. Cell Biol.* 2, 267–74.

Kidder, G. M., and Conlon, R. A. (1985) Utilization of cytoplasmic poly(A)$^+$ RNA for protein synthesis in preimplantation mouse embryos. *J. Embryol. Exp. Morphol.* 89, 223–34.

Kidder, G. M., and McLachlin, J. R. (1985) Timing of transcription and protein synthesis underlying morphogenesis in preimplantation mouse embryos. *Dev. Biol.* 112, 265–75.

Konarska, M. M., Padgett, R. A., and Sharp, P. A. (1984) Recognition of cap structure in splicing *in vitro* of mRNA precursors. *Cell* 38, 731–6.

Kramer, A., Keller, W., Appel, B., and Luhrmann, R. (1984) The 5' terminus of the RNA moiety of U1 small nuclear ribonucleoprotein particles is required for splicing of messenger RNA precursors. *Cell* 38, 299–307.

La Marca, M. J., and Wassarman, P. M. (1979) Program of early development in the mammal: changes in absolute rates of synthesis of ribosomal proteins during oogenesis and early embryogenesis in the mouse. *Dev. Biol.* 73, 103–19.

Levinson, J., Goodfellow, P., Vadeboncoeur, M., and McDevitt, H. (1978) Identification of stage-specific polypeptides synthesized during murine preimplantation development. *Proc. Natl. Acad. Sci. U.S.A.* 75, 3332–6.

Levy, I. L., Stull, G. B., and Brinster, R. L. (1978) Poly(A) and synthesis of polyadenylated RNA in the preimplantation mouse embryo. *Dev. Biol.* 64, 140–8.

Lewin, B. (1975) Units of transcription and translation: sequence components of heterogeneous nuclear RNA and mRNA. *Cell* 4, 77–93.

– (1980) Complexity of mRNA populations. In *Gene Expression. II. Eukaryotic Chromosomes*, ed. B. Lewin, pp. 694–727. New York: Wiley.

Li, G.C., and Werb, Z. (1982) Correlation between synthesis of heat shock proteins and development of thermotolerance in Chinese hamster fibroblasts. *Proc. Natl. Acad. Sci. U.S.A.* 79, 3268–72.

Lynn, D. A., Angerer, L. M., Bruskin, A. M., Klein, W. H., and Angerer, R. (1983) Localization of a family of mRNAs in a single cell type and its precursors in sea urchin embryos. *Proc. Natl. Acad. Sci. U.S.A.* 80, 2656–60.

McAlister, L., and Finkelstein, D. (1980) Heat shock proteins and thermal resistance in yeast. *Biochem. Biophys. Res. Commun.* 93, 891–24.

Manes, C., Byers, M. J., and Carver, A. S. (1981) Mobilization of genetic information in the early rabbit trophoblast. In *Cellular and Molecular Aspects of Implantation,* eds. S. R. Glasser and D. W. Bullock, pp. 113–24. New York: Plenum.

Matheson, R. C., and Schultz, G. A. (1980) Histone synthesis in preimplantation rabbit embryos. *J. Exp. Zool.* 213, 337–49.

Maxson, R. E., Mohun, T., Gormexano, G., Childs, G., and Kedes, L. (1983) Distinct organizations and patterns of expression of early and late histone gene sets in sea urchin. *Nature* 301, 120–5.

Maxson, R. E., and Wu, R. S. (1976) A simple method for measuring specific radioactivities of ribonucleoside triphosphates using RNA polymerase. *Eur. J. Biochem.* 62, 551–4.

Merz, E. A., Brinster, R. L., Brunner, S., and Chen, H. Y. (1981) Protein degradation during preimplantation development of the mouse. *J. Reprod. Fertil.* 61, 415–18.

Mescher, A., and Humphreys, T. (1974) Activation of maternal mRNA in the absence of poly(A) formation in fertilized sea urchin eggs. *Nature* 249, 138–9.

Moore, G. P. M. (1975) The RNA polymerase activity of the preimplantation mouse embryo. *J. Embryol. Exp. Morphol.* 34, 291–8.

Morange, M., Diu, A., Bensaude, O., and Babinet, C. (1984) Altered expression of heat shock proteins in embryonal carcinoma and mouse early embryonic cells. *Mol. Cell. Biol.* 4, 730–5.

Newrock, K. M., Cohen, L. H., Hendricks, M. B., Donnelly, R. J., and Weinberg, E. S. (1978) Stage-specific mRNAs coding for subtypes of H2A and H2B histones in the sea urchin embryo. *Cell* 14, 327–36.

Olds, P. J., Stern, S., and Biggers, J. D. (1973) Chemical estimates of the RNA and DNA contents of the early mouse embryo. *J. Exp. Zool.* 186, 39–46.

Osborn, J. C., and Moor, R. M. (1982) Cell interactions and actin synthesis in mammalian oocytes. *J. Exp. Zool.* 220, 125–9.

Petzoldt, U., Hoppe, P. C., and Illmensee, K. (1980) Protein synthesis in enucleated fertilized and unfertilized mouse eggs. *Wilhelm Roux Arch. Dev. Biol.* 189, 215–19.

Piko, L., and Clegg, K. B. (1982) Quantitative changes in total RNA, total poly(A), and ribosomes in early mouse embryos. *Dev. Biol.* 89, 362–78.

Piko, L., Hammons, M. D., and Taylor, K. D. (1984) Amounts, synthesis, and some properties of intracisternal A-particle-related RNA in early mouse embryos. *Proc. Natl. Acad. Sci. U.S.A.* 81, 488–92.

Preisa, A., Rosenberg, U., Kienlin, A., Seifert, E., and Jackle, H. (1985) Molecular genetics of Kruppel, a gene required for segmentation of the *Drosophila* embryo *Nature* 315, 27–32.

Richter, J. D., Smith, L. D., Anderson, D. M., and Davidson, E. H. (1984) Interspersed poly(A) RNAs of amphibian oocytes are not translatable. *J. Mol. Biol.* 173, 227–41.

Roccheri, M. C., Bernardo, M. G., and Guidice, G. (1981a) Synthesis of heat shock proteins in developing sea urchin. *Dev. Biol.* 83, 173–7.

Roccheri, M. C., Sconzo, G., DiBernardo, M. G., Albanese, I., DiCarlo, M., and Giudice, G. (1981b) Heat shock proteins in sea urchin embryos. Territorial and intracellular regulation. *Acta Embryol. Morph. Exp. N.S.* 2, 91−9.

Sawicki, J. A., Magnuson, T., and Epstein, C. J. (1982) Evidence for expression of the paternal genome in the two-cell mouse embryo. *Nature* 294, 450−1.

Schlafke, S., and Enders, A. C. (1967) Cytological changes during cleavage and blastocyst formation in the rat. *J. Anat.* 102, 13−32.

Schultz, G. A., Clough, J. R., Braude, P. R., Pelham, H. R. B., and Johnson, M. H. (1981) A reexamination of messenger RNA populations in the preimplantation mouse embryo. In *Cellular and Molecular Aspects of Implantation,* eds. S. R. Glasser and D. W. Bullock, pp. 137−54. New York: Plenum.

Schultz, G. A., Clough, J. R., and Johnson, M. H. (1980) Presence of cap structures in the messenger RNA of mouse eggs. *J. Embryol. Exp. Morphol.* 56, 139−56.

Schultz, G. A., Manes, C., and Hahn, W. E. (1973) Estimation of the diversity of transcription in early rabbit embryos. *Biochem. Genet.* 9, 247−59.

Sittman, D. B., Graves, R. A., and Marzluff, W. F. (1983) Structure of a cluster of mouse histone genes. *Nucleic Acids Res.* 11, 6679−97.

Slater, I., Gillespie, D., and Slater, D. W. (1973) Cytoplasmic adenylation and processing of maternal RNA. *Proc. Natl. Acad. Sci. U.S.A.* 70, 406−11.

Sonenberg, N., Morgan, M. A., Merrick, W. C., and Shatkin, A. J. (1978) A polypeptide in eukaryotic initiation factors that cross-links specifically to the 5′-terminal cap in mRNA. *Proc. Natl. Acad. Sci. U.S.A.* 75, 4843−7.

Ulberg, L. C., and Sheean, L. A. (1973) Early development of mammalian embryos in elevated temperatures. *J. Reprod. Fertil.* Suppl. 19, 155−61.

Van Blerkom, J. (1981) The structural relation and post-translational modification of stage-specific proteins synthesized during early preimplantation development in the mouse. *Proc. Natl. Acad. Sci. U.S.A.* 78, 7629−33.

− (1985) Post-translational regulation of early development in the mammal. In *Differentiation and Proliferation,* ed. C. Venizale. pp. 67−86. New York: Van Nostrand Reinhold.

Van Blerkom, J., Barton, S. C., and Johnson, M. H. (1976) Molecular differentiation in the preimplantation mouse embryo. *Nature* 259, 319−21.

Van Blerkom, J., and Brockway, G. O. (1975) Qualitative patterns of protein synthesis in the preimplantation mouse embryo. *Dev. Biol.* 44, 148−57.

Van Blerkom, J., and Runner, M. N. (1984) Mitochondrial reorganization during resumption of arrested meiosis in the mouse oocyte. *Am. J. Anat.* 171, 335−55.

Voellmy, R., Bromley, P., and Kocher, H. P. (1983) Structural similarities between corresponding heat-shock proteins from different eucaryotic cells. *J. Biol. Chem.* 258, 3516−22.

Wilt, F. H. (1973) Polyadenylation of maternal RNA of sea urchin eggs after fertilization. *Proc. Natl. Acad. Sci. U.S.A.* 70, 2345−9.

Winkler, A. M., Bruening, G., and Hershey, J. W. B. (1983) An absolute requirement for the 5′ cap structure for mRNA translation in sea urchin eggs. *Eur. J. Biochem.* 137, 227−32.

Wittig, S., Hensse, S., Keitel, C., Elsner, C., and Wittig, B. (1983) Heat shock gene expression is regulated during teratocarcinoma cell differentiation and early embryonic development. *Dev. Biol.* 96:507−14.

Wu, R. S., and Wilt, F. H. (1974) The synthesis and degradation of RNA containing polyriboadenylate during sea urchin embryogeny. *Dev. Biol.* 41, 352−70.

Wudl, L., and Chapman, V. (1976) The expression of β-glucuronidase during preimplantation development of mouse embryos. *Dev. Biol.* 48, 104−9.

Young, R. J. (1977) Appearance of 7-methylguanosine-5′-phosphate in the RNA of 1-cell embryos three hours after fertilization. *Biochem. Biophys. Res. Commun.* 76, 32−9.

Young, R. J., Stull, G. B., and Brinster, R. L. (1973) RNA in mouse ovulated oocytes. *J. Cell Biol.* 59, 372a.

Young, R. J., and Sweeney, K. (1979) Adenylation and ADP-ribosylation in the mouse 1-cell embryo. *J. Embryol. Exp. Morphol.* 49, 139−52.

Young, R. J., Sweeney, K., and Bedford, J. M. (1978) Uridine and guanosine incorporation by the mouse one-cell embryo. *J. Embryol. Exp. Morphol.* 44, 133−48.

Zeller, R., Nyffenegger, T., and DeRobertis, E. M. (1983) Nucleocytoplasmic distribution of snRNPs and stockpiled snRNA-binding proteins during oogenesis and early development in *Xenopus laevis*. *Cell* 32, 425−34.

Zimmerman, J. L., Petri, W., and Meselson, M. (1983) Accumulation of a specific subset of *D. melanogaster* heat shock mRNAs in normal development without heat shock. *Cell* 32, 1161−70.

9 Metabolic aspects of the physiology of the preimplantation embryo

PETER L. KAYE

CONTENTS

I. Introduction

Fertilization and preimplantation development of eutherian mammals occur in the oviduct and uterine lumen, when the embryo is often described as a closed system. Largely successful attempts to devise a totally defined medium in which normal preimplantation development may occur in vitro have reinforced this concept. However, the mammalian embryo must communicate with the maternal physiology via the fluid of the reproductive tract. It is clear that this channel is used in the case of nutrients, which has been demonstrated by studies of embryonic metabolism in vitro.

The early mammalian embryo has several advantages for the study of metabolism. Development of several species can occur in a completely defined medium in vitro (for a review, see Mastroianni and Biggers 1981), enabling definition of the effects of environmental perturbations on the embryo. Although the tissue mass is small, with many techniques it is now possible to assay one or a few embryos, diminishing animal and chemical requirements and facilitating rapid manipulation and assay of labile compounds. Thus, it is feasible in many cases to obtain estimates of metabolite levels and their embryonic and maternal variations. This should permit further studies of the effects of nutritional and environmental manipulation of the mother's environment and toxicological insults on embryonic and reproductive tract metabolism. Finally, it is possible to study the metabolic bases and ramifications of the first differentiations that occur during this time, as discussed in other chapters in this volume.

There is wide species diversity in developmental patterns from fertilization to implantation. Except where comparative data are illuminating or innovative, this review deals predominantly with the mouse, for which most detailed information exists. Comparative aspects are dealt with elsewhere (see Chapter 3).

II. Carbohydrate metabolism

This review covers progress in the last decade. Earlier work has been reviewed (Biggers and Stern 1973; Wales 1973b, 1975; Biggers and Borland 1976).

A. Lactate and pyruvate

The first cell division requires pyruvate. This may be supplied in vivo by the cumulus cells (Leese and Barton in press) and/or oviductal secretions. Division occurs in the ampullary region of the oviduct. Using the precise microhistochemical procedures of Lowry, it was shown that rises in oviductal pyruvate and lactate levels occurred shortly after ovulation. This rise and a prolonged increase in isthmic lactate levels specific to mated mice coincided with the predicted embryonic reqirements for these two substrates. The observation that this latter change occurred only in pregnant oviducts implies acknowledgment by the maternal system that an embryo is present (Neider and Corder 1983).

Pyruvate enters embryos by a combination of diffusion and a facilitated mechanism (Leese and Barton 1984), as will be described later. Lactate appears to enter by rapid diffusion (Wales and Whittingham 1974).

Fates of these substrates include oxidation and conversion to metabolic intermediates, including glutamate, alanine, aspartate, and those of the mitochondrial tricarboxylic acid cycle. Some 5–10 percent is incorporated into acid-insoluble material, presumably (but not necessarily) protein. The importance of this incorporation decreases as development proceeds (Wales 1973b).

B. Glucose

Glucose supports development as a sole energy source only after the third cell division. Thereafter, metabolic diversity rapidly accelerates, and a variety of energy sources may support development. Blastocysts require high concentrations of glucose to hatch from the zona pellucida and exhibit trophoblast outgrowth in vitro (Wordinger and Brinster 1976).

Although they do not require glucose as an energy source, early cleavage-stage mouse embryos possess a specific facilitated glucose-transport system (Wales and Brinster 1968; Gardner and Kaye 1984).

Using ultramicrofluorometric procedures, Leese and Barton (1984) measured the disappearance of unlabeled glucose and pyruvate from the incubation medium surrounding embryos. The technique requires as little as one embryo and may be used for viability studies or measurements of developmental changes in the same embryo. Pyruvate uptake by mouse embryos remained near 1 pmol/h from egg to morula but was reduced to one-third that level in the blastocyst stage (Leese and Barton 1984). Glucose uptake, barely detectable in eggs, increased slightly after fertilization but remained less than 0.5 pmol/h to the morula stage, increasing 10-fold in the blastocyst. Thus, premorula stages prefer pyruvate to glucose as a substrate, but this preference is reversed in the blastocyst. The uptake rates measured are similar to those determined earlier (Barbehenn, Wales, and Lowry 1974).

Glucose uptake by bovine blastocysts correlates with their development after embryo transfer and emphasizes the potential of these noninvasive techniques for viability assays in embryo transfer and in vitro fertilization (Renard, Philippon, and Menezo 1980).

Transport of glucose into mouse blastocysts has a low affinity (Dabich and Acey 1982; Gardner and Kaye 1984). If the glucose levels reported in sheep uterine fluid (0.25 mM) (Wales 1973a) apply to the mouse, then the maternal system may regulate embryonic glucose uptake (and metabolic) rates via the availability of substrate in the reproductive tract. Moreover, glucose uptake by blastocysts in vitro is stimulated by insulin. The effects of insulin include at least an increase in maximal uptake rate (Gardner and Kaye 1984). This may be explained by recruitment of transport systems to the plasma membrane from cytoplasmic pools. If a similar situation occurs in vivo, then the maternal system may additionally regulate embryonic glucose metabolism via an endocrine mechanism.

C. Glycogen

During early development, glycogen is the major product of incorporated glucose. Levels increase 10-fold between the 1-cell and 2-cell stages (Ozias and Stern 1973), but this is not reflected in incorporation of radiolabeled glucose into acid-soluble glycogen (Brinster 1969). Moreover, rapid glycogen synthesis occurs during development from 2-cell to blastocyst in vitro (Brinster 1969), but freshly collected embryos show almost constant glycogen levels until the late blastocyst stage (Ozias and Stern 1973). The glycogen levels in vitro were reduced if glucose was omitted from the medium (Ozias and Stern 1973), reflecting the kinetic results mentioned earlier, but reduction to the uterine level of 0.25 mM did not reduce glucose incorporation to in vivo levels (Pike and Wales 1982).

The answers to this puzzle lie in the form of glycogen synthesized by the early cleavage stages and maternal control of glycogen metabolism in the morula/blastocyst. Glycogen may be extracted from tissues in two forms: an acid-insoluble glucoprotein precursor called desmoglycogen (Whelan 1976), and the typical acid-soluble polysaccharide form. Unless precautions are taken to release the glucose from desmoglycogen with glycosidases, this fraction will be lost from the total glycogen extract made with acid. Treatment of embryos with amyloglucosidases before acid extraction showed that the incorporation of [^{14}C]glucose into total glycogen during the 2-cell-to-8-cell period increased from 34 percent to 93 percent of the total incorporation. Extraction of normally obtained acid-insoluble fraction with amyloglucosidase revealed that during the 1-cell-to-8-cell period 43 percent of the glucose accumulated as desmoglycogen, representing 56−66 percent of the acid-insoluble fraction (Pike and Wales 1982). This explained the discrepancy between the early high glycogen estimations by Ozias and Stern (1973) using an enzymic assay for glycogen and the low incorporation data of Brinster (1969) using acid extraction.

Using glycogen and desmoglycogen synthesis rates, Pike and Wales (1982) calculated theoretical glycogen levels in embryos cultured from 1-cell stage to blastocysts and obtained results very close to those measured by Ozias and Stern (1973) at each stage. During the 1-cell and 2-cell stages, desmoglycogen synthesis predominates over glycogen synthesis, but from the 8-cell stage glycogen synthesis increases rapidly to reach 60 percent of the glucose accumulated by late blastocysts, whereas desmoglycogen synthesis remains relatively constant. It appears that the early mouse embryo represents an ideal system in which to investigate the function of desmoglycogen in cells. Unfortunately, this avenue has not been pursued, despite the considerable groundwork these studies provide.

The high levels of glycogen in cultured embryos remain unexplained. During development, in vitro glucose incorporation increases, particularly between the 8-cell and morula/early blastocyst stages. Much of the glucose (20−60%) is stored as glycogen. Until the morula/early blastocyst stage, most of this is incorporated into desmoglycogen. At that time a dramatic increase in soluble-glycogen synthesis predominates, so that only about 2 percent of glucose incorporated by late blastocysts over 5 h is desmoglycogen, but 58 percent is soluble glycogen.

Pulse-chase experiments revealed little degradation of the glycogen accumulated in the later stages, unless complete energy starvation occurred. Because enzyme activities favor synthesis, the comparatively low glycogen levels in uterine blastocysts probably arise from the maternal effects on glycogen catabolism (Edirisinghe, Wales, and Pike 1984a).

This hypothesis was tested by combining an in vitro pulse-labeling period with chase in vitro or in vivo after transfer to pseudopregnant or ovariectomized recipients. During 24-h chase in the uteri of day-4 pseudopregnant mice, the label in acid-soluble glycogen was reduced to 20 percent of the initial pulse value, compared with 73 percent in embryos chased in vitro. If the chase time was increased to 48 h (by which time blastocysts have implanted in vivo), only 2 percent of the label incorporated by morulae remained. Thus, the pseudopregnant uterine environment results in a net degradation of acid-soluble glycogen in blastocysts (Edirisinghe, Wales, and Pike 1984b). When morulae/early blastocysts were transferred to the uteri of untreated ovariectomized mice for a 24-h chase, the level of label retained was between those for embryos transferred to pseudopregnant mice and embryos chased in vitro. Treatment of ovariectomized mice with progesterone reduced the amount of label retained to the level observed in embryos chased in pseudopregnant uteri. Estrogen alone, or estrogen priming, did not alter the effects of progesterone, but when given in combination antagonized the effect of progesterone (Edirisinghe and Wales 1984).

Thus, the progesterone-dominated environment in utero results in a net degradation of glycogen pools, accounting for the differences observed between the glycogen levels of uterine blastocysts and blastocysts obtained by culture in the presence of glucose. Progesterone treatment of mice in implantational delay also decreased the glycogen content of blastocysts below that for controls (Ozias and Weitlauf 1971).

Because glycogen breakdown occurred in unstimulated uteri and was increased by progesterone treatment, it is unlikely that progesterone acts directly on the embryos. Progesterone may alter the nutritional state of the genital tract, resulting in glycogen breakdown. There is evidence that this may be the case. Reduced lactate and pyruvate levels result in increased glycogen breakdown in vitro (Pike and Wales 1982), and isthmic oviduct levels of lactate fall dramatically in the period when embryonic glycogen levels are increasing in vitro (Neider and Corder 1983). In addition, oxygen levels in the rat uterine lumen (Yochim and Mitchell 1968) and rabbit oviduct (Mastroianni and Jones 1965) are around 5 percent, compared with the 20 percent used for most culture. Low levels of O_2 in the pregnant mouse oviduct/uterus could result in greater glycogen breakdown in vivo. High O_2 levels are responsible for the Pasteur effect in which phosphofructokinase (PFK) is inhibited by high levels of ATP, citrate, and other metabolites that accumulate in cells exposed to high oxygen. Such inhibition was proposed as the block on glycolysis in precompaction embryos (Barbehenn et al. 1974). ATP does accumulate in blastocysts in vitro, but not in vivo (Quinn and Wales 1973c; Spielman et al. 1984), suggesting that a Pasteur effect may operate in vitro but not in vivo. Further support for this proposal comes from autoradiographic studies showing that mouse trophectodermal cells accumulate large amounts of glycogen as compared with inner cells (Edirisinghe, Wales, and Pike 1984c) and in the rabbit have higher mitochondrial concentrations reflecting greater aerobic metabolism (Benos et al. 1985). These cells would presumably be exposed to higher O_2 levels and would accumulate more glycogen than inner cells in utero. Maternal progesterone may reduce oviduct/uterine energy substrate (e.g., lactate, pyruvate, and O_2) levels that are manifested metabolically in blastocysts by absence of a Pasteur effect, lowered ATP levels, and decreased glycogen synthesis. Determination of the levels of O_2, lactate, pyruvate, and other energy substrates in the pregnant mouse genital tract could confirm whether or not maternal regulation of embryonic metabolism does occur via environmental regulation.

These effects of progesterone are intriguing in view of the reported antiimplantation effects of passive immunization of inbred mice with monoclonal mouse antibodies to progesterone administered 109–130 h after mating. Furthermore, passive immunization during preimplantation development, or specifically 32 h after mating, arrested embryo cleavage (Heap, Rider, and Feinstein 1984).

D. O_2 metabolism

Greater numbers of mouse embryos of all stages develop in vitro from the 1-cell stage to morula/early blastocyst under 5 percent O_2 than the usual 20 percent (Quinn and Harlow 1978). This is approximately the O_2 level found in pregnant rat uterine lumen (Yochim and Mitchell 1968). The range of O_2 tensions compatible with greatest development to blastocysts was narrower for 1-cell and 2-cell embryos than later stages. This was reflected by lower cell numbers in blastocysts developing under 20 percent O_2. It was suggested that an O_2 gradient in the blastocyst may induce cells on the outside to divide more slowly and form the trophectoderm, whereas inside blastomeres would divide rapidly and form the inner cell mass (Harlow and Quinn 1979). This hypothesis has not been tested further. However, trophectodermal cells accumulate large amounts of glycogen (Edirisinghe et al. 1984c). This is consistent with aerobic metabolism that a higher O_2 tension would fuel. Similarly, these cells have higher mitochondrial concentrations in the rabbit (Benos et al. 1985), consistent with greater aerobic metabolism.

E. Periimplantation metabolism

Because of the difficulties in manipulation, and lack of a suitably defined culture medium, knowledge of the metabolic requirements of the mammalian periimplantation period is limited.

During somitogenesis (days $10-12$), glucose is converted to lactate even in the presence of high O_2 levels by rat embryos growing in serum. Glucose oxidation is much lower and relatively constant during this period (Tanimura and Shepard 1970). Recently, the metabolism of glucose over 3 h by $6-9.5$-day mouse embryos in the same completely defined medium used for preimplantation embryos was examined (Clough and Whittingham 1983). The glucose oxidation rate was independent of the glucose concentration in the medium, was inhibited by lactate, and occurred predominantly via the pentose phosphate pathway. Most of the glucose was converted to lactate. This aerobic glycolysis is unusual, being restricted to a few cell types, including muscle, gut mucosa, retinal, and tumor cells. This curious metabolic profile may be related to the rapid growth rate of the embryo, but our understanding of its significance awaits more data.

III. Lipid metabolism

A. Triacylglycerols

Various radioactive substrates are converted into embryonic lipids (Wales 1975). During culture of mouse embryos from the 2-cell stage to morulae, the major lipid derived from [^{14}C]glucose was the glycerol of triacylglycerols (Flynn and Hillman 1978). Eight-cell embryos obtained from culture of 2-cell embryos incorporated about 300 fmol of [^{14}C] palmitate over 2 h, 83 percent into the triacylglycerol fraction (Flynn and Hillman 1980). This represents significant energy storage. Each mole of palmitate could yield 129 moles of ATP via complete oxidation (i.e., about 40 pmole ATP equivalents stored per 2 h in this study). Glycogen storage of energy over the same time at the same stage represents about 7 pmole ATP equivalents (Pike and Wales 1982). Thus, the mouse embryo may store most of its energy as triacylglycerols, provided a fatty acid source is available. As with glycogen, triacylglycerol metabolism in utero is of great interest.

Unfertilized eggs and 2-cell embryos oxidized palmitate at about 75 fmol/h, increasing to 400 fmol/h in the late blastocyst, about 10 percent the rate of glucose, pyruvate, and lactate oxidation (Flynn and Hillman 1980).

Little is known of embryonic lipid metabolism in other species. In the rabbit, pyruvate can be replaced in the culture medium by myristic, palmitic, stearic, oleic, linoleic, propionic, and acetic acids for development from 1-cell to morula. These probably act as energy sources, because the 1-cell rabbit embryo has an active tricarboxylic acid cycle and oxidative phosphorylation system, which are both essential (Kane and Buckley 1977). Arachidonic acid was toxic, and butyric and valeric acids did not support growth (Kane 1979).

B. Membrane lipids

In mouse embryos, about 4 percent of the palmitate entered the polar lipids, 75 percent as choline phosphatides and 10 percent as glycolipid (Flynn and Hillman 1980). These observations are supported by studies of phospholipid synthesis from choline (Pratt 1980). Choline was incorporated into phosphatidylcholine as early as the 2-cell stage. Incorporation increased 9–13-fold during the 8-cell stage and compaction, remaining elevated in the blastocyst. This indicates that the embryo is capable of new membrane-component synthesis from the 2-cell stage. In morulae and blastocysts, choline transport was concentrative, specific, and saturable, with a K_m of $5\mu M$. Simple diffusion also occurred. Less than half of the transported choline was converted to phosphocholine, perhaps due to regulation of choline kinase or an active phosphatase.

Another major membrane lipid is cholesterol, which increases from 0.34 pmol per egg to 1.08 pmol per blastocyst (Pratt 1982). Uptake and conversion to lipid of the precursor mevalonate increased linearly from the 2-cell to the 8-cell stage, then, like phospholipid synthesis, accelerated to 10 times that level in mature blastocysts. The labeled lipid products were cholesterol, lanosterol, squalene, and cholesterol esters. From the 2-cell stage to the early blastocyst, more mevalonate appears as lanosterol than cholesterol, but from compaction onward the proportion found as cholesterol increases and in the mature blastocyst exceeds that in lanosterol. Chase experiments show conversion of lanosterol to cholesterol during compaction. Because inhibition of compaction by cytochalasin had no effect, this metabolic switch is not tied to formation of cell contacts, which is prevented by cytochalasin-induced inhibition of microfilament growth (Johnson and Maro 1984). Cholesterol could be derived from exogenous low-density lipoproteins (LDL), which inactivate hydroxymethylglutaryl-CoA reductase (HMG-CoA-reductase), inhibiting cholesterol synthesis. In the mouse embryo, because the cholesterol content can be accounted for by embryonic synthesis, exogenous cholesterol is probably not utilized or available as LDL in oviductal fluid. Development of embryos in a medium devoid of lipid supports this conclusion.

Because HMG-CoA-reductase could be measured only in the blastocyst (perhaps because of insensitivity of the assay procedure), this is the earliest stage at which the complete sterol pathway can be shown to operate. In addition to membrane sterols, products of this pathway may be steroid hormone precursors.

C. Glycoproteins
The later results of Pratt (1982) showing HMG-CoA-reductase activity only in blastocysts cast some doubt on earlier conclusions from morphological studies of the effects of HMG-CoA-reductase inhibitors (Pratt, Keith, and Chakraborty 1980). Surani, Kimber, and Osborn (1983) showed that mevalonate reversed inhibition of compaction by HMG-CoA-reductase inhibitors, but cholesterol did not. Inhibition of squalene cyclization had no effect on development. Thus, the effects of inhibition of HMG-CoA-reductase on development (Pratt et al. 1980) result from lack of nonsterol isoprenoids or dolichols, not cholesterols. Dolichol is required as a sugar carrier in glycoprotein formation. HMG-CoA-reductase inhibition did not affect incorporation of amino acids into protein, but did inhibit incorporation of sugars into glycoproteins. Membrane glycoproteins are involved in cell interactions and morphogenesis and are synthesized by embryos (Fishel and Surani 1980; Cooper and MacQueen 1983).

IV. Nitrogen metabolism
Although early studies demonstrated a need for a fixed nitrogen source to obtain maximum development of mouse embryos in vitro, it has recently been found that addition of protein to an otherwise amino-acid-free, macromolecule-free medium for culture of 2-cell embryos from F_1 hybrid mice to blastocysts has no effect on subsequent development in vivo after transfer (Caro and Trounson 1984). Apparently normal development of mouse embryos to blastocysts in vitro is independent of nitrogen. The albumin found necessary in earlier studies probably served a protective rather than nutritive role. But a note of caution must be sounded before accepting this conclusion. Few measurements other than that by Caro and Trounson (1984) have been made of the effects of protein or amino acid supplements on complete development of cultured embryos transferred to pseudopregnant hosts. These usually show only about 30–50 percent live births from embryos transferred, even in controls.

Thus, it may be that a nitrogen requirement exists but is not revealed by this experimental approach. Certainly, observations of protein metabolism in vitro reveal both mouse and rabbit embryos to be active in protein turnover and amino acid transport.

A. Amino acid transport

Amino acid transport has been characterized in several different systems with different properties and reactivities (Table 9.1) (for review, see Christensen 1984). Preimplantation studies have concentrated on the zwitterionic systems that are subdivided on the basis of Na^+ requirement.

Although experimental efforts concentrate on uptake or exchange of individual amino acids from a medium lacking other amino acids (in order to analyze particular systems), the real situation is the substantially more complex one of an embryo bathing in a broth of many amino acids and ions. Here uptake of each amino acid will be the net result of the complex kinetics of all transport systems operating, with competition for transporters by amino acids on both sides of the membrane.

Early studies reported increased incorporation of radioactive amino acids into protein by embryos as development proceeded (for review, see Biggers and Stern 1973). Later studies attempted to analyze amino acid transport in terms of the newly emerging nomenclature (Table 9.1). The known kinetic constants for transport in preimplantation mouse embryos gathered from these are presented in Table 9.2 and will be discussed, with evidence of other systems, later.

In the isolated oocyte, highly exchangeable systems resembling L and ASC transport leucine (see Table 9.1 for definition of transport systems). The Na^+-independent L system had a relatively constant affinity ($K_m \approx 60\mu M$), but the maximal transport rate (V_{max}) increased with growth of the oocyte. Alanine uptake was partially (30%) dependent on Na^+ and became increasingly so as the oocytes enlarged. Competition by phenylalanine, an L-specific passenger, together with serine, an ASC passenger, suggested that this Na^+-dependent uptake occurred by an ASC system, because N-methylaminoisobutyric acid (MeAIB), believed to be specifically transported by the A system, had no effect. The exchange of preloaded alanine for external alanine was increased by the presence of Na^+ in the medium. Because ASC is reportedly a highly exchangeable system and A is not so, it was concluded that the oocyte possessed transport systems resembling L and ASC (Colonna et al. 1983).

Studies of unfertilized and fertilized eggs (Holmberg and Johnson 1979) confirmed the presence of the Na^+-independent exchangeable L system, with a K_m value for methionine similar to that reported for leucine in the oocyte. Another system specific for and with high affinity for glycine and sarcosine that is highly Na^+-dependent, nonexchangeable, and concentrative has been reported in mouse eggs and precom-

Table 9.1. *Amino acid transport systems*

System	Amino acid reactivity	Selective substrate	Features
Zwitterionic Na$^+$-dependent			
A	Most, but especially those with small branched side chains, e.g., methionine, alanine, proline, serine, threonine	MeAIB[a]	Ubiquitous, highly concentrative, little exchange; subject to hormonal and adaptive regulation
ASC	Many, especially those with small side chains, e.g., threonine, alanine, cysteine	L-cysteine	Concentrative, active in exchange; may be involved in metabolic regulation
gly	Glycine, sarcosine		Cotransports 2 Na$^+$; may be a precursor of A system; subject to hormonal control
N	Glutamine, asparagine, histidine	Histidine	Subject to adaptive control; so far, found only in hepatocytes and hepatoma cells
Zwitterionic Na$^+$-independent			
L (L2)	Prefers those with branched apolar side chains, e.g., leucine, isoleucine, phenylalanine, methionine	BCH	Highly exchangeable, ubiquitous; L2 may or may not be identical with L
L1	High-affinity variant of L		Develops in hepatocytes in culture
Anionic			
X$^-_{AG}$	Aspartate, glutamate	Cysteate and cysteine sulfinate	Na$^+$-dependent; wide distribution
x^-_G	Glutamate		Na$^+$-independent
Cationic			
ly$^+$	Arginine, lysine		Na$^+$-independent; ubiquitous

Note:

[a] Abbreviations: MeAIB = *n*-methylaminoisobutyric acid; BCH = 2-amino-norbornyl-2-carboxylic acid.

Table 9.2. *Kinetic parameters of amino acid transport in mouse oocytes and embryos*

Stage	K_m (μM)	V_{max} (fmol/min/embryo)	Amino acid and system	Reference
Oocyte	66	45	Leucine (L)	Colonna et al. (1983)
Egg	56	30	Methionine (L)	Holmberg and Johnson (1979)
	220	98	Methionine (preloaded with Met) (L2)	Miller (1984)
2-cell	89	15	Glycine (*gly*)	Hobbs and Kaye (1985)
4-cell	33	10	Leucine (L)	Borland and Tasca (1974)
	63	25	Methionine (L)	Borland and Tasca (1974)
Early blastocyst	50	26	Leucine (L)	Borland and Tasca (1974)
	62	46	Methionine (L)	Borland and Tasca (1974)
	890	10	Methionine (A)	Borland and Tasca (1974)
	607	40	Glycine (A?)	Hobbs and Kaye (1985)
	15	7	Glycine (*gly*)	Hobbs and Kaye (1985)
Late blastocyst	22	133	Methionine (?)	Miller (1984)
ICM	660	28	Methionine (L2)	Miller (1984)

paction embryos (Hobbs and Kaye 1985). This *gly*-system may persist into the blastocyst, where glycine is also taken up by the A system (Kaye et al. 1982).

Study of methionine exchange mechanisms suggests that in addition to these L, ASC, and *gly* systems, unfertilized eggs possess an additional L2 system, on the basis of decreased uptake affinity and differing competition behavior of amino-acid-loaded and nonloaded eggs (Miller 1984). There may also be a Na^+-dependent A-like system, because methionine continues to accumulate rather than reach equilibrium with the medium. Miller has proposed that the system with K_m of 60μM, the same as that previously reported for the L system in eggs, is L1-like and that the true L (or L2) system has a higher K_m (250μM). The evidence for the extra two transport systems requires confirmation. In late 4-cell embryos, leucine and methionine are both accumulated by the Na^+-independent L system, with an unchanged affinity. V_{max} values for oocytes and eggs were similar but appeared to decrease in 4-cell embryos, perhaps reflecting decreased access of amino acids to the membrane.

In cleavage-stage rabbit embryos, transport differs from that in mice. Eggs possess concentrative, exchangeable, Na^+-requiring systems for methionine uptake. Competition studies suggest the presence of L, A, and ASC systems. Fertilization stimulates methionine uptake by tripling the affinity of its transporter, an increase that persists in the morula and suggests transporter modifications. A further large increase in methionine uptake by morulae 69–76 h after luteinizing hormone administration may be due to loss or segregation of an exchange system at compaction (Miller and Schultz 1983).

In early mouse blastocysts, a third Na^+-dependent system transporting methionine was observed, with affinity and V_{max} one-quarter those of the L system (Borland and Tasca 1974). Na^+ appeared to increase the affinity of the transporter for methionine (Borland and Tasca 1975). This system was described as A-like, on the basis of competition by glycine and alanine and the presence of a methylaminoisobutyric acid (MeAIB, see Table 9.1) transporting system in the blastocyst (Kaye et al. 1982), also previously observed for aminoisobutyric acid (AIB) by Van Winkle and Dabich (1977). Blastocysts did not exchange preloaded methionine for external methionine, in contrast to morulae and earlier-stage embryos. Preloading with methionine depleted eggs of L-system passengers, whereas blastocysts were depleted of only A-system passengers. This also suggested that if the L system persists in the blastocyst, it no longer exchanges with the external medium (Kaye et al. 1982).

The process of compaction and the extra compartments this permeability seal subsequently creates in the blastocyst further complicate an already complex system. Methionine transport was inhibited by ouabain and K^+-free medium in late morulae, but these effects were not present in early blastocysts (Borland and Tasca 1975). Because ouabain inhibits Na^+-K^+ ATPase, these observations implicate ATPases or Na^+ pumps in methionine uptake by morulae, but during blastocoel formation these pumps become localized on the blastocoelic cell surfaces (i.e., separated from the external medium). In cytochalasin-collapsed blastocysts, ouabain gained access to these inner surfaces and inhibited Na^+-dependent alanine uptake (DiZio and Tasca 1977). The amino acid exchange properties of cytochalasin-collapsed blastocysts resemble those of the precompaction embryos and indicate that the trophectoderm is analogous to gut epithelium, where concentrative systems are located on the outer mucosal membrane and exchange systems on the inner basolateral membrane. In the blastocyst, such an arrangement would transport amino acids into the blastocoel, where they could serve as an embryonic pool and supply intracellular Na^+ for the ATPases important in blastocoel formation (Kaye et al. 1982).

Segregation of transport systems to different cell membranes in blastocysts has been studied in the rabbit (Miller and Schultz 1983) and mouse (Miller 1984). Mouse inner cell masses (ICMs) possessed both saturable and nonsaturable methionine-uptake systems. The K_m was much higher than for the whole embryo, and there was free exchange, suggesting an L-like system. However, competition for uptake by alanine and a partial Na^+ dependence indicate the presence in ICM cells of an A- or ASC-like system as well. About 70 percent of the methionine entering mouse blastocysts was taken up in the cavity (Miller 1984), confirming the earlier hypothesis (Kaye et al. 1982) that transport would occur into the blastocoel.

The rabbit blastocyst showed weak methionine efflux to the medium, but the kinetics resembled those of uptake into cavity only; so the rabbit blastocyst has an asymmetric transporter distribution also. Because most amino acids compete, it is likely that systems resembling L, A, and ASC (Guidotti et al. 1975) exist in the rabbit (Miller and Schultz 1983).

Interacting transporters may regulate the whole range of internal amino acid pools purely by a combination of the various kinetic parameters and external substrate levels. However, in many cells transport is subject to external regulation (e.g., various hormones may stimulate A- or gly-system transport). In addition, the absence of amino acids from the environment may increase amino acid transport by so-called adaptive regulation, resulting in provision of more transporters to the membrane (Guidotti et al. 1975).

With so little information it is difficult to confidently predict how such effects might be utilized during embryonic development. Glycine transport was stimulated in vitro by the peptide hormones insulin and epidermal growth factor (EGF) in blastocysts but not 2-cell embryos. This suggested that glycine enters blastocysts via a hormone-responsive A system

(Hobbs and Kaye 1985). Because the A system may be a differentiative variant of the more restricted, but Na$^+$-dependent, *gly* system (Christensen 1973), this could represent differentiation of transport systems to allow hormonal and/or nutritive regulation of amino acid transport, at about the time the blastocyst enters the uterus.

B. Amino acid pools

i. In embryos. Increased sensitivity of amino acid quantitation enabled measurement of absolute amino acid pools in eggs and embryos of the mouse (Schultz et al. 1981) and rabbit (Miller 1984) (Table 9.3). Seventeen amino acids were detected (proline is not detected by the method). Major components were taurine and glycine. Many of the zwitterionic amino acids (especially those with affinity for the A system) increased in concentration during development, mostly at the expense of glycine, but there was a drop in the total pool.

The rabbit egg pool was composed mainly of aspartate, glutamate, glycine, alanine, and taurine. A 50 percent increase in the total pool occurred in the morula, predominantly because of 4-fold increases in concentrations of glycine and serine and 5- to 10-fold increases in the cationic amino acids. Taurine decreased to about one-quarter its concentration in eggs. Because the volume of the embryo changes little over this time, these changes represent accumulations. The enormous increase in blastocyst volume was accompanied by increases in total amino acids to 200 nmol. The cavity contains 97 percent of this, of which glycine makes up 70 percent. There was 10–30 times as much of most amino acids in the cavity as in the cellular component, except for the anionic amino acids, of which about one-third is intracellular.

ii. Contribution of oviduct/uterine fluids. Growing embryos in amino-acid-free medium containing protein reduced the amino acid pools, but protein levels remained close to those of freshly collected embryos until the midblastocyst stage (Sellens, Stein, and Sherman 1981). This and the amino acid transport systems discussed earlier indicate that the mouse embryo relies on the oviduct/uterine fluid to maintain its endogenous amino acid pools. However, sufficient amino acids are available for protein synthesis within the total embryonic pool (or from whole protein in the medium), because protein levels are maintained close to those found in vivo.

Amino acids accumulated in fluid perfused through the rabbit oviduct to reach concentrations about 27 percent of those in plasma. Treatment of the rabbits with hCG increased levels of neutral amino acids (Leese, Aldridge, and Jeffries 1979). The rat uterine horn contains transport

Table 9.3. *Amino acid compositions in embryos and tract fluids (%)*

| | Egg | | Rabbit oviduct fluid[b] | | Morula | | Rabbit uterine fluid[b] | | Blastocyst | | |
| | | | | | | | | | | Rabbit[b] | |
	Mouse[a]	Rabbit[b]	Day 1	Day 3	Mouse[a]	Rabbit[b]	Day 3	Day 6	Mouse[a]	Cells	Cavity
Taurine	50.0	37.3	7.1	5.8	17.0	6.7	4.8	5.4	39.0	n.d.	0.7
Aspartate	3.5	14.6	1.9	1.4	5.8	15.2	0.9	0.5	9.7	3.4	0.2
Threonine	0.9	0.8	12.0	10.8	4.5	2.4	1.9	2.2	2.0	4.1	3.2
Glutamine	0.7	1.0	n.d.[c]	n.d.	4.9	n.d.	n.d.	2.2	4.5	1.9	1.7
Serine	1.3	1.9	5.6	9.3	5.4	5.8	6.2	3.1	4.3	6.4	6.3
Glutamate	6.8	13.3	8.7	10.2	17.7	16.1	12.0	2.3	16.2	22.3	1.2
Glycine	30.0	12.6	37.8	33.2	27.4	30.4	64.1	75.7	4.4	29.3	67.1
Alanine	4.5	10.1	13.4	15.5	7.6	7.4	3.8	4.9	4.5	10.0	8.7
Valine	5.5	1.5	2.6	2.9	4.0	2.0	1.0	0.7	1.9	3.2	1.2
Methionine	0.1	0.5	1.3	2.4	0.6	2.2	0.4	0.4	1.2	2.3	0.6
Isoleucine	0.2	0.7	1.0	1.0	0.9	0.5	0.3	0.4	1.0	2.0	0.5
Leucine	0.4	1.8	1.7	2.0	1.2	2.0	0.7	0.8	1.5	4.1	0.8
Tyrosine	<0.1	0.3	1.3	1.2	<0.1	0.7	0.4	0.4	1.5	1.7	0.6
Phenylalanine	0.2	0.7	0.9	0.8	<0.1	0.9	0.3	0.3	1.4	1.9	0.4
Histidine	0.1	1.1	1.3	1.5	1.4	1.0	0.5	1.3	2.7	2.3	0.8
Lysine	0.6	0.5	2.8	2.0	1.8	3.3	0.6	0.9	2.4	4.4	4.2
Arginine	<0.1	1.0	1.6	2.0	<0.1	3.4	0.6	0.7	1.5	2.6	1.8
Pool (pmol)	13.2	22.4	1.6×10^6	0.9×10^6	7.5	34.8	0.5×10^6	1.6×10^6	7.7	6×10^3	20×10^3
Concentration (mM)	66	20	—[d]	—	37.5	31.6	28.8	25.5	25.7	—	12.1

Note:

[a]Data from Schultz et al. (1981). [b]Data from Miller (1984). [c]n.d., not detected. [d]Not determined.

systems that actively secrete amino acids into the lumen when estrogen-primed (Walters, Hazelwood, and Lawrence 1979).

Glycine was the most abundant amino acid in rabbit oviducal flushings (Leese et al. 1979; Miller 1984), representing, together with alanine, threonine, glutamate, serine, and taurine, 85 percent of the total pool, which on day 1 was about equal to that of uterine flushings on day 6 (Table 9.3). The glycine concentration of rabbit uterine fluid increases almost fivefold between day 1 and day 3 of pregnancy, whereas most other amino acid concentrations change little. Taurine, high in fluid of the pregnant uterus on day 1, decreases to one-third that level by day 3. Comparison of these results with earlier figures from nonpregnant uteri (Gregoire, Gongsakdi, and Rakoff 1961) shows that pregnancy leads to increased concentrations of amino acids in uterine fluid, presumably to support the increases in embryonic pools. When compared with embryonic pools, as discussed earlier, this gives some insight into the relationship between the amino acid composition of the fluid and the embryo. The changes in glycine and taurine in tract fluids are reflected in changes in the embryo concentrations, where glycine increases about fourfold by the morula stage and taurine decreases to about 30 percent of its egg value. Precise definition of the link between tract fluids and embryonic pools would require quantitative analysis of changes in embryonic pools in vitro in the presence of various combinations of amino acids and the amino acid levels in luminal fluids during pregnancy and pseudopregnancy.

C. Protein metabolism

Few measurements of embryonic protein turnover exist. Electrophoretic analyses of proteins labeled with precursors throughout development reveal different labeling patterns that suggest turnover of diverse species.

In mouse embryos, half-lives of proteins labeled for 1 h with [^{3}H]-methionine decreased from 18.2 h in the 1-cell embryo to 13.1 h in the blastocyst. The 25 percent increase in turnover rate was apparent by the early 2-cell stage, preceding a major increase in protein synthesis (Merz et al. 1981) that may in turn be associated with destruction of maternal mRNA and activation of the embryonic genome (Flach et al. 1982).

Recently [^{125}I]BSA was used to measure whole protein uptake by mouse embryos in vitro. Uptake was low in the cleavage stages, but increased in the blastocyst (Pemble and Kaye 1984, in press; Kay, Pemble, and Hobbs in press). These results support earlier electron-microscopic observations of uptake of peroxidase and ferritin by rat and rabbit embryos (Schlafke and Enders 1973; Hastings and Enders 1974), and immunocytochemical demonstrations of uptake of exogenous BSA by mouse embryos in vivo (Glass 1963). Blastocysts loaded with [^{125}I]BSA lost radioactivity within 2 h. Because the [^{125}I]label is attached to internal tyrosine residues, it is likely that this represents protein hydrolysis. Pinocytosed protein is hydrolyzed within the lysosomes of embryonic rat yolk sac, and the released labeled amino acids are reutilized by the embryo in protein synthesis (Freeman and Lloyd 1983). A similar arrangement may exist in the mouse blastocyst, with the trophectodermal cells processing pinocytosed protein for use by the whole embryo. Certainly, because protein levels in the mouse uterus (Pratt 1977) are high, pinocytosis could represent a significant contribution to embryonic nitrogen metabolism.

V. Interconversions

Little is known of the interconversions of various precursors and their subsequent fate. Quinn and Wales (1973a, 1973b) observed synthesis of alanine, glutamate, and aspartate from lactate and pyruvate in early mouse embryos, and from glucose in day-6 rabbit blastocysts in vitro. As expected, glucose has also been found incorporated into lipids (Flynn and Hillman 1978), nucleic acids (Murdoch and Wales 1973; Pike, Kaye, and Wales 1977), and proteins (Schneider et al. 1976).

Not all the amino acid taken up by embryos is incorporated into protein or remains as the simple amino acid. Of the acid-soluble ^{14}C accumulated as glycine by blastocysts over 2 h, 22 percent did not chromatograph with glycine, but resembled serine and alanine, and a further 30 percent was unidentified (Hobbs and Kaye 1985). This implies intermediate formation of pyruvate, a key precursor of various compounds, including fatty acids. Glycine is also a precursor of purines and pyrimidines; pyrimidine biosynthesis occurs in mouse embryos in culture (Troike and Brinster 1981). Lysine and leucine were also converted to unidentified compounds at significant rates by mouse embryos (Epstein and Smith 1973). Both these amino acids and glycine are ketogenic and could contribute to fatty acid synthesis, as described earlier.

VI. Blastocoel formation

With establishment of tight junctions between the outer cells of the morula, a permeability seal is formed, isolating the inner cells from the maternal environment, and blastocoel formation is possible (Ducibella et al. 1975; Hastings and Enders 1975). Blastocoel fluid begins to accumulate in an eccentric position about 8 h after this permeability seal is established in the mouse. In early blastocysts, the ICM is covered by trophectodermal processes on all sides, including the juxtacoelic (Fleming et al. 1984).

Because the rabbit forms a larger blastocyst than other laboratory species, expanding by several orders of magnitude, it has become the model for studies of blastocoel formation. Only a summary of this area follows, because a thorough review has recently appeared (Benos et al. 1985).

In 1976, Borland, Biggers, and Lechene used electron-probe microanalysis to demonstrate parallel increases in the concentrations of Na, Cl, K, Ca, and S during rabbit blastocyst expansion. The accumulation of a fluid nearly isosmotic to the bathing medium in the range 230–370 mOsm (Borland et al. 1977) suggested that fluid transport occurs by osmosis. Expansion continued against a sucrose gradient, with increased accumulation of Na and Cl, not by K, Mg, Ca, S, and P, implying passive movement of water associated with movement of Na and Cl. These ionic movements presumably arise from Na^+-K^+ ATPase localized to the juxtacoelic membrane of the rabbit blastocyst (Benos et al. 1985). Ouabain can be selectively applied to either trophectodermal surface, because tight junctions are impermeable to it (Borland and Tasca 1974). At the abcoelic membrane, ouabain had no effect on Na^+ or Cl^- levels, but at the juxtacoelic surface it decreased Na^+ and Cl^- and increased K^+ levels. This suggested that a constant leakage of K^+ into the blastocoel occurred that is normally returned to the cell in exchange for Na^+, generating the osmotic gradient for H_2O to follow.

The ATPase is central to fluid expansion in the rabbit blastocyst. It is possible to use labeled ouabain as a marker for Na^+-K^+ ATPase because it binds to the outer K^+-binding surface of the membrane ATPase. The number of pump sites so measured increased biphasically, with peaks on days 5 and 7 (Benos 1981). The later increase coincided with growth of endodermal cells around the inside of the blastocoel (Benos et al. 1985). These results were confirmed by the observation that incorporation of [35S]methionine into electrophoretically isolated Na^+ K^+ ATPase increased 90-fold between days 4 and 5, remained elevated for one day, and then decreased (Benos et al. 1985).

The continued and increased activity of the ATPases requires an increased supply of intracellular Na^+. Between days 6 and 7 the rabbit blastocyst acquires an amiloride-sensitive Na^+-uptake system (Benos 1981). Because embryos cultured from five days neither express this Na^+-uptake system nor expand as rapidly, this system probably provides greater quantities of Na^+ for the increased activity of the increasing numbers of ATPases. During the same period, a furosemide-sensitive Na^+ Cl^- cotransport system develops that could provide additional intracellular Na^+ (Benos and Biggers 1983). In addition to this, a *gly* system similar to that observed in mouse embryos (Hobbs and Kaye 1985) may exist. In the mouse, this system appears to cotransport two Na^+ (Hobbs and Kaye in press). The enormous levels of glycine in the rabbit blastocyst fluid and uterine fluid (Miller 1984) (Table 9.3) suggest that such a system would be significant in Na^+ flow. Because the amiloride-sensitive Na^+-

uptake system does not appear in vitro, it thus may require intervention by a maternally derived factor.

An alternative process for formation of the blastocoel in the mouse was proposed by Wiley (1984) to explain some of the cytoplasmic events that have been observed early in blastocoel formation. This model asserts that blastocoel formation commences with "seeding" water derived from mitochondrial oxidation of lipid droplets. These and other organelles accumulate in a cortical distribution in the morula. The model attributes this arrangement to drift of these organelles in osmotic flows caused by ionic currents resulting from the asymmetric distribution of Na^+-K^+-ATPase ion pumps in the apposed blastomeric membranes (Vorbrodt et al. 1977). The constant morula volume as cavitation commences is said to result from balancing the volume of disappearing vesicles against the generated metabolic water and paracellular losses through immature tight junctions. As these junctions mature, paracellular loss of water is prevented, and the blastocoel expands rapidly because of the transcellular movement of water into the blastocoel. The concept of "metabolic" water is difficult to assess because of the basic lack of knowledge of embryonic metabolism. However, this model is not basically different from that proposed for the rabbit (Benos et al. 1985), because both depend primarily on Na^+-K^+-ATPase activity. Determination of the details awaits further research.

VII. Conclusion

Studies during the last decade have expanded the metabolic map proposed by Wales (1973b) into new areas. Mouse embryos are capable of biosynthesis of a number of lipids, including many of those required for membrane assembly. There is good evidence for biosynthesis of other membrane and cytoskeletal components such as glycoproteins. In terms of energy metabolism, the mouse embryo may store considerable quantities of triacylglycerols in addition to glycogen, while relying on oxidation of glucose, lactate, and pyruvate and some fatty acid derived from tract fluids for energy production. Nitrogen metabolism is yielding to new technologies that also show that the embryo is in a dynamic relationship with its environment. The simple structure of the preimplantation embryo suits this research. Modern techniques of amino acid quantitation permit examination of amino acid pools in a few embryos or a small volume of fluid. This should enable analysis of the fluctuations in the amino acid compositions of tract secretions during early pregnancy and of the variation of embryonic pools and transport that these fluid compositions reflect. The studies of blastocoel formation tie the complex area of membrane transport to metabolism and fluid expansion. It is not unreasonable to expect that the next decade will see the isolation of various transport proteins and an understanding of the regulation of their expression resulting from recombinant DNA studies.

The most difficult area for study is metabolic interconversions, where, in addition to high sensitivity, techniques require high resolution. Nevertheless, innovative application of techniques such as liquid chromatography and NMR should enable advances here.

The strong thread through this review is the evidence for embryo-maternal interactions. The embryo is dependent on maternal physiology for a regulated supply of nutrients, such as pyruvate, lactate, glucose, amino acids, and O_2, and the maternal physiology appears to recognize conception and alter nutrient levels appropriately. Evidence for direct effects of hormones on embryonic metabolism is lacking, but likely candidates may include substances such as early pregnancy factor (EPF) (Morton 1984), platelet-activating factor (O'Neill 1985), the inhibitory influence of uterine flushings (O'Neill and Quinn 1983), the acid-soluble contaminant of some BSA preparations that stimulates growth of rabbit blastocysts in vitro (Kane 1985), or any of the growth factors isolated from serum now known to have growth-regulatory effects in many other systems.

The use of quantitative microanalytic procedures as noninvasive tests of metabolic performance should also prove useful in the assessment of embryo viability in in vitro fertilization and embryo transfer programs.

Acknowledgments

The author's work referred to was supported by a grant from the National Health and Medical Research Council of Australia. I am grateful to Marrianne Hamlet, Fay McElligott, and Margaret Muncaster for their expert preparation of this manuscript.

References

Barbehenn, E. K., Wales, R. G., and Lowry, O. H. (1974) The explanation for the blockade of glycolysis in early mouse embryos. *Proc. Natl. Acad. Sci. U.S.A.* 71, 1056–60.

Benos, D. J. (1981) Ouabain binding to preimplantation rabbit blastocysts. *Dev. Biol.* 83, 69–78.

Benos, D. J., and Biggers, J. D. (1983) Sodium and chloride co-transport by preimplantation rabbit blastocysts. *J. Physiol. (Lond.)* 342, 23–33.

Benos, D. J., Biggers, J. D., Balaban, R. S., Mills, J. W., and Overstrom, E. G. (1985) Developmental aspects of sodium dependent transport processes of preimplantation rabbit embryos. In *Regulation and Development of Membrane Transport Processes*, ed. J. S. Graves, pp. 211–35. New York: Wiley.

Biggers, J. D., and Borland, R. M. (1976) Physiological aspects of growth and development of the preimplantation mammalian embryo. *Annu. Rev. Physiol.* 38, 95–116.

Biggers, J. D., and Stern, S. (1973) Metabolism of the preimplantation mammalian embryo. *Adv. Reprod. Physiol.* 6, 1–60.

Borland, R. M., Biggers, J. D., and Lechene, C. P. (1976) Kinetic aspects of rabbit blastocoele fluid accumulation: an application of electron probe microanalysis. *Dev. Biol.* 50, 201–11.

— (1977) Fluid transport by preimplantation rabbit blastocysts *in vitro. J. Reprod. Fertil.* 51, 131–5.

Borland, R. M., and Tasca, R. J. (1974) Activation of a Na$^+$-dependent amino-acid transport system in preimplantation mouse embryos. *Dev. Biol.* 30, 169–82.

— (1975) Na$^+$-dependent amino-acid transport in preimplantation mouse embryos. II. Metabolic inhibitors and nature of cation requirement. *Dev. Biol.* 46, 192–201.

Brinster, R. L. (1969) Incorporation of carbon from glucose and pyruvate into the preimplantation mouse embryo. *Exp. Cell Res.* 58, 153–8.

Caro, C. M., and Trounson, A. (1984) The effect of protein on preimplantation mouse embryo development. *J. In Vitro Fertil. Embryo Trans.* 1, 183–7.

Christensen, H. N. (1973) On the development of amino-acid transport systems. *Fed. Proc.* 32, 19–28.

— (1984) Organic ion transport during seven decades. The amino acids. *Biochim. Biophys. Acta* 779, 255–69.

Clough, J. R., and Whittingham, D. G. (1983) Metabolism of ^{14}C-glucose by postimplantation mouse embryos *in vitro. J. Embryol. Exp. Morphol.* 74, 133–42.

Colonna, R., Cecconi, S., Buccione, R., and Magnia, F. (1983) Amino-acid transport systems in growing mouse oocytes. *Cell Biol. Int. Rep.* 7, 1007–15.

Cooper, A. R., and MacQueen, H. A. (1983) Subunits of laminin are differentially synthesized in mouse eggs and early embryos. *Dev. Biol.* 96, 467–71.

Dabich, D., and Acey, R. A. (1982) Transport of glucosamine aldohexoses by preimplantation mouse blastocysts. *Biochim. Biophys. Acta* 684, 146–8.

DiZio, S. M., and Tasca, R. J. (1977) Sodium dependent amino-acid transport in preimplantation mouse embryos. III. Na$^+$-K$^+$-ATPase-linked mechanism in blastocysts. *Dev. Biol.* 59, 198–205.

Ducibella, T., Albertini, D. F., Anderson, E., and Biggers, J. D. (1975) The preimplantation mammalian embryo: characterization of intercellular junctions and their appearance during development. *Dev. Biol.* 45, 231–50.

Edirisinghe, W. D., and Wales, R. G. (1984) Effect of parenteral administration of oestrogen and progesterone on the glycogen metabolism of mouse morulae-early blastocysts *in vivo. J. Reprod. Fertil.* 72, 67–73.

Edirisinghe, W. R., Wales, R. G., and Pike, I. L. (1984a) Synthesis and degradation of labelled glycogen pools in preimplantation mouse embryos during short periods of *in vitro* culture. *Aust. J. Biol. Sci.* 37, 137–46.

— (1984b) Degradation of biochemical pools labelled with [^{14}C]glucose during culture of 8-cell and morula-early blastocyst stage mouse embryos *in vitro* and *in vivo. J. Reprod. Fertil.* 72, 59–65.

— (1984c) Studies on the distribution of glycogen between the inner cell mass and trophoblast cells of mouse embryos. *J. Reprod. Fertil.* 71, 533–8.

Epstein, C. J., and Smith, S. A. (1973) Amino-acid uptake and protein synthesis in preimplantation mouse embryos. *Dev. Biol.* 33, 171–84.

Fishel, S. B., and Surani, M. A. H. (1980) Evidence for the synthesis and release of a glycoprotein by mouse blastocysts. *J. Reprod. Fertil.* 59, 181–5.

Flach, G., Johnson, M. H., Braude, P. R., Taylor, A. S., and Bolton, V. N. (1982) The transition from maternal to embryonic control of preimplantation mouse development. *EMBO Journal* 1, 681–6.

Fleming, T. P., Warren, P. D., Chisholm, J. C., and Johnson, M. H. (1984) Trophectodermal processes regulate the expression of totipotency within the inner cell mass of the mouse expanding blastocyst. *J. Embryol. Exp. Morphol.* 84, 63–90.

Flynn, T. J., and Hillman, N. (1978) Lipid synthesis from [U-^{14}C]glucose in preimplantation mouse embryos in culture. *Biol. Reprod.* 19, 922–6.

— (1980) The metabolism of exogenous fatty acids by preimplantation mouse embryos developing *in vitro. J. Embryol. Exp. Morphol.* 56, 157–68.

Freeman, S. J., and Lloyd, J. B. (1983) Evidence that protein ingested by the rat visceral yolk sac yields amino acids for synthesis of embryonic protein. *J. Embryol. Exp. Morphol.* 73, 307–15.

Gardner, H. G., and Kaye, P. L. (1984) Effects of insulin on preimplantation mouse embryos. In *Proceedings of the 16th Annual Meeting of the Australian Society for Reproductive Biology*, Vol. 16, p. 107 (abstract).

Glass, L. E. (1963) Transfer of native and foreign serum antigens to oviductal mouse eggs. *Am. Zool.* 3, 135–56.

Gregoire, A. T., Gongsakdi, D., and Rakoff, A. E. (1961) The free amino-acids content of female genital tract. *Fertil. Steril.* 12, 322–7.

Guidotti, G. C., Gazola, C. C., Borghetti, F. F., and Franchi-Gazola, R. (1975) Adaptive regulation of amino-acid transport across the cell membrane in avian and mammalian tissues. *Biochim. Biophys. Acta* 406, 264–79.

Harlow, G. M., and Quinn, P. (1979) Foetal and placental growth in the mouse after preimplantation development *in vitro* under oxygen concentrations of 5 and 20%. *Aust. J. Biol. Sci.* 32, 363–9.

Hastings, R. A., and Enders, A. C. (1975) Junctional complexes in the preimplantation rabbit embryo. *Anat. Rec.* 181, 17–34.

— (1974b) Uptake of exogenous protein by the preimplantation rabbit embryo. *Anat. Rec.* 179, 311–30.

Heap, R. B., Rider, V., and Feinstein, A. (1984) Monoclonal progesterone antibodies and early embryo development. In *Proceedings of the 10th International Congress of Animal Reproduction and Artificial Insemination*, Vol. 8, pp. 14–21.

Hobbs, J. G., and Kaye, P. L. (1985) Glycine transport in mouse eggs and preimplantation embryos. *J. Reprod. Fertil.* 74, 77–86.

— (in press) Glycine and Na$^+$ transport in preimplantation mouse embryos. *J. Reprod. Fertil.*

Holmberg, S. E. M., and Johnson, M. H. (1979) Amino-acid transport in the unfertilized and fertilized egg. *J. Reprod. Fertil.* 56, 223–31.

Johnson, M. H., and Maro. B. (1984) The distribution of cytoplasmic actin in mouse 8-cell blastomeres. *J. Embryol. Exp. Morphol.* 82, 97–117.

Kane, M. T. (1979) Fatty acids as energy sources for culture of one-cell rabbit ova to viable morulae. *Biol. Reprod.* 20, 323–32.

– (1985) A low molecular weight extract of bovine serum albumin stimulates rabbit blastocyst cell division and expansion *in vitro. J. Reprod. Fertil.* 73, 147–50.

Kane, M. T., and Buckley, N. J. (1977) The effects of inhibitors of energy metabolism on the growth of one cell rabbit ova to blastocysts *in vitro. J. Reprod. Fertil.* 49, 261–6.

Kaye, P. L., Pemble, L. B., and Hobbs, J. G. (in press) Protein metabolism in preimplantation mouse embryos. In *New Discoveries and Technologies in Developmental Biology*, ed. H. C. Slavkin. New York: Alan R. Liss.

Kaye, P. L., Schultz, G. A., Johnson, M. H., Pratt, H. P. M., and Church R. B. (1982) Amino-acid transport and exchange in preimplantation mouse embryos. *J. Reprod. Fertil.* 65, 367–80.

Leese, H. J., Aldridge, S., and Jeffries, K. S. (1979) The movement of amino acids into rabbit oviductal fluid. *J. Reprod. Fertil.* 56, 623–6.

Leese, H. J., and Barton, A. M. (1984) Pyruvate and glucose uptake by mouse ova and preimplantation embryos. *J. Reprod. Fertil.* 72, 9–13.

– (in press) The production of pyruvate by isolated mouse cumulus cells. *J. Exp. Zool.*

Mastroianni, L., and Biggers, J. D. (1981) *Fertilization and Embryonic Development in Vitro.* New York: Plenum Press.

Mastroianni, L., and Jones, R. (1965) Oxygen tension within the rabbit fallopian tube. *J. Reprod. Fertil.* 9, 99–102.

Merz, E. A., Brinster, R. L., Brunner, S., and Chen, H. Y. (1981) Protein degradation during preimplantation development of the mouse. *J. Reprod. Fertil.* 61, 415–18.

Miller, J. G. O. (1984) Amino-acid transport in preimplantation mammalian embryos. Ph.D. thesis, University of Calgary, Alberta, Canada.

Miller, J. G. O., and Schultz, G. A. (1983) Properties of amino-acid transport in preimplantation rabbit embryos. *J. Exp. Zool.* 228, 511–25.

Morton, H. (1984) Early pregnancy factor (EPF), a link between fertilization and immunomodulation. *Aust. J. Biol. Sci.* 37, 393–407.

Murdoch, R. N., and Wales, R. G. (1973) Incorporation of [^{14}C]glucose and [^{3}H]uridine into the major classes of RNA in mouse embryos during preimplantation development. *Aust. J. Biol. Sci.* 26, 889–901.

Neider, G. L., and Corder, C. N. (1983) Pyruvate and lactate levels in oviducts of cycling, pregnant and pseudopregnant mice. *Biol. Reprod.* 28, 566–74.

O'Neill, C. (1985) Examination of the causes of early pregnancy associated thrombocytopenia in mice. *J. Reprod. Fertil.* 73, 567–77.

O'Neill, C., and Quinn, P. (1983) Inhibitory influence of uterine secretions on mouse blastocysts decreases at the time of blastocyst activation. *J. Reprod. Fertil.* 68, 269–74.

Ozias, C. B., and Stern, S. (1973) Glycogen levels of preimplantation mouse embryos developing *in vitro. Biol. Reprod.* 8, 467–79.

Ozias, C. B., and Weitlauf, H. M. (1971) Hormonal influences on the glycogen content of normal and delayed implanting mouse blastocysts. *J. Exp. Zool.* 177, 147–52.

Pemble, L. B., and Kaye, P. L. (1984) ^{125}I-BSA uptake by preimplantation mouse embryos. In *Proceedings of the 16th Annual Meeting of the Australian Society for Reproductive Biology*, Vol. 16, p. 105 (abstract).

– (in press) Whole protein uptake and metabolism by mouse blastocyst. *J. Reprod. Fertil.*

Pike, I. L., Kaye, P. L., and Wales, R. G. (1977) Glucose incorporation into nucleic acids of mouse morulae. In *Proceedings of the 10th Annual Meeting of the Australian Society for Reproductive Biology*, Vol. 10, p. 75 (abstract).

Pike, I. L., and Wales, R. G. (1982) Uptake and incorporation of glucose especially into the glycogen pools of preimplantation mouse embryos during culture *in vitro*. *Aust. J. Biol. Sci.* 35, 195–206.

Pratt, H. P. M. (1977) Uterine proteins and the activation of embryos from mice during delayed implantation. *J. Reprod. Fertil.* 50, 1–8.

– (1980) Phospholipid synthesis in the preimplantation mouse embryo. *J. Reprod. Fertil.* 58, 237–48.

– (1982) Preimplantation mouse embryos synthesize membrane sterols. *Dev. Biol.* 89, 101–10.

Pratt, H. P. M., Keith, J., and Chakraborty, J. (1980) Membrane sterols and the development of the preimplantation mouse embryo. *J. Embryol. Exp. Morphol.* 60, 303–19.

Quinn, P., and Harlow, G. M. (1978) The effect of oxygen on the development of preimplantation mouse embryos *in vitro*. *J. Exp. Zool.* 206, 73–80.

Quinn, R., and Wales, R. G. (1973a) Uptake and metabolism of pyruvate and lactate during preimplantation development of the mouse embryo *in vitro*. *J. Reprod. Fertil.* 35, 273–87.

– (1973b) The *in vitro* metabolism of [U-^{14}C]glucose by the preimplantation rabbit embryo. *Aust. J. Biol. Sci.* 26, 653–67.

– (1973c) The effect of culture *in vitro* on the levels of adenosine triphosphate in preimplantation mouse embryos. *J. Reprod. Fertil.* 32, 231–41.

Renard, J.-P., Philippon, A., and Menezo, Y. (1980) *In vitro* uptake of glucose by bovine blastocysts. *J. Reprod. Fertil.* 58, 161–4.

Schlafke, S., and Enders, A. C. (1973) Protein uptake by rat preimplantation stages. *Anat. Rec.* 175, 539–60

Schneider, J. H., Olds, D., Tucker, R. E. and Mitchell, G. E. (1976) Mouse embryos use glucose in protein synthesis. *J. Anim. Sci.* 42, 271 (abstract).

Schultz, G. A., Kaye, P. L., McKay, D. J., and Johnson, M. H. (1981) Endogenous amino-acid pool sizes in mouse eggs and preimplantation embryos. *J. Reprod. Fertil.* 61, 387–93.

Sellens, M. H., Stein, S., and Sherman, M. I. (1981) Protein and free amino-acid content in preimplantation mouse embryos and in blastocysts under various culture conditions. *J. Reprod. Fertil.* 61, 307–15.

Spielman, H., Jacob-Mueller, U., Schulz, P., and Schimmel, A. (1984) Changes in the adenine ribonucleotide content during preimplantation development of mouse embryos *in vivo* and *in vitro*. *J. Reprod. Fertil.* 71, 467–73.

Surani, M. A. H., Kimber, S. J., and Osborn, J. C. (1983) Mevalonate reverses the developmental arrest of preimplantation mouse embryos by compactin, an inhibitor of HMG-CoA-reductase. *J. Embryol. Exp. Morphol.* 75, 205–23.

Tanimura, T., and Shepard, T. G. (1970) Glucose metabolism by rat embryos *in vitro. Proc. Soc. Exp. Biol. Med.* 135, 51–3.

Troike, D. E., and Brinster, R. L. (1981) De novo pyrimidine nucleotide synthesis in the preimplantation mouse embryo. *Exp. Cell Res.* 134, 481–4.

Van Winkle, L. J., and Dabich, D. (1977) Transport of naturally occurring amino acids and alpha aminosobutyric acid by normal and diapausing mouse blastocysts. *Biochem. Biophys. Res. Commun.* 78, 357–63.

Vorbrodt, A., Konwinski, M., Solter, D., and Koprowski, H. (1977) Ultrastructural cytochemistry of membrane-bound phosphatases in preimplantation mouse embryos. *Dev. Biol.* 55, 117–34.

Wales, R. G. (1973a) The uterus of the ewe. II. Chemical analysis of uterine fluid collected by cannulation. *Aust. J. Biol. Sci.* 26, 947–59.

–   (1973b) Biochemistry of the developing embryo. *J. Reprod. Fertil.* Suppl. 8, 117–25.

–   (1975) Maturation of the mammalian embryo: biochemical aspects. *Biol. Reprod.* 12, 66–81.

Wales, R. G., and Brinster, R. L. (1968) The uptake of hexoses by preimplantation mouse embryos *in vitro. J. Reprod. Fertil.* 15, 415–22.

Wales, R. G., and Whittingham, D. G. (1974) Further studies of the accumulation of energy substrates by 2-cell mouse embryos. *Aust. J. Biol. Sci.* 27, 519–21.

Walters, M. R., Hazelwood, R. L., and Lawrence, A. L. (1979) Amino-acid transport from the lumen of the rat uterus. *Biol. Reprod.* 20, 985–90.

Whelan, W. J. (1976) On the origin of primer for glycogen synthesis. *Trends Biochem. Sci.* 1, 13–15.

Wiley, L. M. (1984) Cavitation in the preimplantation embryo: Na/K-ATPase and the origin of nascent blastocoel fluid. *Dev. Biol.* 105, 330–42.

Wordinger, R. J., and Brinster, R. L. (1976) Influence of reduced glucose levels on the *in vitro* hatching of the mouse blastocyst. *Dev. Biol.* 53, 294–6.

Yochim, J. M., and Mitchell, J. A. (1968) Intrauterine oxygen tension in the rat during progesterone: its possible relation to carbohydrate metabolism and the regulation of nidation. *Endocrinology* 83, 706–13.

10 Role of cell surface molecules in early mammalian development

JEAN RICHA AND DAVOR SOLTER

CONTENTS

Introduction

Multicellular organisms exist through the cooperative interactions of single cells that become programmed during development to carry out specific functions. This programming occurs in a sequential and orderly fashion and, at least in the earliest events that can be studied in the mammalian embryo, may be brought about by cues transmitted from cell to cell by cell surface molecules.

In the mouse embryo the first such morphogenetic event, compaction, occurs at the late 8-cell stage, when the individual spherical blastomeres flatten against each other and maximize their cell contacts (Ducibella and Anderson 1975; Lehtonen 1980) (see Chapter 2). Concurrent with this change in shape is the development of more specialized contacts such as desmosomes and focal tight junctions (Ducibella et al. 1975; Magnuson, Jacobson, and Stackpole 1978; Ducibella and Anderson, 1979), and the start of intercellular communication via gap junctions (Lo and Gilula

293

1979; Lo 1980; Goodall and Johnson 1982, 1984). Moreover, surface and cytoplasmic polarity become evident: On the surface, short microvilli become restricted to an apical pole (Calarco and Epstein 1973; Ducibella et al. 1975; Ziomek and Johnson, 1980; Reeve and Ziomek 1981), while the nucleus occupies the basal region of the blastomeres (Reeve and Kelly 1982), and the intervening cytoplasm becomes increasingly filled with various organelles (Ducibella et al. 1977; Reeve 1981; Fleming 1984). Compaction is followed, after 24 h, by cavitation, which marks the onset of the blastocyst stage. Here the embryo consists of an outer single layer of cells, the trophectoderm, surrounding a fluid-filled blastocoelic cavity and an aggregate of cells called the inner cell mass.

If compaction is impeded, further development is inhibited; individual blastomeres accumulate fluid-filled vacuoles and degenerate without formation of the blastocyst (Ducibella and Anderson 1979). Inhibition of compaction can be caused by disruption of the cytoskeleton (cytochalasin D, colcemid), perturbation of intercellular contacts (concanavalin A, tunicamycin), or absence of extracellular calcium (Pratt et al. 1982). The role of cell surface molecules in this process was confirmed by the demonstration that several heterologous antisera raised against mouse embryonal carcinoma (EC) cells also inhibit compaction (Kemler et al. 1977; Johnson et al. 1979; Ducibella 1980; Ogou, Okada, and Takeichi 1982).

Changes in expression of cell surface molecules during preimplantation development have been demonstrated using antibodies of various specificities; these studies have been reviewed extensively (Jacob 1979; Solter and Knowles 1979; Wiley 1979; Johnson and Calarco 1980; Webb 1983). The work presented herein is a description of stage-specific embryonic antigens (SSEAs) of known molecular structure. In addition, we address the established or suggested roles in development of several cell surface and cytoskeletal components.

II. Structurally defined embryonic cell surface molecules

At present, oligosaccharides, detected using various probes (i.e., monoclonal antibodies or lectins), are the only embryonic cell surface structures whose chemical compositions and sequences are known (Table 10.1). This knowledge provides the basis for experimental approaches to elucidate their functions.

A. Lacto-series carbohydrates

Several oligosaccharide molecules belonging to this group have been detected on mammalian embryos. The antigen that has been studied in most detail is stage-specific embryonic antigen 1 (SSEA-1), detected by a monoclonal antibody raised against F9 embryonal carcinoma cells (Solter and Knowles 1978). SSEA-1 first appears on the 8-cell-stage pre-implantation mouse embryo, although not all blastomeres are positive. The pattern of negative and positive blastomeres persists through the morula stage but changes at the blastocyst stage, where positive expression of the antigen becomes restricted to the inner cell mass (ICM) (Solter and Knowles 1978). When ICMs are grown in vitro and prevented from attaching, an outside layer of endodermal cells is formed (Hogan and Tilly 1978; Solter and Knowles 1978). Some of these endodermal cells are positive and some negative for SSEA-1 expression (Solter and Knowles 1979). When endoderm is removed by repeated immunosurgery (Strickland, Reich, and Sherman 1976), most of the resulting ectodermal cells express SSEA-1 (Solter and Knowles 1979).

In postimplantation embryos, SSEA-1 is expressed only on the visceral endoderm and the embryonic ectoderm (Fox et al. 1981). In the adult mouse, specific cells in only a few tissues, such as brain, kidney, epididymis, and uterine epithelium, are positive (Fox et al. 1981, 1982a, 1982b). In addition, SSEA-1 is present on EC cells but not on their differentiated derivatives (Solter et al. 1979; Knowles et al. 1980), which has made it a useful marker for monitoring the differentiation of EC cells in culture. However, when EC cells differentiate into cells that correspond to visceral endoderm, the expression of SSEA-1 persists (Hogan, Barlow, and Tilly 1983).

SSEA-1 is expressed in different tissues of different species on normal and tumor cells. Monoclonal antibodies raised against human myelocytic leukemia cells (Huang et al. 1983a), adenocarcinoma of the colon or stomach (Brockhaus et al. 1982), or different lung carcinomas (Huang et al. 1983b) detect an antigenic determinant that is identical with or very similar to SSEA-1. Human EC cells do not express this antigen, although it is expressed by some of their differentiated derivatives (Andrews et al. 1982, 1984; Kannagi et al. 1983a). Two additional antigens of the lacto series have been detected on mouse embryos. These are the human blood group precursor antigens I and i, defined by naturally occurring human monoclonal antibodies (Feizi and Kabat 1972). I and i antigens are found on glycoproteins and glycolipids of human erythrocytes (for review, see Hakomori 1981) and other cell types (Childs, Kapadia, and Feizi 1980). Indirect immunofluorescence studies of unfertilized eggs and preimplantation embryos showed staining of all stages with anti-I but not with anti-i antibodies (Feizi et al. 1982; Knowles, Rappaport, and Solter 1982). A similar pattern of reactivity was observed on undifferentiated cells in teratocarcinoma cell line PSMB. By day 6 of development, parietal and visceral endoderm of the mouse embryo showed i and I staining (Kapadia, Feizi, and Evans 1981), whereas ectodermal cells reacted only weakly or not at all.

Table 10.1. *Structurally defined antigens detected by monoclonal antibodies*

Antibody	Antigen	Structure[a]	Distribution[b]	References
IIC3	—[c]	Galβ GalNAc	Mouse EC cells, lymphoid cells; mouse morulae, blastocysts, and primitive endoderm; blastocyst outgrowths only after neuraminidase treatment	Marticorena et al. (1983)
5D4	—	Galα	Mouse EC cells, unfertilized and fertilized eggs, less on early cleavage-stage embryos	Stern et al. (1983)
C6	—	Galβ1→4GlcNAcβ	Mouse EC cells; preimplantation mouse embryos (following neuraminidase treatment)	Fenderson et al. (1983)
A5	—	Galβ1→4GlcNAcβ		
Anti-i	i	Galβ1→4GlcNAcβ1→3Galβ1→4GlcNAcβ	Primitive endoderm of mouse embryos	Kapadia et al. (1981)
Anti-I	I	Galβ1→4GlcNAcβ1 $\searrow$ 6 $\qquad\qquad\qquad$ GlcNAcβ Galβ1→4GlcNAcβ1 $\nearrow$ 3	Preimplantation mouse embryo	Feizi et al. (1982), Knowles et al. (1982)

480	SSEA-1 X-hapten	Galβ1→4GlcNAcβ ↑1,3 Fucα	Mouse EC cells; mouse embryos, 8-cell to blastocyst, ectoderm, and visceral endoderm; differentiated human EC cells	Solter and Knowles (1978), Fox et al. (1981), Gooi et al. (1981), Hakomori et al. (1981)
75.12	Y-hapten	Galβ1→4GlcNAcβ ↑1,2 ↑1,3 Fucα Fucα	Some mouse EC cell lines; not on early mouse embryos	Balineau et al. (1983)
M1 : 22 : 25 : 8	Forssman	GalNAcα1→3GalNAcβ1→3Galα1→4Galβ	Mouse EC cells and early mouse blastocysts	Willison and Stern (1978)
MC 631	SSEA-3	SSEA-4 SSEA-3 NeuAcα2→3Galβ1→3GalNAcβ1→3Galα1→4Galβ	Unfertilized and fertilized mouse eggs; preimplantation stages of mouse embryo; human EC cells	Shevinsky et al. (1982), Kannagi et al. (1983a, 1983b)
MC 813-70	SSEA-4			

Note:

[a] In some cases the complete structure of the antigenic determinant has not been identified.

[b] Only the distribution of antigens or mouse embryos and EC cells is included.

[c] No identifying name.

SSEA-1 and Ii antigens contain repeated N-acetyllactosamine units, Galβ1→GlcNAc, characteristic of type-2 blood group precursor chains (Watkins 1980), in contrast to Galβ1→3GlcNAc typical of type-1 isomers. Structural studies of the Ii antigens have revealed that their antigenic determinants consist of linear (i) (Neimann, Watanabe, and Hakomori 1978) or branched (I) oligosaccharide sequences (Feizi et al. 1979; Watanabe et al. 1979) (Table 10.1) and that loss of the terminal Gal residue results in loss of Ii activity (Neimann et al. 1978; Watanabe et al. 1979). Furthermore, both I and i oligosaccharide chains can be converted into blood group H, A, or B structures by the addition of fucose in α1→2 linkage to the terminal Gal residue (H antigen), followed by N-acetylgalactosamine (A antigen) or galactose (B antigen) in α1→3 linkage (Feizi et al. 1982). SSEA-1 can be derived from the blood group antigens I or i by α1→3 fucosylation of the N-acetylglucosamine (Gooi et al. 1981; Hakomori et al. 1981) (Table 10.1). As expected in inhibition or direct binding assays, monoclonal antibodies detecting SSEA-1 react with purified lacto-N-fucopentaose III (Kannagi et al. 1982).

The appearance of SSEA-1 at the 8-cell-stage embryo suggests that this antigen might play a role in compaction. A monoclonal antibody to SSEA-1 and its F(ab) or F(ab')$_2$ fragments has been ineffective in inhibiting compaction of mouse embryos (D. Solter, unpublished observations). On the other hand, it has been shown recently that monovalent (Bird and Kimber 1984) or multivalent (Fenderson, Zehavi, and Hakomori 1984) lacto-N-fucopentaose III can reverse the compaction of 8-cell-stage embryos. The inability of the antibody to inhibit compaction could be due to the antibody binding site being distinct from the functional part of the antigen, or the agglutinating effect of the multivalent antibody reversing its inhibiting action, the monovalent F(ab) fragments having too low an avidity to functional sites.

Carbohydrate molecules of the type described earlier can participate in cell-cell interactions in several ways. They can bind to specific receptor molecules (endogenous lectins) or act as substrates for cell surface glycosyl transferases. Receptors that specifically recognize the fucosyl α1→3 N-acetylglucosamine linkage have been described on the surface of human hepatocytes (Prieels et al. 1978). It is not known if such receptors exist in embryos or EC cells, although endogenous lectins that specifically bind molecules with high fucose content (like fucoidan) have been described (Grabel et al. 1981, 1983) on mouse EC cells.

It has been suggested that multivalent lactosaminoglycans, which are abundant on the surfaces of EC cells, might mediate the EC cell adhesion by binding to galactosyl transferase on the cell surface (Shur 1983). Similar mechanisms might be involved in compaction of the mouse morula (Shur 1984), because an excess of UDP-Gal can prevent compaction and blastocyst formation (Shur, Oettgen, and Bennett 1979). Galactosyl

Table 10.2. *Binding of several antibodies to parental and mutant F9 cells*

Antibody[a]	Cell lines[b]		
	F9	OT F9	SOT F9
Anti-SSEA-1	8,500[c]	9,000	120
Anti-H	300	250	3,500
Anti-I-Ma	8,600	8,400	9,000
Anti-Forssman	5,000	5,000	4,800

Note:
[a]Anti-SSEA-1, anti-H, and anti-Forssman are monoclonal antibodies, and anti-I-Ma is a human monospecific antiserum.
[b]F9 and OT F9 are two different clones of the original F9 cell line. SOT F9 cells were derived by mutagenizing OT F9 and selecting for SSEA-1-negative cells (Rosenstraus 1983).
[c]Data are given as counts per minute bound to 10^5 cells. See Solter and Knowles (1978) for details of indirect binding assay.

transferase has been detected, using immunohistochemical methods, on the surfaces of preimplantation mouse embryos, particularly at the 8-cell to early blastocyst stages (Sato, Muramatsu, and Berger 1984).

The presence of the SSEA-1 antigenic determinant suggested the presence of another glycosyl transferase, namely $\alpha 1 \rightarrow 3$ fucosyl transferase, on the cell surfaces of embryos and EC cells. This enzyme was indeed detected on the surfaces of F9 cells, where its specific activity decreased when the stem cells were induced to differentiate into parietal endoderm by retinoic acid and dibutyryl cAMP (Muramatsu and Muramatsu 1983). Two further lines of evidence indicate an association between SSEA-1 and surface fucosyl transferase. Campbell and Stanley (1983) isolated from Chinese hamster ovary (CHO) cells a mutant cell line that expresses SSEA-1 and has fucosyl transferase activity, whereas the parental line and the revertants are negative for both. In addition, Rosentraus (1983) isolated a mutant cell line from F9 cells that is SSEA-1-negative. We have shown (Table 10.2) that the mutant cell line is H-blood-group-antigen-positive, whereas the parental line is negative. This result can be explained by loss of $\alpha 1 \rightarrow 3$ fucosyl transferase in the mutant cells that in turn allows $\alpha 1 \rightarrow 2$ fucosyl transferase access to the common substrate.

Thus, several glycosyl transferases are present on the surfaces of embryonic cells and are able to alter cell surface carbohydrates. Glycosyl transferases and their acceptor substrates might play an important role during development by constantly changing patterns of cell surface molecules, thus providing a mechanism for new and specific cellular interactions.

B. Globo-series carbohydrates

Another class of carbohydrate antigens detected on mouse embryos and on human and mouse EC cells belong to globo-series oligosaccharide chains. These include globoside (GalNAcβ1→3Galα1→4Galβ1→4Glc-Cer), Forssman antigen (GalNAcα1→3 globoside), SSEA-3 (Galβ1→3 globoside), and SSEA-4 (NeuAcα2→3Galβ1→3 globoside).

Affinity-purified antibodies to globoside first bind to 2- to 4-cell embryos and reach peak binding levels at the morula stage. Purified antibodies to Forssman antigen first bind to the late morula, and the binding increases at the early blastocyst stage (Willison et al. 1982). A similar binding pattern was observed when monoclonal antibody to Forssman antigen (M1/22.25) was used (Willison and Stern 1978). The same antibody was subsequently used to analyze the lineage relationship of Forssman-antigen-positive cells in postimplantation embryos (Stinnakre et al. 1981). Additional antigens of this class include SSEA-3 and -4 detected on early mouse embryos and human (but not mouse) EC cells by monoclonal antibodies MC631 and MC813.70, respectively. SSEA-3 is present on mouse oocytes and becomes restricted first to the ICM at the blastocyst stage and later to the primitive endoderm (Shevinsky et al. 1982). It is also present on cells of the visceral but not the parietal endoderm (Fox et al. 1984). SSEA-4 has a distribution on mouse oocytes and early cleavage-stage embryos similar to that of SSEA-3 (Kannagi et al. 1983a). When oocytes, fertilized eggs, and cleavage-stage embryos were pretreated with neuraminidase, their reactivities with anti-SSEA-3 MC631 were not affected, whereas those with anti-SSEA-4 MC813.70 were abolished (Kannagi et al. 1983a). Detailed analysis of MC631 and MC813.70 reactivities with various purified glycolipids (Kannagi et al. 1983a, 1983b) demonstrated that both react with the following structure, which represents a unique embryonic antigen with at least two known epitopes (*a* and *b*):

$$\overbrace{\text{NeuAc}\alpha2 \rightarrow 3\text{Gal}\beta1 \rightarrow \underbrace{3\text{GalNAc}\beta1 \rightarrow 3\text{Gal}\alpha1 \rightarrow 4\text{Gal}\beta1 \rightarrow 4\text{Glc}\beta1 \rightarrow 1\text{-Cer}}_{b}}^{a}$$

MC813.70 recognized the terminal *a* structure, whereas MC631 recognized the internal *b* structure. These antigenic determinants are expressed on human EC cells (Andrews et al. 1982) and are associated with cell surface glycoproteins (Shevinsky et al. 1982) and glycolipids (Kannagi et al. 1983a).

The biosynthetic pathway of globo-series antigens and of previously

Table 10.3. *Effects of retinoic acid on F9 cell antigen expression*

Culture conditions	F9 cell binding (cpm) by	
	Anti-SSEA-1[a]	Anti-SSEA-3[a]
Tissue culture dish		
−RA[b]	5,600	140
+RA	1,900	100
Bacteriological dish		
−RA	6,700	210
+RA	5,900	3,100

Note:
[a]RIA conditions were as described previously (Solter and Knowles 1978; Solter et al. 1979).
[b]RA, retinoic acid.

described lacto-series antigens initiates from a common precursor, lacto-sylceramide (Galβ1→4Glcβl→1Cer). The addition of αGal by the action of α1→4-Gal transferase begins the globo series, and the addition of βGlcNAc by the action of β1→3-GlcNAc transferase initiates the lacto series (for details, see Kannagi et al. 1983a). The same branch point leads to the divergence of glycolipids of P and ABO blood groups, respectively (Watkins 1980). Competition for substrates and the activation of various glycosyl transferases during development and during differentiation of EC cells very likely underlies the switch from one glycolipid series to another and the appearance of new and more complex glycolipids. This is clearly exemplified by the analysis of cell surface properties of differen-tiated cells derived from the F9 cell line exposed to retinoic acid under various in vitro conditions. Hogan, Taylor, and Adamson (1981) demon-strated that F9 cells grown as attached monolayers in the presence of retinoic acid differentiate into parietal endoderm, whereas those grown as floating aggregates differentiate into visceral endoderm. We examined the expression of SSEA-1 and -3 in cells grown under these various conditions (Table 10.3) and demonstrated that SSEA-3, a marker of visceral endoderm (Fox et al. 1984), is induced only in aggregates in the presence of retinoic acid.

C. Monosaccharides and disaccharides

The presence of different monosaccharides and dissacharides has been detected on embryos and EC cells using either monoclonal antibodies or various lectins or a combination of both. It is very likely that the identified sugars represent only one part of the antigenic determinant or lectin binding site. Identification of these sugars was based on binding of antibody or lectin and by partial or complete inhibition of binding using the appropriate sugar. In most cases the sugar concentration necessary to inhibit binding was very high, further suggesting that the antigenic site was only partially identified.

A rat monoclonal antibody, 5D4, recognized surface antigens shared by murine EC cells and preimplantation embryos. Indirect evidence suggests that the determinants recognized are carbohydrates and may include α-linked galactose residues (Stern et al. 1983).

Another monoclonal antibody, IIC3, raised against F9 teratocarcinoma cells (Marticorena et al. 1983) detects an antigen expressed on F9 as well as other EC cells (PCC3, PCC4, and Nulli) and some differentiated cells. Indirect immunofluorescence studies reveal a weak binding of IIC3 to 8- to 12-cell mouse embryos and stronger binding to morulae and blastocysts. ICMs are faintly reactive, but after two days in culture the primitive endoderm cells react very strongly. Trophoblastic outgrowths show few fluorescent cells, although after neuraminidase treatment the attached blastocysts show considerable fluorescence with IIC3. The binding of this antibody is completely inhibited by β-D-Gal at 5 mg/ml or by GalNAc at 20 to 40 mg/ml. Fucose and GlcNAc have no effect on IIC3 binding.

Additional carbohydrate antigens are recognized on preimplantation mouse embryos and on multipotent EC cell lines by two IgM class monoclonal antibodies C6 and A5 (Fenderson, Hahnel, and Eddy 1983). Detection of these antigens requires prior treatment of the embryos, but not of EC cells, with neuraminidase. A sugar hapten inhibition assay, used to characterize the determinants recognized by these antibodies, identified N-acetyllactosamine (Galβ1→4GlcNAc) as inhibitor of the binding of both antibodies to F9 cells (Fenderson et al. 1983). However, the high concentration (0.1 M) of the hapten required to inhibit antibody binding suggested that N-acetyllactosamine is again only part of the antigenic determinant for both antibodies.

At the present time it is difficult to determine the relationship between the antibodies described earlier and those detecting antigens of the lacto and globo series. For example, antibodies C6 and A5 detect an N-acetyllactosamine-containing structure following neuraminidase treatment of embryos but react directly with untreated EC cells. A monoclonal antibody to N-acetyllactosamine (D. Solter, unpublished results) has been found to bind to EC cells but not to embryos. N-acetyllactosamine is the terminal structure of both I and i antigens present on preimplantation embryos (Feizi et al. 1982; Knowles et al. 1982), so that it is very likely that these antigens on preimplantation embryos are further modified, probably by the addition of neuraminic acid. It is also possible that the presence of neuraminic acid on other neighboring structures prevents binding of antibodies to terminal N-acetyllactosamine. These questions are unlikely to be definitively resolved in the near future, because the amount of available embryonic material is prohibitively small for isolation and sequence analysis of glycolipid antigens. Thus, analysis is limited to the use of indirect methods, and results must be interpreted cautiously.

I and i antigen can be fucosylated at penultimate as well as internal GlcNAc, leading to formation of SSEA-1-type (X-determinant-type) antigens (Fukushi et al. 1984). However, probably only unbranched type-2 chain (i antigen) is fucosylated in embryos and EC cells (Kannagi et al. 1982; Fenderson et al. 1986). Additional fucosylation of the terminal Gal converts SSEA-1 into Y hapten (Table 10.1). The existence of Y hapten on mouse embryos is controversial, because some monoclonal antibodies (e.g., 75.12) recognizing this determinant do not bind to preimplantation mouse embryos (Blaineau et al. 1983), whereas others (e.g., AH-6) (Abe, McKibbin, and Hakomori 1983) bind to morulae and blastocysts (Fenderson et al. 1986). Similar discrepancies have been noted using several monoclonal antibodies that presumably react with lacto N-III-fucopentaose (SSEA-1); most of these antibodies do not react with preimplantation embryos, but all react with EC cells, and lacto-N-III-fucopentaose inhibited this reactivity (Solter and Knowles 1986). It is at present unclear whether these findings reflect subtle differences in antigenic structure or the effect of surrounding molecules.

D. Lectin receptors

Lectin-saccharide interactions might be involved in cell adhesion among teratocarcinoma cells (Grabel et al. 1981, 1983) as well as in other systems (Harrison and Chesterton 1980; Barondes 1981). Several lectin receptors have been identified on preimplantation mouse embryos, and a possible role in development has been suggested.

Lectin receptors found on preimplantation mouse embryos include the receptors of concanavalin A (ConA), peanut agglutinin (PNA), wheat germ agglutinin (WGA), *Dolichos biflorus* agglutinin (DBA), and *Ricinus communis* agglutinin (RCA). The specificity of each lectin depends on the sugar residue to which it binds: ConA binds to Man or Glu (Poretz and Goldstein 1970), PNA preferentially recognizes Galβ1→3GalNAc sequence (Lotan et al. 1975), and WGA binds GlcNAc and NeuAc (Goldstein, Hammerström, and Sundblad 1975). DBA recognizes terminal GalNAc (Etzler and Kabat 1970), and RCA binds to Gal (Nicolson, Blaustein, and Etzler 1974). DBA receptors are expressed on cleavage-stage embryos (Fujimoto et al. 1982), and PNA labeling, using FITC-PNA, is uniform until the 16-cell stage (Handyside 1980), when it increases and becomes polarized. RCA, ConA, and WGA bind to mouse eggs and cleavage-stage embryos in a uniform manner (Brownell 1977). However, ConA receptors become polarized at the 8- to 16-cell stage (Ziomek and Johnson 1980), paralleling the polarization of surface microvilli.

Although the presence of these lectin receptors suggests the involvement of endogenous lectins in cell-cell interactions, it remains difficult to correlate lectin interactions and specific developmental functions, because more than one molecule might bind a certain lectin, and a given lectin might display different affinities depending on the structure of the carbohydrate chain. Moreover, lectins of various origins but presumably with the same sugar specificity can display considerable differences in binding (both in intensity and distribution) to early mouse embryos (Sato and Muramatsu 1985; Sato et al. in press).

Several studies have suggested a possible role for lectins in compaction. First, single 8-cell blastomeres can spread over PNA-, WGA-, or ConA-coated agarose beads (Kimber and Surani 1982). However, sugar-containing molecules at physiological concentrations did not inhibit compaction of intact 8-cell embryos (Kimber and Surani 1982), suggesting that the sugar-lectin interaction alone may not account for compaction of 8-cell embryos or that the affinity/avidity of endogenous lectins for the added molecules is fairly low. Second, WGA can induce compaction and cavitation-like events in 2-cell mouse embryos (Johnson 1986): The increased apposition of the blastomeres can be detected 1 h following exposure to the lectin, and after 12–18 h a blastocoel-like cavity is evident between the blastomeres, and junctional complexes can be seen at the periphery of the cavity. Third, in zonae-free mouse embryos cultured in ConA at 20 μg/ml, blastocyst formation is inhibited; cell division is retarded, compaction is prevented, and fluid accumulation occurs within but not between blastomeres (Reeve 1982).

These findings suggest that endogenous lectins may exist on the cell surfaces of the embryos and may play a role in compaction. The lectin receptors could also represent, at least in part, the antigenic determinants defined by monoclonal antibodies. Comparative studies on the distribution of lectin receptors and structurally defined antigens are lacking, except for that of monoclonal antibody IIC3 versus WGA and RCA-I (Marticorena et al. 1983). Simultaneous monitoring of lectin receptors and corresponding carbohydrate cell surface antigens could be helpful in determining the role of cell surface molecules in development.

III. Functionally defined embryonic molecules

Unlike the previously described antigens, this group of cell surface molecules, also detected by immunological approaches, consists of glycoproteins with only partially identified structures but a known involvement in cell-cell adhesion, a function postulated only for the structurally defined antigens. These glycoproteins are expressed on embryos, EC cells, and adult cells, where they mediate cell-cell adhesion by an apparently similar, if not identical, mechanism. Antibodies raised against cells presenting these glycoproteins can inhibit cell-cell adhesion. The presence of Ca^{2+} renders these glycoproteins more resistant to proteolytic enzymes such as trypsin. Molecules of this type have been repeatedly described by several groups of investigators, and it is quite likely that uvomorulin, cadherin, and cell CAM 120/80 are the same or very similar molecules. Uvomorulin (UM) is involved in compaction of 8-cell-stage mouse embryos and the calcium-dependent aggregation of EC cells. Its identification was achieved by producing rabbit anti-EC-cell serum that interfered with the compaction of preimplantation embryos and the aggregation of EC cells (Hyafil et al. 1980). Indirect immunofluorescence analysis indicated that this cell surface molecule was expressed on F9 EC cells but not on PYS-2 parietal endoderm cells (Vestweber and Kemler 1984).

UM is uniformly distributed on unfertilized mouse eggs and preimplantation-stage mouse embryos (Hyafil et al. 1983), becomes restricted to ICM cells of the blastocyst (Hyafil et al. 1983; Vestweber and Kemler 1984), and reappears on all epithelial cells, independent of their germ-layer origin, of 12-day-old embryos (Vestweber and Kemler 1984). In adult mouse tissues, UM is detected on epithelial cells from tongue, uterus, kidney, trachea, and liver (Vestweber and Kemler 1984) and also on a very restricted region on the intermediate junction of intestinal epithelial cells (Boller, Vestweber, and Kemler 1985). It is a nonintegral membrane protein of molecular weight 120,000 (Peyrieras et al. 1983; Vestweber and Kemler 1984), of which a tryptic fragment of 84,000 (UMt) was purified from EC cell membranes (Hyafil et al. 1980). This molecule reverses the effects of rabbit antibodies on mouse morulae and EC cells. A rat monoclonal antibody (DE1) raised against UMt immunoprecipitates the native molecule (Peyrieras et al. 1983) and binds the UMt only in the presence of calcium (Hyafil, Babinet, and Jacob 1981). Analysis of the role of calcium, in conjunction with UM in early mouse embryogenesis, suggests that the effects of calcium on compaction are mediated through conformational changes in UM. On binding to calcium, UM changes its conformation to favor trypsin resistance and DE1 recognition. This conformational change maintains compaction. Decompaction occurs when UM undergoes conformational changes in the absence of calcium (Hyafil et al. 1983).

A similar immunological approach led to the identification of cadherin (Yoshida-Noro, Suzuki, and Takeichi 1984). Cadherin was identified by a rabbit antiserum raised against F9 cells and represents a portion of a calcium-dependent site for cell-cell adhesion between teratocarcinoma cells (Takeichi et al. 1981; Ogou et al. 1982; Yoshida and Takeichi 1982) and for compaction of 8-cell-stage mouse embryos (Ogou et al. 1982). It is resistant to trypsin treatment in the presence of calcium but sensitive in the absence of calcium (Takeichi et al. 1983). A rat monoclonal antibody, ECCD-1, raised against F9 cells, inhibited cell-cell adhesion of mouse hepatocytes and teratocarcinoma cells (Ogou et al. 1983; Takeichi et al. 1983). The binding of ECCD-1 to the cell surface is calcium-dependent, and the bound antibody can be removed by 1 μM EGTA (Yoshida-Noro et al. 1984). When tested on preimplantation mouse embryos, ECCD-1 inhibited compaction of 8- to 16-cell-stage embryos. However, cell proliferation proceeded at a normal rate, generating blastocysts deficient in ICM (Takeichi et al. 1983; Shirayoshi, Okada, and Takeichi 1983). Furthermore, ECCD-1 recognizes multiple cell surface proteins, the major one of which has a molecular weight of 124,000 (Ogou, Yoshida-Noro, and Takeichi 1983), formerly reported to be of 140,000 (Yoshida and Takeichi 1982). In addition, the antigens reacting with DE1, a monoclonal antibody raised against UMt (Hyafil et al. 1981), were exactly the same as those recognized by ECCD-1 (Yoshida-Noro et al. 1984). However, DE1 itself does not cause disruption of cell-cell adhesion (Hyafil et al. 1981). Thus, DE1 and ECCD-1 probably recognize different sites on the same molecule.

A monoclonal antibody (DECMA-1) with properties very similar to those of ECCD-1 has been recently described (Vestweber and Kemler 1985). It disrupts monolayers of epithelial cells, inhibits aggregation of teratocarcinoma cells, and inhibits compaction of mouse preimplantation embryos. Inhibition of compaction is not complete; embryos compact after a delay of 24 h and form normal blastocysts. In this respect, DECMA-1 is apparently different from ECCD-1, because in the presence of the latter, the embryos form blastocysts without inner cell mass. Vestweber and Kemler (1985) demonstrated that DECMA-1 binds to mouse and dog ovomorulin, and following proteolytic cleavage the antibody binds to a 26,000-d fragment. It is interesting to note that two other monoclonal antibodies, anti-arc-1 and rrl, which both affect the cell-cell interaction of Madin-Darby canine kidney (MDCK) epithelial cells, also bind to the same 26,000-d fragment of mouse and dog ovomorulin (Vestweber and Kemler 1985). It is very likely that the 26,000-d fragment contains at least one of the functional adhesion sites of the ovomorulin molecule.

Recently, an antiserum against material shed into serum-free medium by MCF-7 human carcinoma cells led to the identification of a cell surface glycoprotein involved in embryonic and adult cell-cell adhesion (Damsky et al. 1983). This glycoprotein has a molecular weight of 120,000; an 80,0000-d (GP80) component is shed into the medium or can be generated by proteolysis in the presence of calcium; hence the name of the molecule, cell-CAM 120/80. It is expressed on cleavage-stage mouse embryos and mediates compaction at the 8-cell stage. As embryo development proceeds, expression of cell-CAM 120/80 becomes restricted to the ICM, where it mediates cell-cell adhesion (Richa et al. 1985). It is also expressed on human and mouse EC cells and on certain human and mouse epithelial tissues (Damsky et al. 1983). A rabbit antiserum raised against the purified GP80 (anti-GP80) was used to analyze different cell lines by immunofluorescence assay. The distribution of the molecule recognized by anti-GP80 was restricted to areas of cell-cell contact in cultured cells with epithelial morphology and was absent from cultured fibroblasts. This pattern is consistent with the role of this molecule in cell-cell adhesion.

Cell adhesion molecules sharing the common property of calcium dependence have been identified in embryonic chicken neural retina (Magnani, Thomas, and Steinberg 1981; Thomas and Steinberg 1981; Cook and Lilien 1982; Grunwald, Pratt, and Lilien 1982) and in embryonic chicken liver (L-CAM) (Bertolotti, Rutishauser, and Edelman 1980; Brackenbury, Rutishauser, and Edelman 1981; Gallin, Edelman, and Cunningham 1983). In the neural retina system, glycoproteins of 130,000 and 70,000 d have been immunoprecipitated with adhesion-blocking antisera (Grunwald et al. 1982). In the liver, a monoclonal antibody that blocks cell adhesion recognizes L-CAM as a cell surface glycoprotein of 124,000 d. In the presence of calcium, the molecule can be released from membranes by trypsin as a soluble 81,000-d fragment (Gallin et al. 1983). These cell surface glycoproteins resemble UM, cadherin, and cell-CAM 120/80 in terms of molecular weight, calcium sensitivity, and involvement in calcium-dependent cell-cell adhesion.

It is quite possible that several molecules with similar antigenic properties and possibly shared functional roles exist in epithelial cells of different species. The recent demonstration that polyclonal and monoclonal (DE1) antibodies to uvomorulin recognize additional cell surface proteins (Peyreiras et al. 1985) supports such a notion. The question remains as to the role of such molecules in the compaction of mouse embryos. It is unlikely that they are active in inducing and regulating compaction, because these molecules are obviously present on mouse zygotes and cleavage-stage embryos well before the start of compaction (Ogou et al. 1982; Damjanov, Damjanov, and Damsky 1986). The presence of adhesion-permissive molecules is apparently necessary but not sufficient for compaction of late 8-cell-stage embryos, and identification of molecules and mechanisms that trigger this important event in normal development awaits further investigation.

IV. Cytoskeleton and cell surface

Most of the evidence for the roles of various cytoskeleton elements in preimplantation development has been derived using two major approaches. The first consists of treating cleavage-stage embryos with cytoskeleton-disrupting drugs and subsequently monitoring the development of the embryos. The second approach relies on indirect immunofluorescence or electron microscopy to localize the cytoskeletal elements throughout the preimplantation stages. These elements include microfilaments, microtubules, intermediate filaments, myosin, α-actinin, and spectrin.

Treatment of embryos with cytochalasin B or cytochalasin D, which disrupt the organization of microfilaments, results in an arrest of cleavage that prevents further development (Surani, Barton, and Burling 1980; Pratt, Chakraborty, and Surani 1981; Pratt et al. 1982). However, DNA synthesis is unaffected (Surani et al. 1980), and surface antigens appear at the correct developmental time (Pratt et al. 1981; Petzoldt et al. 1983; Petzoldt 1986). Thus, intact microfilaments are necessary for normal cell division but not for differentiation, analogous to the case in ascidian embryos (Whittaker 1973; Satoh 1979) and in *Caenorhabditis elegans* (Laufer, Bazzicalupo, and Wood 1980). At the 8-cell stage, cytochalasin B or D also disrupts and inhibits compaction (Ducibella and Anderson 1975; Surani et al. 1980; Pratt et al. 1982), rendering the embryo an aggregate of blastocyst-like vesicles (Pratt et al. 1981). These dual effects of cytoskeletal inhibitors on cell division and cell shape make them inadequate probes to analyze the role of microfilaments in preimplantation development, particularly compaction (see Chapter 2).

The role of microtubules in development has been investigated by using colchicine and colcemid to inhibit the polymerization of tubulin (Wilson 1975). These drugs arrest the cleavage of 2- and 4-cell-stage embryos (Surani et al. 1980; Pratt et al. 1982), cause decompaction of 8-cell embryos (Surani et al. 1980), and prevent cavitation of morulae (Wiley and Eglitis 1980). Other studies using colchicine and colcemid (Ducibella and Anderson 1975; Ducibella 1980; Wiley and Eglitis 1980) suggest that compaction is not microtubule-dependent. However, nocodazole, a drug that induces depolymerization of intracellular microtubules (Hoebeke, Van Nigen, and DeBrabander 1976), does not inhibit compaction of 8-cell embryos but rather accelerates its completion (Maro and Pickering 1984). Thus, the involvement of microtubules in compaction remains controversial, although presumably a great imbalance between polymerized and nonpolymerized tubulin will interfere with the physiological role of this cytoskeletal element.

Proteins of intermediate filaments have been detected in cleavage-stage embryos and blastocysts (Jackson et al. 1980; Lehtonen and Badley 1980; Lehtonen et al. 1983; Oshima et al. 1983), and myosin has been localized in preimplantation embryos as a cytoplasmic cortical band that is absent in regions of cell-cell (Sobel 1983a, 1983b) and cell-substrate contact (Sobel 1984). A spectrin-like protein has been detected as a continuous cortical band (Sobel and Alliegro 1985) from the 2-cell stage onward in preimplantation embryos; this protein is consistently present in the cortical cytoplasm, underlying regions of contact between blastomeres and between cells of the ICM.

These findings suggest the cytoskeletal elements are involved in preimplantation development, particularly compaction. However, the mechanism of their involvement remains unclear. In striated muscle, actin and myosin are organized into highly structured arrays of filaments, and contraction is generated by the sliding of the actin filaments with respect to the myosin filaments (Huxley 1969). In nonmuscle systems, these proteins have been implicated in the mechanisms responsible for cell shape change and cell mobility (for review, see Korn 1978 and Weatherbee 1981). In the mouse embryo, a cortical layer of microfilaments has been identified in the 2-cell stage (Opas and Soltynska 1978) and the 8-cell stage (Ducibella et al. 1977), in addition to the peripheral localization of cytoplasmic actin (Lehtonen and Badley 1980; Johnson and Maro 1984). Cortical myosin (Sobel 1983a, 1983b, 1984) might interact with the actin microfilaments to effect the changes in the shape of blastomeres during compaction. The anchoring of actin filaments to the membrane could be achieved by the α-actinin present at the periphery (Lehtonen and Badley 1980) of the blastomeres. A similar role for α-actinin in nonmuscle cells was suggested by studies in which this protein was detected at the ends of microfilament bundles (Lazarides and Burridge 1975; Lazarides 1976) and in adhesion plaques (Badley et al. 1978). Spectrin might also play an anchoring role, because it is thought to complex with actin (Ungewickell et al. 1979; Fowler and Taylor 1980) and to be linked to an integral membrane protein (band 3) of erythrocytes (Bennett and Stenbuck 1979).

From the data, it is clear that an intact cytoskeleton is essential for several morphogenetic processes during preimplantation development. Cytoplasmic and membrane polarization of 8-cell-stage blastomeres and subsequent compaction depend on and may be regulated by cytoskeletal elements (Fleming, Cannon, and Pickering 1986) and their proper intracellular architecture. The role of stage-specific embryonic antigens acting as membrane anchoring elements could also be visualized in this context. A highly organized cytoplasmic architecture and the appropriate movement and redistribution of organelles have been observed even in oocyte and fertilized eggs, and it seems likely that the cytoskeleton is instrumental in these processes (Van Blerkom 1985). In contrast, the resilience of the embryo to the transitory disturbances of the cytoskeleton is striking. Complete depolymerization of microtubules and microfilaments is essential for nuclear transfer (McGrath and Solter 1983); yet the embryo is able to recover and develop normally. Redistribution of cytoplasmic components of the egg by centrifugation or by physical mixing is also completely reversible (J. McGrath and D. Solter, unpublished results). At the present time, it is not known whether the precise cytoplasmic localization of various components is essentially irrelevant for successful development or whether mechanisms exist for rapid and accurate restructuring of the cytoplasmic components following any disturbance.

V. Concluding remarks

We have described some known surface components of early mouse embryos that are likely to be involved in cellular interactions and changes in cell shape and early morphogenetic movements. The presence and proper functioning of most of these components are essential for normal development, but we are still far from understanding their function and mutual interactions. The genes encoding cell surface molecules in early embryos or the enzymes necessary for their synthesis have not yet been cloned, and little is known about their control and activation.

Identification of these molecules is, however, the first essential and sometimes the most difficult step. Problems posed by the limited amounts of embryonic material available are gradually being overcome. Micromethods in protein analysis and in recombinant DNA technology are fast approaching the point where analysis of gene expression governing the cell surface in early mammalian embryos will be possible.

Acknowledgments

Original work presented in this chapter was supported in part by grants CA-10815 and CA-25875 from the National Cancer Institute, by grants HD-12487 and HD-17720 from the National Institute of Child Health and Human Development, and by grant PCM-81/18801 from the National Science Foundation. We thank Barbara B. Knowles for helpful comments.

References

Abe, K., McKibbin, J. M., and Hakomori, S. (1983) The monoclonal antibody directed to difucosylated type 2 chain (Fucα1→2Galβ1→4[Fucα1→3]GlcNAc; Y determinant). *J. Biol. Chem.* 258, 11793−7.

Andrews, P. W., Damjanov, I., Simon, D., Banting, G. S., Carlin, C., Dracopoli, N. C., and Fogh, J. (1984) Pluripotent embryonal carcinoma clones derived from the human teratocarcinoma cell line Tera-2: differentiation *in vivo* and *in vitro*. *Lab Invest.* 50, 147−62.

Andrews, P. W., Goodfellow, P. N., Shevinsky, L. H., Bronson, D. L., and Knowles, B. B. (1982) Cell-surface antigens of a clonal human embryonal carcinoma cell line: morphological and antigen differentiation in culture. *Int. J. Cancer* 29, 523−31.

Badley, R. A., Lloyd, C. W., Woods, A., Carruthers, L., Allock, C., and Rees, D. A. (1978) Mechanisms of cellular adhesion. III. Preparation and preliminary characterization of adhesions. *Exp. Cell Res.* 117, 231−44.

Barondes, S. H. (1981) Lectins: their multiple endogenous cellular functions. *Annu. Rev. Biochem.* 5, 207–31.

Bender, B. L., Jaffe, R., Carlin, B., and Chung, A. E. (1981) Immunolocalization of entactin, a sulfated basement membrane component, in rodent tissues, and comparison with GP-2 (laminin). *Am. J. Pathol.* 103, 419–26.

Bennett, V., and Stenbuck, P. J. (1979) The membrane attachment protein for spectrin is associated with band 3 in human erythrocyte membranes. *Nature* 280, 468–73.

Bertolotti, R., Rutishauser, V., and Edelman, G. H. (1980) A cell surface molecule involved in aggregation of embryonic liver cells. *Proc. Natl. Acad. Sci. U.S.A.* 77, 4831–5.

Bird, J. M., and Kimber, S. J. (1984) Oligosaccharides containing fucose linked α(1–3) and α(1–4) to N-acetylglucosamine cause decompaction of mouse morulae. *Dev. Biol.* 104, 449–60.

Blaineau, C., LePendu, J., Danielle, A., Connan, F., and Avner, P. (1983) The glycosidic antigen recognized by a novel monoclonal antibody, 75.12, is developmentally regulated on mouse embryonal carcinoma cells. *EMBO Journal* 2, 2217–22.

Boller, K., Vestweber, D., and Kemler, R. (1985) Cell-adhesion molecule uvomorulin is localized in the intermediate junctions of adult intestinal epithelial cells. *J. Cell Biol.* 100, 327–32.

Brackenbury, R., Rutishauser, U., and Edelman, G. M. (1981) Distinct calcium-independent and calcium-dependent adhesion systems of chicken embryo cells. *Proc. Natl. Acad. Sci. U.S.A.* 78, 387–91.

Brockhaus, M., Magnani, J. L., Herlyn, M., Blaszczyk, M., Steplewski, Z., Koprowski, H., and Ginsburg, V. (1982) Monoclonal antibodies directed against the sugar sequence of lacto-N-fucopentaose III are obtained from mice immunized with human tumors. *Arch. Biochem. Biophys.* 217, 647–51.

Brownell, A. G. (1977) Cell surface carbohydrates of preimplantation embryos as assessed by lectin binding. *J. Supramol. Struct.* 7, 223–34.

Calarco, P. G., and Epstein, C. J. (1973) Cell surface changes during preimplantation development in the mouse. *Dev. Biol.* 32, 208–13.

Campbell, C., and Stanley, P. (1983) Regulatory mutations in CHO cells induce expression of the mouse embryonic antigen SSEA-1. *Cell* 35, 303–9.

Childs, R. A., Kapadia, A., and Feizi, T. (1980) Expression of blood group I and i active carbohydrate sequences on cultured human and animal cell lines assessed by radioimmunoassays with monoclonal cold agglutinins. *Eur. J. Immunol.* 10, 379–84.

Cook, J. H., and Lilien, J. (1982) The accessibility of certain proteins on embryonic chick neural retina cells to iodination and tryptic removal is altered by calcium. *J. Cell Sci.* 55, 85–103.

Damjanov, I., Damjanov, A., and Damsky, C. H. (1986) Developmentally regulated expression of the cell-cell adhesion glycoprotein cell-CAM 120/80 in preimplantation mouse embryos and extraembryonic membranes. *Dev. Biol.* 116, 194–202.

Damsky, C. H., Knudsen, K. A., Dorio, R., and Buck, C. A. (1981) Manipulation of cell-cell and cell-substratum interaction in mouse mammary tumor epithelia. *J. Cell Biol.* 89, 173–84.

Damsky, C. H., Richa, J., Solter, D., Knudsen, K., and Buck, C. A. (1983) Identification and purification of a cell surface glycoprotein mediating intercellular adhesion in embryonic and adult tissue. *Cell* 34, 455–66.

Ducibella, T. (1980) Divalent antibodies to mouse embryonal carcinoma cells inhibit compaction in the mouse embryo. *Dev. Biol.* 79, 356–66.

Ducibella, T., Albertini, D. F., Anderson, E., and Biggers, S. D. (1975) The preimplantation mammalian embryo: characterization of intercellular junctions and their appearance during development. *Dev. Biol.* 45, 231–50.

Ducibella, T., and Anderson, E. (1975) Cell shape and membrane changes in the eight-cell mouse embryo: prerequisites for morphogenesis of the blastocyst. *Dev. Biol.* 47, 45–58.

– (1979) The effects of calcium deficiency on the formation of the zonula occludens and blastocoel in the mouse embryo. *Dev. Biol.* 73, 46–58.

Ducibella, T., Ukena, T., Karnovsky, M., and Anderson, E. (1977) Changes in cell surface and cortical cytoplasmic organization during embryogenesis in the preimplantation mouse embryo. *J. Cell Biol.* 74, 153–67.

Etzler, M. E., and Kabat, E. A. (1970) Purification and characterization of a lectin (plant haemagglutinin) with blood group A specificity from *Dolichos biflorus*. *Biochemistry* 9, 869–77.

Feizi, T., Childs, R. A., Watanabe, K., and Hakomori, S.-I. (1979) Three types of blood group I specificity among monoclonal anti-I autoantibodies revealed by analogues of branched erythrocyte glycolipid. *J. Exp. Med.* 149, 975–80.

Feizi, T., and Kabat, E. A. (1972) Immunochemical studies on blood groups. LIV. Classification of anti-I and anti-i sera into groups based on reactivity patterns with various antigens related to the blood group A, B, H, Le^a, Le^b and precursor substances. *J. Exp. Med.* 135, 1247–58.

Feizi, T., Kapadia, A., Gooi, H. C., and Evans, M.J. (1982) Human monoclonal autoantibodies detect changes in expression and polarization of the Ii antigens during cell differentiation in early mouse embryos and teratocarcinomas. In *Teratocarcinoma and Embryonic Cell Interactions,* eds. T. Muramatsu, G. Gachelin, A. A. Moscona, and Y. Ikawa, pp. 201–15. Tokyo: Japan Scientific Societies Press/Academic Press.

Fenderson, B. A., Hahnel, A. C., and Eddy, E. M. (1983) Immunohistochemical localization of two monoclonal antibody-defined carbohydrate antigens during early murine embryogenesis. *Dev. Biol.* 100, 318–27.

Fenderson, B. A., Holmes, E. H., Fukushi, Y., and Hakomori, S.-I. (1986) Coordinate expression of X and Y haptens during murine embryogenesis. *Dev. Biol.* 114, 12–21.

Fenderson, B. A., Zehavi, U., and Hakomori, S.-I. (1984) A multivalent lacto-N-fucopentaose III-lysyllysine conjugate decompacts preimplantation mouse embryos, while the free oliogosaccharide is ineffective. *J. Exp. Med.* 160, 1591–6.

Fleming, T. P. (1984) Maturation and polarization of the endocytotic apparatus in outer blastomeres of the preimplantation mouse embryo. *J. Embryol. Exp. Morphol. Suppl.* 82, 153.

Fleming, T. P., Cannon, P. M., and Pickering, S. J. (1986) The cytoskeleton, endocytosis and cell polarity in the mouse preimplantation embryo. *Dev. Biol.* 113, 406–19.

Fowler, V., and Taylor, D. L. (1980) Spectrin plus band 4.1 cross-link actin. Regulation by micromolar calcium. *J. Cell Biol.* 85, 361–76.

Fox, N., Damjanov, I., Knowles, B. B., and Solter, D. (1982a) Teratocarcinoma antigen is secreted by epididymal cells and coupled to maturing sperm. *Exp. Cell Res.* 137, 485–8.

– (1984) Stage-specific embryonic antigen 3 as a marker of visceral extraembryonic endoderm. *Dev. Biol.* 103, 263–6.

Fox, N., Damjanov, I., Martinez-Hernandez, A., Knowles, B. B., and Solter, D. (1981) Immunohistochemical localization of the early embryonic antigen (SSEA-1) in postimplantation mouse embryos, and fetal and adult tissues. *Dev. Biol.* 83, 391–8.

Fox, N., Shevinsky, L., Knowles, B. B., Solter, D., and Damjanov, I. (1982b) Distribution of murine stage specific embryonic antigens in the kidneys of three rodent species. *Exp. Cell Res.* 140, 331–9.

Fujimoto, H., Muramatsu, T., Urushihara, H., and Yanagisawa, K. O. (1982) Receptors to *Dolichos biflorus* agglutinin. A new cell surface marker common to teratocarcinoma cells and preimplantation mouse embryos. *Differentiation* 22, 59–61.

Fukushi, Y., Hakomori, S., Nudelman, E., and Cochran, N. (1984) Novel fucolipids accumulating in human adenocarcinoma. II. Selective isolation of hybridoma antibodies that differentially recognize mono-, di-, and trifucosylated type 2 chain. *J. Biol. Chem.* 259, 4681–5.

Gallin, W. J., Edelman, G., and Cunningham, B. (1983) Characterization of L-CAM, a major cell adhesion molecule from embryonic liver cells. *Proc. Natl. Acad. Sci. U.S.A.* 80, 1038–42.

Goldstein, I. J., Hammarström, S., and Sundblad, G. (1975) Precipitation and carbohydrate-binding specificity studies on wheat germ agglutinin. *Biochim. Biophys. Acta* 405, 53–61.

Goodall, H., and Johnson, M. H. (1982) The use of carboxy-fluorescein diacetate to study the formation of permeable channels between mouse blastomeres. *Nature* 295, 524–6.

–(1984) The nature of intercellular coupling within the preimplantation mouse embryo. *J. Embryol. Exp. Morphol.* 79, 53–76.

Gooi, H. C., Feizi, T., Kapadia, A., Knowles, B. B., Solter, D., and Evans, M. J. (1981) Stage specific embryonic antigenic (SSEA-1) involves the 3-fucosylated type 2 precursor chain. *Nature* 292, 156–8.

Grabel, L. B., Glabe, C. G., Singer, M. S., Martin, G. R., and Rosen, S. D. (1981) A fucan specific lectin on teratocarcinoma stem cells. *Biochem. Biophys. Res. Commun.* 102, 1165–71.

Grabel, L. B., Singer, M. S., Martin, G. R., and Rosen, S. D. (1983) Teratocarcinoma stem cell adhesion: the role of divalent cations and a cell surface lectin. *J. Cell Biol.* 96, 1532–7.

Grunwald, G. B., Pratt, R. S., and Lilien, J. (1982) Enzymatic dissection of embryonic cell adhesive mechanisms. III. Immunological identification of a component of the calcium-dependent adhesive system of embryonic chick neural retina cells. *J. Cell Sci.* 55, 69–83.

Hakomori, S.-I. (1981) Blood group ABH and Ii antigens of human erythrocytes: chemistry, polymorphism, and their developmental change. *Semin. Haematol.* 18, 39–62.

314 JEAN RICHA AND DAVOR SOLTER

Hakomori, S.-I., Nudelman, E., Levery, S., Solter, D., and Knowles, B. B. (1981) The hapten structure of developmentally regulated glycolipid antigen (SSEA-1) isolated from human erythrocytes and adenocarcinomas. A preliminary note. *Biochem. Biophys. Res. Commun.* 100, 1578–86.

Handyside, A. H. (1980) Distribution of antibody- and lectin-binding sites on dissociated blastomeres from mouse morulae: evidence for polarization at compaction. *J. Embryol. Exp. Morphol.* 60, 99–116.

Harrison, F. L., and Chesterton, C. J. (1980) Factors mediating cell-cell recognition and adhesion. *FEBS Lett* 122, 157–65.

Hoebeke, J., Van Nigen, G., and DeBrabander, M. (1976) Interaction of nocodazole (R17934), a new antitumoral drug, with rat brain tubulin. *Biochem. Biophys. Res. Commun.* 69, 319–42.

Hogan, B. L. M., Barlow, D. P., and Tilly, R. (1983) F9 teratocarcinoma cells as a model for the differentiation of parietal and visceral endoderm in mouse embryo. *Cancer Surv.* 2, 115–40.

Hogan, B. L. M., Taylor, A., and Adamson, E. (1981) Cell interactions modulate embryonal carcinoma cell differentiation into parietal or visceral endoderm. *Nature* 291, 235–7.

Hogan, B., and Tilly, R. (1978) In vitro development of inner cell masses isolated immunosurgically from mouse blastocysts. I. Inner cell masses from 3.5 day p.c. blastocysts incubated for 24 hr before immunosurgery. *J. Embryol. Exp. Morphol.* 45, 93–105.

Huang, L. C., Brockhaus, M., Magnani, J. L., Cuttitta, F., Rosen, S., Minna, J. C., and Ginsburg, V. (1983b) Many monoclonal antibodies with an apparent specificity for certain lung cancers are directed against a sugar sequence found in lacto-*N*-fucopentaose III. *Arch. Biochem. Biophys.* 220, 318–20.

Huang, L. C., Civin, C. I., Magnani, J. L., Shaper, J. H., and Ginsburg, V. (1983a) My-1, the human myeloid-specific antigen detected by mouse monoclonal antibodies, is a sugar sequence found in lacto-*N*-fucopentaose III. *Blood* 61, 1020–3.

Huxley, H. E. (1969) The mechanism of muscular contraction. *Science* 164, 1356–66.

Hyafil, F., Babinet, C., Huet, C., and Jacob, F. (1983) Uvomorulin and compaction. In *Teratocarcinoma Stem Cells*, eds. L. Silver, G. R. Martin, and S. Strickland, pp. 197–207. Cold Spring Harbor, N.Y.: Cold Spring Harbor Laboratory.

Hyafil, F., Babinet, C., and Jacob, F. (1981) Cell-cell interactions in early embryogenesis: a molecular approach to the role of calcium. *Cell* 26, 447–54.

Hyafil, F., Morello, D., Babinet, C., and Jacob, F. (1980) A cell surface glycoprotein involved in the compaction of embryonal carcinoma cells and cleavage stage embryos. *Cell* 21, 927–34.

Jackson, B. W., Grund, C., Schmid, E., Burki, K., Franke, W. W., and Illmensee, K. (1980) Formation of cytoskeletal elements during mouse embryogenesis. I. Intermediate filaments of the cytokeratin type and desmosomes in preimplantation embryos. *Differentiation* 17, 161–79.

Jacob, F. (1979) Cell surface and early stages of mouse embryogenesis. *Curr. Top. Dev. Biol.* 13, 117–35.

Johnson, L. V. (1986) Wheat germ agglutinin induces compaction- and cavitation-like events in 2-cell mouse embryos. *Dev. Biol.* 113, 1–9.

Johnson, L. V. and Calarco, P. G. (1980) Mammalian preimplantation development: the cell surface. *Anat. Rec.* 196, 201–19.

Johnson, M. H., Chakraborty, J., Handyside, A. H., Willison, K., and Stern, P. (1979) The effect of prolonged decompaction on the development of the preimplantation mouse embryo. *J. Embryol. Exp. Morphol.* 54, 241–61.

Johnson, M. H., and Maro, B. (1984) The distribution of cytoplasmic actin in mouse 8-cell blastomeres. *J. Embryol. Exp. Morphol.* 82, 97–117.

Kannagi, R., Cochran, N. A., Ishigami, F., Hakomori, S.-I., Andrews, P. W., Knowles, B. B., and Solter, D. (1983a) Stage-specific embryonic antigens (SSEA-3 and -4) are epitopes of a unique globo-series ganglioside isolated from human teratocarcinoma cells. *EMBO Journal* 2, 2355–61.

Kannagi, R., Levery, S. B., Ishigami, F., Hakomori, S.-I., Shevinsky, L. H., Knowles, B. B., and Solter, D. (1983b) New globoseries glycosphingolipids in human teratocarcinoma reactive with the monoclonal antibody directed to a developmentally regulated antigen, stage-specific embryonic antigen 3. *J. Biol. Chem.* 258, 8934–42.

Kannagi, R., Nudelman, E., Levery, S. B., and Hakomori, S.-I. (1982) A series of human erythrocyte glycosphingolipids reacting to the monoclonal antibody directed to a developmentally regulated antigen, SSEA-1. *J. Biol. Chem.* 257, 14865–74.

Kapadia, A., Feizi, T., and Evans, M. J. (1981) Changes in the expression and polarization of blood group I and i antigens in post-implantation embryos and tertocarcinomas of mouse associated with cell differentiation. *Exp. Cell Res.* 131, 185–95.

Kemler, R., Babinet, C., Eisen, H., and Jacob, F. (1977) Surface antigen in early differentiation. *Proc. Natl. Acad. Sci. U.S.A.* 74, 4449–52.

Kimber, S. J., and Surani, M. A. H. (1982) Spreading of blastomeres from eight-cell mouse embryos on lectin coated beads. *J. Cell Sci.* 56, 191–206.

Kleinman, H. K., Klebe, R. J., and Martin, G. R. (1981) Role of collagenous matrices in the adhesion and growth of cells. *J. Cell Biol.* 88, 473–85.

Knowles, B. B., Pan, S. H., Solter, D., Linnenbach, A., Croce, C., and Huebner, K. (1980) SV40-DNA transformed mouse teratocarcinoma cells. II. Expression of H-2, laminin and SV40 T and TASA upon differentiation. *Nature* 288, 615–18.

Knowles, B. B., Rappaport, J., and Solter, D. (1982) Murine embryonic antigen (SSEA-1) is expressed on human cells and structurally related human blood group antigen I is expressed on mouse embryos. *Dev. Biol.* 93, 54–8.

Korn, E. D. (1978) Biochemistry of actomyosin-dependent cell motility (a review). *Proc. Natl. Acad. Sci. U.S.A.* 75, 588–99.

Laufer, J. S., Bazzicalupo, P., and Wood, W. B. (1980) Segregation of developmental potential in early embryos of *Caenorhabditis elegans*. *Cell* 19, 569–77.

Lazarides, E. (1976) Actin, α-actinin, and tropomyosin interaction in the structural organization of actin filaments in nonmuscle cells. *J. Cell Biol.* 68, 202–19.

Lazarides, E., and Burridge, K. (1975) α-Actinin: immunofluorescent localization of a muscle structural protein in nonmuscle cells. *Cell* 6, 289–98.

Lehtonen, E. (1980) Changes in cell dimensions and intercellular contacts during cleavage-stage cell cycles in mouse embryonic cells. *J. Embryol. Exp. Morphol.* 58, 231–49.

Lehtonen, E., and Badley, R. A. (1980) Localization of cytoskeletal proteins in preimplantation mouse embryos. *J. Embryol. Exp. Morphol.* 55, 211–25.

Lehtonen, E., Lehto, V.-P., Vartio, T., Badley, R. A., and Virtanen, I. (1983) Expression of cytokeratin polypeptides in mouse oocytes and preimplantation embryos. *Dev. Biol.* 100, 158–65.

Lo, C. W. (1980) Gap junctions and development. In *Development in Mammals*, Vol. 4, ed. M. H. Johnson, pp. 39–80. Amsterdam: Elsevier/North Holland.

Lo, C. W., and Gilula, N. B. (1979) Gap junctional communication in the preimplantation mouse embryo. *Cell* 18, 399–409.

Lotan, R., Skutelsky, E., Danon, D., and Sharon, N. (1975) The purification, composition and specificity of the anti-T lectin from peanut (*Arachis hypogaea*). *J. Biol. Chem.* 250, 8518–23.

McGrath, J., and Solter, D. (1983) Nuclear transplantation in the mouse embryo by microsurgery and cell fusion. *Science* 220, 1300–2.

Magnani, J. L., Thomas, W. A., and Steinberg, M. S. (1981) Two distinct adhesion mechanisms in embryonic neural retina cells. I. A kinetic analysis. *Dev. Biol.* 81, 96–105.

Magnuson, T., Jacobson, J. B., and Stackpole, C. W. (1978) Relationship between intercellular permeability and junction organization in the preimplantation mouse embryo. *Dev. Biol.* 67, 214–24.

Maro, B., and Pickering, S. J. (1984) Microtubules influence compaction in preimplantation mouse embryos. *J. Embryol. Exp. Morphol.* 84, 217–32.

Marticorena, P., Hogan, B., DiMeo, A., Artzt, K., and Bennett, D. (1983) Carbohydrate changes in pre- and peri-implantation mouse embryos as detected by a monoclonal antibody. *Cell Differ.* 12, 1–10.

Muramatsu, H., and Muramatsu, T. (1983) A fucosyltransferase in teratocarcinoma cells. Decreased activity accompanying differentiation to parietal endoderm cells. *FEBS Lett.* 163, 181–4.

Neimann, H., Watanabe, K., and Hakomori, S.-I. (1978) Blood group i and I activities of "lacto-*N*-*nor*hexaosylceramide" and its analogues: the structural requirements for i-specificities. *Biochem. Biophys. Res. Commun.* 81, 1286–93.

Nicolson, G. L., Blaustein, J., and Etzler, M. E. (1974) Characterization of two plant lectins from *Ricinus communis* and their quantitative interactions with murine lymphoma. *Biochemistry* 13, 196–203.

Ogou, S.-I., Okada, T. S., and Takeichi, M. (1982) Cleavage stage mouse embryos share a common cell adhesion system with teratocarcinoma cells. *Dev. Biol.* 92, 521–8.

Ogou, S.-I., Yoshida-Noro, C., and Takeichi, M. (1983) Calcium-dependent cell-cell adhesion molecules common to hepatocytes and teratocarcinoma stem cells. *J. Cell Biol.* 97, 944–8.

Opas, J., and Soltynska, M. S. (1978) Reorganization of the cortical layer during cytokinesis in mouse blastomeres. *Exp. Cell Res.* 113, 208–11.

Oshima, R. G., Howe, W. E., Klier, F. G., Adamson, E. D., and Shevinsky, L. H. (1983) Intermediate filament protein synthesis in preimplantation murine embryos. *Dev. Biol.* 99, 447–55.

Petzoldt, U. (1986) Expression of two surface antigens and paternal glucose-phosphate isomerase in polyploid one-cell mouse eggs. *Dev. Biol.* 113, 512–16.

Petzoldt, U., Bürki, K., Illmensee, G. R., and Illmensee, K. (1983) Protein synthesis in mouse embryos with experimentally produced asynchrony between chromosome replication and cell division. *Roux's Arch. Dev. Biol.* 192, 138–44.

Peyrieras, N., Hyafil, F., Louvard, D., Ploegh, H. L., and Jacob, F. (1983) Uvomorulin: A nonintegral membrane protein of early mouse embryo. *Proc. Natl. Acad. Sci. U.S.A.* 80, 6274–7.

Peyrieras, N., Louvard, D., and Jacob, F. (1985) Characterization of antigens recognized by monoclonal and polyclonal antibodies directed against uvomorulin. *Proc. Natl. Acad. Sci. U.S.A.* 82, 8067–71.

Poretz, R. D., and Goldstein, I. J. (1970) An examination of the topography of the saccharide binding site of concanavalin A and of the forces involved in complexation. *Biochemistry* 9, 2890–6.

Pratt, H. P. M., Chakraborty, J., and Surani, M. A. H. (1981) Molecular and morphological differentiation of the mouse blastocyst after manipulations of compaction with cytochalasin D. *Cell* 26, 279–92.

Pratt, H. P. M., Ziomek, C. A., Reeve, W. J. D., and Johnson, M. H. (1982) Compaction of the mouse embryo: an analysis of its components. *J. Embryol. Exp. Morphol.* 70, 113–32.

Prieels, J.-P., Pizzo, S. V., Glasgow, L. R., Paulson, J. C., and Hill, R. L. (1978) Hepatic receptor that specifically binds oligosaccharides containing fucosyl α1→3N-acetylglucosamine linkages. *Proc. Natl. Acad. Sci. U.S.A.* 75, 2215–19.

Reeve, W. J. D. (1981) Cytoplasmic polarity develops at compaction in rat and mouse embryos. *J. Embryol. Exp. Morphol.* 62, 351–67.

– (1982) Effect of concanavalin A on the formation of the mouse blastocyst. *J. Reprod. Immunol.* 4, 53–64.

Reeve, W. J. D., and Kelly, F. P. (1982) Nuclear position in the cells of the mouse early embryo. *J. Embryol. Exp. Morphol.* 75, 117–39.

Reeve, W. J. D., and Ziomek, C. A. (1981) Distribution of microvilli on dissociated blastomeres from mouse embryos: evidence for surface polarization at compaction. *J. Embryol. Exp. Morphol.* 62, 339–50.

Richa, J., Damsky, C. H., Buck, C. A., Knowles, B. B., and Solter, D. (1985) Cell surface glycoproteins mediate compaction, trophoblast attachment and endoderm formation during early mouse development. *Dev. Biol.* 108, 513–21.

Rosenstraus, M. J. (1983) Isolation and characterization of an embryonal carcinoma cell line lacking SSEA-1 antigen. *Dev. Biol.* 99, 318–23.

Sato, M., and Muramatsu, T. (1985) Reactivity of five *N*-acetylgalactosamine-recognizing lectins with preimplantation embryos, early postimplantation embryos, and teratocarcinoma cells of the mouse. *Differentiation* 29, 29–38.

Sato, M., Muramatsu, T., and Berger, E. G. (1984) Immunological detection of cell surface galactosyltransferase in preimplantation mouse embryos. *Dev. Biol.* 102, 514–18.

Sato, M., Yonezawa, S., Uehara, H., Arita, Y., Sato, E., and Muramatsu, T. (in press) Distinct distribution of receptors for two fucose-recognizing lectins in embryos and adult tissue of the mouse. *Differentiation*.

Satoh, N. (1979) On the "clock" mechanism determining the time of tissue-specific enzyme development during ascidian embryogenesis. I. Acetylcholine-esterase development in cleavage arrested embryos. *J. Embryol. Exp. Morphol.* 54, 131–9.

Shevinsky, L. H., Knowles, B. B., Damjanov, I., and Solter, D. (1982) Monoclonal antibody to murine embryos defines a stage-specific embryonic antigen expressed on mouse embryos and human teratocarcinoma cells. *Cell* 30, 697–705.

Shirayoshi, Y., Okada, T. S., and Takeichi, M. (1983) The calcium-dependent cell-cell adhesion system regulates inner cell mass formation and cell surface polarization in early mouse development. *Cell* 35, 631–8.

Shur, B. D. (1983) Embryonal carcinoma cell adhesion: the role of surface galactosyltransferase and its 90K lactosaminoglycan substrate. *Dev. Biol.* 99, 360–72.

– (1984) The receptor function of galactosyltransferase during cellular interactions. *Mol. Cell. Biochem.* 61, 143–58.

Shur, B. D., Oettgen, P., and Bennett, D. (1979) UDP-galactose inhibits blastocyst formation in the mouse. *Dev. Biol.* 73, 178–81.

Sobel, J. S. (1983a) Cell-cell contact modulation of myosin organization in the early mouse embryo. *Dev. Biol.* 100, 207–13.

– (1983b) Localization of myosin in the preimplantation mouse embryo. *Dev. Biol.* 95, 227–1.

– (1984) Myosin rings and spreading in mouse blastomeres. *J. Cell Biol.* 99, 1145–50.

Sobel, J. S., and Alliegro, M. A. (1985) Changes in the distribution of a spectrin-like protein during development of the preimplantation mouse embryo. *J. Cell Biol.* 100, 333–6.

Solter, D., and Knowles, B. B. (1978) Monoclonal antibody defining a stage-specific mouse embryonic antigen (SSEA-1). *Proc. Natl. Acad. Sci. U.S.A.* 75, 5565–9.

– (1979) Developmental stage-specific antigens during mouse embryogenesis. *Curr. Top. Dev. Biol.* 13, 139–65.

– (1986) Cell surface antigens of germ cells, embryos, and teratocarcinoma stem cells. In *Principles and Management of Testicular Cancer*, ed. N. Javadpour, pp. 88–98. New York: Thieme.

Solter, D., Shevinsky, L., Knowles, B. B., and Strickland, S. (1979) The induction of antigenic changes in a teratocarcinoma stem cell line (F9) by retinoic acid. *Dev. Biol.* 70, 515–21.

Stern, P. L., Gilbert, P., Heath, J. K., and Furth, M. (1983) A monoclonal antibody which detects a cell surface antigen on murine embryonal carcinoma and early mouse embryo stages may recognize a carbohydrate determinant involving α-linked galactose. *J. Reprod. Immunol.* 5, 145–60.

Stinnakre, M. G., Evans, M. J., Willison, K. R., and Stern, P. L. (1981) Expression of Forssman antigen in the post-implantation mouse embryo. *J. Embryol. Exp. Morphol.* 61, 117–31.

Strickland, S., Reich, E., and Sherman, M. I. (1976) Plasminogen activator in early embryogenesis: enzyme production by trophoblast and parietal endoderm. *Cell* 9, 231–40.

Surani, M. A. H., Barton, S. C., and Burling, A. (1980) Differentiation of 2-cell and 8-cell mouse embryos arrested by cytoskeletal inhibitors. *Exp. Cell Res.* 125, 275–86.

Takeichi, M., Atsumi, T., Yoshida, C., Uno, K., and Okada, T. S. (1981) Selective adhesion of embryonal carcinoma cells and differentiated cells by Ca^{++}-dependent sites. *Dev. Biol.* 87, 340–50.

Takeichi, M., Yoshida-Noro, C., Ogou, S., Shirayoshi, Y., Okada, T. S., and Wartiovaara, J. (1983) A cell-cell adhesion molecule involved in embryonic cellular interactions as studied by using teratocarcinoma cells. In *Teratocarcinoma Stem Cells*, eds. L. Silver, G. R. Martin, and S. Strickland, pp. 163–71. Cold Spring Harbor, N.Y.: Cold Spring Harbor Laboratory.

Thomas, W. A., and Steinberg, M. S. (1981) Two distinct adhesion mechanisms in embryonic neural retina cells. II. An immunological analysis. *Dev. Biol.* 81, 106–14.

Ungewickell, E., Bennett, P. M., Calvert, R., Ohanian, V., and Gratzer, W. B. (1979) In vitro formation of a complex between cytoskeletal proteins of the human erythrocyte. *Nature* 811–14.

Van Blerkom, J. (1985) Extragenomic regulation and autonomous expression of a developmental program in the early mammalian embryo. *Ann. N.Y. Acad. Sci.* 442, 58–72.

Vestweber, D., and Kemler, R. (1984) Rabbit antiserum against a purified surface glycoprotein decompacts mouse preimplantation embryos and reacts with specific adult tissues. *Exp. Cell Res.* 152, 169–78.

— (1985) Identification of a putative cell adhesion domain of uvomorulin. *EMBO Journal* 4, 3393–8.

Watanabe, K., Hakomori, S.-I., Childs, R. A., and Feizi, T. (1979) Characterization of a blood group I-active ganglioside. Structural requirements for I and i specificities. *J. Biol. Chem.* 254, 3221–8.

Watkins, W. M. (1980) Biochemistry and genetics of the ABO, Lewis, and P blood group systems. *Adv. Hum. Genet.* 10, 1–135.

Weatherbee, J. A. (1981) Membranes and cell movement: interactions of membranes with the proteins of the cytoskeleton. *Int. Rev. Cytol.* Suppl. 12, 113–76.

Webb, C. A. G. (1983) Glycoproteins on gametes and early embryos. In *Development in Mammals*, Vol. 5, ed. M. H. Johnson, pp. 155–85. Amsterdam: Elsevier.

Whittaker, J. R. (1973) Segregation during ascidian embryogenesis of egg cytoplasmic information for tissue-specific enzyme development. *Proc. Natl. Acad. Sci. U.S.A.* 70, 2096–100.

Wiley, L. M. (1979) Early embryonic cell surface antigens as developmental probes. *Curr. Top. Dev. Biol.* 13, 167–97.

Wiley, L. M., and Eglitis, M. A. (1980) Effect of colcemid on cavitation during mouse blastocoele formation. *Exp. Cell Res.* 127, 89–101.

Willison, K. R., Karol, R. A., Suzuki, A., Kundu, S. K., and Marcus, D. M. (1982) Neutral glycolipid antigens as developmental markers of mouse teratocarcinomas and early embryos: an immunologic and chemical analysis. *J. Immunol.* 129, 603–9.

Willison, K. R., and Stern, P. L. (1978) Expression of a Forssman antigenic specificity in the preimplantation mouse embryo. *Cell* 14, 785–93.

Wilson, L. (1975) Action of drugs on microtubules. *Life Sci.* 17, 303–10.

Yoshida, C., and Takeichi, M. (1982) Teratocarcinoma cell adhesion: identification of a cell-surface protein involved in calcium-dependent cell aggregation. *Cell* 28, 217–24.

Yoshida-Noro, C., Suzuki, N., and Takeichi, M. (1984) Molecular nature of the calcium-dependent cell-cell adhesion in mouse teratocarcinoma and embryonic cells studied with a monoclonal antibody. *Dev. Biol.* 101, 19–27.

Ziomek, C. A., and Johnson, M. H. (1980) Cell surface interaction induces polarization of mouse 8-cell blastomeres at compaction. *Cell* 21, 935–42.

11   Cell-lineage-specific gene expression in development

EILEEN D. ADAMSON

CONTENTS

I. Introduction

This chapter describes the identified genes that successively become activated during mammalian development, especially post implantation (see Chapters 8 and 10 for preimplantation studies). The genes to be discussed include those that code for secretory products, matrix components, growth factors and their receptors, cell surface proteins, cytoskeletal proteins, and proto-oncogenes. The aim is to trace the ontogeny and appearance in specific cell lineages of gene products that have been identified by either cloned molecular probes or specific antibodies.

The usefulness of this attempt is threefold. First, identification of cell types by their characteristic products is the major means by which the experimental embryologist can study model differentiating systems. Second, identification of tumor cells and their tissues of origin may play

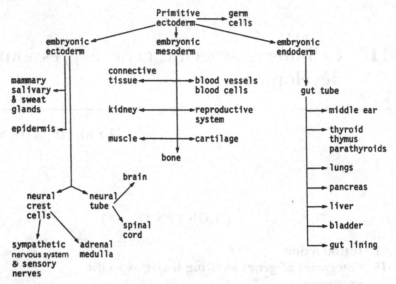

Figure 11.1. Principal tissue lineages derived from primitive ectoderm. For further details, see Chapter 5.

a significant role in diagnosis, prognosis, and therapy for the many human neoplasms that feature oncodevelopmentally activated genes. Third, analysis of gene regulation is the first stage in understanding the mechanisms by which differentiative decisions occur. Clearly, such decisions involve positional information, cell-cell interactions, and diffusible products. Such interactions will not be discussed here except to note that gene products that are growth factors, for instance, will have profound effects on responsive cell types.

The largest category of genes addressed here includes those that are expressed either in a small set of cell lineages or at a wide range of levels that may identify individual cell lineages if quantitative analyses are applied. It is usually necessary, therefore, to trace the antecedents of a cell type under study by a set of quantified gene products. The reader is referred to textbooks of embryology for detailed descriptions of the origins of cell types in the postimplantation mouse embryo (Green 1966). The cell types of the extraembryonic lineages are described in Chapter 4, and Figure 11.1 indicates the major derivatives of the embryonic lineages.

Teratocarcinoma cells have also proved invaluable for biochemical analyses of some of the events of embryonic differentiation. For complete descriptions of the properties and potentials of teratocarcinoma cells, see reviews by Martin (1975, 1978, 1980), Strickland (1981), and Hogan, Barlow, and Tilly (1983) (also see Chapter 15). Figure 11.2 illustrates two examples of teratocarcinoma cells that can be stimulated to differentiate into pathways leading to all three embryonic germ layers.

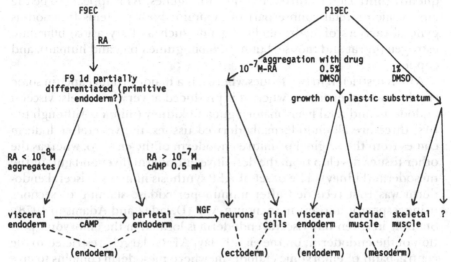

Figure 11.2. Some teratocarcinoma model systems. EC, embryonal carcinoma cells; RA, retinoic acid; DMSO, dimethyl sulfoxide; NGF, nerve growth factor.

Further details can be found in publications by Hogan, Taylor, and Adamson (1981), Grover and Adamson (1985), Hogan and associates (1983), McBurney and associates (1982), and Edwards and McBurney (1983). The differentiated cell types were identified by markers that were first identified in embryos. Eventually, in several cases, teratocarcinoma cells provided the means of cloning genes expressed at early stages in development (Wang and Gudas 1983; Levine, La Rosa, and Gudas 1984; Wang, La Rosa, and Gudas 1985; Trevor and Oshima 1985).

II. Categories of genes showing lineage-specific expression

A. Serum proteins and growth factors

i. Alpha-fetoprotein and albumin. Alpha-fetoprotein (AFP) (reviewed by Ruoslahti and Seppala 1979) is a serum glycoprotein of 75,000 d coded for by a single gene (25 kb) on chromosome 5 in the mouse. The AFP gene is 13.5 kb downstream from the albumin gene, with which it shares similar structure and function. AFP and albumin were probably derived from a single ancestral gene that gave rise to threefold repeats of four exons and three protein domains, because both genes have 15 coding blocks or exons with 14 intervening sequences (Eiferman et al. 1981). They may be activated together in specific tissues, but they are subsequently controlled by different regulatory genes. AFP appears to be the embryonic or fetal counterpart of albumin; both proteins function as general carriers of hydrophobic ligands such as fatty acids, bilirubin, estrogen (in rat and mouse, but not rabbit, guinea pig, and human), and copper.

AFP is restricted to two tissues where it is a major product and (in some species) two other tissues where it is produced at very low levels: visceral endoderm and fetal liver (major); gut and kidney (minor). Although the first three are all endodermally derived tissues, the visceral endoderm comes from the original primitive endoderm of the embryo, whereas the other tissues develop from the definitive endoderm (liver and gut) or the mesoderm (kidney). The onset of AFP synthesis in mouse visceral endoderm was first recorded after immunoperoxidase staining of sections using a specific antiserum to mouse AFP (Dziadek and Adamson 1978). Staining in seven-day visceral endoderm is located in the embryonic portion of the endoderm; on the eighth day, AFP is largely restricted to the central band of embryonic endoderm, where mesoderm conjoins to give rise to the visceral yolk sac.

Extraembryonic visceral endoderm is also capable of expressing the AFP gene, but only after it has been removed from the negative influence of the extraembryonic ectoderm and cultured to form epithelial vesicles (Dziadek 1978). Thus, AFP expression appears to be switched off under certain conditions. For example, visceral embryonic endoderm that grows out after the excision of extraembryonic endoderm from the seven-day egg cylinder (Hogan and Tilly 1981) switches to a parietal type that synthesizes large amounts of basement membrane material, but no AFP. Although the cell of origin cannot be identified as visceral endoderm with certainty, visceral endoderm taken from a six- or seven-day egg cylinder can give rise to both parietal and visceral endoderm in a chimeric embryo (Gardner 1982).

Albrechtsen and Nørgaard-Pedersen (1978) confirmed that AFP was confined to the visceral layer of the yolk sac endoderm in rats, as shown earlier by Gitlin, Ketzes, and Boesman (1967). AFP was also found in the endodermal structures (sinuses of Duval) within the placenta. AFP seems to be actively absorbed by other tissues and cell types, many of which are endocytotic, and therefore the demonstration of staining for AFP is not sufficient to demonstrate its synthesis in that tissue. In situ hybridization using cloned cDNA probes can be used on sections or cells to detect AFP mRNA (Figure 11.3).

AFP is present at high levels in amniotic fluid and in serum of fetuses, but this decreases 10^4- to 10^5-fold to very low levels in the adult. Whereas the half-life of albumin is long (about 19 days) (Peters, 1975), that of human AFP is 4 to 6 days, and the liver probably accounts for most of this degradation. The half-life of the mRNA of AFP, however,

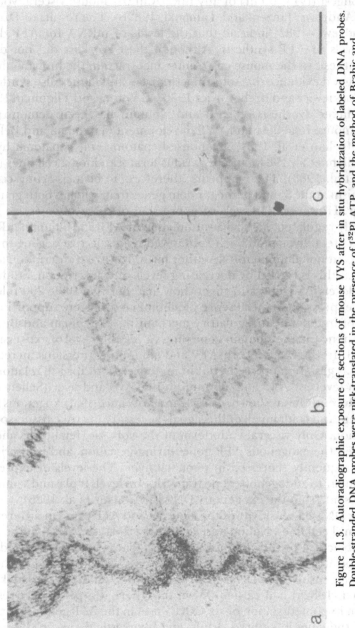

Figure 11.3. Autoradiographic exposure of sections of mouse VYS after in situ hybridization of labeled DNA probes. Double-stranded DNA probes were nick-translated in the presence of [^{32}P] ATP, and the method of Brahic and Haase (1978) was used with minor modifications: (a) AFP probe; (b) v-fos probe; (c) control v-rel probe. AFP mRNA is readily detected in the endoderm portion of the VYS, whereas levels of fos transcripts in (b) are barely detectably higher than the control levels in (c).

is the longest (t½ = 30 hr) of any mRNA in the mouse visceral yolk sac (VYS) (Andrews, Janzen, and Tamaoki, 1982b). The results of Dziadek and Andrews (1983) indicate that the levels of mRNA for AFP determine levels of AFP synthesis. Although these workers did not detect AFP synthesis in the mouse gastrointestinal tract, in the human embryo there is good evidence that intestine, stomach, and kidney also synthesize AFP (Mackiewicz and Breborowicz 1980; cf. Young and Tilghman 1984).

Coordinate synthesis of AFP and albumin has been demonstrated during mouse fetal liver and VYS development (Tilghman and Belayew 1982; Meehan et al. 1984), in mouse-hepatoma—rat-hepatoma hybrid cells (Sellem et al. 1984), and in various teratocarcinoma cell lines (Buc-Caron et al. 1983). There appears, therefore, to be a cis-acting control mechanism at the 5' end of the tandem genes that switches both genes on together. However, the levels of expression of the two genes are independently regulated. Parallel accumulation of AFP and albumin mRNAs in mouse fetal liver until birth is followed by a selective decline in AFP mRNA, but not albumin mRNA, after birth. In vitro transcription assays show that this is caused by decreasing levels of transcription. Studies of mRNA levels in regenerating liver show that there is a 10- to 200-fold rise in AFP expression in the absence of albumin expression, supporting the notion of independent regulatory mechanisms (Tilghman and Belayew 1982). In the rat, the albumin gene shows a very low level of expression in the VYS. In the mouse, both AFP and albumin expressions increase in parallel as gestation proceeds, but in rat there is a reciprocal relationship between levels of AFP and albumin in the VYS and liver (Sellem et al. 1984). Recently it was shown, using transgenic mice, that exogenous DNA encoding AFP contains all of the signals necessary to evoke tissue-specific expression. Only visceral endoderm of the yolk sac, fetal liver, and gut expressed the exogenous AFP gene during gestation, and transcription was subsequently repressed in neonatal liver. The levels of expression were similar to endogenous gene transcription levels if plasmid sequences were absent from the exogenous DNA (Krumlauf et al. 1985).

Using DNase I sensitivity of the albumin and AFP genes in different rat tissues and cell lines, Nahon and associates (1984) showed that the chromatin structure of the albumin and AFP genes could be involved with tissue-specific gene expression, because cells with active genes showed increased sensitivity to DNase I in those genes. Andrews, Dziadek, and Tamaoki (1982a) had earlier shown that there is a positive correlation between hypomethylation of six CCGG sites in the AFP gene and expression of the gene in embryo, adult, and neoplastic tissues in the mouse. Studies of six methylation sites in the rat albumin gene showed that methylation changes at the sites analyzed are not responsible for the increases in gene activity during rat liver development (Vedel et al. 1983). However, undermethylation of one site at the 5' end of the rat albumin gene is correlated with the activation of gene expression in hepatoma cells (Ott et al. 1982).

Other studies have made use of the visceral endoderm differentiated from F9 teratocarcinoma cells. The AFP gene is active on the third day after retinoic acid treatment of F9 aggregates and reaches high levels by the eighth day (Grover, Oshima, and Adamson 1983b). Albumin and transferrin are also synthesized by embryoid bodies at much lower levels. Several pieces of evidence suggest that the AFP gene can be activated in the absence of overt visceral endoderm morphology, but high levels of AFP production are achieved only when a well-organized epithelial layer of polarized visceral endoderm cells is established on a thin basement membrane (Grover, Andrews, and Adamson 1983a; E. D. Adamson et al., unpublished observations).

F9 aggregate cultures do not activate the AFP gene spontaneously. Scott and Tilghman (1983) constructed an AFP minigene and used this as a marker to show that this exogenous gene transfected into F9 cells is activated in parallel with the endogenous gene when retinoic acid is added (Scott et al. 1984), suggesting that there is tissue-specific cis-regulation of the AFP gene. They were also able to show that although the AFP gene is demethylated during the course of differentiation, it is only one of many genes (both active and inactive) found to be demethylated. The genes examined include actin, H-2, histone H4, and albumin, and these are similarly affected by 5-azacytidine treatment of cells, a procedure that does not lead to AFP gene activation or differentiation in F9 cells (Young and Tilghman 1984).

AFP is one of the very few genes that is expressed in a limited number of cell types during embryogenesis, and although this has proved extremely useful in distinguishing between the endodermal products derived from teratocarcinoma cells, it poses a question about its precise role in endoderm function and indeed in embryogenesis. The ability of AFP to act as a carrier of smaller ligands is clearly one essential function, but there seems to be no reason why parietal endoderm should not also synthesize AFP or why albumin should not serve this function.

ii. Transferrin, apolipoprotein Al, and α_1-antitrypsin.

Transferrin is a serum β_1-globulin glycoprotein (80,000 d) whose major function is transport of iron to tissues, where it binds to specific cell surface receptors. Transferrin also binds copper and zinc and appears to have an additional stimulatory effect on growth, in addition to providing iron for the synthesis of such proteins as hemoglobin, myoglobin, and cytochromes. Apolipoprotein Al (28,000 d) is one of several components of high-density lipoproteins (HDL) in serum. All of these components are synthesized by the liver in adult animals. α_1-Antitrypsin is a serum protein that binds to and neutralizes trypsin-like or other serine proteases.

There is a single gene for apolipoprotein Al in the mouse (Meehan et al. 1984) and for transferrin in rat (Levin et al. 1984) and human (Uzan et al. 1984).

Several serum proteins that are liver-specific in adults are synthesized by both the VYS and the fetal liver but are regulated differently. Both organs synthesize high levels of albumin, transferrin, α_1-antitrypsin (Meehan et al. 1984), and apolipoproteins (Shi and Heath 1984). However, at least two products of the liver at later developmental stages, anti-α_1-chymotrypsin and major urinary proteins, are not expressed by the VYS or other tissues in the fetus (Meehan et al. 1984). Furthermore, whereas albumin and α_1-antitrypsin mRNAs are produced by the VYS at one-tenth of the level of the liver in the 13.5-day fetus, apolipoprotein Al and transferrin productions are several times higher than in fetal liver. The expressions of α_1-antitrypsin and apolipoprotein Al increase 10- to 20-fold in the fetal liver between the 13th and 18th days of gestation. Expression in the VYS, in contrast, does not change during this time. The cellular source of transferrin, apolipoprotein Al, and α_1-antitrypsin is predominantly the endodermal layer of the VYS, as shown by immuno-peroxidase staining (Adamson 1982b; Shi and Heath 1984) or manual dissection of the two layers followed by specific mRNA detection (Meehan et al. 1984).

The production of transferrin and its regulation in the fetus are more complicated than for other serum proteins. Because transferrin is a necessary component of serum-free-defined media for in vitro cultures of nearly all cells (Bottenstein et al. 1979), it is likely also to be indispensable for the growth and development of embryos. In the mouse, maternal tranferrin appears not to cross the placenta, and so at least by the 17th day of gestation the fetus is its own source (Renfree and McLaren 1974). Transferrin is synthesized as early as the 7th day of gestation in the mouse, as detected by specific immunoprecipitation of metabolically la-beled 80-kd proteins on gels. Immunoperoxidase staining indicates that it may be produced even earlier, and its major source is the visceral endo-derm of the egg cylinder (Adamson 1982b). At the 13th day, the endo-derm of the VYS is the major source; at the 15th day the liver and VYS are equally important, and later the liver predominates, although low levels are synthesized by some other tissues. Rates of transferrin synthesized and secreted have been measured by enzyme-linked immunosorbent assays (ELISA) (Meek and Adamson 1985). Several lines of evidence suggest that transferrin may be involved in cell proliferation and dif-ferentiation and particularly in nerve control of muscle differentiation (Markelonis et al. 1982; Matsuda, Spector, and Strohman 1984). It is possible, therefore, that synthesis in fetal tissues might lead to autocrine stimulatory effects or effects on adjacent tissues.

The relative lack of tissue specificity in the production of transferrin by embryonic cells is mirrored by results with teratocarcinoma cells, although, for the most part, cells that synthesize transferrin have been identified as visceral endoderm-like on the basis of other criteria (e.g., PSA5E and 1H5). Undifferentiated embryonal carcinoma (EC) cells synthesize barely detectable levels of transferrin, except for F9 cells, which synthesize up to 6ng/10^6 cells per day (Adamson and Hogan 1984). When EC cells are stimulated to differentiate by the conditions of culture or by retinoic acid, transferrin synthesis may be stimulated (Adamson 1982b; Adamson and Hogan 1984). Differentiation toward parietal endoderm does not lead to transferrin synthesis (e.g., F9 cultured with retinoic acid and dibutyryl cAMP, F9ACc19, PYS-2). In contrast, differentiation toward visceral-like endoderm in OC15, PC13, F9, P19, and P19S 1801A1 cells leads to activation of the transferrin gene (Adamson 1982b, in press; Adamson and Hogan 1984). In addition, the 1H5 cell line cultured as monolayers synthesizes undetectable amounts, but the gene is activated when cysts form spontaneously, and even greater activity is induced when retinoic acid is also added (Adamson et al. 1985b).

Metallothionein (MT) (reviewed by Karin 1985) is another example of a gene that is highly active in both visceral endoderm and fetal liver (Andrews, Adamson, and Gedamu 1984) and is therefore considered at this point even though it is not a serum protein. The metaleothioneins are a group of cytoplasmic proteins (6,000 d) that bind heavy metals such as Cu^{2+}, Cd^{2+}, and Zn^{2+} and whose expression in the adult animal is inducible in vitro and in vivo. Their functions are thought to be chiefly in detoxification by chelating these metals. Although the MT promoter can be regulated by metals and by glucocorticoids in the liver, in the VYS the gene is not inducible. Tissue- and stage-specific expression of MT occurs in fetal rats (Anderson et al. 1983; Piletz et al. 1983; Templeton, Banerjee, and Cherian 1985) and mice (Ouellette 1982; Andrews et al. 1984). In fetal mouse liver, the time course of expression is different from that in the rat. High expression is observed from day 12 and this rises to day 16 and then falls slightly. At the same time, the pregnant mother's liver expresses five times more MT mRNA than normal (10-fold less than fetal liver). Fetal brain and placenta have undetectable amounts of MT mRNA, small amounts are found in the amnion, and very high levels occur in the visceral and parietal endoderm. Because the expression of this gene is of some importance in considering the effects of DNA constructs containing the MT promoter in transgenic mice, studies of endogenous expression of MT have been well documented.

Teratocarcinoma cells are variable in their levels of expression of metallothionein. Some stem cells, such as OC15, have little or none; other have high levels (PC13); some (F9) have moderate levels, and these levels increase during differentiation toward visceral endoderm or parietal endoderm. In general, the distribution among differentiated cell types matches that in the embryo. It is high in the endoderm cell lines, PS HR9 and 1H5, but rather low in PSA5E (Andrews et al. 1984). The latter, however, is inducible by 5-azacytidine and sodium butyrate (G. K. Andrews and E. D. Adamson, unpublished observation).

The visceral endoderm and fetal liver originate from very different embryological lineages and yet have many common products (Table 11.1). The obvious differences in mode of expression and regulation of the serum protein genes serve to illustrate that these tissues may be thought of as functional equivalents adapted to their particular stages of development. This view is further strengthened by including the visceral mesoderm and its capacity for hematopoiesis, thus making the VYS more clearly an organ that provides early liver functions, as well as other unique functions.

iii. Epidermal growth factor. General reviews of several growth factors have recently been published (Li 1984). Epidermal growth factor (EGF) is a 53-amino-acid polypeptide (6,045 d) synthesized and stored in the mouse (but not rat or human) male submandibular salivary gland and found also in human urine (isolated as urogastrone). The physiological role for EGF is unknown, but it does occur in body fluids and tissues (summarized in Table 1 in Adamson 1983). Experimentally, it affects epidermal maturation, eye opening, and tooth eruption in neonatal animals. It also inhibits gastric acid secretion. In vitro, it acts as a mitogen for a wide range of cell types (reviewed by Herschman 1985). Among many effects on cells, EGF induces the rapid appearance of transferrin receptors in fibroblasts in vitro (Wiley and Kaplan 1984). Because the gene for EGF has been cloned, it is now possible to search for fetal tissues that synthesize it. One such survey has failed to detect EGF mRNA in the fetus, but precursor EGF mRNA is first detected in neonatal submandibular salivary glands and kidneys (C. G. Webb, personal communication). When the salivary glands are removed from adult mice, serum levels eventually return to normal; therefore, other tissues could synthesize EGF. The EGF detected in fetal tissues and fluids appears to be maternally derived, and if there is an embryonic equivalent it is sufficiently different not to be recognized by nucleic acid probes or antibodies (Nexó et al. 1980).

iv. Transforming growth factors α and β (TGFα and TGFβ). TGFαs are polypeptides that are closely similar to EGF and bind and act through the EGF receptor. The first one to be recognized was sarcoma growth factor (SGF), a product of Moloney-murine-sarcoma-virus-transformed fibroblasts, but a similar polypeptide is also produced by human and rat cancer cells. The rat protein has been purified (5,616 d), sequenced, synthesized, molecularly cloned, and shown to have 35−40

Table 11.1. Markers characteristic of extraembryonic tissues, early ectoderm, and fetal liver[a]

Marker	Early ectoderm	Trophoblast/placenta	Parietal endoderm/PYS	Visceral endoderm/VYS	Fetal liver (definitive endoderm)
Serum proteins					
AFP	−	−	−	++	++
Transferrin	?		−	+	+
Transferrin-R	?	−/+	?	?	+
Apolipoprotein A1	−	−	−	+	+
α₁-Antitrypsin	−	−	−	+	+
Albumin	−	−	−	+	+
Growth factors and receptors					
EGF	−	−	−	−	−
EGF-R	?	+	−	+	+
TGFβ	?	+	?	?	+
IGFs	?	+	?	+	+
IGF-R	?	+	?	−	?
NGF	?	+	?	?	?
Ins-R	?	+	?	?	+
Proto-oncogenes					
c-ras	+	+	+	++	++
c-myc	+	+/−	?	+	+
c-fos	−	+	+	+	+
c-fms	−	+	+	+	?

Table 11.1 (cont.)

Marker	Early ectoderm	Trophoblast/ placenta	Parietal endoderm/PYS	Visceral endoderm/VYS	Fetal liver (definitive endoderm)
Cell surface antigens					
MHC-Ag	+ (later)	−	?	+	+
I	+	+	+	+	?
i	−	−	− (or slight)	+	?
SSEA-1	+	+/?	−	+ (later)	?
SSEA-3	+	−/+	−	+	?
Alkaline phosphatase	+	−/+	−	−	−
Matrix proteins					
Fibronectin	+	+	−	+ low	+
Laminin	− → + low	+	++ high	+ low	+
Collagen IV	− → + low	−/+	++ high	+ (later)	+
Collagen I	−	−/+	−	+ (later)	+
Collagen III	+		−		+
Intermediate filament proteins					
Endo A/TROMA-1	+ → −	+	+	+	+
TROMA-3	?	+	+	−	−
Endo B	+ → −	+	+	+	+
Vimentin	− → +	−/+	+	−	+
Others					
Metallothionein	?	?/−	++	++	++
Oncomodulin	?	+	+	−	−

Note:
[a] Abbreviations are as in the text; PYS = parietal yolk sac.

percent sequence and structural homology with EGF (Anzano et al. 1983; Marquardt et al. 1984; Tam et al. 1984; Lee et al. 1985). It is also like EGF in that it is able to stimulate anchorage-independent growth (e.g., in soft agar) of normal untransformed fibroblasts when used in conjunction with TGFβ. TGFβ (approximately 25,000 d) is composed of two disulfide-linked polypeptide chains of 12,000 d. It binds to a receptor distinct from that for EGF and will stimulate growth in soft agar in the presence of either TGFα or EGF. TGFβ is produced by a wide range of transformed cell lines and normal tissues, such as bovine kidney, placenta, liver, muscle, and brain, and by blood platelets, which also produce an EGF-like protein and platelet-derived growth factor (PDGF). TGFβ activity is produced by human fetal fibroblasts and 15-day mouse fetal fibroblasts in a latent form that can be released by acidification of conditioned medium (Lawrence et al. 1984).

When TGFs are extracted from mouse and rat embryos, the product is an EGF-like component that competes with EGF for its receptor but is antigenically distinct (Nexó et al. 1980; Proper, Bjornson, and Moses 1982). There is evidence that TGFα is also produced as a larger precursor form (Derynck et al. 1984) that may exist as a transmembrane protein and that a 4.5-kb mRNA coding for TGFα is present in normal rat tissues (Lee et al. 1985). It remains to be seen whether or not endogenous TGF-like growth factors stimulate growth and fetal development. Some products of teratocarcinoma cells in serum-free culture are TGF-like, and so it seems likely that the embryo produces and responds to such growth factors, but which cell types are capable of doing so is unknown.

EGF-like growth factors with properties similar to those produced by sarcoma-virus-transformed cell lines appear to occur in normal rodent embryos (Matrisian, Pathak, and Magun 1982; Twardzik, Ranchalis, and Todaro 1982; Twardzik 1985) and human placenta (Stromberg et al. 1982). It is important to determine the sources of these factors, because they could be the embryonic equivalent of EGF. The production of TGFα by embryos could be associated with activated proto-oncogenes in fetal tissues. TGFα appears to have activities similar to those of EGF in neonatal mice (Tam 1985), but its ability to interact with TGFβ imparts great flexibility to its actions, because both antagonistic and synergistic responses have been described in in vitro studies (reviewed by Massagué 1985). The effects of TGFβ also interact with those of PDGF and insulin-like growth factors (IGF), suggesting a role in the control of growth and differentiation in embryonic and fetal tissues.

v. Platelet-derived growth factor. Like TGFβ, PDGF consists of two subunits of 11,000 to 12,000 d held together by sulfide bonds. However, in PDGF, the two chains are dissimilar to each other and to TGFβ, and one of them has been shown to be closely similar to the oncogene product of the simian sarcoma virus, p28^{sis} (Doolittle et al. 1983; Waterfield et al. 1983; Chiu et al. 1984). As its name suggests, PDGF is derived from platelets and is a mitogen for connective-tissue-derived cells and glial cells. PDGF is a "competence" factor, acting on quiescent cells to stimulate their advance into the G_1 phase of the cell cycle (see Section II.C).

A PDGF-like growth factor has been detected in goat colostrum (Brown and Blakely 1984), and similar factors may be produced by early embryo cells and teratocarcinoma cells, as well as transformed and tumor cell lines (Graves, Owen, and Antoniades 1983; Bowen-Pope, Vogel, and Ross 1984b; Nister et al. 1984). PDGF appears to be mainly a product of mesoderm-derived cell types (megakaryocytes, smooth-muscle cells, endothelial cells) as well as glial cells (Di Corleto and Bowen-Pope 1983). Developmental regulation of PDGF secretion by rat smooth-muscle cells occurs, because two-week-old rat cells secrete substantial amounts, whereas three-month-old rat cells do not (Seifert, Schwartz, and Bowen-Pope 1984).

vi. Somatomedins. Somatomedins (Sms) or insulin-like growth factors (IGF) have been reviewed by Adamson (1983) and Perdue (1984). Sms (7,000 d) are a family of serum peptide growth factors that are growth-hormone-dependent (GH-dependent) and are thought to be the mediators of GH action. They stimulate the incorporation of sulfate into cartilage, have insulin-like action on nonskeletal tissues, and stimulate mitosis in a wide variety of cultured cells. SmC and IGF-I are very similar or identical basic peptides. IGF-II, SmA, and multiplication-stimulating activity (MSA) are neutral or less basic, with 62 percent homology to IGF-I and with a lower degree of homology to proinsulin.

The human and rat IGF-II coding sequences have been cloned by two groups and shown to be transcribed as larger precursors (Bell et al. 1984; Dull et al. 1984; Yang et al. 1985a, 1985b). The liver is believed to be the source of the Sm peptides found in plasma, but other cell types release Sms in culture. For example, human fetal lung fibroblasts, WI-38, release IGF-I, and this is stimulated by GH. IGF-II synthesis by fetal fibroblasts in culture is stimulated by placental lactogen, but not GH. IGF-I is produced by fetal mouse cells (D'Ercole and Underwood 1981) and by multiple tissue sites (D'Ercole, Stiles, and Underwood 1984), and this is potentiated by EGF (Richman et al. 1985). Sm-like activity has recently been shown to be synthesized by human embryonic fibroblast cells (WI-38) as well as day-13 mouse embryo palates (Atkinson, Bala, and Hollenberg 1985). Production of this growth factor, measured by radioimmunoassay, is stimulated by a number of other factors, with EGF > insulin > human placental lactogen > human growth hormone >> human prolactin.

It is possible that SmC-like activity is an active promoter of fetal development. Because a number of other factors regulate its production, this may confer a means for fine tuning by combinations of growth-factor activities that are present at different locations and at different stages. IGFs are also produced by the placenta and kidney, but as yet the specific cell type of origin has not been identified. IGF-II is produced by the allantois and amnion during mouse development (J.K. Heath, personal communication) and therefore may be an important influence during early gestation. In support of this notion, IGFs in human amniotic fluid are specifically regulated during normal pregnancy, with IGF-I constant at 20 ng/ml and IGF-II at 114 ng/ml at the earliest time studied until 33 weeks, when levels start to fall (Merimee, Grant, and Tyson 1984).

vii. Insulin and glucagon. Insulin is detectable in the human fetal pancreas at the 7th week of gestation, and glucagon by the 15th week. Insulin secretion appears to be absolutely restricted to the β cell of the islets of Langerhans in the pancreas. Glucagon is synthesized by the α cells in the same structures. Apart from insulinomas, the synthesis of these two hormones is restricted to these highly specialized cell types, and no teratocarcinoma cell with similar ability has been reported.

viii. Growth hormone, prolactin, placental lactogen, and chorionic gonadotropin. GH is synthesized by the pituitary gland and is not likely to play a significant role in fetal growth. Placental lactogen (or chorionic somatomammotropin) may do so, because maternal plasma concentrations are 1,000 times greater than normal nonpregnant GH levels.

The placenta is the source of several peptide hormones that are homologues of hormones synthesized in other endocrine tissues. One set of genes code for at least four different polypeptides related to GH. This family of genes is thought to play a crucial role in normal fetal development, but their precise roles are not known. The placental hormones can substitute for the pituitary hormones prolactin and GH, because removal of the pituitary during the second half of rodent pregnancy does not block pregnancy. The major placental hormone is placental lactogen, reviewed by Talamantes and associates (1980). Recently a newly discovered gene encoding prolactin-related protein has been found to be 95 percent homologous, in parts of the gene, to proliferin, first found in proliferating mouse tumor cells (Linzer and Nathans 1985; Linzer et al. 1985). Two different rodent placental lactogen proteins have been described, and these have prolactin-like activities. Amino acid sequences of mouse placental lactogens distinguish them from other members of the GH family, but their specific targets and functions are unknown. The trophoblast giant cell appears to be the source of prolactin-related protein in rodents (Tabarelli, Kofler, and Wick 1983; Soares, Julian, and Glasser 1985).

Human placental lactogen shares 90 percent homology with GH and is synthesized, together with human chorionic gonadotropin (hCG), in the syncytiotrophoblast cells of the human placenta (Boime et al. 1982; Hoshina, Boothby, and Boime 1982). The synthesis of hCG is highest in the first trimester and it is detected in serum two weeks after fertilization. Its β subunit is unique to hCG and has been used as a marker of placental-type cells such as BeWo choriocarcinoma cells. Prolactin is synthesized by the pituitary, but is found in amniotic fluid at 100 times higher concentrations than in maternal serum, and several tissues respond to it by secreting Sm.

ix. Nerve growth factor (NGF). NGF is certainly involved in the development and maintenance of function of sympathetic and some sensory neurons. Two identical subunits (13,000 d) or β-NGF form the active component of a larger complex found in male mouse salivary gland, guinea pig prostate gland, and a few other sources in large amounts. However, the actual source of NGF relevant to the nervous system still is not certain, suggesting that NGF is produced locally to stimulate neuron development. Evidence shows that glial cells and peripheral tissues innervated by sensory and sympathetic neurons produce NGF. Certain neuroblastomas also produce NGF (Schwartz and Costa 1978). It is not, therefore, a product of a specific cell lineage.

x. Teratocarcinoma-cell-derived growth factors. Because teratocarcinoma cells resemble cell types found in early embryos, their growth-factor products could point to endogenous embryonal growth factors. TGF activity is readily detected in EC-cell-conditioned medium, but several other growth factors also appear to be present. A PDGF-like activity was detected in PSA1G cells (Gudas, Singh, and Stiles 1983) and F9 cells (Rizzino and Bowen-Pope 1985), although there is no evidence yet for v-sis-related transcription. A mixture of growth factors produced by PSA1 cells elutes from sizing columns in four peaks, one of which inhibits the differentiation of Friend leukemia cells stimulated with dimethyl sulfoxide. At least one factor is autocrine in type, and none of them is related to PDGF, EGF, or TGFα (Jakobovits, Banda, and Martin 1985). However, a growth factor (6 kd) produced by PC13 A2 undifferentiated cells is thought to be EGF-like (Stern and Priddle 1984). When EC cells differentiate, their growth-factor production declines, and at the same time they lose tumorigenicity. They do, however, start to produce other growth-promoting factors such as transferrin (Adamson 1982b), IGF-II, and extracellular-matrix components that stimulate the growth and/or differentiation of EC cells. Mixed cell cultures therefore grow better than pure cell types, and it seems likely that cross-feeding also occurs between stem cells and their differentiated products in the embryo (Heath and Isacke 1984).

xi. Growth factors in development. Various growth factors (reviewed by Adamson 1983) are present in the embryo, and the timing of the appearance and spatial distribution of these factors may be critical for normal development. One might speculate that early embryonic cells require constant but controlled growth-factor stimulation in order to proliferate, and different combinations of factors are progressively required for differentiation and morphogenesis, whereas in the adult, growth factors appear to be needed predominantly for tissue and wound repair. It is therefore important to identify which embryonic cell types produce growth factors and the timing of their expression. With the increasing availability of monospecific antibodies and cloned probes, this task will be easier.

B. Growth-factor receptors

i. EGF receptors. The EGF receptor is a single transmembrane polypeptide of 170,000 d (Downward at al. 1984; Lin et al. 1984; Ullrich et al. 1984). The receptor appears very early in post-implantation development, thus indicating a possible general role during development, because they are not specific to any cell lineage. EGF receptors occur on all mouse fetal tissues except parietal endoderm from at least the 12th day of gestation, and the numbers of binding sites increase during gestation (Adamson and Meek 1984). The phosphokinase activity associated with the EGF receptor is first detectable in 11-day mouse embryos (Hortsch et al. 1983). The amnion and also the liver of the late-gestation fetus and the newborn express the highest levels of EGF receptor in the mouse. Developing myoblasts, epidermal cells, palatal epithelium, and lung also express and utilize EGF receptors during development.

EGF receptors were first detected in developing mouse tissues when teratocarcinoma stem cells (PC13 and OC15) were allowed to differentiate toward visceral endoderm-type cells in vitro (Rees, Adamson, and Graham 1979). Stem cells do not have detectable receptors, but their differentiated products do. Approximately 70,000 and 30,000 receptors per cell, respectively, appear with average binding affinities ($K_a = 1.3 \times 10^9 M^{-1}$). F9 cells, however, even when unstimulated, express a low level of EGF-binding activity that doubles after four days of stimulation of aggregate cultures with retinoic acid. PC13 cells treated similarly express a much larger increase in EGF receptors as they differentiate, but the time course is similar (Adamson and Hogan 1984). Murine stem cells therefore do not appear to express EGF receptors unless they have become partially differentiated, although it is possible that EGF receptors are present but are masked or continuously down-regulated by endogenous EGF-like ligands. Human teratocarcinoma stem cells that are thought to represent an earlier stage in development also express a low level of EGF receptors (Carlin and Andrews 1985).

ii. TGF receptors. TGFα receptors appear to be identical with EGF receptors, and therefore the embryo can, in theory, respond to TGFα. ^{125}I-labeled purified TGFβ binds to a variety of cultured cells of both epithelial and mesenchymal origin, including embryonic fibroblasts. However, nothing is known about the distribution of TGFβ receptors in the developing embryo.

iii. IGF and IGF-like receptors. The human insulin receptor is a 350,000-d tetramer consisting of two α and two β subunits (Ullrich et al. 1985). It has structural similarities to human EGF receptor and homology with members of the *src* family of oncogene products. When PC13 teratocarcinoma cells differentiate, previously undetectable insulin receptors appear in the endoderm-like cells (Heath, Bell, and Rees 1981). F9 cells differ in expressing insulin receptors and responding to insulin by increased growth (Nagarajan and Anderson 1982). Noting that insulin must be added to stimulate serum-free growth of PC13 stem cells, Heath (1983) concluded that insulin may be acting through high-affinity IGF receptors that bind insulin at lower affinities. Rat IGF, MSA, can substitute for insulin in serum-free growth, but binding of MSA to PC13 cells is not completed by insulin. This conflict is further complicated by the finding that insulin and IGF requirements for growth of EC cells are abrogated if lipids are added to the medium (Heath 1983).

Although insulin is not considered important in fetal development, receptors are probably present on embryonic tissues, as in EC cells. If so, IGFs (particularly IGF-II), which probably are produced in the embryo, may act by binding to insulin receptors. Alternatively, there may be separate IGF receptors. Basic Sm receptors have been measured during the development of rat placenta, and possible receptor polypeptides of 330 kd and 140 kd were detected after gel electrophoresis of affinity-labeled receptors (Bhaumich and Bala 1985). IGF-I receptors are very similar to insulin receptors in having two dissimilar subunits (130 kd and 90 kd). The placenta is a rich source of these receptors, and the human placenta is reported to have two IGF-I receptors with different affinities (Jonas and Harrison 1985).

iv. Transferrin receptors. Most cultured cells, including teratocarcinoma cells, express transferrin receptors (Karin and Mintz 1981). In vivo, stable adult differentiated cell types do not express this cell surface

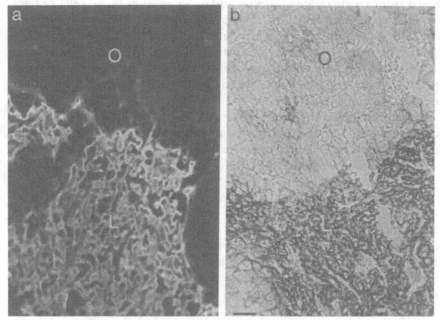

Figure 11.4. Sections of mouse placenta to show (a) transferrin receptors by immunofluorescent reaction and (b) alkaline phosphatase histochemical reaction. O = outer placental zone or spongiotrophoblast; L = inner labyrinthine zone; bar = 50 μm. [Reproduced from Müller et al. (1983c), with permission.]

glycoprotein, whereas stem cells that continue to divide do so (Omary and Trowbridge 1981; Trowbridge and Omary 1981). Although no systematic study has been performed, two of the developing cell types studied are low in transferrin receptor. One is the giant trophoblast cell produced when blastocysts are allowed to grow out in vitro. The inner cell mass and endoderm are both stained with antibodies to the transferrin receptor (Adamson in press). In contrast, there is low expression of transferrin receptors on the outer part of the mouse placenta. The spongiotrophoblast has a low or undetectable level of transferrin receptors, as revealed by fluorescent antibody staining, and this distribution is similar to that of alkaline phosphatase (Figure 11.4). Both of these cell surface markers are particularly strongly expressed in the labyrinthine portion of the mouse placenta.

v. PDGF receptors. PDGF receptors have been reviewed by Bowen-Pope, Seifert, and Ross (1985). The distribution of this receptor (170–185 d) is more restricted, although a systematic study of fetal tissues has not yet been done. Receptors are present on mesodermal-type cells such as vascular smooth-muscle cells, glial cells, chondrocytes, and fibroblasts from a wide range of species and on some white blood cells, but not on epithelial, endothelial, or skeletal muscle cells. There is strong evidence that the receptor phosphorylates itself and acts as a tyrosine phosphokinase on other cellular substrates (reviewed by Sefton and Hunter 1984). The receptor may therefore exhibit homologies with oncogene products, but these have not yet been identified, because the receptor has not been sequenced or cloned.

Teratocarcinoma stem cells (F9 and PC13) do not bind ^{125}I-PDGF but do secrete PDGF-like factors into the medium. When they differentiate by induction with retinoic acid to parietal-endoderm-like cells, they become able to bind and respond to PDGF (Rizzino and Bowen-Pope 1985).

vi. NGF receptors. NGF receptors occur mainly on sympathetic neurons, because NGF injected into newborn rodents results in a great increase in the size of the sympathetic ganglia (Levi-Montalcini and Booker 1960). With the development of monoclonal antibodies to the rabbit NGF receptor, purification and characterization of the receptor is now possible (Morgan and Bradshaw 1985). The time of onset of NGF in the embryo is not known, but it could shed light on the developmental role of NGF.

C. Proto-oncogenes

Proto-oncogenes (reviewed by Varmus 1984 and Müller and Verma 1984) are normal cellular genes that are homologues of retroviral transforming genes, although, in some cases, a retroviral homologue has not been identified. Because each of these viral genes is responsible for converting normal cells into tumorigenic cells, their functions must concern some aspects of growth control. It is suspected that their normal cellular counterparts (from which they are thought to have originated millions of years ago) also have important growth functions. Table 11.2 shows a classification of some of the proto-oncogenes that have been most studied.

In view of the growth-associated functions of the oncogenes, the embryo might express these genes at different stages of normal development. Indeed, such a variation in expression has been observed, but the significance for development is not yet clear. Variations of c-*onc* expression during development have been recorded by Slamon and Cline (1984) as accumulated transcripts in whole embryos. Among class-I c-*oncs*, c-*abl* is expressed in mouse embryos, rising to a peak at day 10. The c-*src* is active in many tissues throughout development; it peaks at day 14 and declines thereafter. The c-*src* is thought to have a function in brain development and has been found in chick embryonic and adult brain at very high levels, whereas in other tissues such as muscle, initially high embryonic transcript levels fall during development (Gessler and Barnekow 1984). Levy

Table 11.2. *Summary of proto-oncogene characteristics*

Gene	c-*onc* Gene product [a]	Location	Function
Class I: tyrosine protein kinase (PK)			
src	pp60[c-src]	Plasma membrane	Tyrosine PK
fps	p98[c-fps]	Plasma membrane? & cytoplasm	Tyrosine PK
abl	p150[c-abl]	Cytoplasmic and membrane-associated?	Tyrosine PK not normally detected
Class-I-related			
erbB	EGF receptor	Plasma membrane	EGF receptor
fms	CSF-1 receptor	Plasma membrane	CSF-1 receptor
mos	?	Cytoplasm?	Tyrosine PK; serine/threonine PK?
Class II: GTP binding			
Ha-*ras*	p21[c-Ha-ras]	Plasma membrane	GTP binding (no autokinase)
Ki-*ras*	p21[c-Ki-ras]	Plasma membrane	GTP binding
Class III: secreted growth factors			
sis	PDGF	Secreted & cytoplasmic	Growth factor
Class IV: nuclear			
fos	p55[c-fos]	Nuclear	?
myc	64K	Nuclear matrix	DNA binding
myb	75K	Nuclear	?

Note:
[a]PK, protein kinase; CSF, colony-stimulating factor.

and associates (1984) showed that chick heart also expresses high levels of c-*src*, and these are highest during organogenesis (stages 21–32). The human fetus has a similar distribution of c-*src* protein, with high levels in cerebral cortex, spinal cord, and heart. A role for c-*src* in electrogenic tissue function appears to be suggested by these data. A stronger correlation of c-*src* expression with differentiation of mature neuronal function was suggested by immunoperoxidase staining of developing chick cerebellum with rabbit antibodies (Fults et al. 1985).

Closely related to type-I oncogenes are c-*erb*, c-*fms*, and c-*mos*. The c-*erb* oncogene also has its maximum expression at the same time as c-*src*, and it remains high for the rest of gestation, but this is largely c-*erbA* expression, because c-*erbB* is not detectable in high amounts in any fetal tissues (Slamon and Cline 1984). This result conflicts with the presence of the EGF receptor (homologous with c-*erbB*) in fetal tissues from the 11th day of gestation. The c-*fms* product has recently been shown to be closely similar to if not identical with the CSF-1 receptor (Sherr et al. 1985), and its expression in embryos occurs in a tissue- and stage-specific manner, as will be described later. A third member of class-I-related oncogenes, c-*mos*, was thought to be inactive in all tissues, but recently transcripts have been detected in rat and mouse embryos as well as in testes and ovaries (Propst and Vande Woude 1985).

The *ras* genes are highly active in all fetal tissues throughout development and in teratocarcinoma cells (Müller 1983), with very little variation in transcript levels. High levels of transcripts hybridizing with v-*sis* are present at early stages in development (8-day peak), falling to one-fifth this level at day 10, and rising gently again during the rest of gestation (Slamon and Cline 1984). One may therefore suspect an early production of PDGF-like growth factors in embryos; indeed, PDGF, or a similar growth factor, has been detected in conditioned medium from EC cells (Rizzino and Bowen-Pope 1985).

The distribution of transcript levels of c-*fos* and c-*fms* has been traced in greater detail, because they show tissue-specific expression during development. They are detected predominantly in extraembryonic tissues, each in a different pattern. The c-*fos* oncogene is quite strongly expressed in mouse VYS endoderm and mesoderm, but even more in amnion, and both VYS and amnion express increasing amounts as gestation proceeds. High and progressively increasing levels of transcripts of c-*fms* are expressed in the placenta as gestation proceeds. The labyrinth (inner) and the spongiotrophoblast (outer) portions of the placenta have similar levels of c-*fms*, whereas c-*fos* is expressed more in the spongiotrophoblast (Adamson, Müller, and Verma 1983; Müller 1983; Müller et al. 1983a; Müller, Verma, and Adamson 1983c). Human extraembryonic tissues also express high levels of c-*fos* and c-*fms*, with the former highest in amnion and chorion, and the latter highest in placenta (Müller et al. 1983b).

Transcripts of c-*fos* are translated in these extraembryonic tissues, and the protein products have been detected in the nuclei of amnion cells in primary culture (Curran et al. 1984) and in sections of tissues by immunohistochemical staining (Adamson, Meek, and Edwards 1985a). The levels of staining of c-*fos* protein with anti-*fos* antibodies correlate with the levels of mRNA. When antibody titers are high, it is possible to stain nuclei in tissues that have very low levels of mRNA, and one may conclude that although the c-*fos*-reactive protein is ubiquitous, some cell lineages such as amnion express up to 100-fold more than others (e.g., liver). The functions of c-*fos* and c-*fms* products in development remain to be determined.

The function of c-*myc* is more clearly related to proliferation than are those of the other nuclear c-*onc*s. It is expressed in several cell lines during rapid growth, but declines when differentiation occurs (Campisi et al. 1984; Gonda and Metcalf 1984; Greenberg and Ziff 1984). The c-*myc* transcripts are present at high levels in the human placenta at four to five weeks, especially in cytotrophoblast, and decline with time (Pfeifer-

Ohlsson et al. 1984), whereas strong *myc* mRNA signals are detected later in some epithelial structures of the rapidly growing fetus, such as skin and gut in the late first trimester (Pfeifer-Ohlsson et al. 1985).

Both c-*myc* and c-*fos* are expressed transiently during the processes that follow the progression of quiescent embryonic fibroblasts stimulated by PDGF to pass from G_0 to G_1 and hence to cell division (Armelin et al. 1984; Cochran et al. 1984; Greenberg and Ziff 1984; Kruijer et al. 1984; Müller et al. 1984). It is clear from this kind of study that several oncogene products may be involved in a single growth-related process, for example, c-*sis*, c-*fos*, and c-*myc*. In seeming contrast, c-*fos* expression generally increases when differentiation occurs (Gonda and Metcalf 1984; Mitchell et al. 1985; S.A. Edwards and E. D. Adamson, unpublished data) and may even be a crucial component in differentiation (Müller and Wagner 1984; Rüther, Wagner, and Müller 1985).

Little is known about lineage specificity for most oncogenes, with the exceptions of c-*fos* and c-*fms*. It is premature to predict that high expression of c-*fms* may be a useful marker of the trophectoderm cell lineage in view of the limited amount of data. Levels of transcripts, however, are highest in the placenta and occur at similar levels in the human choriocarcinoma cell line BeWo. However, a mouse teratocarcinoma cell line that is thought to be trophoblastic, 3TDM-1, expresses only modest levels of c-*fos* and c-*fms* (Adamson et al. 1983). A human teratocarcinoma stem cell line, HT-H, that differentiates spontaneously in culture gives rise to a morphologically different cell type that expresses readily detectable c-*fms* mRNA, whereas the parent cells do not (M. Izhar and M. Fukuda, personal communication). It might be assumed that differentiation to trophoblast cells has occurred if other supporting markers are found.

Further studies are needed to shed light on the functions of the proto-oncogenes in general and to determine if they have special roles in embryonic development. One possible explanation for the high oncogene expression in extraembryonic tissues is the undermethylated state of the DNA in these tissues (Manes and Menzel 1981; Chapman et al. 1984). The undermethylated state of specific oncogenes in extraembryonic tissues may lead to the production of growth factors (such as TGFα) that are utilized in embryonic development. Another possibility is that the constant exposure of the extraembryonic tissues to growth factors in the surrounding fluids, together with receptors that can bind and ingest the factors, leads to constitutive activation of interrelated oncogenes. These hypotheses are likely to be tested in the near future using combinations of cloned probes and antibodies that are now becoming available.

D. Cell surface markers

A battery of specific surface marker molecules (as reviewed by Jacob 1977) that are totally characteristic of each individual cell type is highly desirable for many experimental approaches.

Because histocompatibility antigens are generally ubiquitous, their absence from murine embryos and embryonal carcinoma cells is itself a useful marker. H-2 antigens make a transient appearance on the cell surface of the trophoblast cell before implantation (Searle et al. 1976). H-2 antigens begin to be detected on all tissues of the embryo by mid-gestation (Kirkwood and Billington 1981) but remain absent from the labyrinth or syncytiotrophoblast and placental giant cells. However, the less differentiated spongiotrophoblast does express H-2 antigens (Jenkinson and Owen 1980). Murine teratocarcinoma stem cells also lack histocompatibility antigens, and these appear when differentiation occurs (Croce et al. 1981).

Numerous other cell surface differences have been detected between different embryonic stages and cell layers. The blood group antigens and other developmentally regulated antigens, such as SSEA-1 and SSEA-3, are discussed in Chapter 10.

Alkaline phosphatase expression on cell surfaces of teratocarcinoma stem cell lines decreases as differentiation occurs. Alkaline phosphatase is expressed increasingly in mouse embryos from the 2-cell stage onward, predominantly on inward-facing cell membranes (Mulnard and Huygens 1978). It disappears from trophoblast and primary endoderm but remains strongly expressed on primary germ cells. The mature mouse placenta expresses alkaline phosphatase asymmetrically, because the labyrinthine zone stains strongly, whereas the spongiotrophoblast stains weakly (Figure 11.4).

E. Extracellular-matrix components

Fibronectin, laminin, collagens, and proteoglycans play immense but poorly understood roles in the life of an embryonic cell. They surround, attach to, hold, and shape cells as they divide, migrate, and organize into complex organs. It is important, therefore, to trace the cellular origins of these secretory deposits and to attempt to understand their regulatory mechanisms within the cells and also the responses that are made as cells develop (reviewed by Adamson 1982a and Trelstad 1984).

Quantitative differences in the matrix components synthesized (especially by extraembryonc tissues) are useful in characterizing the differentiated products of cells in culture. Fibronectin (FN) (450,000 d) is synthesized to some extent by nearly all cells tested, but early embryos start to show deposits of FN only at the blastocyst stage, when the primitive endoderm cells accumulate it on their basal surfaces (Wartiovaara, Leivo, and Vaheri 1979). Parietal endoderm cells may synthesize FN transiently, but it is located mainly near the trophoblast margin (Semoff, Hogan, and Hopkins 1982). Trophoblast and visceral endoderm cells make much more, and it becomes incorporated into the basement membrane of visceral endoderm and the newly forming mesodermal layers. All embryonic basement membranes then stain with antibodies in immunohistochemical tests, and these deposits of FN are clearly important in cell mobility (Boucaut et al. 1984; Bronner-Fraser 1984). Soluble FN is present in amniotic fluid and serum, and cells in vitro may attach to the substratum using fibronectin in calf serum.

Laminin and type-IV collagen are the main structural proteins of basement membranes, and both are seen intracellularly in the primitive endoderm cell layer (Adamson and Ayers 1979; Leivo et al. 1980). However, one of the three major subunits of laminin (B1) is synthesized even by unfertilized eggs, whereas deposits of laminin are detected at the 8- to 16-cell stage (Wu et al. 1983), when B2 and A subunits are also synthesized (Cooper and MacQueen 1983). Type-III collagen is the next to appear as immunohistochemically stained deposits at the 16-cell stage (Sherman et al. 1980). Type-I collagen is not detected until about the 8th day, when type-I and -III collagens are seen in the head and heart mesenchyme, somites, chorion, amnion, and VYS (Leivo et al. 1980). Embryos lacking collagen type-I expression by virtue of a retrovirus gene inserted 5′ to the collagen gene die shortly after the 12th day, because the aorta is structurally weak and ruptures (Schnieke, Harbers, and Jaenisch 1983). Type-III collagen is synthesized in a largely fixed ratio with type-I collagen by fibroblasts (Liau, Yamada, and deCrombrugghe 1985) even when both levels are reduced, as in the transformed state. They are not tightly linked, however, and one gene can be stimulated exclusively (Setoyama, Liau, and deCrombrugghe 1985).

Parietal endoderm is an excellent source of basement-membrane components, because an extremely thick deposit (Reichert's membrane) is progressively laid down until the 13th day of gestation (Hogan, Cooper, and Kurkinen 1980). Both mouse (Timpl et al. 1979) and rat (Wewer 1982) yolk sac tumors that synthesize large amounts of basement membrane have been used to characterize the components and to make antibodies (Chang et al. 1983; Engvall et al. 1983). Recently cDNA probes to laminin and type-IV collagen (Kurkinen et al. 1983; Wang and Gudas, 1983; Marotti, Brown, and Strickland 1985), to a chondroitin sulfate proteoglycan core protein (Bourdon et al. 1985), and to fibronectin (Oldberg, Linney, and Ruoslahti 1983) have been prepared; so it will soon be possible to determine the origin of matrix components and to study their regulatory mechanisms. Model systems based on the ability of teratocarcinoma cells to differentiate into either parietal or visceral endodem will be important in distinguishing the crucial steps that characterize the choice of differentiative pathway: parietal endoderm characterized by little or no FN, much laminin, proteoglycans, and type-IV collagen synthesis; visceral endoderm characterized by more FN and much less laminin and collagen type-IV synthesis (summarized by Hogan, Barlow, and Tilly

1983 and Wu et al. 1983). Earlier studies to characterize matrix protein production by embryonal carcinoma cells and their differentiated products (Wartiovaara et al. 1978; Adamson, Gaunt, and Graham 1979; Wolfe et al. 1978; Cooper et al. 1981; Hogan, Taylor, and Cooper, 1982) are now being extended using cDNA probes. For example, retinoic acid stimulates the transcriptional rate of laminin and type-IV collagen genes while decreasing that of fibronectin when F9 cells differentiate into parietal endoderm (Grover and Adamson 1985; Marotti et al. 1985; Wang et al. 1985; A. Grover and E. D. Adamson, unpublished observations).

F. Intermediate-filament proteins

Cytoskeletal proteins include tubulin (56,000 d), which is in microtubules and mitotic spindle, actin (43,000 d), which is found in microvilli and under the plasma membrane, and intermediate-filament (8–10 nm diameter) proteins that maintain the shape of the cell and may have a crucial role in regulatory activities in the cell.

Intermediate filaments have been shown to be produced from a family of related genes that are differentially expressed in different tissues. Intermediate-filament proteins have been characterized biochemically and by reaction with specific antibodies as the following types: desmins (53,000 d), which are present in muscle; vimentins (57,000 d), in fibroblasts, endothelial and smooth-muscle cells, and most cells in vitro; neuronal intermediate filaments (68,000, 145,000, and 200,000 d); glial fibrillar acidic (GFA) protein (51,000 d) in glial cells and astrocytes and cytokeratins in epithelial cells (Altmannsberger et al. 1981). Intermediate filaments are sparse in the mesoderm cells of 8.5-day embryos, but these consist of vimentin, with no cytokeratins or desmin. Primitive ectoderm cells do not express vimentin and stain weakly for cytokeratins; cytokeratin expression is then switched off when ectoderm delaminates to form the mesoderm cells of the primitive streak at day 8.5. Desmoplakins are present in both proximal endoderm and ectoderm (Franke et al. 1982), but not in primary mesoderm cells (Franke et al. 1983). Although cytokeratins are considered unique to epithelial tissues and vimentin is characteristic of endothelial and mesenchymal cells, after epithelial cells are adapted to culture they express vimentin as well as cytokeratins. The mesothelial cell contains both vimentin and a pair of cytokeratin polypeptides in vivo; it is also a motile cell (Rheinwald et al. 1984). Interestingly, parietal endoderm also expresses both vimentin and cytokeratins (Lane et al. 1983; Lehtonen et al. 1983). These cells are also highly motile, and although they lay down a basement membrane (Reichert's membrane), it is multilayered and the cells responsible do not form junctional complexes; instead, they leave intercellular gaps. This is in contrast with visceral endoderm cells, which express cytokeratins, but not vimentin, and form a normal epithelium.

The cytokeratin group of intermediate-filament proteins (see Møll et al. 1982 for review) appears early in the developing mouse and can be recognized as specialized cytokeratins by means of antibodies TROMA-1, -2, and -3 (Brûlet et al. 1980; Brûlet and Jacob 1982) and by rabbit polyclonal antibodies to Endo A and B (Oshima 1982). Members of this family of cytokeratins are present as intermediate filaments in the trophoblast, parietal and visceral endoderm of mouse embryos, PYS, and PF HR9 teratocarcinoma cells (Chung et al. 1977). TROMA-1 recognizes the same antigen as antibodies to Endo A, whereas Endo B is related but distinct (Oshima 1981, 1982). These cytokeratins are not found in EC cells either as mRNA or as protein and are undetectable in the inner cell mass of blastocysts and in mouse embryos earlier than the 4- to 8-cell stage. They are synthesized increasingly (Oshima et al. 1983) as gestation proceeds and finally come to be located solely in the trophectoderm and visceral- and parietal-like endoderm cells in mouse embryos.

Fetal epidermal cells stain for TROMA-1, -2, and -3 up to day 12 of gestation, as do most epithelial cells, including gut, lung, pancreas, kidney, thymus, and uterus (but not neuroepithelium). The liver expresses TROMA-1 and -2, but not TROMA-3. However, by day 14, the skin epithelium no longer stains with TROMA antibodies, although these antibodies stain most epithelia derived from the mesoderm and endoderm (Kemler, Brûlet, and Jacob 1981). As fetal skin matures, it comes to express the epidermal-type cytokeratins found in adult skin. TROMA-3 reacts only with parietal endoderm (weakly) and distinguishes this kind of endoderm from visceral endoderm, which does not react (Boller and Kemler 1983). Cytokeratins have proved to be excellent markers for identifying the differentiated products of murine teratocarcinoma cells (Grover et al. 1983a, 1983b; Kahan and Adamson 1983; Müller and Wagner 1984).

The genes for several of the cytokeratins have been cloned in order to study the regulation of their expression. The mRNA for Endo A is present in trophectoderm, but not inner cell mass, at the blastocyst stage, and by 7.5 days it accumulates in visceral endoderm and in the amnion (Duprey et al. 1985). When F9 cells differentiate into parietal endoderm cells after induction with retinoic acid, the expression of a 1.5-kb (EndoB) and a 2.0-kd (Endo A) mRNA species can be detected after 48 h of treatment (Tabor and Oshima 1982; Singer, Trevor, and Oshima 1986). It is now clear that there are two distinct families of cytokeratin gene products. One subfamily (type II or B) is more basic than the other subfamily (type I or A). At least 20 members of this family have been identified, and they always occur in pairs of one A and one B polypeptide together (or, more likely, tetramers of coils with a highly conserved central helix portion). Simple epithelia in human tissues are characterized by cytokeratins of 52,000 d (type II) and 44,000 and 40,000 d (type I), numbered 8, 18, and 19, respectively (Sun et al. 1983). Mouse extraembryonic endoderm and trophoblast contain a comparable set (type II = Endo A = TROMA-1 = 55,000 d = No. 8; type I = Endo B = 50,000 d = No. 18) and this set is also the main cytokeratin pair that occurs in adult mouse liver (Trevor and Oshima 1985). As epithelia become more complex, for example, the embryonic epidermis changes into a keratinizing stratified epithelium, the cytokeratins change and become characteristic of their tissue type. This set of genes then provides an excellent source of stable tissue-specific markers.

G. Heat-shock proteins

Heat-shock proteins (HSP) are synthesized from a set of highly evolutionarily conserved genes that are activated after cells recover from stresses such as elevated temperatures or toxic chemicals. Similar proteins are found in normal mouse embryos and teratocarcinôma cells. A 70,000 d HSP is newly synthesized in the 2-cell mouse embryo and is constitutively expressed also in F9 cells (Bensaude et al. 1983; Bensaude and Morange 1983; Morange et al. 1984). Stem cells and early embryos express 70-kd and 89-kd HSPs, and beginning at the blastocyst stage, embryos respond to heat shock by synthesizing a 68-kd, but not a 105-kd protein as do some EC cells. The PCC7S1009 cell line does not respond to hyperthermia, however, unless it is stimulated to differentiate by retinoic acid. F9 cells express 89-kd, 70-kd, and 59-kd HSPs at high levels in the absence of stress and start to synthesize 70-kd, 68-kd, and 24-kd HSPs at increased levels when arsenite is added (Bensaude and Morange 1983). Because the 70-kd HSP is also represented among the masked mRNAs in *Xenopus* oocytes, and similar 70-kd proteins are synthesized in preblastula stages of *Xenopus* (Bravo and Knowland 1979) and 16-cell rabbit embryos (Schultz and Tucker 1977), this protein may have important functions. Whatever these are, however, HSPs are useful as a means of determining the stage of an embryonic cell type, because several are developmentally regulated.

The 89-kd HSP of F9 cells was found to decrease after stimulation of differentiation by retinoic acid and dibutyryl cAMP, and the relevant sequence was detected in a cDNA library by hybrid selection and in vitro translation (Levine et al. 1984). The 89-kd HSP is associated with a tyrosine phosphokinase activity in avian-sarcoma-virus-transformed cell lines (Opperman, Levinson, and Bishop 1981) and may play a role in tumorigenicity.

H. Oncomodulin

Oncomodulin is a calcium-binding protein (11,500 d) that has potential as a marker to distinguish parietal endoderm from visceral endoderm. Although both tissues in the rat conceptus stain with antibodies to the protein, the visceral endoderm merely absorbs it, because metabolically labeled oncomodulin synthesis cannot be demonstrated. Rat parietal yolk sac, placenta (particularly the outer spongiotrophoblast zone), and amnion also synthesize oncomodulin, as assayed by RIA of tissue extracts (L. Brewer and J. MacManus, personal communication).

III. Conclusions and summary

There is some evidence that the active genes that characterize a particular cell type are switched on as a set. For example, both VYS endoderm and fetal liver (Table 11.1) are responsible for activating a very similar set of protein products such as albumin, AFP, transferrin, α_1-anti-trypsin, Endo-A and Endo-B cytoskeletal proteins, and metallothionein. In the case of the tandem genes, albumin and AFP, the evidence is quite strong. However, it is clear that once activated, each of the genes in this set is subject to individual modulation. This is particularly obvious for albumin synthesis, which is sustained throughout life in the adult liver, whereas AFP is rapidly shut down after birth.

Another set of genes that may be activated together are those responsible for the formation of an epithelium. The components of the basement membrane must be made, and desmosomal plaques and tight junctions formed, that will interact with the cytoskeleton. However, even these are subject to individual modulation, because the parietal endoderm makes an excessive amount of basement membrane (Reichert's membrane), with little evidence of desmosomes or tight junctions. The presence of vimentin together with prekeratin-type intermediate-filament proteins is a further difference that emphasizes the uniqueness of this early epithelial cell type. The mesothelial cell found in peritoneal cavities is similarly a specialized epithelial cell type.

This chapter has reviewed a selection of the earlier genes to be activated. Some of these are concerned with growth control and tissue morphogenesis or are associated with clinically useful markers. One marker product that has not yet been mentioned and that may be considered as represented by at least two tissue-specific types is plasminogen activator. Other tissue-specific markers are characteristic of the function of mature tissues. They include myosin, actin, tropomysin, myoglobin, steroid-hormone-producing enzymes, nerve-conduction-associated products, enzymes, and so forth. A large list of isoenzymic forms that have tissue-specific distributions should also be mentioned, such as aldolase, creatine phosphokinase, enolase, and lactic dehydrogenase. Some of these change isoenzymic forms during development and may be useful in the identification of cell types in differentiating cell systems (Adamson 1976; Fletcher et al. 1978).

Study of cell lineages in embryonic development initially traces the antecedents and descendants of each cell type during the formation of an adult. Identification of markers that characterizes these lineages enables the establishment of teratocarcinoma models that further help progress in these studies. The gene products that I have discussed have included secreted proteins such as serum proteins and growth factors, cell surface markers such as growth-factor receptors, oncogenes, cytoskeletal proteins, and extracellular-matrix proteins. Most of these are products of differentiated cell types and are chiefly useful as lineage markers. However, detailed analyses of how these gene products are controlled in a tissue-specific manner may give clues to the mechanisms of gene control in specific lineages and hence how a cell develops. Certain genes may play a more direct role in controlling differentiation, and candidates for this role include growth factors, proto-oncogenes, and homeotic genes. Other unknown genes probably exist that are finally responsible for turning on differentiation, and these will be increasingly sought in future studies.

Acknowledgments

I thank Drs. R. Oshima, K. Trevor, L. Shevinsky, and S. Edwards for their suggestions and comments. This work was supported by grants CA-28427, HD-18772, and P30-CA-30199 from the NIH.

References

Adamson, E. D. (1976) Isoenzyme transitions of creatine phosphokinase, aldolase, and phosphoglycerate mutase in differentiating mouse cells. *J. Embryol. Exp. Morphol.* 35, 355–67.
–(1982a)The effect of collagen on cell division, cellular differentiation and embryonic development. In *Collagen in Health and Disease*, eds. J. B. Weiss and M. I. V. Jayson, pp. 218–43. Edinburgh: Churchill Livingstone.
–(1982b) The location and synthesis of transferrin in mouse embryos and teratocarcinoma cells. *Dev. Biol.* 91, 227–34.
–(1983) Growth factors in development. In *The Biological Basis of Reproductive and Developmental Medicine*, ed. J. B. Warshaw, 307–36. New York: Elsevier.
–(in press) Extraembryonic tissues as sources and sinks of humoral factors in development: teratocarcinoma model systems. *Cell Endocrinol.*
Adamson, E. D., and Ayers, S. E. (1979) The localization and synthesis of collagen in mouse embryos. *Cell* 16, 953–65.
Adamson, E. D., Gaunt, S. J., and Graham, C. F. (1979) The differentiation of teratocarcinoma stem cells is marked by the types of collagen synthesized. *Cell* 17, 469–76.

Adamson, E. D., and Hogan, B. L. M. (1984) Expression of EGF receptor and transferrin by F9 and PC13 teratocarcinoma cells. *Differentiation* 27, 152−7.

Adamson, E. D., and Meek, J. (1984) The ontogeny of epidermal growth factor receptors during mouse development. *Dev. Biol.* 103, 62−71.

Adamson, E. D., Meek, J., and Edwards, S. A. (1985a) Product of the cellular oncogene, c-*fos*, observed in mouse and human tissues using an antibody to a synthetic peptide. *EMBO Journal* 4, 941−7.

Adamson, E. D., Müller, R., and Verma, I. (1983) Expression of c-*onc* genes, c-*fos* and c-*fms*, in developing mouse tissues. *Cell Biol. Int. Rep.* 7, 557−9.

Adamson, E. D., Strickland, S., Tu, M., and Kahan, B. (1985b) A teratocarcinoma-derived endoderm stem cell line (1H5) that can differentiate into extra-embryonic endoderm cell types. *Differentiation,* 29, 68−76.

Albrechtsen, R., and Nørgaard-Pedersen, B. (1978) Immunofluorescent localisation of alpha-fetoprotein synthesis in the endodermal sinus of rat placenta. *Scand. J. Immunol.* 8, 193−9.

Altmannsberger, M., Osborn, M., Schauer, A., and Weber, K. (1981) Antibodies to different intermediate filament proteins. *Lab. Invest.* 45, 427−34.

Anderson, R. E., Piletz, J. E., Birren, B. W., and Herschman, H. R. (1983) Levels of metallothionein messenger RNA in fetal, neonatal and maternal rat liver. *Eur. J. Biochem.* 131, 496−500.

Andrews, G. K., Adamson, E. D., and Gedamu, L. (1984) The ontogeny of expression of murine metallothionein: comparison with the α-fetoprotein gene. *Dev. Biol.* 103, 294−303.

Andrews, G. K., Dziadek, M., and Tamaoki, T. (1982a) Expression and methylation of the mouse α-fetoprotein gene in embryonic, adult and neoplastic tissues. *J. Biol. Chem.* 257, 5148−53.

Andrews, G. K., Janzen, R. G., and Tamaoki, T. (1982b) Stability of α-fetoprotein messenger RNA in mouse yolk sac. *Dev. Biol.* 89, 111−16.

Anzano, M. A., Roberts, D. D., Smith, J. M., Sporn, M. B., and DeLarco, J. E. (1983) Sarcoma growth factor from conditioned medium of virally-transformed cells is composed of both type α and type β transforming growth factors. *Proc. Natl. Acad. Sci. U.S.A.* 80, 6264−8.

Armelin, H. A., Armelin, M. C. S., Kelly, K., Stewart, T., Leder, P., Cochran, B. H., and Stiles, C. D. (1984) Functional role for c-*myc* in mitogenic response to platelet-derived growth factor. *Nature* 310, 655−60.

Atkinson, P. R., Bala, R. M., and Hollenberg, M. D. (1985) Somatomedin-like activity from cultured embryo-derived cells: partial characterization and stimulation of production by epidermal growth factor (urogastrone). *Can. J. Biochem. Cell Biol.* 62, 1335−42.

Bell, G. I., Merryweather, J. P., Sanchez-Pescador, R., Stempien, M. M., Priestly, L., Scott, J., and Rall, L. B. (1984) Sequence of a cDNA clone encoding human preproinsulin-like growth factor. *Nature* 310, 775−7.

Bensaude, O., Babinet, C., Morange, M., and Jacob, F. (1983) Heat shock proteins, first major products of zygotic gene activity in mouse embryo. *Nature,* 305, 331−3.

Bensaude, O., and Morange, M. (1983) Spontaneous high expression of heat shock proteins in mouse embryonal carcinoma cells and ectoderm from day 8 mouse embryo. *EMBO Journal* 2, 173–7.

Bhaumich, B., and Bala, R. M. (1985) Ontogeny and characterization of basic somatomedin receptors in rat placenta. *Endocrinology* 116, 492–8.

Boime, I., Boothby, M., Hoshina, M., Daniels-McQueen, S., and Darnell, R. (1982) Expression and structure of human placental hormone genes as a function of placental development. *Biol. Reprod.* 26, 73–91.

Boller, K., and Kemler, R. (1983) *In vitro* differentiation of embryonal carcinoma cells characterized by monoclonal antibodies against embryonic cell markers. In *Teratocarcinoma Stem Cells*, eds. L. M. Silver, G. R. Martin, and S. Strickland, pp. 39–50. Cold Spring Harbor, N.Y.: Cold Spring Harbor Laboratory.

Bottenstein, J., Hayashi, I., Hutchings, S., Masui, H., Mather, J., McClure, D. B., Ohasa, S., Rizzino, A., Sato, G., Serrero, G., Wolfe, R., and Wu, R. (1979) The growth of cells in serum-free hormone-supplemented media. *Methods Enzymol.* 58, 94–109.

Boucaut, J.-C., Darribere, T., Poole, T. J., Aoyama, H., Yamada K. M., and Thiery, J. P. (1984) Biologically active synthetic peptides as probes of embryonic development: a competitive peptide inhibitor of fibronectin function inhibits gastrulation in amphibian embryos and neural crest cell migration in avian embryos. *J. Cell Biol.* 99, 1822–30.

Bourdon, M. A., Oldberg, A., Pierschbacher, M., and Ruoslahti, E. (1985) Molecular cloning and sequence analysis of a chondroitin sulfate proteoglycan cDNA. *Proc. Natl. Acad. Sci. U.S.A.* 82, 1321–5.

Bowen-Pope, D. F., Seifert, R. A., and Ross, R. (1985) The platelet-derived growth factor receptor. In *Control of Animal Cell Proliferation*, Vol. 1, eds. A. L. Boynton and H. L. Leffert, pp. 281–314. New York: Academic Press.

Bowen-Pope, D. G., Vogel, A., and Ross, R. (1984) Production of PDGF-like molecules and reduced expression of PDGF-R accompany transformation by a wide spectrum of agents. *Proc. Natl. Acad. Sci. U.S.A.* 81, 2396–400.

Brahic, M., and Haase, A. T. (1978) Detection of viral sequences of low reiteration frequency by *in situ* hybridization. *Proc. Natl. Acad. Sci. U.S.A.* 75, 6125–9.

Bravo, R., and Knowland, J. (1979) Classes of proteins synthesized in oocytes, eggs, embryos and differentiated tissues of *Xenopus laevis*. *Differentiation* 13, 101–8.

Bronner-Fraser, M. (1984) Latex beads as probes of a neural crest pathway: effects of laminin, collagen, and surface charge on bead translocation. *J. Cell Biol.* 98, 1947–50.

Brown, K., and Blakeley, D. M. (1984) Partial purification and characterization of a growth factor present in goat's colostrum: similarities with platelet-derived growth factor. *Biochem. J.* 219, 609–17.

Brûlet, P., Babinet, C., Kemler, R., and Jacob, F. (1980) Monoclonal antibodies against trophectoderm-specific markers during mouse blastocyst formation. *Proc. Natl. Acad. Sci. U.S.A.* 77, 4113–14.

Brûlet, P., and Jacob, F. (1982) Molecular cloning of a cDNA sequence encoding a trophectoderm-specific marker during mouse blastocyst formation. *Proc. Natl. Acad. Sci. U.S.A.* 79, 2328–32.

Buc-Caron, M. H., Darmon, M., Poiret, M., Sellem, C., Sala-Trepat, J. M., and Erdos, T. (1983) Analysis of α-fetoprotein and albumin gene expression in teratocarcinoma cells. In *Teratocarcinoma Stem Cells*, eds. L. M. Silver, G. R. Martin, and S. Strickland, pp. 411–20. Cold Spring Harbor, NY: Cold Spring Harbor Laboratory.

Campisi, J., Gray, H. E., Pardee, A. B., Dean M., and Sonenshein, G. E. (1984) Cell-cycle control of c-*myc* but not c-*ras* expression is lost following chemical transformation. *Cell* 36, 241–7.

Carlin, C. R., and Andrews, P. W. (1985) Human embryonal carcinoma cells express low levels of functional receptor for epidermal growth factor. *Exp. Cell Res.* 159, 17–26.

Chapman, V. M., Forrester, L., Sanford, J., Hastie, N., and Rossant, J. (1984) Cell lineage specific undermethylation of mouse repetitive DNA. *Nature* 304, 284–96.

Chiu, I.-M., Reddy, E. P., Givol, D., Robbins, K. C., Tronick, S. R., and Aaronson, S. A. (1984) Nucleotide sequence analysis identifies the human c-*sis* proto-oncogene as a structural gene for platelet-derived growth factor. *Cell* 37, 123–9.

Chung, A. E., Estes, L. E., Shinozuka, H., Braginski, J., Lorz, C., and Chung, C. A. (1977) Morphological and biochemical observations on cells derived from the in vitro differentiation of the embryonal carcinoma cell line PCC4-F. *Cancer Res.* 37, 2072–81.

Chung, A. E., Jaffe, R., Bender, B., Lewis, M., and Durkin, M. (1983) Monoclonal antibodies against the GP-2 subunit of laminin. *Lab. Invest.* 49, 576–81.

Cochran, B. H., Zullo, J., Verma, I. M., and Stiles, C. D. (1984) Expression of the c-*fos* gene and of a *fos*-related gene is stimulated by platelet-derived growth factor. *Science* 226, 1080–2.

Cooper, A. R., Kurkinen, M., Taylor, A., and Hogan, B. L. M. (1981) Studies on the biosynthesis of laminin by murine parietal endoderm cells. *Eur. J. Biochem.* 119, 189–97.

Cooper, A. R., and MacQueen, H. A. (1983) Subunits of laminin are differentially synthesized in mouse eggs and early embryos. *Dev. Biol.* 96, 467–71.

Croce, C. M., Linnenbach, A., Huebner, K., Parnes, J. R., Margulies, D. H., Apella, E., and Seidman, J. G. (1981) Control of expression of histocompatibility antigens (H-2) and β₂-microglobulin in F9 teratocarcinoma cells. *Proc. Natl. Acad. Sci. U.S.A.* 78, 5754–8.

Curran, T., Miller, A. D., Zokas, L., and Verma, I. M. (1984) Viral and cellular *fos* proteins: a comparative analysis. *Cell* 36, 259–68.

D'Ercole, A. J., Stiles, A. D., and Underwood, L. E. (1984) Tissue concentrations of somatomedin C: further evidence of multiple sites of synthesis or autocrine mechanisms of action. *Proc. Natl. Acad. Sci. U.S.A.* 81, 935–9.

D'Ercole, A. J., and Underwood, L. E. (1981) Growth factors in fetal growth and development. In *Fetal Endocrinology*, Vol. 1, eds. M. J. Novy and J. A. Resko, pp. 155–82. New York: Academic Press.

Derynck, R., Roberts, A. B., Winkles, M. E., Chen, E. Y., and Goeddel, D. V. (1984) Human transforming growth factor-α: precursor structure and expression in *E. coli. Cell* 38, 287–97.

Di Corleto, P. E., and Bowen-Pope, D. R. (1983) Cultured endothelial cells produce a platelet-derived growth factor-like protein. *Proc. Natl. Acad. Sci. U.S.A.* 80, 1919–23.

Doolittle, R. F., Hunkapillar, M. W., Hood, L. E., Devore, S. G., Robbins, K. C., Aaronson, S. A., and Antoniades, H. N. (1983) Simian sarcoma virus oncogene, v-*sis*, is derived from the gene (or genes) encoding a platelet-derived growth factor. *Science* 221, 275–6.

Downward, J., Yarden, Y., Mayes, E., Scrace, G., Totty, N., Stockwell, P., Ullrich, A., Schlessinger, J., and Waterfield, M. D. (1984) Close similarity of epidermal growth factor receptor and v-*erb* B oncogene protein sequences. *Nature* 307, 521–7.

Dull, T. J., Gray, A., Hayflick, J. S., and Ullrich, A. (1984) Insulin-like growth factor II precursor gene organization in relation to insulin gene family. *Nature* 310, 777–81.

Duprey, P., Morello, D., Vasseur, M., Babinet, C., Condamine, H., Brûlet, P., and Jacob, F. (1985) Expression of the cytokeratin endoA gene during early mouse embryogenesis. *Proc. Natl. Acad. Sci. U.S.A.* 82, 8535–9.

Dziadek, M. (1978) Modulation of alphafoetoprotein synthesis in the early postimplantation mouse embryo. *J. Embryol. Exp. Morphol.* 46, 135–46.

Dziadek, M., and Adamson, E. D. (1978) Localisation and synthesis of alpha-foetoprotein in post-implantation mouse embryos. *J. Embryol. Exp. Morphol.* 43, 289–313.

Dziadek, M., and Andrews, G. K. (1983) Tissue specificity of alpha-fetoprotein messenger RNA expression during mouse embryogenesis. *EMBO Journal* 2, 549–54.

Edwards, M. K. S., and McBurney, M. W. (1983) The concentration of retinoic acid determines the differentiated cell types formed by a teratocarcinoma cell line. *Dev. Biol.* 98, 187–91.

Eiferman, F. A., Young, P. R., Scott, R. W., and Tilghman, S. M. (1981) Intragenic amplification and divergence in the mouse α-fetoprotein gene. *Nature* 294, 713–18.

Engvall, E., Krusius, T., Wewer, U., and Ruoslahti, E. (1983) Laminin from rat yolk sac tumor: isolation, partial characterization and comparison with mouse laminin. *Arch. Biochem. Biophys.* 222, 649–56.

Faulk, W. P., and Galbraith, G. M. P. (1979) Trophoblast transferrin and transferrin receptors in the host-parasite relationship of human pregnancy. *Proc. R. Soc. Lond. Biol.* 204, 83–97.

Fletcher, L., Rider, C. C., Taylor, C. B., Adamson, E. D., Luke, B. M., and Graham, C. F. (1978) Enolase isoenzymes as markers of differentiation in teratocarcinoma cells and normal tissues of mouse. *Dev. Biol.* 65, 211–24.

Franke, W. W., Grund, C., Jackson, B. W., and Illmensee, K. (1983) Formation of cytoskeletal elements during mouse embryogenesis. IV. Ultrastructure of primary mesenchymal cells and their cell-cell interactions. *Differentiation* 25, 121–41.

Franke, W. W., Grund, C., Kuhn, C., Jackson, B. W., and Illmensee, K. (1982) Formation of cytoskeletal elements during mouse embryogenesis. III. Primary mesenchymal cells and the first appearance of vimentin filaments. *Differentiation* 23, 43–59.

Fults, D. W., Towle, A. C., Lauder, J. M., and Maness, P. F. (1985) pp60$^{c\text{-}src}$ in the developing cerebellum. *Mol. Cell. Biol.* 5, 27–32.

Gardner, R. L. (1982) Investigation of cell lineage and differentiation in the extraembryonic endoderm of the mouse embryo. *J. Embryol. Exp. Morphol.* 68, 175–98.

Gessler, M., and Barnekow, A. (1984) Differential expression of the cellular oncogenes c-*src* and c-*yes* in embryonal and adult chicken tissues. *Bioscience Reports* 4, 757–70.

Gitlin, D., Ketzes, J., and Boesman, M. (1967) Cellular distribution of serum fetoprotein in organs of fetal rat. *Nature* 215, 534.

Gonda, T. J., and Metcalf, D. (1984) Expression of *myb, myc,* and *fos* proto-oncogenes during the differentiation of a murine myeloid leukemia. *Nature* 310, 249–51.

Graves, D. T., Owen, A. J., and Antoniades, H. N. (1983) Evidence that a human osteosarcoma cell line which secretes a mitogen similar to PDGF requires growth factors present in platelet-poor plasma. *Cancer Res.* 43, 83–7.

Green, E. L., ed. (1966) *Biology of the Laboratory Mouse*, 2nd ed. New York: McGraw-Hill.

Greenberg, M. E., and Ziff, E. B. (1984) Stimulation of 3T3 cells induces transcription of the c-*fos* proto-oncogene. *Nature* 311, 433–7.

Grover, A., and Adamson, E. D. (1985) Roles of extracellular matrix components in differentiating teratocarcinoma cells. *J. Biol. Chem.* 260, 12252–8.

Grover, A., Andrews, G., and Adamson, E. D. (1983a) Role of laminin in epithelium formation by F9 aggregates. *J. Cell Biol.* 97, 137–44.

Grover, A., Oshima, R. G., and Adamson, E. D. (1983b) Epithelial layer formation in differentiating aggregates of F9 embryonal carcinoma cells. *J. Cell Biol.* 96, 1690–6.

Gudas, L. J., Singh, J. P., and Stiles, C. D. (1983) Secretion of growth regulatory molecules by teratocarcinoma stem cells. In *Teratocarcinoma Stem Cells*, eds. L. M. Silver, G. R. Martin, and S. Strickland, pp. 229–36. Cold Spring Harbor, N.Y.: Cold Spring Harbor Laboratory.

Heath, J. K. (1983) Regulation of murine embryonal carcinoma cell proliferation and differentiation. *Cancer Surveys* 2, 141–64.

Heath, J., Bell, S., and Rees, A. R. (1981) Appearance of functional insulin-receptor during the differentiation of embryonal carcinoma cells. *J. Cell Biol.* 91, 293–7.

Heath, J. K., and Isacke, C. (1983) Reciprocal control of teratocarcinoma proliferation. *Cell Biol. Int. Rep.* 7, 561–2.

–(1984) PC13 embryonal carcinoma derived growth factor. *EMBO Journal* 3, 2957–62.

356 EILEEN D. ADAMSON

Hershman, H. R. (1985) The EGF receptor. In *Control of Animal Cell Proliferation*, Vol. 1, eds. A. L. Boynton and H. L. Leffert, pp. 169–200. New York: Academic Press.

Hogan, B. L. M., Barlow, D. P., and Tilly, R. (1983) F9 teratocarcinoma cells as a model system for the differentiation of parietal and visceral endoderm in the mouse embryo. *Cancer Surveys* 2, 115–40.

Hogan, B. L. M., Cooper, A. R., and Kurkinen, M. (1980) Incorporation into Reichert's membrane of laminin-like extracellular proteins synthesized by parietal endoderm cells of the mouse embryo. *Dev. Biol.* 80, 289–300.

Hogan, B. L. M., Taylor, A., and Adamson, E. D. (1981) Cell interactions modulate embryonal carcinoma cell differentiation into parietal or visceral endoderm. *Nature* 291, 235–7.

Hogan, B. L. M., Taylor, A., and Cooper, A. R. (1982) Murine parietal endoderm cells synthesize heparan sulphate and 170 K and 145 K sulphated glycoproteins as components of Reichert's membrane. *Dev. Biol.* 90, 210–14.

Hogan, B. L. M., and Tilly, R. (1981) Cell interactions and endoderm differentiation in cultured mouse embryos. *J. Embryol. Exp. Morphol.* 62, 379–94.

Hortsch, M., Schlessinger, J., Gootwine, E., and Webb, C. (1983) Appearance of functional EGF receptor kinase during rodent embryogenesis. *EMBO Journal* 2, 1937–41.

Hoshina, M., Boothby, M., and Boime, I. (1982) Cytological localization of chorionic gonadotropin α and placental lactogen mRNAs during development of the human placenta. *J. Cell Biol.* 93, 190–8.

Jacob, F. (1977) Mouse teratocarcinoma and embryonic antigens. *Immunol. Rev.* 33, 3–32.

Jakobovits, A., Banda, M. J., and Martin, G. R. (1985) Embryonal carcinoma-derived growth factors: specific growth-promoting and differentiation-inhibiting activities. In *Cancer Cells*, Vol. 3, *Growth Factors and Transformations*, eds. J. Feramisco, B. Ozanne, and C. D. Stiles, pp. 393–9. Cold Spring Harbor, N.Y.: Cold Spring Harbor Laboratory.

Jenkinson, E. J., and Owen, V. (1980) Ontogeny and distribution of major histocompatibility complex (MHC) antigens on mouse placenta trophoblast. *J. Reprod. Immunol.* 2, 173–81.

Jonas, H. A., and Harrison, L. C. (1985) The human placenta contains two distinct binding and immunoreactive species of insulin-like growth factor-I receptors. *J. Biol. Chem.* 260, 2288–94.

Kahan, B., and Adamson, E. D. (1983) A teratocarcinoma-derived bipotential cell line with primitive endoderm properties. In *Teratocarcinoma Stem Cells*, eds. L. M. Silver, G. R. Martin, and S. Strickland, pp. 131–41. Cold Spring Harbor, N.Y.: Cold Spring Harbor Laboratory.

Karin, M. (1985) Metallothioneins: proteins in search of function. *Cell* 41, 9–10.

Karin, M., and Mintz, B. (1981) Receptor mediated endocytosis of transferrin in developmentally totipotent mouse teratocarcinoma stem cells. *J. Biol. Chem.* 256, 3245–52.

Kemler, R., Brûlet, P., and Jacob, F. (1981) Monoclonal antibodies as a tool for the study of embryonic development. In *The Immune System*, Vol. 1, eds. C. M. Steinberg and I. Lefkovits, pp. 102–9. Basel: S. Karger.

Kirkwood, K. J., and Billington, W. D. (1981) Expression of serologically detectable H-2 antigens on mid-gestation mouse embryonic tissues. *J. Embryol. Exp. Morphol.* 61, 207–19.

Kruijer, W., Cooper, J. A., Hunter, T., and Verma, I. M. (1984) PDGF induces rapid but transient expression of the c-*fos* gene. *Nature* 312, 711–16.

Krumlauf, R., Hammer, R. E., Tilghman, S. M., and Brinster, R. L. (1985) Developmental regulation of α-fetoprotein genes in transgenic mice. *Mol. Cell. Biol.* 5, 1639–48.

Kurkinen, M., Barlow, D. P., Helfman, D. M., Williams, J. G., and Hogan, B. L. M. (1983) Isolation of cDNA clones for basal lamina components: type IV procollagen. *Nucleic Acids Res.* 11, 6199–209.

Lane, E. B., Hogan, B. L. M., Kurkinen, M., and Garrels, J. I. (1983) Co-expression of vimentin and cytokeratins in cells of the early mouse embryo. *Nature* 303, 701–4.

Lawrence, D. A., Pircher, R., Kryceve-Martinerie, C., and Jullien, P. (1984) Normal embryo fibroblasts release transforming growth factors in a latent form. *J. Cell. Physiol.* 121, 184–8.

Lee, D. C., Rose, T. M., Webb, N. R., and Todaro, G. J. (1985) Cloning and sequence analysis of a cDNA for rat transforming growth factor-α. *Nature* 313, 489–92.

Lehtonen, E., Lehto, V.-P., Passivuo, R., and Virtanen, I. (1983) Parietal and visceral endoderm differ in their expression of intermediate filaments. *EMBO Journal* 2, 1023–8.

Leivo, I., Vaheri, A., Timpl, R., and Wartiovaara, J. (1980) Appearance and distribution of collagens and laminin in the early mouse embryo. *Dev. Biol.* 76, 100–14.

Levi-Montalcini, R., and Booker, B. (1960) Destruction of the sympathetic ganglia in mammals by an antiserum to the nerve growth protein. *Proc. Natl. Acad. Sci. U.S.A.* 46, 384–91.

Levin, M. J., Tuil, D., Uzan, G., Dreyfus, J.-C., and Kahn, A. (1984) Expression of the transferrin gene during development of non-hepatic tissues: high levels of transferrin mRNA in fetal muscle and adult brain. *Biochem. Biophys. Res. Commun.* 122, 212–17.

Levine, R. A., LaRosa, G. J., and Gudas, L. J. (1984) Isolation of cDNA clones for genes exhibiting reduced expression after differentiation of murine teratocarcinoma stem cells. *Mol. Cell. Biol.* 4, 2142–50.

Levy, B. T., Sorge, L. K., Meymandi, A., and Maness, P. F. (1984) pp60$^{c\text{-}src}$ kinase is in chick and human embryonic tissues. *Dev. Biol.* 104, 9–17.

Li, C. H., ed. (1984) *Hormonal Proteins and Peptides. Growth Factors*, Vol. 12. New York: Academic Press.

Liau, G., Yamada, Y., and deCrombrugghe, B. (1985) Coordinate regulation of the levels of type III and type I collagen mRNA in most but not all mouse fibroblasts. *J. Biol. Chem.* 260, 531–6.

Lin, C. R., Chen, W. S., Kruijer, W., Stolarsky, L. S. Weber, W., Evans, R. M., Verma, I. M., Gill, G. N., and Rosenfeld, M. G. (1984) Expression cloning of human EGF receptor complementary DNA: gene amplification and three related messenger RNA products in A431 cells. *Science* 224, 843–8.

Linzer, D. I. H., Lee, S. J., Ogren, L., Talamantes, F., and Nathans, D. (1985) Identity of proliferin mRNA and protein in mouse placenta. *Proc. Natl. Acad. Sci. U.S.A.* 82, 4356–9.

Linzer, D. I. H., and Nathans, D. (1985) A new member of the prolactin-growth hormone gene family expressed in mouse placenta. *EMBO Journal* 4, 1419–23.

McBurney, M. W., Jones-Villeneuve, E. M. V., Edwards, M. K. S., and Anderson, P. J. (1982) Control of muscle and normal differentiation in a cultured embryonal carcinoma cell line. *Nature* 229, 165–7.

Mackiewicz, A., and Breborowicz, J. (1980) The *in vitro* production of alpha-fetoprotein variants by human fetal organs. *Onco. Dev. Biol. Med.* 1, 251–61.

Manes, C., and Menzel, P. (1981) Demethylation of CpG sites in DNA of early rabbit trophoblast. *Nature* 293, 589–90.

Markelonis, G. J., Bradshaw, R. A., Oh,T. H., Johnson, J. L., and Bates, O. J. (1982) Sciatin is a transferrin-like polypeptide. *J. Neurochem.* 39, 315–20.

Marotti, K. R., Brown, G. D., and Strickland, S. (1985) Two-stage hormonal control of type IV collagen mRNA levels during differentiation of F9 teratocarcinoma cells. *Dev. Biol.* 108, 26–31.

Marquardt, H., Hunkapillar, M. W., Hood, L. E., and Todaro, G. J. (1984) Rat transforming factor type I: structure and relation to epidermal growth factor. *Science* 223, 1079–81.

Martin, G. R. (1975) Teratocarcinomas as a model system for the study of embryogenesis and neoplasia. *Cell* 5, 229–43.

–(1978) Advantages and limitations of teratocarcinoma stem cells as models of development. In *Development in Mammals*, Vol. 3, ed. M. H. Johnson, pp. 225–65. Amsterdam: Elsevier/North Holland.

–(1980) Teratocarcinomas and mammalian embryogenesis. *Science* 208, 768–76.

Massagué, J. (1985) The transforming growth factors. *Trends Biochem. Sci.* 10, 237–41.

Matrisian, L. M., Pathak, M., and Magun, B. E. (1982) Identification of an epidermal growth factor-related transforming growth factor from rat fetuses. *Biochem. Biophys. Res. Commun.* 104, 761–9.

Matsuda, R., Spector, D., and Strohman, R. C. (1984) There is selective accumulation of a growth factor in chicken skeletal muscle. *Dev. Biol.* 103, 267–75.

Meehan, R. R., Barlow, D. P., Hill, R. E., Hogan, B. L. M., and Hastie, N. D. (1984) Pattern of serum protein gene expression in mouse visceral yolk sac and foetal liver. *EMBO Journal* 3, 1881–5.

Meek, J., and Adamson, E. D. (1985) Transferrin in fetal and adult mouse tissues: synthesis, storage, and secretion. *J. Embryol. Exp. Morphol.* 86, 205–18.

Merimee, T. J., Grant, M., and Tyson, J. E. (1984) Insulin-like growth factors in amniotic fluid. *J. Clin. Endocrinol. Metab.* 59, 752–67.

Mitchell, R. L., Zokas, L., Schreiber, R. D., and Verma, I. M. (1985) Rapid induction of the expression of proto-oncogene *fos* during human monocytic differentiation. *Cell* 20, 209–17.

Møll, R., Franke, W. W., Schiller, D. L., Geiger, B., and Krepler, R. (1982) The catalog of human cytokeratins: patterns of expression in normal epithelia, tumors, and cultured cells. *Cell* 31, 11−24.

Morange, M., Diu, A., Bensaude, O., and Babinet, C. (1984) Altered expression of heat shock proteins in embryonal carcinoma and mouse early embryonic cells. *Mol. Cell. Biol.* 4, 730−5.

Morgan, C. J., and Bradshaw, R. A. (1985) Production of a monoclonal antibody directed against nerve growth factor receptor from sympathetic membranes. *J. Cell Biochem.* 27, 121−32.

Müller, R. (1983) Differential expression of cellular oncogenes during murine development and in teratocarcinoma cell lines. In *Teratocarcinoma Stem Cells*, eds. L. M. Silver, G. R. Martin, and S. Strickland, pp. 451−68. Cold Spring Harbor, N.Y.: Cold Spring Harbor Laboratory.

Müller, R., Bravo, R., Burckhardt, J., and Curran, T. (1984) Induction of c-*fos* gene and protein by growth factors precedes activation of c-*myc*. *Nature* 312, 716−20.

Müller, R., Slamon, D. J., Adamson, E. D., Tremblay, J. M., Müller, D., Cline, M. J., and Verma, I. M. (1983a) Transcription of cellular oncogenes c-*ras*^{Ki} and c-*fms* during mouse development. *Mol. Cell. Biol.* 3, 1062−9.

Müller, R., Tremblay, J. M., Adamson, E. D., and Verma, I. M. (1983b) Tissue and cell type-specific expression of two human c-*onc* genes. *Nature* 304, 454−6.

Müller, R., and Verma, I. M. (1984) Expression of cellular oncogenes. *Curr. Top. Microbiol.* 112, 73−115.

Müller, R., Verma, I. M., and Adamson, E. D. (1983c) Expression of c-*onc* genes: c-*fos* transcripts accumulate to high levels during development of mouse placenta, yolk sac, and amnion. *EMBO Journal* 2, 679−84.

Müller, R., and Wagner, E. (1984) Differentiation of F9 teratocarcinoma stem cells after transfer of c-*fos* proto-oncogenes. *Nature* 311, 438−42.

Mulnard, J., and Huygens, R. (1978) Ultrastructural localization of nonspecific alkaline phosphatase during cleavage and blastocyst formation in the mouse. *J. Embryol. Exp. Morphol.* 44, 121−31.

Nagarajan, L., and Anderson, W. B. (1982) Insulin promotes the growth of F9 embryonal carcinoma cells apparently by acting through its own receptor. *Biochim. Biophys. Res. Commun.* 106, 974−80.

Nahon, J. -L., Gal, A., Erdos, T., and Sala-Trepat, J. M. (1984) Differential DNase I sensitivity of the albumin and α-fetoprotein genes in chromatin from rat tissues and cell lines. *Proc. Natl. Acad. Sci. U.S.A.* 81, 5031−5.

Nexó, E., Hollenberg, M. D., Figueroa, A., and Pratt, R. M. (1980) Detection of epidermal growth factor-urogastrone and its receptors during mouse fetal development. *Proc. Natl. Acad. Sci. U.S.A.* 77, 2782−5.

Nister, M., Heldin, C. -H., Wasteson, A., and Westermark, B. (1984) A glioma-derived analog to platelet-derived growth factor: demonstration of receptor-competing activity and immunological cross-reactivity. *Proc. Natl. Acad. Sci. U.S.A.* 81, 926−30.

Oldberg, A., Linney, E., and Ruoslahti, E. (1983) Molecular cloning and nucleotide sequence of a cDNA clone coding for the cell attachment domain in human fibronectin. *J. Biol. Chem.* 258, 10193−6.

Omary, M. B., and Trowbridge, I. S. (1981) Biosynthesis of the human transferrin receptor in cultured cells. *J. Biol. Chem.* 256, 12988–92.

Opperman, H., Levinson, W., and Bishop, J. M. (1981) A cellular protein that associates with the transforming protein of Rous sarcoma virus is also a heat shock protein. *Proc. Natl. Acad. Sci. U.S.A.* 78, 1067–71.

Oshima, R. G. (1981) Identification and immunoprecipitation of cytoskeletal proteins from murine extra-embryonic endodermal cells. *J. Biol. Chem.* 256, 8124–33.

–(1982) Developmental expression of murine extra-embryonic endodermal cytoskeletal proteins. *J. Biol. Chem.* 257, 3414–21.

Oshima, R. G., Howe, W. E., Klier, F. G., Adamson, E. D., and Sheinsky, L. H. (1983) Intermediate filament protein synthesis in preimplantation murine embryos. *Dev. Biol.* 99, 447–55.

Ott, M.-O., Sperling, L., Cassio, D., Levilliers, J., Sala-Trepat, J., and Weiss, M. C. (1982) Undermethylation at the 5' end of the albumin gene is necessary but not sufficient for albumin production by rat hepatoma cells in culture. *Cell* 30, 825–33.

Ouellette, A. J. (1982) Metallothionein mRNA expression in fetal mouse organs. *Dev. Biol.* 92, 240–6.

Perdue, J. F. (1984) Chemistry, structure, and function of insulin-like growth factors and their receptors: a review. *Can. J. Biochem. Cell. Biol.* 62, 1237–45.

Peters, T. (1975) Serum albumin. In *The Plasma Proteins*, Vol. 1, 2nd ed., ed. F. W. Putnam, pp. 133–81. New York: Academic Press.

Pfeifer-Ohlsson, S., Goustin, A. S., Rydnert, J., Wahlstrom, T., Bjersing, L., Stehelin, D., and Ohlsson, R. (1984) Spatial and temporal pattern of cellular *myc* oncogene expression in developing human placenta: implications for embryonic cell proliferation. *Cell* 38, 585–96.

Pfeifer-Ohlsson, S., Rydnert, J., Goustin, A. S., Larsson, E., Betsholtz, C., and Ohlsson, R. (1985) Cell-type specific pattern of *myc* proto-oncogene expression in developing human embryos. *Proc. Natl. Acad. Sci. U.S.A.* 82, 5050–4.

Piletz, J. E., Anderson, R. D., Birren, B. W., and Herschman, H. R. (1983) Metallothionein synthesis in foetal, neonatal, and maternal rat liver. *Eur. J. Biochem.* 131, 489–95.

Proper, J. A., Bjornson, C. L., and Moses, H. L. (1982) Mouse embryos contain polypeptide growth factor(s) capable of inducing a reversible neoplastic phenotype in nontransformed cells in culture. *J. Cell. Physiol.* 110, 169–74.

Propst, F., and Vande Woude, G. F. (1985) Expression of c-*mos* proto-oncogene transcripts in mouse tissues. *Nature* 315, 516–18.

Rees, A. R., Adamson, E. D., and Graham, C. F. (1979) Epidermal growth factor receptors increase during the differentiation of embryonal carcinoma cells. *Nature* 281, 309–11.

Renfree, M. B., and McLaren, A. (1974) Foetal origin of transferrin in mouse amniotic fluid. *Nature* 252, 159–60.

Rheinwald, T. G., O'Connell, T. M., Connell, N. D., Rybak, S. M., Allen-Hoffmann, B. L., LaRocca, P. J., Wu, Y. -J., and Rehwoldt, S. M. (1984) Expression of specific keratin subsets and vimentin in normal human epithelial cells: a function of cell type and conditions of growth during serial culture. In *Cancer Cells, The Transformed Phenotype,* Vol. 1, eds. A. J. Levine, G. F. Vander Woude, W. C. Topp, and J. D. Watson, pp. 217–28. Cold Spring Harbor, N.Y.: Cold Spring Harbor Laboratory.

Richman, R. A., Denedret, M. R., Florini, J. R., and Toly, B. A. (1985) Hormonal regulation of somatomedin secretion by fetal rat hepatocyte in primary culture. *Endocrinology* 116, 180–8.

Rizzino, A., and Bowen-Pope, D. F. (1985) Production of PDGF-like growth factors by embryonal carcinoma cells and response to PDGF by endoderm-like cells. *Dev. Biol.* 110, 15–22.

Ruoslahti, E., and Seppala, M. (1979) α-Fetoprotein in cancer and fetal development. *Adv. Cancer Res.* 29, 275–346.

Rüther, U., Wagner, E. F., and Müller, R. (1985) Analysis of the differentiation-promoting potential of inducible c-*fos* gene introduced into embryonal carcinoma cells. *EMBO Journal* 4, 1775–81.

Schnieke, A., Harbers, K., and Jaenisch, R. (1983) Embryonic lethal mutation in mice induced by retrovirus insertion into the α₁(I) collagen gene. *Nature* 304, 315–20.

Schultz, G. A., and Tucker, E. B. (1977) Protein synthesis and gene expression in preimplantation rabbit embryo. In *Development in Mammals,* Vol. 1, ed. M. H. Johnson, pp. 69–97. Amsterdam: North Holland.

Schwartz, J. P., and Costa, E. (1978) Regulation of nerve growth factor content in a neuroblastoma cell line. *Neuroscience* 3, 473–80.

Scott, R. W., and Tilghman, S. M. (1983) Transient expression of a mouse α-fetoprotein minigene: deletion analyses of promoter function. *Mol. Cell. Biol.* 3, 1295–309.

Scott, R. W., Vogt, T. F., Croke, M. E., and Tilghman, S. M. (1984) Tissue specific activation of a cloned α-fetoprotein gene during differentiation of a transfected embryonal carcinoma cell line. *Nature* 310, 562–7.

Searle, R. F., Sellens, M. H., Elson, J., Jenkinson, E. J., and Billington, W. D. (1976) Detection of alloantigen during preimplantation development and early trophoblast differentiation in the mouse by immunoperoxidase labeling. *J. Exp. Med.* 143, 348–59.

Sefton, B. M., and Hunter, T. (1984) Tyrosine protein kinases. In *Advances in Cyclic Nucleotide and Protein Phosphorylation Research,* eds. P. Greengard and G. A. Robison, pp. 195–226. New York: Raven Press.

Seifert, R. A., Schwartz, S. M., and Bowen-Pope, D. F. (1984) Developmentally regulated production of platelet-derived growth factor-like molecules. *Nature* 311, 669–71.

Sellem, C. H., Frain, M., Erdos, T., and Sala-Trepat, J. M. (1984) Differential expression of albumin and α-fetoprotein genes in fetal tissues of mouse and rat. *Dev. Biol.* 102, 51–60.

Semoff, S., Hogan, B. L. M., and Hopkins, C. R. (1982) Localisation of fibronectin, laminin-entactin and entactin in Reichert's membrane by immuno-electron microscopy. *EMBO Journal* 1, 1171–5.

Setoyama, C., Liau, G., and deCrombrugghe, B. (1985) Pleiotropic mutants of NIH 3T3 cells with altered regulation in the expression of both type I collagen and fibronectin. *Cell* 41, 201–9.

Sherman, M. I., Gay, R., Gay, S., and Miller, E. J. (1980) Association of collagen with preimplantation and peri-implantation mouse embryos. *Dev. Biol.* 74, 470–8.

Sherr, C. J., Rettenmier, C. W., Sacca, R., Roussel, M. F., Look, A. T., and Stanley, E. R. (1985) The c-*fms* proto-oncogene product is related to the receptor for the mononuclear phagocyte growth factor, CSF-1. *Cell* 41, 665–76.

Shi, W.-K., and Heath, J. K. (1984) Apolipoprotein expression by murine visceral yolk sac endoderm. *J. Embryol. Exp. Morphol.* 81, 143–52.

Silver, L. M., Martin, G. R., and Strickland, S., eds. (1983) *Teratocarcinoma Stem Cells*, Cold Spring Harbor, N.Y.: Cold Spring Harbor Laboratory.

Singer, P.A., Trevor, K., and Oshima, R. G. (1986) Molecular cloning and characterization of the Endo B cytokeratin expressed in preimplantation mouse embryos. *J. Biol. Chem.* 261, 538–47.

Slamon, D. J., and Cline, M. J. (1984) Expression of cellular oncogenes during embryonic and fetal development of the mouse. *Proc. Natl. Acad. Sci. U.S.A.* 81, 7141–5.

Soares, M. J., Julian, J. A., and Glasser, S. R. (1985) Trophoblast giant cell release of placental lactogens: temporal and regional characteristics. *Dev. Biol.* 107, 520–6.

Stern, P., and Priddle, J. D. (1984) Preliminary characterization of a murine embryonal carcinoma cell-derived growth promoting activity. *Cell Biol. Int. Rep.* 8, 579–85.

Strickland, S. (1981) Mouse teratocarcinoma cells: prospects for the study of embryogenesis and neoplasia. *Cell* 24, 277–8.

Stromberg, K., Pigott, D. A., Ranchalis, J. E., and Twardzik, D. R. (1982) Human term placenta contains transforming growth factors. *Biochem. Biophys. Res. Commun.* 106, 354–64.

Sun, T. T., Eichner, W. G., Nelson, S. C. G., Tseng, R. A., Weiss, M., Jarvinen, M., and Woodcock-Mitchell, J. (1983) Keratin classes: molecular markers for different types of epithelial differentiation. *J. Invest. Dermatol.* 81, 109S–14S.

Tabarelli, M., Kofler, R., and Wick, G. (1983) Placental hormones. I. Immunofluorescence studies of the localization of chorionic gonadotropin, placental lactogen and prolactin in human and rat placenta and in the endometrium of pregnant rats. *Placenta* 4, 379–88.

Tabor, J. M., and Oshima, R. G. (1982) Identification of mRNA species that code for extra-embryonic endodermal cytoskeletal proteins in differentiated derivatives of murine embryonal carcinoma cells. *J. Biol. Chem.* 257, 8771–4.

Talamantes, F., Ogren, L., Markoff, E., Woodard, S., and Madrid, L. (1980) Phylogenetic distribution, regulation of secretion and prolactin-like effects of placental lactogen. *Fed. Proc.* 39, 2582–7.

Tam, J. P. (1985) Physiological effect of transforming growth factor in the new born mouse. *Science* 229, 633–5.

Tam, J. P., Marquardt, H., Rosberger, D. F., Wong, T. W., and Todaro, G. J. (1984) Synthesis of biologically active rat transforming growth factor α. *Nature* 309, 376–8.

Templeton, D. M., Banerjee, D., and Cherian, M. G. (1985) Metallothionein synthesis and location in relation to metal storage in rat liver during gestation. *Can. J. Biochem. Cell Biol.* 63, 16–22.

Tilghman, S. M., and Belayew, A. (1982) Transcriptional control of the murine albumin/α-fetoprotein locus during development. *Proc. Natl. Acad. Sci. U.S.A.* 79, 5254–7.

Timpl, R., Rohde, H., Gehron-Robey, P., Rennard, S. I., Foidart, J. M., and Martin, G. R. (1979) Laminin–a glycoprotein from basement membranes. *J. Biol. Chem.* 254, 9933–7.

Trelstad, R. L., ed. (1984) *The Role of Extracellular Matrix in Development.* New York: Alan R. Liss.

Trevor, K., and Oshima, R. G. (1985) Preimplantation mouse embryos and liver express the same type I keratin gene product. *J. Biol. Chem.* 260, 15885–91.

Trowbridge, I. S., and Omary, M. B. (1981) Human cell surface glycoprotein related to cell proliferation is the receptor for transferrin. *Proc. Natl. Acad. Sci. U.S.A.* 78, 3039–43.

Twardzik, D. R. (1985) Differential expression of transforming growth factor-α during prenatal development of the mouse. *Cancer Res.* 45, 5413–16.

Twardzik, D. R., Ranchalis, J. E., and Todaro, G. J. (1982) Mouse embryonic transforming growth factors related to those isolated from tumor cells. *Cancer Res.* 42, 590–3.

Ullrich, A., Bell, J. R., Chen, E. Y., Herrara, R., Petruzelli, L. M., Dull, T. J., Gray, A., Coussens, L., Liao, Y. -C. Tsubokawa, M., Mason, A., Seeburg, P. H., Grunfeld, C., Rosen, O. M., and Ramachandran, J. (1985) Human insulin receptor and its relationship to the tyrosine kinase family of oncogenes. *Nature* 313, 756–61.

Ullrich, A., Coussens, L., Hayflick, J. S., Dull, T. J., Gray, A., Tam, A. W., Lee, J., Yarden, Y., Liberman, T. A., Schlessinger, J., Downward, J., Mayes, E. L. V., Whittle, N., Waterfield, M. D., and Seeburg, P. H. (1984) Human epidermal growth factor receptor cDNA sequence and aberrant expression of the amplified gene in A431 epidermoid carcinoma cells. *Nature* 309, 418–25.

Uzan, G., Frain, M., Park, L., Desmond, C., Maessens, G., Sala-Trepat, J., Zakin, M. W., and Kahn, A. (1984) Cloning and sequence analysis of cDNA for human transferrin. *Biochem. Biophys. Res. Commun.* 119, 273–81.

Varmus, H. E. (1984) The molecular genetics of cellular oncogenes. *Annu. Rev. Genet.* 18, 553–612.

Vedel, M., Gomez-Garcia, M., Sala, M., and Sala-Trepat, J. M. (1983) Changes in methylation pattern of albumin and α-fetoprotein in developing rat liver and neoplasia. *Nucleic Acids Res.* 11, 4335–54.

Wang, S. -Y. and Gudas, L. J. (1983) Isolation of cDNA clones specific for collagen IV and laminin from mouse teratocarcinoma cells. *Proc. Natl. Acad. Sci. U.S.A.* 80, 5880–4.

Wang, S. -Y., La Rosa, G. J., and Gudas, L. J. (1985) Molecular cloning of gene sequences transcriptionally regulated by retinoic acid and dibutyryl cyclic AMP in cultured mouse teratocarcinoma cells. *Dev. Biol.* 107, 75–86.

Wartiovaara, J., Leivo, I., and Vaheri, A. (1979) Expression of the cell surface-associated glycoprotein, fibronectin, in the early mouse embryo. *Dev. Biol.* 69, 247–57.

Wartiovaara, J., Leivo, I., Virtanen, I., Vaheri, A., and Graham, C. F. (1978) Cell surface and extracellular matrix glycoprotein fibronectin: expression in embryogenesis and in teratocarcinoma differentiation. *Ann. N.Y. Acad. Sci.* 312, 132–41.

Waterfield, M. D., Scrace, G. T., Whittle, N., Stroobant, P., Johnsson, A., Wasteson, A., Westermark, B., Heldin, C. -H., Huang, J. S., and Devel, T. F. (1983) Platelet-derived growth factor is structurally related to the putative transforming protein p28^{sis} of simian sarcoma virus. *Nature* 304, 35–9.

Wewer, U. (1982) Characterization of a rat yolk sac carcinoma cell line. *Dev. Biol.* 93, 416–21.

Wiley, H. S., and Kaplan, J. (1984) EGF rapidly induces a redistribution of transferrin-receptor pools in human fibroblasts. *Proc. Natl. Acad. Sci. U.S.A.* 81, 7456–60.

Wolfe, J., Mautner, V., Hogan, B., and Tilly, R. (1978) Synthesis and retention of fibronectin (LETS protein) by mouse teratocarcinoma cells. *Exp. Cell. Res.* 118, 63–71.

Wu, T.-C., Wan, Y. -J., Chung, A. E., and Damjanov, I. (1983) Immunohistochemical localization of entactin and laminin in mouse embryos and fetuses. *Dev. Biol.* 100, 496–505.

Yang, Y. W. -H., Rechler, M. M., Nissley, S. P., and Coligan, J. E. (1985a) Biosynthesis of rat insulin-like growth factor II: II. Localization of mature rat insulin-like growth factor II (7434 daltons) to the amino terminus of the 20 kilodalton biosynthetic precursor by radiosequence analysis. *J. Biol. Chem.* 260, 2578–82.

Yang, Y. W., Romanus, Y. A., Liu, T.-Y., Nissley, S. P., and Rechler, M. M. (1985b) Biosynthesis of rat insulin-like growth factor II: I. Immunochemical demonstration of a 20 kilodalton biosynthetic precursor of rat insulin-like growth factor II in metabolically labeled BRL-3A rat liver cells. *J. Biol. Chem.* 260, 2570–77.

Young, P.R., and Tilghman, S. M. (1984) The induction of α-fetoprotein synthesis in differentiating F9 teratocarcinoma cells is accompanied by a genome-wide loss of DNA methylation. *Mol. Cell. Biol.* 4, 898–907.

12   X-chromosome regulation in oogenesis and early mammalian development

VERNE M. CHAPMAN

Contents

365

I. Introduction

The X chromosome in placental mammals is subject to a unique system of developmental regulation. This regulation involves coordinate activation and inactivation of the entire chromosome during the course of female development. The inactivation event occurs during early embryogenesis, and, once established, the inactive condition is somatically heritable in a cell lineage. The inactivation event in the embryonic lineages typically occurs at random with respect to the parental origin of X chromosomes such that the developing embryo and resulting adult is a mosaic with respect to X-chromosome expression. The primordial germ cells of the developing fetus appear to be similar to somatic cells with respect to single active X expression, and the mature X is reactivated in oogonial stages prior to the onset of meiotic prophase.

The molecular mechanisms involved in the onset of inactivation and the maintenance of inactive X chromosomes are not well established, but there is an increasing amount of evidence that DNA modification, probably in the form of cytosine methylation, is involved. Moreover, there is ample evidence to suggest that DNA methylation is also associated with changes in chromatin structure and that the critical methylation changes in the inactive X chromosome occur in gene promoter regions and other regulatory sites (Wolf and Migeon 1985).

This review is primarily concerned with the elements of X-chromosome regulation that take place in the early embryo. The issues of primary interest include (1) the reactivation of X chromosomes during oogenesis, (2) the expression of X chromosomes during early embryogenesis, (3) the basis for preferential expression of the maternal or oocyte-derived X chromosome (X^M) in trophectoderm and yolk sac endoderm lineages, and (4) the timing of the onset of random inactivation in the embryonic lineages. These elements of X-chromosome regulation occur during early embryonic stages, and there is evidence to indicate that preferential expression in extraembryonic lineages is a consequence of X-chromosome imprinting during gametogenesis. Thus, it is important to include an overview of X-chromosome regulation and expression during the process of oogenesis and establish the relationships between these events and the expression of X chromosomes in early cleavage embryos.

In the final portion of this review, I shall try to develop some general models of X-chromosome regulation that apply to the reactivation and inactivation processes. My primary interest in this section will be to extend earlier models by several workers in this area to account for more recent data. More important, I hope that this will serve as a focus for future experimental work that will help clarify the molecular mechanisms that underlie X-chromosome regulation during development.

The topic of X-chromosome regulation has received considerable attention, and the reader is invited to examine reviews by West (1982), Gartler and Riggs (1983), Gartler and Cole (1981), Lyon (1974), Russell (1978), and Monk and Harper (1978), where the reader will find additional data and perspectives on X-chromosome regulation that may not be included in this chapter.

II. X-chromosome expression in oocytes

Several studies indicate that both X chromosomes are functional during oogenesis. Most of the evidence comes from the phenotypic expression of ubiquitously expressed enzymes associated with metabolism, such as glucose-6-phosphate dehydrogenase (G6PD), phosphoglycerate kinase (PGK), and hypoxanthine phosphoribosyl transferase (HPRT). Indirect evidence that both X chromosomes are expressed comes from comparisons of HPRT, G6PD, and PGK activity levels in oocytes from X/X versus X/O females (Epstein 1969, 1972). Each of these enzymes shows the expected twofold difference in specific activity between two-X and one-X oocytes. More direct evidence suggests that both X chromosomes are expressed in the same oocyte. These conclusions come from studies of oocytes from female fetuses that are heterozygous for allelic variants of X-chromosome genes, particularly those gene products such as G6PD and HPRT that are dimeric structures as active enzymes. Oocytes from fetuses heterozygous for G6PD and HPRT show heteropolymer charge forms that are not present in mosaic diploid somatic tissues (Figure 12-1A). Heteropolymer charge forms have been observed in G6PD heterozygotes in humans (Gartler, Andina, and Gant 1975; Migeon and Jelalian 1977) and in mice (Kratzer and Chapman 1981), whereas HPRT has been examined only in mice (Figure 12-1B). Collectively, these data demonstrate that both X chromosomes are functional in oocytes up to the time of ovulation and fertilization. The two-X-active condition of oocytes could have arisen either from a reactivation event during the course of germ cell differentiation or from failure of cells of the germ line to undergo inactivation during embryogenesis.

Experimentally, it is difficult to demonstrate that 100 percent of the embryonic cells have undergone inactivation and carry a single active X chromosome. However, it is possible to examine the functional expression of X chromosomes in oogonia at the earliest stages of germ cell migration into the ovary and ask whether or not the heterodimer form of G6PD is expressed in a manner similar to that of mature oocytes.

The initial work was done in human fetal material that was heterozygous for G6PD electrophoretic-variant alleles (Migeon and Jelalian 1977). The age of the fetus and the stage of the germ cells present were estimated from fetus size. Migeon and Jelalian (1977) found one G6PD A/B fetus

368 VERNE M. CHAPMAN

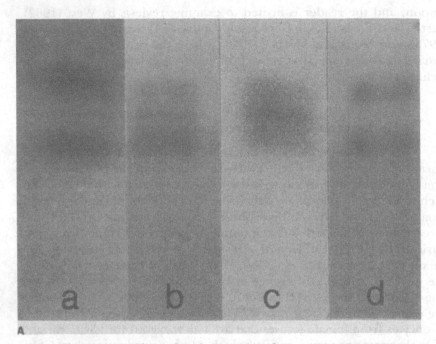

Figure 12.1A. Electrophoretic expression of *M. caroli* G6PD A/B heterozygous germ cells. Lanes a, b, and c are germ cells from ovaries of fetuses at days 11, 12, and 14, respectively; lane d is ovulated oocytes from mature females. [Reproduced from Kratzer and Chapman (1981), with permission.]

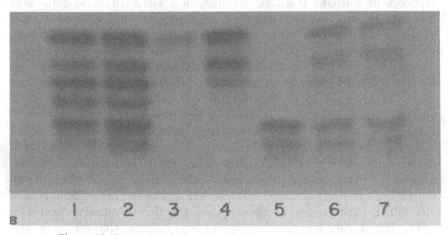

Figure 12.1B. Isoelectric-focusing phenotypes of HPRT: (1, 2) ovaries from HPRT A/B fetuses (15 days); (3) fibroblasts HPRT B; (4) brain from HPRT B; (5) brain from HPRT A; (6, 7) mixtures of HPRT A and HPRT B homogenates. See Chapman et al. (1982) for methods.

that was estimated to be eight weeks on the basis of mean foot length and crown−rump length. Germ cells would be expected to be premeiotic in such a fetus. A G6PD heteropolymer band was detectable in the ovary from this sample, but the relative activity in this band was considerably less than the activity present in the two homodimeric bands. These results suggested that two X chromosomes may be active in at least some germ cells prior to the onset of meiotic prophase. A similar study was conducted by Gartler and associates (1975), who observed that a G6PD heteropolymer was detectable at 13 weeks in germ cells from a G6PD A/B heterozygous fetus. However, these workers failed to observe a G6PD heteropolymer in germ cells recovered from a 12-week fetus, suggesting that a single X chromosome was functioning in the oogonia from this sample. These experiments rely on the electrophoretic phenotypes of G6PD detected on cellulose acetate. The method is relatively simple, but it is difficult to assess whether the differences in results between these two reports are matters of differences in limits of isozyme resolution or consequences of limited sampling. It is difficult to draw final conclusions about the state of X-chromosome expression in human oogonia and primordial germ cells from the results of two experimental observations, but both reports clearly indicate that the proportion of germ cells with two active X chromosomes is relatively small at a time when most of the germ cells are in premeiotic stages. Whether or not the positive result of Migeon and Jelalian represents a minor portion of germ cells that had initiated meiosis or X-chromosome reactivation just prior to meiotic prophase is impossible to judge.

Similar approaches to studying X-chromosome expression in the mouse have been limited to the expression of G6PD heteropolymers in *Mus caroli* germ cells (Kratzer and Chapman 1981). The study of G6PD in *M. caroli* germ cells used crosses of G6PD B females with G6PD A males. Gestational age was determined from the day of mating and by the morphological appearance of fetuses. Germ cells were collected from pooled ovaries of a litter, and the proportion of germ cells was estimated by alkaline phosphatase staining. The results are summarized in Table 12.1, which shows the percentages of germ cells present as oogonia, the change in germ cells per ovary, and a summary of G6PD heterodimer/homodimer ratios at various days of development. The data demonstrate that about 10^3 germ cells are present in the ovary as oogonia on day 10 and that there is a 10-fold increase in the number of germ cells by mitotic divisions during the next 3 days. There is also initiation of meiosis between days 10 and 11 in *M. caroli*, which is 1 or 2 days earlier than in *M. musculus*, with a substantial increase in the proportion of cells in meiotic prophase by day 12. The G6PD heterodimer is detected at day 11, which is coincident with the appearance of the first meiotic prophase, but the initial activity levels of the G6PD heterodimer in pooled oocytes are

Table 12.1. *Germ cell differentiation and G6PD heterodimer expression in* M. caroli *fetuses*

Day of development	Percentage oogonia	Number of germ cells/ovary	Dimeric ratio: G6PD A/B heterodimers G6PD A/A + B/B homodimers
10	100	1,200−1,800	0
11	88	4,000−4,500	0−1.0
12	12	10,000−12,000	1.5−2.0
13	1	10,000−15,000	2.0

Source: Data from Kratzer and Chapman (1981).

considerably less than those observed for the homodimers. However, as the proportion of germ cells in meiotic prophase increases during days 12 through 14, there is an increase in heterodimer activity to a level twice that observed for the two separate homodimers.

Female mice that carry the X/autosome translocation T(X; 16) have the intact X inactive in all somatic cells. Random inactivation probably occurs in T(X; 16), but inactivation of translocated X chromosomes would result in an unbalanced expression of autosome 16. Functional monosomy for either all or part of chromosome 16 would lead to a selection against these cells and expression of the nonmosaic X chromosome in somatic tissues. Nonmosaic X-chromosome expression in T(X; 16) has been characterized by using electrophoretic variants of X-linked phosphoglycerate kinase (*Pgk-1*) (Johnston 1981; McMahon, Fosten, and Monk 1981). These studies demonstrated that the inactive X-chromosome allele is not expressed in either somatic tissues or in germ cells present in 12.5-day fetal ovaries. However, the somatically inactive *Pgk-1*[a] allelic form is detectable in germ cells by 13.5 days and in subsequent days of development. Surprisingly, the relative expression of the *Pgk-1*[a] form does not increase progressively during this period as the proportion of germ cells entering meiosis increases. Moreover, the relative ratio of PGK A and B isozymes in adult oocytes still remains less than 1 : 1 (McMahon and Monk 1983). These results could be a consequence of somatic cell enzyme contamination of the germ cell preparations, but it is also possible that the relative activities of these two X-chromosome gene products reflect a difference in activities between previously active and inactive *Pgk-1* genes. Collectively, the biochemical evidence indicates that at least two X-chromosome genes, *G6PD* and *Hprt*, are fully active and that a third locus, *Pgk-1*, is at least partially active in oocytes. The relative ratios of homopolymers and heteropolymers for G6PD and HPRT suggest that these genes are equally functional, but whether or not this is true for all X-chromosome loci remains to be determined. The isozyme markers further indicate that a single X chromosome is expressed in oogonia and that the inactive X chromosome is reactivated either just prior to or at the time of entering meiotic prophase.

These conclusions are consistent with the results of analyzing quantitative changes of HPRT in germ cells (Monk and McLaren 1981). A comparison of HPRT activities in XX, XO, and XY germ cells of mice on gestational days 11.5 through 15.5 showed substantial increases in HPRT activity in female germ cells during this period. Moreover, the relative activities of HPRT in germ cells of XX females compared with XO females were roughly equivalent in oogonia, but changed as they entered meiosis. By 13.5 days, XX female germ cells had twice the activity found in XO germ cells. Similar dosage differences in HPRT activities between XX and XO females have been reported by other workers (Mangia, Abbo-Halbasch, and Epstein 1975; Andina 1978).

Independent evidence for the single active X chromosome in primordial germ cells prior to meiosis comes from cytogenetic studies that use a variety of techniques to demonstrate that the inactive X chromosome is different from the active X chromosome. These techniques have been well described in a recent review by West (1982), and they will be only summarized here. In brief, the inactive X chromosome can be cytologically detected by differential staining of condensed heterochromatin in interphase, by differential staining after hypotonic potassium chloride treatment of metaphase chromosomes (Kanda 1973), and by asynchronous replication of inactive X chromosomes. The latter can be detected by differential incorporation of a pulse of either [^{3}H]-thymidine or 5-bromodeoxyuridine (Takagi 1974; Mukherjee 1976). Early studies of oocytes suggested that these cells did not have a condensed chromosome and that both X chromosomes were active. More recently, these techniques have been applied to the study of X chromosomes in oogonia by Gartler, Rivest, and Cole (1980). They used the hypotonic potassium chloride technique to demonstrate that mouse oogonia from fetal gonads at days 12 and 13 have a differentially staining and presumably inactive X chromosome. Moreover, they foud that the frequency of differential X-chromosome staining found in oogonia did not differ from the frequencies of differentially staining X chromosomes found in somatic cells of female mouse embryos.

Overall, three sources of data, cytogenetic, dosage of X-chromosome gene products, and allozyme expression, all indicate that both X chromosomes are functional in oocytes and that the premeiotic germ cells have a single active X chromosome. Thus, the evidence is consistent with reactivation of an inactive X chromosome at about the time germ cells enter meiotic prophase. None of the available techniques is sufficiently sensitive to pinpoint the cell-cycle events correlated with reactivation, but the results with G6PD expression in *M. caroli* germ cells, as well as those from other studies, demonstrate that the functional expression of both X chromosomes increases in concert with the recruitment of oogonia into meiotic prophase, which occurs over a two-day period in the mouse. These findings suggest that the reactivation is closely tied to other features of germ cell differentiation.

The occurrence of X-chromosome reactivation in somatic tissues is relatively uncommon. In vivo, functional expression of heteropolymers for G6PD and HPRT in tissues of heterozygous females has not been reported. Our laboratory has screened tissues from several hundred heterozygous females in the course of X-chromosome genetic analysis. None of these has shown clear evidence of heteropolymers (Chapman, Kratzer, and Quarantillo 1983). Reactivation of X chromosomes has also been extensively studied in cells grown in culture by several laboratories (Migeon 1972; Kahan and DeMars 1975). These studies typically demonstrate that the inactive X chromosome does not spontaneously reactivate at frequencies greater than 1×10^{-8} or 1×10^{-7}. Reactivation of some X-chromosome genes can be achieved by treating hybrid cells (in which one of the parent cells was a female cell carrying an HPRT mutant on the active X) with the drug 5-azacytidine (5-azaC) and selecting for HPRT expression. The 5-azaC treatment will induce a high frequency of reactivation of HPRT on the inactive X chromosome that can rescue cells in the selective medium containing hypoxanthine, aminopterin, and thymidine (HAT). Analyses of these reactivated HPRT$^+$ cells indicate that other nonselected X-chromosome genes may be reactivated as well, but not all cells reactivate the same nonselected loci (Mohandas, Sparkes, and Shapiro 1981; Graves 1982; Lester, Korn, and DeMars 1982). These results indicate that drug-induced reactivation can occur for inactive X-chromosome genes but that the reactivation may be localized to either one gene or a few marker loci.

In contrast to drug-induced gene reactivation, the X-chromosome reactivation process in oogenesis is apparently global, which leads to several questions about germ cell reactivation and the mechanisms that are responsible for achieving it. Although biochemical and cytogenetic data indicate that oogonia and presumably primordial germ cells have an inactive X chromosome, the data do not exclude the possibility that the chromatin structure of their inactive X chromosome and/or the state of DNA methylation of inactive X-chromosome genes in germ cells are fundamentally different from those of the inactive X chromosomes of other somatic tissues, thus allowing reactivation of the entire X chromosome during meiotic prophase. It also is not known whether X-chromosome reactivation is triggered by extrinsic factors present in the differentiating gonad or whether the reactivation is an active process maintained by gene products synthesized in oocytes. Developing experimental approaches to the activation process will require some knowledge of the state of the inactivated X chromosome in primordial germ cells.

III. X-chromosome expression in preimplantation embryos

Both X chromosomes remain functional in oocytes up to the time of ovulation, which is followed immediately by fertilization. The mammalian embryo undergoes the initial cleavages relatively slowly, and there is ample evidence from a variety of sources that the embryonic genome is transcriptionally active by the 2-cell stage (see Chapter 8) and that it is responsible for directing critical aspects of cell metabolism and possibly cell differentiation in very early mammalian embryogenesis. Thus, there is no reason a priori for assuming that X-chromosome genes should not be functional during the preimplantation period of development. The major issues of concern are whether or not the sperm-derived or paternal X chromosome (X^P) is delayed either in its expression or in postfertilization activation and whether or not the early embryo manifests any evidence of X-chromosome dosage compensation. The issue of paternal X-chromosome function during early embryogenesis is related to the observation that two of the early cell lineages that separate during early embryogenesis, the trophectoderm and yolk sac endoderm, express only the X chromosome from the oocyte, X^M. Experiments designed to identify maternal or uterine environmental factors responsible for this nonmosaic expression suggest that the expression of X^M in the extraembryonic lineages is a consequence of nonrandom inactivation. The possible occurrence of nonrandom inactivation suggests that the sperm- and oocyte-derived X chromosomes are different at the time of fertilization. Thus, it is important to demonstrate that X^P is expressed before trophectoderm and primitive endoderm lineages are normally differentiated and that the apparent nonrandom inactivation in these extraembryonic lineages is not due to a delay in expression or activation of the sperm-derived X chromosome until after these lineages become differentiated.

A. Cytogenetic expression of X chromosomes in early embryos

The early stages of mammalian embryogenesis do not show the typical cytogenetic manifestations of X-chromosome dosage compensation, namely, asynchronous X-chromosome replication or the presence of condensed heterochromatin (Plotnick, Klinger, and Kosseff 1971; Takagi 1974; Mukherjee 1976). Asynchronous X-chromosome replication was not detectable in preimplantation mouse embryos before the blastocyst or 40–50-cell stage with either [^{3}H]-thymidine labeling techniques (Mukherjee 1976) or quinacrine mustard fluorescence of BUdR incorporated into DNA (Takagi 1974). Overtly, these results suggested that an inactive X chromosome was not present during the first four or five cleavages following fertilization. On the other hand, the data do not exclude the possibility that the unusual cell-cycle properties of early embryos (Mukherjee 1976) mask possible differences between active and inactive X chromosomes, especially during DNA replication. Moreover, the lack of inactive heterochromatin could reflect decreased levels of DNA methylation that occur in satellite sequences of germ cells and probably persist through early cleavage stages (Sanford, Chapman, and Rossant 1985). Thus, the appearance of asynchronous X-chromosome replication may be a useful indicator of when cells manifest the properties of somatic cells that have a single active X chromosome, but it may not be a good indicator of either X-chromosome expression in preimplantation embryos or the onset of X-chromosome dosage compensation.

B. Quantitative determination of X-chromosome gene products in early embryos

The expression of quantitative dosage differences of X-chromosome gene products in XX oocytes compared with XO suggested that a similar approach could be used to determine if both X chromosomes were expressed in female embryos by comparison with male embryos. Three laboratory groups developed microquantitative enzyme assay techniques to measure the activity levels of either α-galactosidase (α-gal) or HPRT in single preimplantation embryos (Adler, West, and Chapman 1977; Monk and Kathuria 1977; Kratzer and Gartler 1978a; Monk 1978; Monk and Harper 1978). The experimental approach used in each of these studies was based on determining whether the frequency distributions of enzyme activities among individual embryos were unimodal or bimodal. A bimodal distribution would indicate that female embryos have twice the dosage of an X-linked gene product as males and therefore have both X chromosomes active. Both HPRT and α-gal are suitable for these tests because both enzymes increase in activity substantially during the preimplantation period. Biochemical assessments of rates of synthesis for these enzymes in preimplantation embryos are difficult, if not prohibitively expensive, but the high rate of general protein synthesis makes it likely that these enzyme activity changes represent newly synthesized enzyme proteins. We cannot eliminate a role for mRNA stored in the egg at fertilization in producing some portion of the enzyme activity change (Harper and Monk 1983). However, there is little evidence that the early cleavage stages utilize stored mRNA to a major extent in the mouse embryo (see Chapter 8). Each of the three laboratory groups was able to demonstrate bimodality in the distribution of embryo HPRT or α-gal activity levels at various stages of embryogenesis, but there are some critical concerns that must be borne in mind. In brief, analysis of mixed populations that are composed of presumptive subpopulations whose mean activities differ by twofold will show significant departures from a normal distribution only if the coefficient of variation of the enzyme assay is sufficiently small, probably on the order of 15 to 20 percent. A second concern is the sample sizes tested. Most of the statistical procedures used for tests of normality require large sample sizes to adequately choose between alternative distributions (normal versus bimodal). Each of these studies has some limitations in addressing the foregoing concerns, but collectively they do suggest that a dosage difference exists in the levels of two separate X-linked gene products. This dosage difference is consistent with the expression of both X chromosomes in females at this stage of development.

A direct test for HPRT activity levels in male and female embryos was reported by Epstein and associates (1978a, 1978b). These workers separated blastomeres at the 2-cell stage and cultured the resulting twin embryos in vitro through the blastocyst stage. One member of successfully grown pairs was karyotyped to determine the sex of the embryo, and the remaining twin or half-blastocyst was pooled with four additional embryos of the same sex to measure HPRT activity levels. The absolute levels of activity varied between samples by as much as twofold even though the apparent stages of development and the cell numbers were comparable. However, the female embryos had an average of twice as much activity as males.

The observed twofold dosage of HPRT in female embryos clearly demonstrates that an X-chromosome gene product is higher in female embryos during cleavage stages of development, and it indicates that both X chromosomes are functioning before X inactivation occurs in the entire embryo.

C. Qualitative evidence for expression of X^P and X^M in early embryos

The use of allelic variants of X-chromosome gene products provides a more direct method for identifying the onset of X^P expression during early embryogenesis. Embryos are collected from crosses of mice that differ for electrophoretic alleles of an X-linked enzyme and are analyzed for heterozygous patterns of expression.

Electrophoretic variants of *Pgk-1* were used to determine the timing of X^P expression in early embryos (Krietsch et al. 1982). The PGK electrophoretic phenotype was determined in single embryos between the 1-cell stage and the egg-cylinder stage at day 6. Expression of the paternal allele of *Pgk-1* was not detectable until day 6. Analyses of PGK activity levels in early embryos indicate that the levels of PGK are relatively high in eggs at the time of fertilization and that the active enzyme persists in cleaving embryos until the morula stage (Kozak and Quinn 1975; Krietsch et al. 1982). After the blastocyst stage, the activity levels increase 10-fold or more by the egg-cylinder stage, indicative of new enzyme synthesis.

The failure to find paternal expression of *Pgk-1* before day 6 could be a consequence of either a delay in the expression of *Pgk-1* or a total lack of X^P expression in early embryos. Alternatively, the techniques used may have been insufficiently sensitive to detect a very low level of newly synthesized *Pgk-1* before day 6. The relatively high levels of oocyte PGK activity in early stages are similar to the expressions of other glycolytic enzymes, notably glucose phosphate isomerase (GPI), an autosomal gene product, and G6PD, an X-linked gene product. Studies of paternal GPI expression clearly show that a paternal gene product can be detected as early as the 4-cell stage even though the oocyte GPI levels are as much as 10- to 20-fold greater than the levels produced by embryos (A. Peterson, personal communication). Thus, it is possible that the levels of newly synthesized PGK are equally low and below the lower limits of the PGK assay.

More recently, electrophoretic variation for HPRT has been found in several wild-derived or feral populations of house mice (Chapman et al. 1983). We have made crosses of mice that vary for *Hprt* allozyme forms and have examined the isoelectric-focusing phenotypes of HPRT in pooled and single embryos (V. M. Chapman and G. Johnson, unpublished data). We have observed that morula and blastocyst-stage embryos show an HPRT phenotype similar to the HPRT phenotype observed in HPRT A/B oocytes (Figure 12.2A). These pooled embryo results are consistent with the expression of HPRT A/B heterodimers in early embryos and hence activity of both X^M and X^P. Moreover, the pattern shows the preponderance of the X^M form of activity that is predicted from a mixture of male and female embryos. Preliminary studies also indicate that heterozygous HPRT expression is detected in pooled embryos from the 8-cell stage (Figure 12.2B). Single-embryo samples were analyzed at the blastocyst stage to determine if the heterozygous pattern was limited to a subset of presumptive female embryos. The data clearly showed that 50 percent of all embryos expressed the heterozygous pattern, which is consistent with the segregation of XX HPRT A/B and XY HPRT B or A embryos (Figure 12.2C).

These results confirm the conclusions reached in studies of bimodal enzyme activity distribution among populations of embryos (Epstein et al. 1978a, 1978b) suggesting that both X chromosomes are expressed during early embryogenesis. More important, demonstration of the heterozygous pattern at the 8-cell stage indicates that X^P is active in the embryo before differentiation of the trophectoderm and primitive endoderm. Thus, the preferential expression of X^M in extraembryonic lineages appears to result from preferential inactivation of a functioning X^P in these lineages.

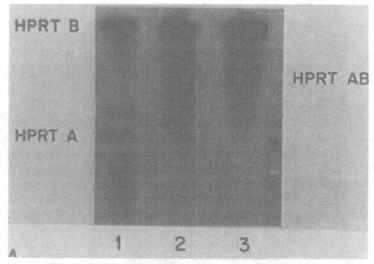

Figure 12.2A. Isoelectric-focusing phenotype of HPRT in pooled embryos from a cross of HPRT B females and HPRT A males. Lane 1 is HPRT A/B hemolysate; lane 2 is pooled morulae; lane 3 is pooled early blastocysts at day 3.5.

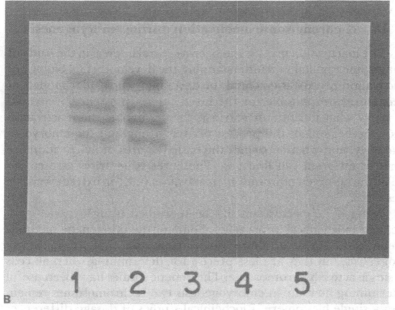

Figure 12.2B. Isoelectric-focusing phenotypes of HPRT in pooled 8-cell-stage embryos from HPRT B X HPRT A. Lanes 1, 2, and 3 are 8-cell-stage embryos from Ha/ICR X *M. castaneus*; lane 4 is the phenotype from 4-cell-stage embryos from Ha/ICR X *M. spretus*; lane 5 is HPRT B control from L cells.

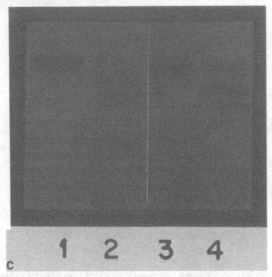

Figure 12.2C. HPRT expression in single blastocysts from a cross of HPRT B X HPRTA. Lanes 1 and 3 are HPRT B (presumptive males); lanes 2 and 4 are HPRT A/B presumptive females.

IV. X-chromosome inactivation during embryogenesis

The inactivation process has been a central interest in the study of X-chromosome regulation. Understanding the developmental biology of the inactivation process is essential for developing molecular models of the mechanisms responsible for the inactivation. In essence, we would like to know what intrinsic or extrinsic factors initiate the inactivation process, whether or not the process occurs throughout the embryo simultaneously, and whether or not the basic features of inactivation are the same in different cell lineages. The latter issue arises as a consequence of the observed nonrandom inactivation of X^P in extraembryonic lineages.

The timing of X inactivation has been studied using genetic, cytogenetic, and biochemical approaches. Some of the cytogenetic methods have been useful for determining when inactivation first occurs during embryogenesis, but they are less precise for determining when all cells have a single active X chromosome. The genetic studies have been useful for determining how late in embryogenesis two X chromosomes remain active in a single blastomere. Biochemical studies of dosage differences in X-chromosome products provide some insight into the number of active X chromosomes present in the whole embryo or in specific lineages, but they are limited by the sensitivity of the assay procedure, the amount of variation in rates of development within a sample of embryos, and the variation in the biochemical assay. In general, no one experimental approach has answered all of the questions about the inactivation process, but all of them combined give a fairly coherent view of the biology of the process. The prevailing view is that inactivation may occur in some cells as early as the morula stage, but that some of the blastomeres found in the embryonic ectoderm of six-day embryos continue to have two active X chromosomes. These observations further suggest that the inactivation process may accompany cell differentiation in the early embryo (Monk and Harper 1978).

A. Timing of X inactivation

Cytogenetic data provide the most direct estimate of the onset of X-chromosome inactivation. Allocyclic or asynchronous replication has been detailed as early as the late morula to early blastocyst stage (DeMars 1967; Takagi 1974) and the late blastocyst stage (Mukherjee 1976). Subsequent data suggested that the allocyclic replication that occurs early during embryogenesis is predominantly X^P (Takagi, Wake, and Sasaki 1978). The cytogenetic data indicated that isocyclic X-chromosome replication continues in a significant proportion of the embryonic ectoderm as late as 6.0 days (Takagi, Sugawara, and Sasaki 1982; Sugawara, Takagi, and Sasaki 1983).

The functional states of both X chromosomes in inner-cell-mass (ICM) cells from 3.5-day to 4.5-day embryos were demonstrated by injection chimera experiments (Gardner and Lyon 1971). These workers isolated single ICM cells from embryos heterozygous for the T(X; 7)1Ct translocation and produced blastocyst injection chimeras. The activities of the intact and translocated X chromosomes could be identified by the expression of alleles in the albino series in the coat color of chimeric mice. The T(X; 7) carried the wild-type allele on the translocated X chromosome, which masked the expression of the autosomal albino allele of donor cells when the translocated X was active. The cells with the translocated X inactive expressed an autosomal albino allele that was distinguishable from the host albino phenotype. Four of nine chimeras showed expression of both translocated and normal X chromosomes with three distinct coat colors. These findings clearly indicated that both X chromosomes from single 4.5-day ICM cells were functional and that irreversible X-chromosome inactivation must occur in these lineages after the fourth day. This suggests that the cytogenetic evidence of X inactivation at the blastocyst stage must be a result of X inactivation in the trophectoderm cells only.

The distributions of quantitative enzyme activity levels in embryo populations have been used to estimate the time of change from two active X chromosomes in female embryos to a single active X chromosome

(Kratzer and Gartler 1978a, 1978b; Monk and Harper 1979). There is some indication that single-embryo activity levels in the embryonic epiblast continue to show bimodal distributions until day 6 (Monk and Harper 1978), but the statistical significance of the bimodality of the distributions is difficult to assess. More convincing evidence for a dosage difference between male and female embryos is provided by a comparison of epiblast HPRT activity with that of adenine phosphoribosyl transferase (APRT), which is the product of an autosomal gene. Activity ratios in karyotypically identified embryos shows a significant difference between female and male epiblasts as late as day 6. These findings suggest that females have higher levels of HPRT activity, but it is difficult to assess whether or not some embryonic cells have undergone X inactivation by 6 days.

Significant bimodal distributions for HPRT activity were reported for whole embryos at the late blastocyst stages by Kratzer and Gartler (1978a, 1978b). These results are consistent with the delay of inactivation in the embryonic portion, but they are not as direct as the HPRT/APRT activity ratios reported earlier (Monk and Harper 1978) for sexed embryo epiblasts.

In summary, the cytogenetic evidence, the results for single-blastomere injection chimeras, and the biochemial data on activity levels of X-linked gene products all support the view that X-chromosome inactivation is not completed in the embryonic lineages until as late as day 6.0 in mouse embryogenesis. On the other hand, the cytogenetic evidence indicates that the inactivation process is initiated in some lineages as early as the 32-cell stage. These findings clearly support the view that the inactivation process occurs first in extraembryonic lineages and last in embryonic progenitors. These results are consistent with the association of X inactivation with some aspect of cytodifferentiation (Monk and Harper 1979).

B. Nonrandom inactivation in extraembryonic lineages

The association of cytodifferentiation and X inactivation in extraembryonic lineages is also accompanied by nonmosaic expression of the maternal X chromosome, X^M, in the trophectoderm and the primitive endoderm lineages of rodents. Nonrandom X-chromosome expression was initially observed by cytogenetic studies of 8.5-day mouse embryos (Takagi and Sasaki 1975). The identities of X^M and X^P were followed in conceptuses using the T(X; 7)1Ct translocation, which carries a segment of chromosome 7 inserted into the X chromosome (Ohno and Cattanach 1962), making it longer than the normal X chromosome. The proportion of cells with an allocyclic replicating X^P was approximately 50 percent in embryonic portions of the conceptus regardless of the parental origin of T(X; 7). By contrast, both the yolk sac and the chorion showed 90 percent or higher allocyclic X^P expression. These results indicated that the pattern of inactivation was different between extraembryonic membranes and embryonic cells of the conceptus.

The initial cytogenetic studies also showed that the asynchronous or allocyclic replicating inactive X chromosomes of early extraembryonic stages could be early replicating rather than late replicating (Takagi 1978). These findings have been extended in recent studies of day-5.3 to day-6.5 embryos (Takagi et al. 1982). These workers observed an early replicating X^P chromosome in proximal endoderm on day 5.3 in 80 percent of cells and a late replicating X^P chromosome in 10 percent of cells. Similar results were observed for the presumptive inactive X^P in extraembryonic ectoderm and ectoplacental cone. The pattern of inactive X-chromosome replication changes in two of these lineages between 6.0 and 6.5 days, so that the allocyclic X chromosome becomes late replicating in the proximal endoderm and extraembryonic ectoderm. The significance of the early replication of inactive X chromosomes in early stages and the transition to late replication is open for speculation, but the functional consequences of early and late replication of inactive X chromosomes are overtly the same for the expression of normal X-chromosome gene elements (Sugawara et al. 1983).

Biochemical evidence that X^M is preferentially expressed in extraembryonic lineages has also been obtained. The basic experimental approach involves the use of allozyme variation for PGK-1, G6PD, and HPRT in crosses that produce heterozygous females. The sex of very early embryos is readily determined by the heterozygous phenotype of the embryonic lineages, which are typically mosaic for X-chromosome expression. The sex of fetuses at later stages can be determined by allozyme phenotype, sex chromatin in the amniotic membrane, or morphology of the gonads.

The initial characterizations of X^M expression in mouse yolk sac used allelic variants of PGK-1 (West et al. 1977). This work demonstrated preferential X^M PGK-1 allozyme expression in total yolk sac. This unbalanced mosaicism of X^M expression was due to nonmosaic X^M expression in the visceral endoderm layer of the yolk sac, whereas the mesodermal yolk sac layer was mosaic for X-chromosome expression. Furthermore, it was possible to show that the preferential expression of X^M was not altered by embryo transfer to genetically different foster mothers (West et al. 1977; Frels, Rossant, and Chapman 1979).

The nonrandomness of parental X-chromosome expression has been extensively studied in both trophectoderm and primitive endoderm lineages of mouse conceptuses using the PGK-1 variation (Frels et al. 1979; Frels and Chapman 1980; Papaioannou and West 1981; Papaioannou et al. 1981; Takagi et al. 1982; McMahon and Monk 1983). The results of these studies demonstrate uniformly that the expression of X^M is nonmosaic in extraembryonic lineages and that even a minor amount of X^P

expression could not be detected either in any of a large number of conceptuses or in pooled samples. These results are at variance with the cytogenetic findings, which report as much as 10 percent of extraembryonic tissues showing allocyclic X^M (Takagi et al. 1982). The available evidence on the relative sensitivity of the PGK-1 fluorescence assay system used by several workers suggests that a 10 percent level of X^P expression would have been detected, especially in the visceral endoderm. The latter results suggest that the allocyclic X^M observed cytogenetically represents either contamination by cell lineages that have undergone random inactivation or the relative error in the cytogenetic methods in estimating the inactive state of X chromosomes.

The general picture that emerges from these studies is that X^M is preferentially expressed in the trophectoderm and primitive endoderm lineages of rodent species. The nonmosaic expression of X^M in these cells probably occurs by a process of nonrandom inactivation of X^P, rather than a mechanism of selective proliferation of X^M active cells from a pool of cells that have undergone random inactivation of X^P and X^M. These conclusions are based on several kinds of evidence. First, the cytogenetic data showed that X^P is already allocyclic by the blastocyst stage (Takagi et al. 1978). Second, the biochemical data clearly showed nonmosaic X^M expression occurring as early as day 5.5 to day 6.0, which further argues that any selective effects would have to occur and be complete within a matter of 1 day (Papaioannou et al. 1981; McMahon and Monk 1983).

Nonrandom inactivation in the extraembryonic lineages suggests that the X chromosomes from the oocyte and sperm differ but that the difference is lost by the time of random X inactivation in the embryonic lineages. Alternatively, it is possible that the X-inactivation mechanisms differ between the various cell lineages and that the X^M/X^P difference(s) is not recognized in embryonic lineages. A difference in the mechanisms of X inactivation between extraembryonic and embryonic lineages has been suggested by Gartler and Riggs (1983) and by Takagi and associates (1982) to account for the unusual early allocyclic replication in extraembryonic lineages and other differences in the inactive X chromosome of these lineages (Kratzer et al. 1983).

The differences in behavior and function of germ cell nuclei during the processes of oogenesis and spermiogenesis offer several opportunities to differentially modify the DNA and chromatin of gamete genomes. Understanding these differences in the respective sperm and oocyte gamete genomes and the postfertilization consequences will probably provide some insight into the nonrandom inactivation in extraembryonic lineages. The occurrence of X-chromosome reactivation in meiosis and the maintenance of two-X-chromosome function during the process of oogenesis were previously described. By contrast, the expression of X chromosomes during spermiogenesis is a matter of speculation, and there is some indication that X chromosomes do not continue to function beyond meiotic prophase (Lifschytz and Lindsley 1974). Study of PGK expression as determined by enzyme synthesis suggests that the sperm-specific autosomal *Pgk-2* is expressed postmeiotically and that the expression of X-chromosome gene *Pgk-1* either is very low or is turned off (Kramer and Erickson 1981).

In general, most of the sperm cell genome appears to diminish in function following meiosis with the exception of gene products that may be special to the process of chromatin condensation (Brock, Trostle, and Meistrich 1980; Erickson et al. 1980; Elsevier, 1982). Many of those sperm-specific proteins are arginine-rich protamines that replace histones in sperm and establish disulfide bridges that appear to stabilize the highly condensed chromatin in sperm heads. Additional DNA-binding nonhistone proteins are synthesized in late spermatid stages during the condensation process (Kistler, Geroch, and Williams-Ashman 1973; Ecklund and Levine 1975). The consequences of sperm DNA condensation during early cleavage stages of embryogenesis are not yet resolved. Available evidence indicates that sperm undergo a rapid decondensation following fertilization and that protamines may be lost during this process (Ecklund and Levine 1975). Cytogenetic analysis of chromosome condensation during the first cleavage indicates that the initial sperm and oocyte pronuclear chromosomes are differentially condensed during mitotic prophase. The female pronuclear chromosomes are actually more condensed than the male chromosomes (Donahue 1972). The difference in condensation is apparently lost by the end of metaphase, but it is possible that some of these differences could persist through early cleavage divisions as a consequence of semiconservation of chromatin structure during DNA and chromatid replication (Seidman, Levine, and Weintraub 1979). Presumably, the extent of differences present at the time of fertilization would be diluted by half with each cell division, with the result that less than 1 percent of the differences would remain by the late blastocyst stage. This mechanism would provide a transient difference between X^M and X^P that could account for the nonrandom inactivation in trophectoderm at the fourth or fifth cleavage and primitive endoderm at the fifth or sixth cleavage.

Even if the proposed semiconservation-dilution model of chromatin structure does not account for the X^M/X^P difference in early embryos, it is interesting to note that the shift from early replicating allocyclic X chromosomes to late replicating allocyclic X chromosomes occurs at about the same stage of development at which random X-chromosome inactivation appears to be initiated in the embryonic lineage (Takagi et al. 1982). This change in the behavior of inactive X chromosomes could indicate either loss of "imprinting" molecules on X chromosomes or changes in the entire genome that shift the relative timing of allocyclic X replication.

Differences in the relative levels of cytosine methylation between the

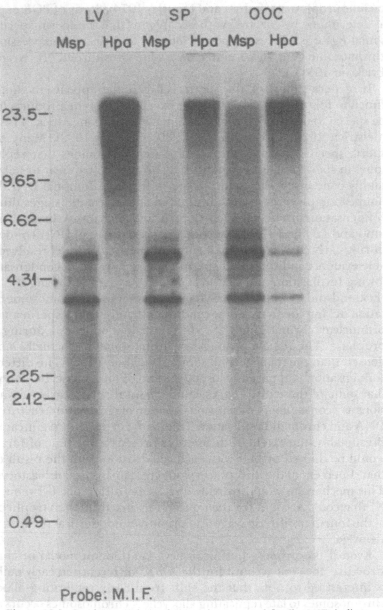

Figure 12.3. Southern analysis of mouse interspersed family (MIF) in liver, sperm, and oocyte DNAs digested with Msp-I and Hpa-II. Msp-I digestion of liver, sperm, and oocyte DNAs produces two prominent DNA sequences at 4.0 and 5.0 kbp, respectively. No comparable MIF bands are observed in liver and sperm when those DNAs are digested with Hpa-II. However, MIF bands are observed in oocyte DNA digested with Hpa-II. These results indicate that MIF CCGG sequences become unmethylated in oogenesis but not in mature sperm. [Reproduced from Sanford et al. (1984), with permission.]

sperm and egg genomes also offer opportunities for imprinting genetic differences that could be maintained during early embryogenesis. We have examined the DNA methylation levels of repetitive sequences in sperm and oocyte DNA using a Southern blot analysis of Hpa-II-digested DNA. We have concentrated on a minor satellite sequence of 5×10^4 copies in the mouse genome and mouse interspersed family (MIF), a dispersed repetitive sequence with $2-3 \times 10^4$ copies in the haploid genome. Both of these sequences identify well-defined fragment sizes when genomic DNA is digested with Msp-1, and both of these sequences are methylated and resistant to Hpa-II digestion in somatic tissues (Chapman et al. 1984). By contrast, Southern blot analysis of Hpa-II-digested oocyte and sperm DNA shows extensive cleavage of minor satellite DNA (Sanford et al. 1984). MIF sequences, on the other hand, are cleaved by Hpa-II in oocyte DNA to a much greater extent than that observed in sperm DNA (Figure 12.3). These results indicate that both sperm and oocyte genomes are equally undermethylated in centromeric satellite sequences but that oocyte DNA may be more undermethylated in dispersed sequences than sperm DNA. It will be important to focus on X-chromosome sequences and determine whether or not differences in DNA methylation are present on the X chromosomes in gametes at the time of fertilization.

Preliminary studies on DNA methylation of repeated elements during the preimplantation period suggest that the undermethylation of satellite sequences present in gametes at the time of fertilization persists in the embryo through the blastocyst stage, but that de novo methylation occurs by day 7 in the embryonic lineages (Sanford et al. 1985). By contrast, the extraembryonic lineages appear to be extensively undermethylated. This has been characterized in repetitive DNA sequences and more recently in a number of single-copy gene elements (see Chapter 4). Random inactivation of X chromosomes occurs in lineages that become de novo methylated, but it is difficult to judge whether the onset of de novo methylation is coincident with X inactivation or occurs separately.

V. X-chromosome expression in embryonal carcinoma (EC) cells

A. Expression of two X chromosomes in undifferentiated XX EC lines

Independent evidence for an association between X-chromosome inactivation and cell differentiation comes from studies of X-chromosome expression in EC cells. Both cytogenetic and biochemical data indicate that some EC cell lines may have two functional X chromosomes. The most direct evidence for expression of both X chromosomes comes from the development of an EC cell line, P10, from a female embryo heterozygous for *Pgk-1* (McBurney and Strutt 1980). Clonally derived populations of these cells showed roughly equal levels of both PGK-1 A and B, which is consistent with these cells having two active X chromosomes. Cytogenetic studies of P10 cells further indicated that the undifferentiated cells did not have an allocyclic X chromosome. Similar cytogenetic results have been reported for the LT-1 cell line derived from an ovarian teratocarcinoma (Takagi and Martin 1984).

The expression of two X chromosomes in undifferentiated cells should produce a doubling of X chromosome gene products compared with similar undifferentiated cell lines that have only a single X chromosome (Martin et al. 1978). A dosage effect should also be evident between undifferentiated and differentiated cell lines if X inactivation accompanies differentiation (McBurney and Adamson 1976; Martin et al. 1978). Martin and associates (1978) demonstrated that three different X-chromosome gene products, G6PD, HPRT, and α-gal, all show relative activities about twofold higher in LT-1 than in three separate undifferentiated EC cell lines that have only a single X chromosome present. By contrast, all four lines have approximately the same levels of activity for several autosomal gene products, including β-glucuronidase and β-galactosidase. These results are somewhat surprising, because the strains of origin for these EC cell lines differ for regulatory and structural alleles of both β-glucuronidase (*Gus*) and β-galactosidase (*Bgs*) and might be expected to produce differences for these enzyme activities. Whether or not the enzyme activities reported for EC cell lines are consequences of either aneuploidy or unusual regulation in EC cells is difficult to determine. The results do suggest that some caution is warranted in evaluating gene dosage from quantitative levels of enzyme activities.

Similar studies of a different series of EC cells further suggest that some XX cell lines may have only one active X chromosome (McBurney and Adamson 1976). For example, the EC cell line C86S1 expressed a late replicating X chromosome, and the levels of X-chromosome gene products did not show a clear dosage difference when compared with XO cells.

B. X inactivation coupled with in vitro differentiation

Undifferentiated EC cell lines can be induced to differentiate either by changes in culture conditions, such as removal from feeder layers, or by chemical inducers, such as retinoic acid. Cytogenetic and gene expression studies indicate that X-chromosome inactivation accompanies this process in XX EC lines (Martin et al. 1978; McBurney and Strutt 1980; Takagi and Martin 1984).

The initial indication that an EC cell line with two X chromosomes active undergoes a change in dosage of X-chromosome gene products was reported by Martin and associates (1978). Their results were based on the relative activity ratios of X-chromosome and autosome-encoded enzymes in XX EC cells and XO EC cell lines. Before differentiation, the ratio was approximately double the ratio observed following differentiation. The results are suggestive of X inactivation, but other interpretations are also possible. More direct evidence for X inactivation comes from cytogenetic analysis of EC cells before and after differentiation (Takagi and Martin 1984). These workers demonstrated that allocyclic replication of X chromosomes did not occur in undifferentiated LT-1 EC cells. However, allocyclic X chromosomes, both early and later replicating, were observable by four to six days following induction of differentiation. The frequency of allocyclic X chromosomes appeared to plateau at about 50 percent around three weeks following induction. The observation that 50 percent of the cells are isocyclic differs from the observed change in replication patterns in normal embryogenesis, where the frequency of isocyclically replicating X chromosomes approaches zero by 6.5 days of development (Rastan 1982; Sugawara et al. 1983). The presence of isocyclic cells in differentiating cultures in vitro may be indicative of a stem cell population that has not differentiated. If that were the case, the expected change in the dosage of X-chromosome gene products should not decrease to a 1 : 1 ratio with XO cells. Alternatively, it would be necessary to argue that functional X inactivation precedes the manifestation of allocyclic replication.

Qualitative biochemical evidence for X inactivation comes from studies of inactivation involving P10 (McBurney and Strutt 1980). Some of the differentiated cultures expressed only one allelic form of PGK-1. However, the differentiating cultures were difficult to analyze clonally, and there is some suggestion that XX EC cell lines may lose an X chromosome (see Chapter 15). Moreover, the predicted changes in relative dosages of PGK activity in undifferentiated and differentiated cultures were not accompanied by corresponding changes in specific activity.

C. X-chromosome reactivation in EC cell/somatic cell hybrids

Reactivation of the inactive X chromosome does not normally accompany the production of somatic cell hybrids; however, recent work demonstrates that somatic cell hybrids involving diploid lymphoid cells and EC cells produce reactivation of the entire inactive X chromosome (Takagi et al. 1983). The reactivation was monitored by the loss of allocyclic replication in 23 separate cloned hybrid lines and by the activation of the PGK-1 allozyme carried on the inactive X chromosome in T(X; 16) translocation hybrids. The reactivation occurred in hybrids with two EC cell lines, OTF9-63 and PSA-1. These results suggest that pluripotent EC cell lines have the ability to reactivate X chromosomes and maintain them in the active condition. Conflicting negative data were obtained by two other studies involving EC cell hybrids with mule cell lines (von Kap-Herr and Mukherjee 1977) or mouse-species hybrid cell lines (Graves and Young 1982). The positive results of Takagi and associates (1983) are compelling, but it is important to derive the basis for the apparent difference between EC cell fusions with diploid cells and aneuploid cells. The fusion of *M. musculus* × *M. caroli* hybrid cells with NG2 EC cells produced a reactivation of the inactive *M. caroli Hprt* gene, but no change in the appearance of late replication. The lack of reactivation could have been a consequence of the unlimited growth of the somatic cells. Moreover, the EC cell hybrids in the work by Takagi and associates (1983) showed EC-cell-like properties, including tumorigenesis. Whether or not EC cell properties were expressed in the hybrids with *M. musculus* × *M. caroli* cells or in the fusion of mule cells with embryonic bodies was not demonstrated.

Experimental reactivation of the entire inactive X chromosome is a highly promising new sytem that may parallel meiotic reactivation. If the mechanism is the same, we shall have new experimental opportunities to identify gene products involved in X-chromosome regulation.

VI. Experimental prospects and models

The focus of research on X-chromosome regulation has generally been on the process of initiating the inactivation and the molecular mechanisms that maintain the mitotically heritable inactive condition (Gartler and Riggs 1983). In this final section, I would like to consider these two topics, but also focus on reactivation.

A. Reactivation

The reactivation of the inactive X chromosome in EC cell/somatic cell hybrids is similar in many respects to the reactivation that occurs during germ cell differentiation. Because EC cell lines are similar in properties to early embryonic cells, the X-chromosomal reactivation mechanism may be a normal element of early embryogenesis and not necessarily limited to female XX cells. Furthermore, the reactivation of inactive X chromosomes in EC cell/somatic cell hybrids suggests an active element in the reactivation process that needs to be continuously present to maintain both X chromosomes active. The element(s) could be either a chromatin- or DNA-binding protein that needs to be synthesized in each cell cycle to maintain both X chromosomes active. A diagram of the expression of reactivation during oogenesis and embryogenesis is shown in Figure 12.4.

The molecular basis of the reactivation event is an open issue, but it is possible to speculate that a germ-cell-specific reactivation mechanism is turned on just prior to meiotic prophase that overrides the somatic

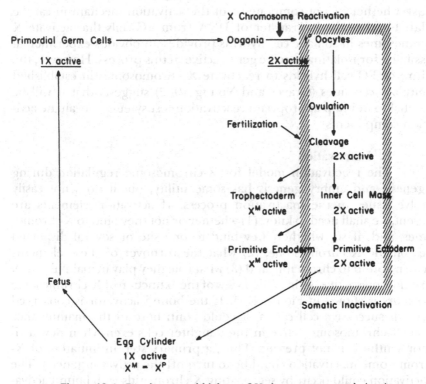

Figure 12.4. Developmental biology of X-chromosome regulation in mammals. The hatched box represents the period of development, in which two X chromosomes are active. The diagram shows that reactivation of X chromosomes occurs when oogonia enter meiotic prophase and that the reactivated X-chromosome condition remains through oogenesis, fertilization, and early embryogenesis.

inactivation. The demethylation of satellite DNA sequences in gametes may be indicative of a generalized demethylation that is a component of the reactivation mechanism (Sanford et al. 1984). However, demethylation per se probably is not sufficient to cause reactivation, because extra-embryonic lineages and EC cells show marked decreases in global genomic DNA methylation at the same time that X inactivation takes place (Chapman et al. 1984).

Once the reactivation mechanism is turned on during oogenesis, it stays on in the oocyte and fertilized egg. Following fertilization, the activation mechanism continues to function until primary differentiation events occur during early embryogenesis. Several molecular mechanisms may be employed, but the reactivation in EC cell hybrids suggests that an enhancer-like element specific to oocytes and early embryos could be responsible and that maintenance of two active X chromosomes requires the active presence of activation elements. Ideally, we would like to ask whether or not components of the activation mechanism can be isolated and studied. Transfer of DNA from EC cells that activate X chromosomes in somatic cell hybrids provides an obvious experimental possibility for isolating specific genes active in this process. However, the failure of EC cell hybrids to reactivate X chromosomes in established aneuploid cell lines (Graves and Young 1982) suggests that it will be essential to develop appropriate reactivation test sytems to evaluate activation components.

B. Inactivation

The reactivation model for X-chromosome regulation during oogenesis and embryogenesis has some utility, but it does not easily resolve issues in the inactivation process. If activation elements are present, we shall need to know (1) whether or not they bind to X chromosomes, and, if so, whether they bind to one site or several dispersed sites on the X chromosome, (2) what the turnover of these elements is when bound to chromatin, and (3) what role they play in maintaining X chromosomes active. A simplified view of the extinction of X-chromosome activators is shown in Figure 12.5. If the bound activator is conserved through successive cell cycles, it could contribute to the maintenance of an X chromosome active in the daughter cells even when new activator synthesis is not present. Thus, a primary step in initiation of X-chromosome inactivation could be to turn off the activator gene(s). The inactivation could occur by segregating X chromatids with bound activator into daughter cells, as shown in the diagram. In this view, inactivation will be a passive process in which the whole X chromosome is excluded from remaining in an active configuration (Ryoji and Worcel 1984). Once the inactive state of the X-chromosome chromatin is established, it can be passively maintained in successive daughter chromatids by allocyclic replication, which will ensure that a different population of DNA-binding proteins will be present during DNA synthesis. The activator model does not exclude the possibility that inactivator elements are produced similar to those described by Gartler and Riggs (1983). Potentially,

Figure 12.5. X-chromosome activator model for regulating X-chromosome expression. Active or functional X chromosomes are indicated as open chromosomes. Inactive X chromosomes are represented by cross-hatched or black chromosomes. Cross-hatched chromosomes represent either imprinting differences and/or inactive but unmodified X chromosomes. (1) X-chromosome reactivation. Reactivation mechanism is turned on, which overrides somatic inactive X condition. Open squares indicate that X-chromosome activity is regulated by "activator" mechanism. (2) Sperm X chromosome differs from oocyte X in chromatin structure and in DNA methylation. These differences may be retained during early cleavage. (3) Trophectoderm differentiation occurs before X^M and X^P difference disappears. (4) Primitive endoderm is dif-

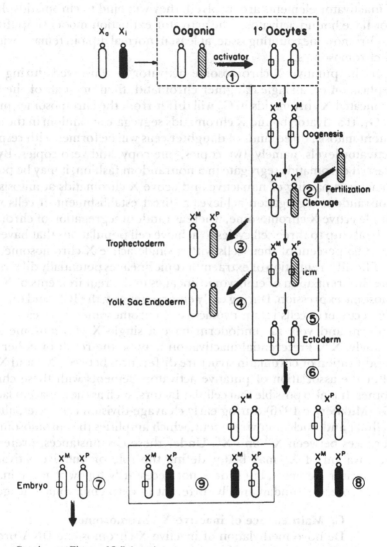

Caption to Figure 12.5 (*cont.*)

ferentiated from ICM X^M, while the X^P difference is still intact. (5) Primitive ectoderm retains the two-active-X condition, but $X^M - X^P$ difference is lost. (6) Activator is turned off. Diagram indicates semiconservation of activator on previously active chromosomes, but newly synthesized chromatids do not have activator. (7) Random assortment of chromatids results in one-fourth of daughter cells with both X chromosomes that retain activator. (8) Random assortment of chromatids results in one-fourth of daughter cells with no activator present. (9) Daughter cells enclosed in box have one active X chromosome, with random but equal expression of cells with either X^M or X^P active. These single-X-active cells could arise from either random daughter chromatid assortment or nonrandom chromatid assortment.

if inactivator elements are involved, they will bind to chromatids that do not have bound activator. The activator extinction model simplifies the X-chromosome counting issue, at least in normal diploid females with two X chromosomes.

If the putative X-chromosome activator is conserved during DNA replication on a single daughter chromatid, then one-half of the newly replicated X chromatids in G_2 will differ from the chromosomes present in G_1. If activator-bound X chromatids segregate at random in the subsequent mitosis, three kinds of daughter cells will be formed with respect to activator levels, namely, two copies, one copy, and zero copies. By contrast, if chromatids segregate in a nonrandom fashion, it may be possible to achieve pairing of nonactive and active X chromatids at mitosis. The nonrandom segregation achieves a direct establishment of cells with a single active X chromosome, whereas random segregation of chromatids will take up to three cell cycles to achieve cell populations that have more than 95 percent of their cells with a single active X chromosome.

The differentiation of extraembryonic lineages potentially differs from the differentiation of embryonic lineages in the requirements of X-chromosome expression. During early embryogenesis, the ICM and the primitive ectoderm cells retain two active X chromosomes, whereas trophectoderm and yolk sac endoderm have a single X chromosome active, namely, X^M. Preferential inactivation may be the result of either DNA modification or chromatin structure differences between X^M and X^P that alter the association of putative activator elements with these chromosomes. It is also possible that cellular factors, such as the apparent lack of a G_1 (Mukherjee 1976) during early cleavage-division cell cycles, alter the cellular and nuclear environment, which amplifies the chromosomal differences between X^M and X^P. Under these circumstances, preferential inactivation of X^P may better define the role of putative activator or inactivator elements, but it does not necessarily follow that the inactivation process is fundamentally different in extraembryonic lineages.

C. Maintenance of inactive X chromosomes

De novo methylation of inactive X-chromosome DNA probably occurs as a secondary step in the inactivation process that acts to extend and stabilize the inactive domain of the X chromosome to prevent random X reactivation or derepression of individual X-chromosome genes. A secondary role of DNA methylation has been suggested by analysis of DNA methylation of inactive X-chromosome genes (Wolf et al. 1984). The secondary role of methylation is also suggested by the observation that inactive X chromosome can be cytogenetically detected in yolk sac endoderm, and functional inactivation in these tissues can be demonstrated with allozyme markers for PGK-1 and HPRT. Nevertheless, the inactive *Hprt* gene functions in DNA-mediated transformation almost as well as the active gene (Table 12.2). By contrast, the inactive *Hprt* gene in

Table 12.2. *DNA-mediated gene transfer from heterozygous HPRT mouse tissues into HPRT hamster cells*

Tissue phenotype	Tissue	HPRT A transformants
B	Yolk sac endoderm[a]	11/42
A = B	Yolk sac mesoderm[a]	3/6
A = B	Adult liver[a]	13/25
B T(X; 16)	Whole fetus[a]	0/32
	Adult brain, liver, kidney[b]	1/59

Note:
[a]Data from Kratzer et al. (1983).
[b]Data from Chapman et al. (1982).

somatic lineages functions only 1 percent as efficiently as the active gene in similar transformation tests (Kratzer et al. 1983). Recent work with an X-chromosome-linked transgenic α-fetoprotein (AFP) minigene further demonstrates that the transgenic AFP minigene is expressed in a correct tissue-specific fashion in visceral yolk sac endoderm, fetal liver, and fetal gut (Krumlauf et al. 1985). Moreover, the X-linked AFP minigene is expressed in yolk sac endoderm even when it is on the inactive X chromosome. By contrast, the AFP minigene is not expressed in fetal liver when it is on the intact, inactive X chromosome of the T(X;16) translocation. These results, coupled with the HPRT transformation data, suggest that the extensive de novo DNA methylation that occurs in the embryonic lineages helps establish a more stable heterochromatic domain on the entire inactive X chromosome that is not present in the inactive X chromosome of extraembryonic lineages, which have a much lower level of genomic DNA methylation. Maintenance of the inactive X condition therefore appears to utilize different molecular mechanisms in embryonic and extraembryonic lineages.

VII. Conclusions

The expression of a single active X chromosome in female mammals first proposed by Lyon (1961) is a well-established phenomenon in several mammalian species. The adaptation of X inactivation as a means of equalizing the X-chromosome gene dosage in males and females continues to provide important opportunities for studying elements of gene regulation that may be broadly applicable to other genetic systems. Moreover, the regulation of the X chromosome is in itself important and deserves special attention. I have attempted to broaden the focus of X-chromosome regulation from the inactivation process to a consideration of reactivation during germ cell differentiation and early embryogenesis in the hope that this perspective may provide new clues and insights and experimental approaches to the issues of X-chromosome regulation.

Acknowledgments

The development of this review was assisted by the helpful discussions of Janet Rossant and Janet Sanford. I would especially like to thank them for permission to include data from their work. I am also grateful to Gerald Johnson for his assistance in the studies of HPRT expression in preimplantation mouse embryos and to Melanie Murawski and Charlotte Schonfeld for technical assistance in these studies. Finally, I would especially like to thank Nancy Holdsworth for valuable assistance in the preparation and organization of the typed manuscript.

Supported in part by USPHS grant GM-24125 from the National Institutes of Health.

References

Adler, D. A., West, J. D., and Chapman, V. M. (1977) Expression of α-galactosidase in preimplantation mouse embryos: Implications for X-chromosome activation. *Nature* 267, 838–9.

Andina, R. J., (1978) A study of X-chromosome regulation during oogenesis in the mouse. *Exp. Cell Res.* 111, 211–18.

Brock, W. A., Trostle, P. K., and Meistrich, M. L. (1980) Meiotic synthesis of testis histones in the rat. *Proc. Natl. Acad. Sci. U.S.A.* 77, 371–5.

Chapman, V., Forrester, L., Sanford, J., Hastie, N., and Rossant, J. (1984) Cell lineage-specific undermethylation of mouse repetitive DNA. *Nature* 307, 284–6.

Chapman, V. M., Kratzer, P. G., and Quarantillo, B. A. (1983) Electrophoretic variation for X-chromosome-linked hypoxanthine phosphoribosyl transferase (HPRT) in wild-derived mice. *Genetics* 103, 785–95.

Chapman, V. M., Kratzer, P. G., Siracusa, L. D., Quarantillo, B. A., Evans, R., and Liskay, R. M. (1982) Evidence for DNA modification in the maintenance of X-chromosome inactivation of adult mouse tissues. *Proc. Natl. Acad. Sci. U.S.A.* 79, 5357–61.

DeMars, R. (1967) The single-active-X: functional differentiation at the chromosome level. *Natl. Cancer Inst. Monogr.* 26, 327–51.

Donahue, R. P. (1972) Cytogenetic analysis of the first cleavage division in mouse embryos. *Proc. Natl. Acad. Sci. U.S.A.* 69, 74–7.

Ecklund, P. S., and Levine, L. (1975) Mouse sperm basic nuclear protein. Electrophoretic characterization and fate after fertilization. *J. Cell Biol.* 66, 251–62.

Elsevier, S. M. (1982) Messenger RNA encoding basic chromosomal proteins of mouse testis. *Dev. Biol.* 90, 1–12.

Epstein, C. J. (1969) Mammalian oocytes: X-chromosome activity. *Science* 163, 1078−9.

−(1972) Expression of the mammalian X-chromosome before and after fertilization. *Science* 175, 1467−8.

Epstein, C. J., Smith, S., Travis, B., and Tucker, G. (1978a) Both X chromosomes function before visible X-chromosome inactivation in female mouse embryos. *Nature* 274, 500−2.

Epstein, C. J., Travis, B., Tucker, G., and Smith, S. (1978b) The direct demonstration of an X-chromosome dosage effect prior to inactivation. In *Genetic Mosaics and Chimeras in Mammals*, ed. L. B. Russell, pp. 261−7. New York: Plenum.

Erickson, R., Kramer, J. M., Rittenhouse, J., and Salkeld, A. (1980) Quantitation of mRNAs during mouse spermatogenesis: protamine-like histone and phosphoglycerate kinase-2 mRNAs increase after meiosis. *Proc. Natl. Acad. Sci. U.S.A.* 77, 6086−90.

Frels, W. I., and Chapman, V. M. (1980) Expression of the maternally derived X chromosome in the mural trophoblast of the mouse. *J. Embryol. Exp. Morphol.* 56, 179−90.

Frels, W. I., Rossant, J., and Chapman, V. M. (1979) Maternal X chromosome expression in mouse chorionic ectoderm. *Dev. Genet.* 1, 123−32.

Gardner, R. L., and Lyon, M. F. (1971) X-chromosome inactivation studies by injection of a single cell into the mouse blastocyst. *Nature* 231, 385−6.

Gartler, S. M., Andina, R., and Gant, N. (1975) Ontogeny of X-chromosome inactivation in the female germ line. *Exp. Cell Res.* 91, 454−6.

Gartler, S. M., and Cole, R. E. (1981) Recent developments in the study of mammalian X-chromosome inactivation. In *Mechanisms of Sex Differentiation in Animals*, eds. C. R. Austin and R. G. Edwards, pp. 113−43. New York: Academic Press.

Gartler, S. M., and Riggs, A. D. (1983) Mammalian X-chromosome inactivation. *Annu. Rev. Genet.* 17, 155−90.

Gartler, S. M., Rivest, M., and Cole, R. E. (1980) Cytological evidence for an inactive X chromosome in murine oogonia. *Cytogenet. Cell Genet.* 28, 203−7.

Graves. J. A. M. (1982) 5-Azacytidine-induced re-expression of alleles on the inactive X chromosome in a hybrid mouse cell line, *Exp. Cell Res.* 141, 99−105.

Graves, J. A. M., and Young, G. J. (1982) X-chromosome activity in heterokaryons and hybrids between mouse fibroblasts and teratocarcinoma stem cells. *Exp. Cell Res.* 141, 87−97.

Harper, M. I., and Monk, M. (1983) Evidence for translation of HPRT enzyme on maternal mRNA in early mouse embryos. *J. Embryol. Exp. Morphol.* 74, 15−28.

Johnston, P. G. (1981) X chromosome activity in female germ cells of mice heterozygous for Searle's translocation T(X; 16)16H. *Genet. Res. Camb.* 37, 317−22.

Kahan, B., and DeMars, R. (1975) Localized derepression on the human inactive X chromosome in mouse-human cell hybrids. *Proc. Natl. Acad. Sci. U.S.A.* 72, 1510−4.

396 VERNE M. CHAPMAN

Kanda, N. (1973) A new differential technique for staining the heteropycnotic X-chromosome in female mice. *Exp. Cell Res.* 80, 463–7.

Kistler, W. S., Geroch, M. E., and Williams-Ashman, H. G. (1973) Specific basic proteins from mammalian testes. Isolation and properties of small basic proteins from rat testes and epididymal spermatozoa. *J. Biol. Chem.* 248, 4532–43.

Kozak, L. P., and Quinn, P. J. (1975) Evidence for dosage compensation of X-linked gene in the 6-day embryo of the mouse. *Dev. Biol.* 45, 65–73.

Kramer, J. M., and Erickson, R. P. (1981) Developmental program of PGK-1 and PGK-2 isozymes in spermatogenic cells of the mouse: specific activities and rates of synthesis. *Dev. Biol.* 87, 37–45.

Kratzer, P. G., and Chapman, V. M. (1981) X chromosome reactivation in oocytes of *Mus caroli. Proc. Natl. Acad. Sci. U.S.A.* 78, 3093–7.

Kratzer, P. G., Chapman, V. M., Lambert, H., Evans, R. E., and Liskay, R. M. (1983) Differences in the DNA of the inactive X-chromosomes of fetal and extraembryonic tissues of mice. *Cell* 33, 37–42.

Kratzer, P. G., and Gartler, S. M. (1978a) HGPRT activity changes in preimplantation mouse embryos. *Nature* 274, 503–4.

−(1978b) Hypoxanthine guianine phosphoribosyl transferase expression in early mouse development. In *Genetic Mosaics and Chimeras in Mammals*, ed. L. B. Russell, pp. 247–60. New York: Plenum.

Krietsch, W. K. G., Fundele, R., Kuntz, G. W. K., Fehlau, M., Burki, K., and Illmensee, K. (1982) The expression of X-linked phosphoglycerate kinase in the early mouse embryo. *Differentiation* 23, 141–4.

Krumlauf, R., Hammer, R. E., Brinster, R., Chapman, V. M., and Tilghman, S. M. (1985) Regulated expression of α-fetoprotein genes in transgenic mice. *Cold Spring Harbor Symp. Quant. Biol.*, 50, 371–8.

Lester, S. C., Korn, N. J., and DeMars, R. (1982) Derepression of genes on the human inactive X chromosome: evidence for differences in locus-specific rates of derepression and rates of transfer of active and inactive genes after DNA-mediated transformation. *Somatic Cell Genet.* 8, 265–84.

Lifschytz, E., and Lindsley, D. L. (1974) Sex chromosome activation during spermatogenesis. *Genetics* 78, 323–31.

Lyon, M. F. (1961) Gene action in the X-chromosome of the mouse (*Mus musculus* L.). *Nature* 190, 372–3.

−(1974) Mechanisms and evolutionary origins of variable X-chromosome activity in mammals. *Proc. R. Soc. Lond. [Biol.]* 187, 243–68.

McBurney, M. W., and Adamson, E. D. (1976) Studies on the activity of the X chromosomes in female teratocarcinoma cells in culture. *Cell* 9, 57–90.

McBurney, M. W., and Strutt, B. J. (1980) Genetic activity of X chromosomes in pluripotent female teratocarcinoma cells and their differentiated progeny. *Cell* 21, 357–64.

McMahon, A., Fosten, M., and Monk, M. (1981) Random X-chromosome inactivation in female primordial germ cells in the mouse. *J. Embryol. Exp. Morphol.* 64, 251–8.

McMahon, A., and Monk, M. (1983) X-chromosome activity in female mouse embryos heterozygous for Pgk-1 and Searle's translocation, T(X;16)16H. *Genet. Res. Camb.* 41, 69–83.

Mangia, F., Abbo-Halbasch, G., and Epstein, C. J. (1975) X-chromosome expression during oogenesis in the mouse. *Dev. Biol.* 45, 366–8.

Martin, G. R., Epstein, C. J., Travis, P., Tucker, G., Yatziv, S., Martin, D. W., Clift, S., and Cohen, S. (1978) X chromosome inactivation during differentiation of female teratocarcinoma stem cells in vitro. *Nature* 271, 329–33.

Migeon, B. R. (1972) Stability of X chromosome inactivation in human somatic cells. *Nature* 239, 87–9.

Migeon, B. R., and Jelalian, K. (1977) Evidence for two active X chromosomes in germ cells of female before meiotic entry. *Nature* 269, 242–3.

Mohandas, T., Sparkes, R., and Shapiro, L. (1981) Reactivation of an inactive human X chromosome: evidence for X inactivation by DNA methylation. *Science* 211, 393–6.

Monk, M. (1978) Biochemical studies on chromosome activity in preimplantation mouse embryos. In *Genetic Mosaics and Chimeras in Mammals*, ed. L. B. Russell, pp. 239–46. New York: Plenum.

Monk, M., and Harper, M., (1978) X-chromosome activity in preimplantation mouse embryos from XX and XO mothers. *J. Embryol. Exp. Morphol.* 46, 53–64.

–(1979) Sequential X chromosome inactivation coupled with celluar differentiation in early mouse embryos. *Nature* 281, 311–13.

Monk, M., and Kathuria, H. (1977) Dosage compensation for an X-linked gene in pre-implantation mouse embryos. *Nature* 270, 599–601.

Monk, M., and McLaren, A. (1981) X-chromosome activity in foetal germ cells of the mouse. *J. Embryol. Exp. Morphol.* 63, 75–84.

Mukherjee, A. B., (1976) Cell cycle analysis and X-chromosome inactivation in the developing mouse. *Proc. Natl. Acad. Sci. U.S.A.* 73, 1608–11.

Ohno, S., and Cattanach, B. M. (1962) Cytological study of an X-autosome translocation in *Mus musculus. Cytogenetics* 1, 129–40.

Papaioannou, V. E., and West, J. D. (1981) Relationship between the parental origin of the X-chromosome, embryonic cell lineage and X-chromosome expression in mice. *Genet. Res. Camb.* 37, 183–97.

Papaioannou, V. E., West, D. D., Bucher, T., and Linke, I. M. (1981) Nonrandom X-chromosome expression early in mouse development. *Dev. Genet.* 2, 305–15.

Plotnick, F., Klinger, H. P., and Kosseff, A. I. (1971) Sex-chromatin formation in preimplantation rabbit embryos. *Cytogenetics* 10, 244–53.

Rastan, S. (1982) Timing of X-chromosome inactivation in postimplantation mouse embryos. *J. Embryol. Exp. Morphol.* 71, 11–24.

Russell, L. B., ed. (1978) *Genetic Mosaics and Chimeras in Mammals, Basic Life Sciences, Vol. 12.* New York: Plenum.

Ryoji, M., and Worcel, A. (1984) Chromatin assembly in Xenopus oocytes: *In vivo* studies. *Cell* 37, 21–32.

Sanford, J. P., Chapman, V. M., and Rossant, J. (1985) DNA methylation in extraembryonic lineages of mammals. *Trends in Genetics* 1, 89–93.

Sanford, J., Forrester, L., Chapman, V., Chandley, A., and Hastie, N. (1984) Methylation patterns of repetitive DNA sequences in germ cells of *Mus musculus. Nucleic Acids Res.* 12, 2823–36.

Seidman, M. M., Levine, A. J., and Weintraub, H. (1979) The asymmetric segregation of parental nucleosomes during chromosome replication. *Cell* 18, 439–49.

Sugawara, O., Takagi, N., and Sasaki, M. (1983) Allocyclic early replicating X chromosome in mice: genetic inactivity and shift into a late replicator in early embryogenesis. *Chromosoma (Berl)* 88, 133–8.

Takagi, N. (1974) Differentiation of X-chromosomes in early female mouse embryos. *Exp. Cell Res.* 86, 127–35.

–(1978) Preferential inactivation of the paternally derived X-chromosome in mice. In *Genetic Mosaics and Chimeras in Mammals*, ed. L. B. Russell, pp. 341–60. New York: Plenum.

Takagi, N., and Martin, G. R. (1984) Studies of the temporal relationship between the cytogenetic and biochemical manifestations of X-chromosome inactivation during the differentiation of LT-1 teratocarcinoma stem cells. *Dev. Biol.* 103, 425–33.

Takagi, N., and Sasaki, M. (1975) Preferential inactivation of the paternally derived X chromosome in the extraembryonic membranes of the mouse. *Nature* 256, 640–2.

Takagi, N., Sugawara, O., and Sasaki, M. (1982) Regional and temporal changes in the pattern of X-chromosome replication during the early postimplantation development of the female mouse. *Chromosoma (Berl.)* 85, 275–86.

Takagi, N., Wake, N., and Sasaki, M. (1978) Cytologic evidence for preferential inactivation of the paternally derived X-chromosome in XX mouse blastocysts. *Cytogenet. Cell Genet.* 20, 240–8.

Takagi, N., Yoshida, M. A., Sugawara, O., and Sasaki, M. (1983) Reversal of X-inactivation in female mouse somatic cells hybridized with murine teratocarcinoma stem cells *in vitro*. *Cell* 34, 1053–62.

von Kap-Herr, C., and Mukherjee, B. B. (1977) Stability of inactive X-chromosome in mouse embryoid body–mule cell and transformed mouse cell–mule heterokaryon. *Exp. Cell Res.* 104, 369–76.

West, J. D. (1982) X-chromosome expression during mouse embryogenesis. In *Genetic Control of Gamete Production and Function*, eds. P. G. Crosignani and B. L. Rubin, pp. 49–91. New York: Academic Press.

West, J. D., Frels, W. I., Chapman, V. M., and Papaioannou, V. E. (1977) Preferential expression of the maternally derived X-chromosome in the mouse yolk sac. *Cell* 12, 873–82.

Wolf, S. F., Jolly, D. J., Lunnen, K. D., Friedman, T., and Migeon, B. R. (1984) Methylation of the hypoxanthine phosphoribosyl transferase locus on the human X-chromosome: implications for X-chromosome inactivation. *Proc. Natl. Acad. Sci. U.S.A.* 81, 2806–10.

Wolf, S. F., and Migeon, B. R. (1985) Clusters of CpG dinucleotides implicated by nuclease hypersensitivity as control elements of housekeeping genes. *Nature* 314, 467–9.

Toward a genetic understanding of development

13 Evidences and consequences of differences between maternal and paternal genomes during embryogenesis in the mouse

M. AZIM H. SURANI

CONTENTS

I. Introduction

The purpose of this chapter is to draw attention to recent investigations that suggest differential roles for paternal and maternal genomes during embryogenesis in the mouse. It appears that while homologous chromosomes are spatially segregated during oogenesis and spermatogenesis, they are subjected to modifications that subsequently evoke different responses at later events throughout development. Hence, differential gene expression during development may be controlled in a major way by specific modifications of homologous chromosomes prior to fertilization.

Recent studies demonstrate that development to term requires the presence of both a maternal genome and a paternal genome (McGrath and Solter 1984a; Surani, Barton, and Norris 1984). Organized growth and development to term cannot be achieved if the embryonic genome lacks the chromosomal complement from either parent, even though such embryonic cells can proliferate, differentiate, and give rise to a full range of cell types under some circumstances. Hence, paternal and maternal genomes must act in concert to allow embryonic development to progress to term. In the account that follows, I hope to show that distinct functions are apparently dependent on the different parental origins of the chromosomes. Aspects of the heritability of this information and its use during development and the probable time of reversal or reintroduction of chromosome modifications during gametogenesis will also be discussed.

II. Genetic constitution and development: roles of paternal and maternal genomes

A. Experimental Approaches

Clear evidence of differences in the influences of paternal and maternal genomes on embryonic development has recently been obtained by altering the genetic constitution of eggs and embryos. Many aspects of this work depend on the ability to carry out micromanipulations of eggs and embryos. Following initial attempts by Lin (1971), many advances in techniques have occurred (Gardner 1968; Modlinski 1975; Hoppe and Illmensee 1977; Tarkowski 1977; Barton and Surani 1983; McGrath and Solter 1983). The ability to extract karyoplasts containing pronuclei (McGrath and Solter 1983) and introduce them into recipient eggs by Sendai-virus-assisted fusion (Graham 1971; McGrath and Solter

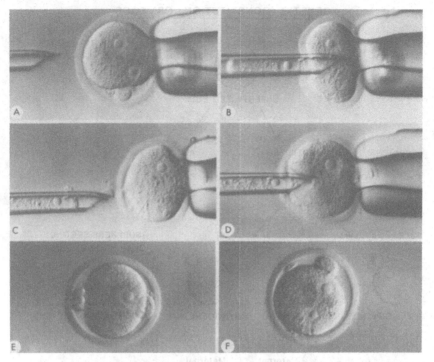

Figure 13.1. Reconstitution of eggs: The egg is held in position by a holding pipette (A), and the donor pronucleus is aspirated into the 25-μm beveled needle (B). The donor karyoplast is held in the microneedle, into which about 5 μl of inactivated Sendai virus is also taken up (C). The contents of the needle are gently released into the perivitelline space of a recipient egg (D). The karyoplast with donor pronucleus is seen on the left-hand side (E), and it eventually fuses with the recipient egg (F). The method is that of McGrath and Solter (1983).

1983; Surani et al. 1984) has allowed rapid progress in this field (Figure 13.1).

With the use of micromanipulative and other techniques, eggs with different genetic constitutions can be prepared (Figure 13.2). If the female pronucleus is removed from a fertilized egg, the resulting haploid egg with a male pronucleus is called androgenetic. A second male pronucleus can be introduced into this haploid egg to produce a biparental heterozygous androgenetic egg. Similarly, if the male pronucleus is removed from a fertilized egg, a haploid gynogenetic egg is produced into which a second female pronucleus can be introduced to obtain a biparental heterozygous gynogenetic egg. Alternatively, eggs that lack both ge-

GENETIC CONSTITUTION OF EGGS

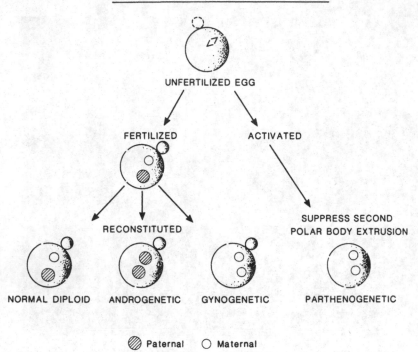

Figure 13.2. Preparation of eggs of different genetic constitutions.

netic and extragenetic contributions from spermatozoa can be prepared by artificial activation of unfertilized eggs (Graham 1974). Such an activated or parthenogenetic egg is haploid, with one female pronucleus. However, if the second polar-body extrusion is prevented, diploid parthenogenetic eggs are obtained that contain two female pronuclei, and in this respect they resemble gynogenetic eggs.

B. Reasons for developmental failure of parthenotes

Almost all the early work was devoted to assessing development of parthenogenetic eggs. In both spontaneously occurring parthenogenetic eggs in the LT/Sv strain of mice (Stevens 1975) and following artificial activation in vitro (Graham 1974), development was initiated, but it did not proceed to term. In our studies on parthenogenetic eggs we have found that the majority of the embryos implant and develop up to day 8 of gestation, but the number of embryos surviving declines rapidly, with only a few giving rise to small 25-somite embryos with very sparse extra-embryonic tissues at day 10 (Kaufman, Barton, and Surani 1977; M. A. H. Surani and S. C. Barton, unpublished data). This midgestation death is a recurring theme in the development of all eggs that lack a paternal genome.

Two reasons for failure of parthenogenetic embryos to develop to term usually have been cited (Graham 1974; Stevens 1975; Whittingham 1980; Kaufman 1981): Because these eggs are essentially homozygous, expression of recessive lethal genes could account for their developmental failure. The other suggested explanation is that lack of an extragenetic contribution from spermatozoa in such eggs could be responsible for their failure to develop. Subsequent studies using both enucleation and nucleus-transfer techniques have shown that both these explanations probably are incorrect.

In one group of studies it appeared that an extragenetic contribution from spermatozoa and/or the cytoplasm of fertilized eggs was crucial for development to term. It was reported that development to term of a few homozygous uniparental androgenetic and gynogenetic eggs was possible after removal of a male or a female pronucleus from fertilized eggs, followed by diploidization of eggs by suppression of the first cleavage division with cytochalasin B (Hoppe and Illmensee 1977). Furthermore, it was reported that if nuclei from the inner cell mass (ICM) of blastocysts resulting from diploid parthenogenetic eggs of the LT/Sv strain of mice were transplanted to enucleated fertilized eggs, development to term also resulted (Hoppe and Illmensee 1982). However, thus far it has not been possible to obtain similar results in other laboratories, and these findings cannot be confirmed (Modlinski 1980; Markert 1982).

An alternative experiment to test the role of extragenetic contributions was designed using a number of inbred and outbred strains of mice (Surani and Barton 1983). Triploid eggs were prepared by suppressing the second polar-body extrusion with cytochalasin B so that the eggs contained two female pronuclei and one male pronucleus (Niemierko 1975; Borsuk 1982) (Figures 13.3 and 13.4). These eggs, when transferred to recipient females, showed abnormal development typical of triploid eggs, with failure of neural tube closure or enlarged head region and inability of the embryo to turn properly (Surani and Barton 1983). When a female pronucleus was removed from such eggs and normal genetic constitution was restored, nearly 40 percent of such manipulated eggs reached term. This clearly demonstrates that a large proportion of eggs are capable of normal development after manipulation. However, when the male pronucleus was removed from such eggs to produce gynogenetic (digynic) eggs with two female pronuclei, development closely resembled that obtained with the genetically similar diploid parthenogenetic eggs. A large number of such eggs implanted, but only a few small 25-somite embryos were detected, and always with extremely sparse trophoblast. The results were incompatible with those reported previously

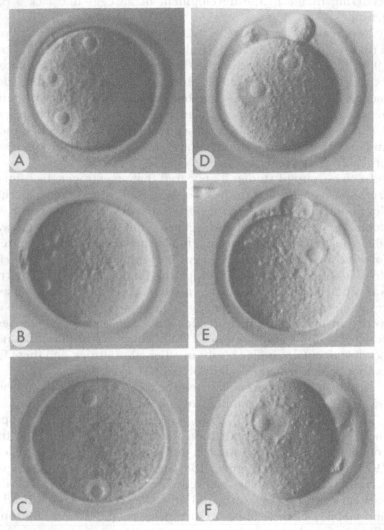

Figure 13.3. Mouse eggs: (A) triploid eggs; (B) digynic egg after removal of male pronucleus; (C) normal diploid egg after removal of female pronucleus; (D) fertilized egg; (E) haploid gynogenetic egg; (F) haploid androgenetic egg.

(Hoppe and Illmensee 1977, 1982) and seemed to eliminate the possibility that an extragenetic component was required for normal embryonic development. This was also confirmed by transfer of pronuclei from normally fertilized to parthenogenetically activated eggs, which allowed development to term (Mann and Lovell-Badge 1984) (Figure 13.5).

MANIPULATION OF TRIPLOID EGGS

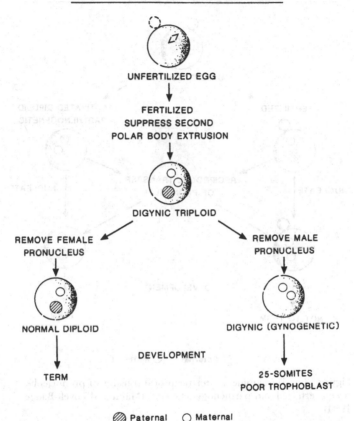

UNFERTILIZED EGG

FERTILIZED
SUPPRESS SECOND
POLAR BODY EXTRUSION

DIGYNIC TRIPLOID

REMOVE FEMALE
PRONUCLEUS

REMOVE MALE
PRONUCLEUS

NORMAL DIPLOID

DIGYNIC (GYNOGENETIC)

DEVELOPMENT

TERM

25-SOMITES
POOR TROPHOBLAST

⬤ Paternal   ◯ Maternal

Figure 13.4. Development of eggs after removal of a male or a female pronucleus from triploid eggs.

The studies described thus far did not rule out the possibility that failure of parthenogenetic development to term could be due to the essentially homozygous genotype of parthenogenetic or gynogenetic (digynic) eggs. However, two different experimental approaches essentially ruled out this possibility. In the first, eggs were activated to produce haploid parthenogenones into which either a second female or a second male pronucleus from a different strain of mouse was introduced (Surani et al. 1984). The reconstituted eggs were thus heterozygous and genetically similar except for the origin of the male or female donor pronucleus (Figure 13.6). The results demonstrated unequivocally that development proceeded to term only when the second pronucleus was male, not when the

MANIPULATION OF ACTIVATED AND FERTILIZED EGGS

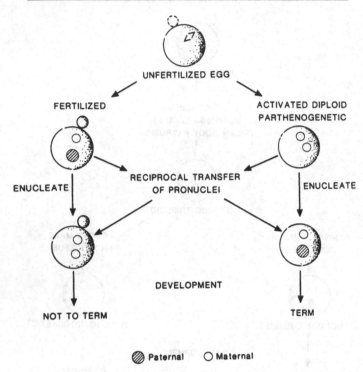

Figure 13.5. Development after reciprocal transfer of pronuclei between fertilized and parthenogenetic eggs (Mann and Lovell-Badge 1984).

donor pronucleus was female. The development of biparental eggs with two female pronuclei was similar to that described for parthenogenetic and gynogenetic (digynic) eggs: Some advanced midgestation development was possible, but always with extremely sparse extraembryonic tissue, especially trophoblast.

In the second experiment, fertilized eggs were used to reconstitute biparental heterozygous gynogenetic or androgenetic eggs by nuclear transfer (Barton, Surani, and Norns 1984; McGrath and Solter 1984a). It was once again shown that both the gynogenetic and androgenetic eggs failed to develop to term (Figure 13.7). Only those eggs that contained a male pronucleus and a female pronucleus developed normally to term.

The results of all these studies demonstrate that there are no functional differences in the properties of cytoplasm of activated parthenogenetic and fertilized eggs; rather, the failure of parthenogenetic eggs is due to lack of a paternal genome. Development to term requires the presence of both male and female genomes in eggs.

RECONSTITUTION OF ACTIVATED AND FERTILIZED EGGS

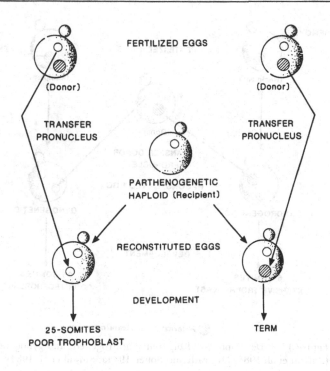

FERTILIZED EGGS

(Donor)   (Donor)

TRANSFER   TRANSFER
PRONUCLEUS   PRONUCLEUS

PARTHENOGENETIC
HAPLOID (Recipient)

RECONSTITUTED EGGS

DEVELOPMENT

25-SOMITES   TERM
POOR TROPHOBLAST

⊘ Paternal   ○ Maternal

Figure 13.6. Development after introduction of a male or a female pronucleus into parthenogenetic or fertilized haploid recipient eggs.

III. Unraveling differences between maternal and paternal genomes

The studies described thus far suggest that both paternal and maternal genomes are necessary for development to term. The purpose now is to extend our understanding of the functional differences between parental chromosomes. It is imperative at this stage to understand the biological nature of the differences, because this will eventually provide the foundation for discovering the molecular basis of the differences.

A. Parental origin of chromosomes and their influence on development: experimental approaches

i. Embryologic. Some 60 to 80 percent of eggs with two maternal genomes (parthenogenetic and gynogenetic) develop to form apparently normal blastocysts, but only about 20 percent of androgenetic eggs do so

RECONSTITUTION OF ANDROGENETIC AND GYNOGENETIC EGGS

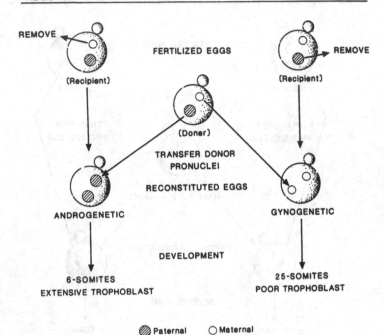

Figure 13.7. Development of biparental androgenetic and gynogenetic eggs (Barton et al. 1984; McGrath and Solter 1984a; Surani et al. 1984).

(M. A. H. Surani, S. C. Barton, and M. L. Norris, unpublished data). Although 25 percent of the androgenetic eggs with YY genetic constitution would not be expected to progress beyond a few cleavage divisions (Morris 1968), the majority of those with XX and XY genetic constitution should be capable of development to the blastocyst stage. The result therefore suggests that differences between maternal and paternal genomes may act during preimplantation development.

There are also marked differences in development of embryos of different parental constitution during the postimplantation period. All embryos that lack chromosomes of paternal origin (parthenogenetic, digynic, and biparental gynogenetic) develop to a relatively advanced stage on day 10 of gestation to form small 25-somite embryos that are characterized by extremely limited development of trophoblast and other extraembryonic tissues (Surani and Barton 1983; Surani et al. 1984). By contrast, an androgenetic embryo develops to form a very poor embryo proper, with only 6–8 somites, but with extensive trophoblast (Barton et al. 1984). The volume of this tissue is similar to that in control embryos (Figure 13.8). Hence, it appears that maternal genome is relatively more important for development of the embryo, whereas the paternal genome appears to be essential for proliferation of the extraembryonic tissues.

On the basis of the results described earlier, one of the reasons for poor development of parthenogenetic and gynogenetic embryos could be the poor development of the extraembryonic placental tissue, especially at the time when the embryo's reliance on it for nutrition becomes acute. Hence, it was of interest to examine the development of the ICM from parthenogenetic and gynogenetic embryos when placed inside normal trophectoderm (TE) vesicles. Previous studies have demonstrated that the ICM, consisting of ectoderm and primitive endoderm cells, contributes to the development of all three germ layers of the fetus and the endoderm of the visceral and parietal yolk sacs, whereas the TE gives rise exclusively to the extraembryonic tissues: trophoblast, ectoplacental cone, and extraembryonic ectoderm (Gardner, Papaioannou, and Barton 1973; Gardner and Rossant 1979; Gardner 1982; Rossant et al. 1983) (see Chapter 4).

First, we examined the development of reconstituted blastocysts from normal ICM and TE (Barton et al. 1985). These control embryos developed normally and reached term, as expected from previous work. However, when normal ICM was introduced into TE vesicles derived from parthenogenetic or heterozygous biparental gynogenetic embryos, the reconstituted embryos developed very poorly, and in most respects their development resembled that obtained with unoperated gynogenones and parthenogenones (Figure 13.9). The trophoblast failed to proliferate, and there was no detectable cellular contribution to the trophoblast from the normal ICM. Therefore, even in the presence of a normal ICM, the trophoblast fails to develop if the cells lack a paternal genome. The embryonic development that then occurred from normal ICM was also extremely poor. Only a few small 25-somite embryos were detected, and these once again resembled those obtained from unoperated parthenogenones and gynogenones (Kaufman et al. 1977; Surani and Barton 1983). Therefore, the presence of parthenogenetic or gynogenetic trophoblast has a marked effect on the development of even a normal ICM. Lack of normal trophoblast proliferation may provide one explanation for failure of embryonic development in the absence of a paternal genome.

Further studies were therefore carried out in which blastocysts were reconstituted from ICM derived from either gynogenetic or parthenogenetic embryos placed inside TE vesicles prepared from blastocysts from fertilized eggs (Figure 13.9). These reconstituted blastocysts developed substantially better than those in the preceding group. The trophoblast proliferated normally, which demonstrates that both parthenogenetic and gynogenetic ICMs can interact appropriately with normal TE and induce its proliferation. Furthermore, all the embryos developed substantially, and indeed some advanced gynogenones and parthenogenones with over 40 somites on day 12 of gestation have now been obtained. This improved development was observed without a direct cellular contribution from TE to the fetus. This result again suggests that lack of tropho-

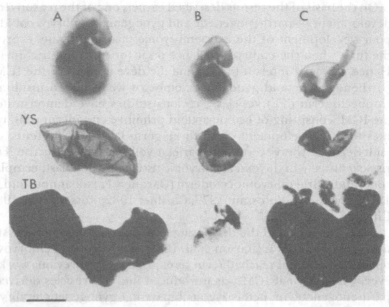

Figure 13.8. Development on day 10 of gestation: (A) control; (B) biparental gynogenetic; (C) biparental androgenetic; YS, yolk sac; TB, trophoblast. Scale: bar = 1 mm. [Reproduced from Surani et al. (1986), with permission; copyright retained by *Cell.*]

blast proliferation is a major cause of developmental failure in the absence of a paternal genome.

However, although some advanced gynogenones and parthenogenones were obtained on day 12 of gestation, they were smaller than the equivalent control embryos and showed some evidence of abnormalities and tissue degeneration. None has reached term. Further improvement in development may occur if normal primitive endoderm and TE cells are provided.

Previous studies have shown that when parthenogenetic embryos are combined with normal embryos, they can develop to term as chimeras (Surani, Barton, and Kaufman 1977; Stevens, Varnum, and Eicher 1977) and sometimes give rise to germ cells of parthenogenetic origin (Stevens 1978). Therefore cells that lack a paternal genome do not suffer from autonomous cell lethality, as also judged by their ability to proliferate and differentiate extensively in ectopic sites (Graham 1974; Iles et al. 1975; Stevens 1975). Furthermore, diploidized, homozygous, gynogenetic embryos can contribute descendants to chimeras, including both their somatic and germ cell lineages; such chimeras have reproduced successfully using their homozygous, uniparental cells as the source of the gametes. (Anderegg and Markert in press). One preliminary finding may also be of significance. We have thus far failed to obtain development to term of

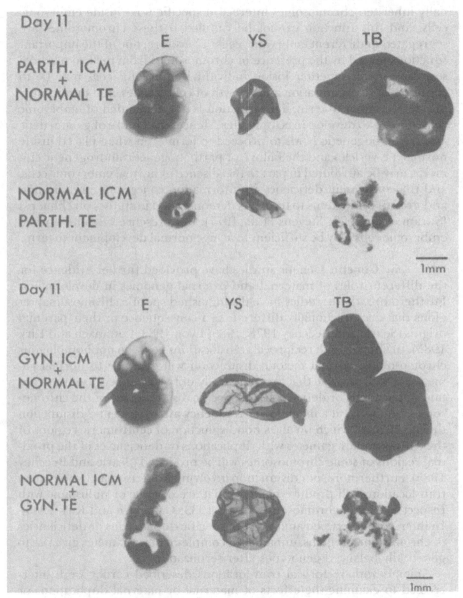

Figure 13.9. Development of reconstituted blastocysts; ICM, inner cell mass; TE, trophectoderm vesicle; E, embryos; YS, yolk sac; TB, trophoblast. Note the extent of development of embryos and extraembryonic tissues in each instance.

chimeras when parthenogenetic and androgenetic embryos were combined, although the former embryos can develop to term as chimeras when combined with normal embryos (Stevens et al. 1977; Surani et al. 1977). This finding, if confirmed, will suggest that paternally and maternally inherited chromosomes interact in specific ways inside embryonic cells, and this function cannot be fulfilled if these chromosomes are segregated in different embryonic cells. Therefore, one of the important functions served by the presence of chromosomes from the two parental sources present together inside individual embryonic cells may be to control spatial organization and growth of cells, tissues, and organs during development to term. This function is not demanded of embryonic cells allowed to develop in ectopic sites. Hence, the failure of gynogenetic and parthenogenetic ICMs to proceed to term even when placed inside normal TE vesicles and the failure of parthenogenetic/androgenetic chimeras may be attributed in part to the absence of normal embryonic cells, and the consequent deficiency in information concerning organization and growth of the fetus to term. As demonstrated in studies on chimeras (Surani et al. 1977; Stevens et al. 1977), the presence of some normal embryonic cells may be sufficient to achive normal development to term.

ii. Genetic. Genetic studies have provided further evidence for the different roles of maternal and paternal genomes in development. Furthermore, these studies have distinguished specific chromosomal regions that are functionally different as a consequence of their parental origins (Searle and Beechey 1978, 1985; Lyon 1983; Cattanach and Kirk 1985). In animals with reciprocal translocations between nonhomologous chromosomes, normal meiotic disjunction will give rise to normal gametes, as well as those that are genetically unbalanced because they contain duplications or deficiencies of specific distal regions of the chromosomes. Similarly, at a much lower frequency after adjacent-2 disjunction during meiosis, which involves nondisjunction of centromeric regions of the chromosomes, gametes with duplications or deficiencies of the proximal regions of some chromosomes will be produced (Searle and Beechey 1985). Furthermore, nondisjunction involving Robertsonian (whole-arm) translocations will produce gametes that are disomic or nullisomic with respect to certain chromosomes (Lyon 1983; Cattanach and Kirk 1985). In intercrosses between animals with specific duplications or deficiencies of chromosomal regions, unbalanced complementary gametes give rise to genetically balanced genotypes after fertilization.

Animals with reciprocal translocations described earlier were intercrossed to examine the effects of maternal or paternal duplications or deficiencies of specific chromosomal regions in genetically balanced embryos. The results (reviewed by Searle and Beechey 1985) showed that maternal duplication of specific regions of chromosomes 2, 6, 7, 8, and 17 failed to complement the corresponding paternal deficiency and this resulted in nonviable embryos or offspring (Figure 13.10). By contrast, the reciprocal combination of paternal duplication with corresponding maternal deficiency of the same chromosomal regions produced fully viable animals. For a specific distal region of chromosome 7, the reverse occurred, because it was paternal duplication of this chromosomal region with maternal deficiency that failed to give rise to normal young. With respect to chromosome 11, the parental origin of a proximal region of the chromosome affected the size of the progeny, rather than viability, as will be described later (Cattanach and Kirk 1985). It was also of interest that there were no functional differences with respect to parental origin of chromosomes 1, 4, 5, 9, 13, 14, and 15 (Lyon 1983; Searle and Beechey 1985). The information on the remaining chromosomes is incomplete or inconclusive.

In a recent study involving chromosome 11, it was shown that mice disomic for maternally derived chromosome 11 were significantly smaller than their littermates, whereas paternally derived disomic-11 mice were substantially larger (Cattanach and Kirk 1985). Using mice with reciprocal translocations, it was demonstrated that this effect was due to a proximal region of chromosome 11. Within the chromosomal region resides the allele for a growth-hormone gene, Ames dwarf (df) (Green 1981), and this region by homology with a region of a human chromosome, may also contain alleles for other growth hormones, one of which acts on the placenta (Cattanach and Kirk 1985). There is, however, no evidence as yet to link the observed effect of the parental origin of the proximal region of chromosome 11 with differential activities of these growth-hormone genes.

Studies involving a distal region of chromosome 2 also revealed contrasting phenotypes depending on the parental origin of the chromosomal region (Cattanach and Kirk 1985). Mice that contained maternal duplication and paternal deficiency of this chromosomal region were long and flat-sided and were essentially inactive (hypokinetic). However, in mice in which this region was paternally duplicated, with corresponding maternal deficiency, the animals were short and square and hyperkinetic.

The complementary phenotypes discussed earlier are dependent on the parental origin of the chromosomes. This is reminiscent of the complementary effects of maternal and paternal genomes on the embryonic and extraembryonic tissues. Duplication of the whole maternal genome in gynogenetic and parthenogenetic eggs produced advanced 25-somite embryos with sparse extraembryonic tissues (Surani and Barton 1983; Surani et al. 1984), whereas duplication of the paternal genome in androgenetic eggs produced very poor embryos but with substantial trophoblast (Barton et al. 1984).

Studies were also carried out on chromosome 17 (Johnson 1974, 1975; Lyon and Glenister 1977), especially with respect to a proximal region of chromosome 17 that contains a deletion mutation, the hairpin tail (T^{hp}), which is an allele of the mouse T/t complex (Bennett 1975). It was shown that the (T^{hp}) mutation is viable in the heterozygote when derived from the father, but lethal at about the midgestation stage when derived from

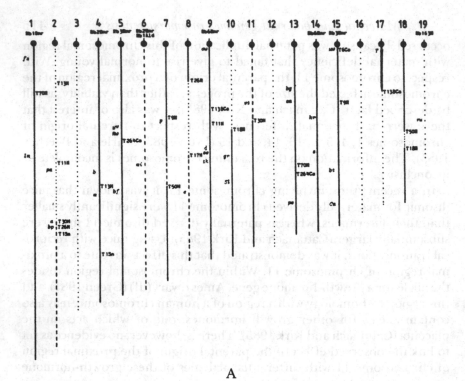

A

B

the mother (Johnson 1974, 1975). The observed effect is clearly due to the nucleus, rather than the egg cytoplasm of $T^{hp}/+$ eggs, because transfer of pronuclei to eggs of another strain of mice gave similar results (McGrath and Solter 1984b). It has been suggested that the affected region of chromosome 17 may contain an activation center that functions only when it is maternally derived, and this may be important for expression of a structural locus (Johnson 1974, 1975). Alternatively, it has been suggested that the maternally inherited locus may be preferentially active, whereas the paternal locus is inactivated (McLaren 1979). A *T*-associated maternal-effect locus (*Tme*) on this chromosome has now been characterized and mapped (Winking and Silver 1984), and it should facilitate a more precise understanding of the mechanism involved in this process.

In studies involving DDK mutant mice it was found that the heterozygous embryos from female DDK mice mated to alien males failed to survive past the implantation stage, whereas the reciprocal cross involving male DDK mice mated to alien females produced viable young (Wakasugi 1974). It has been suggested that the embryonic cytoplasm fails to activate paternal genes from alien mice that are especially important for development of the trophoblast, because the bulk of embryos died at the time of implantation (Wakasugi 1974). No evidence exists thus far to substantiate this view. Whatever the mechanism causing the failure of paternally derived chromosomes from alien mice, the result once again suggests differential modification of the paternal and maternal chromosomes.

Another example in which the behaviors of homologous chromosomes differ concerns preferential inactivation of the paternal X-chromosome (see Chapter 12). In marsupials, the paternal X chromosome is preferentially inactivated in somatic tissues (Sharman 1971). In eutherian mammals, the extraembryonic tissues show preferential paternal X inactivation whereas X inactivation is random in the fetus (Takagi and Sasaki 1975; West et al. 1977; Harper, Fosten, and Monk 1982). The paternal X chromosome is inactivated during the formation of TE and primary endoderm (Chandra and Brown 1975; Monk 1981; Lyon and Rastan 1984), but the imprinting of the paternal X is lost or not recognized in ectoderm cells, in which X inactivation is delayed until day 6 of gestation (Harper et al. 1982; Lyon and Rastan 1984). However, X inactivation is not rigidly dependent on the parental source of X chromosome. For example, in $X^{P}O$ mice, the paternal X chromosome remains active in extraembryonic tissues (Papaioannou and West 1981). Manipulating the *Xce* locus, which may represent the inactivation center, also marginally

Figure 13.10. Mouse autosomes with areas of a maternal duplication with a paternal deficiency (A) or areas of a paternal duplication with a maternal deficiency (B) that result in normal (solid line) or abnormal (broken line) development. Evidence for some chromosomal regions is inconclusive (dotted line). [Reproduced from Searle and Beechey (1985), with permission.]

influences the activity of the paternal X chromosome in the extraembryonic tissue (Rastan and Cattanach 1983). Furthermore, X inactivation has
been detected in the yolk sac in parthenogenetic embryos (Rastan et al.
1980), but the frequency of the event is variable (Endo and Takagi 1981).

iii. Molecular. There is at present no precise understanding at
the molecular level for functional differences between chromosomes
dependent on their parental origins. Recent advances in the field should
help to focus attention on diverse molecular approaches to reveal the
mechanism underlying differential roles for paternal and maternal chromosomes in development.

Attempts were made to determine if newly synthesized proteins in
fertilized, parthenogenetic, and androgenetic preimplantation embryos
were qualitatively distinct from each other (Petzoldt and Hoppe 1980;
Petzoldt et al. 1981). Analysis by one- and two-dimensional polyacrylamide gels failed to show any detectable differences between polypeptides of these embryos (Petzoldt and Hoppe 1980; Petzoldt et al. 1981; M.
A. H. Surani, S. C. Barton, and M. L. Norris, unpublished data). These
results could be interpreted in several ways. First, this type of analysis does
not always readily reveal even major functional differences between differentiated tissues (Van Blerkom, Janzen, and Runner 1982). Alternatively, the results could be interpreted to suggest that the expression of at
least the major structural genes is not influenced by the parental origins of
the genomes; products of regulatory genes probably present in small
amounts may not be distinguished on gels. Finally, differences in gene
expression may be quantitative rather than qualitative, and products of
some genes would need to be measured. The opposite phenotypes observed in studies involving chromosomes 2 and 11 (Cattanach and Kirk
1985), as well as between gynogenetic and androgenetic midgestation
embryos (Barton et al. 1984; Surani et al. 1984), may suggest excesses or
deficiencies of some gene products. Therefore, different approaches,
techniques, and stages of development will need to be considered in
future studies.

An alternative explanation for differential behavior of parental chromosomes may reside in the timing of expression of paternal and maternal
alleles during development. It was reported that some maternal alleles
such as *Pgk* are expressed earlier than paternal alleles (Krietsch et al.
1982). However, no other differences in the timing of expression of major
structural genes have thus far been reported. For example, with respect to
β_2-microglobulin, both parental alleles are expressed as early as the 2-cell
stage (Sawicki, Magnuson, and Epstein 1981). However, further studies
are required, especially regarding expression of genes located within the
homologous chromosomal regions demonstrated to display functional
differences according to parental origin.

Major molecular differences exist between the embryonic and extraembryonic tissues, although the relationship between these differences and the necessity for paternal inheritance of chromosomes for development of extraembryonic tissue is not clear. A substantially higher activity of the cellular homologues of certain viral oncogenes, especially c-*fos*, has been detected in extraembryonic tissue, but not in the embryo, on day 10 of gestation (Müller, Verma, and Adamson 1983). Another major difference is the overall demethylation of DNA in the extraembryonic lineages, whereas the embryonic DNA remains highly methylated (Chapman et al. 1984; Razin et al. 1984; Young and Tilghman 1984; Sanford, Chapman, and Rossant 1985). Evidence in the rabbit suggests that the DNA of early cleaving embryos is highly methylated (Manes and Menzel 1981), which is probably also the case in the mouse (Singer et al. 1979), whereas the DNA of trophoblast is undermethylated (Manes and Menzel 1981) (see Chapter 4).

It should be pointed out that the molecular approaches used thus far to analyze this problem are by no means comprehensive, and indeed virtually no experiments have been specifically designed to elucidate the nature of the differential activities of parental chromosomes at the molecular level. The available information from embryologic and genetic studies should provide a basis for such studies.

B. Heritability of differences between maternal and paternal genomes during development

The evidence presented thus far shows that specific regions of some mammalian chromosomes are functionally distinct as a consequence of their parental origins. However, it is not known if functional differences between parental genomes are heritable and remain intact during the course of development.

Nuclear transplantation studies using embryonic nuclei can potentially resolve a number of key questions in this respect. First, they could provide direct evidence to show if genomic differences with respect to parental origin persist beyond the pronuclear stage and after activation of the embryonic genome at the 2-cell stage. Second, transplantation of embryonic nuclei back to fertilized eggs probably would require reprogramming of the activated embryonic donor nuclei by epigenetic factors in egg cytoplasm before development could resume. Hence, these studies might show that the information with respect to the parental origin of the genomes is heritable and conserved, or they might show that this information is easily obliterated by cytoplasmic and other factors. The results of such studies could therefore reflect on the probable molecular nature of the information and help toward identification of the process of genomic modification.

Previous studies on nuclear transplantation in the mouse have produced contradictory results. Findings from one group of studies showed that nuclei from ICM of normal or parthenogenetic blastocysts were transplanted to enucleated fertilized eggs, development to term resulted in both instances (Illmensee and Hoppe 1981; Hoppe and Illmensee 1982). The latter result suggested that development and reprogramming of donor nuclei obliterate differential roles for parental genomes. However, it is not proved possible to confirm this conclusion, because in subsequent studies such nuclei have failed to support development past a few cleavage divisions (McGrath and Solter 1984c; M. A. H. Surani and S. C. Barton, unpublished data). Only nuclei from 2-cell mouse embryos resulting from fertilized eggs have thus far been confirmed to support development to the apparently normal blastocyst stage, whereas 8-cell totipotent nuclei transplanted into enucleated eggs have failed to progress to the blastocyst stage (McGrath and Solter 1984c). Indeed, it has been argued that with activation of the embryonic genome at the 2-cell stage (Epstein 1975; Van Blerkom and Brockway 1975; Braude et al. 1979; Flach et al. 1982; Bensaude et al. 1983) and with the evident functional differences between paternal and maternal genomes (Barton et al. 1984; McGrath and Solter 1984a; Surani et al. 1984), reprogramming of the donor nucleus once it has progressed past the 2-cell stage may be biologically impossible (McGrath and Solter 1984c). Hence, these conflicting results cannot be easily resolved.

We adopted a different approach to these studies, with the help of haploid embryos, because at least some haploid parthenogenetic embryos can develop normally at preimplantation stages (Surani, Barton, and Norris 1986). Although we found that a small proportion of gynogenetic haploid embryos can develop as far as the blastocyst stage, haploid androgenetic eggs very rarely progress past the 4-cell stage; the reasons for these differences are unclear at present. Nevertheless, we were able to demonstrate that in both haploid gynogenetic and parthenogenetic embryos, activation of the embryonic genome occurred at the 2-cell stage and at the same time as in fertilized eggs.

In our next experiment, we carried out functional tests for haploid embryonic nuclei by transplanting them into fertilized eggs from which either a male or female pronucleus was first removed (Figure 13.11). When 2- and 4-cell haploid nuclei were transplanted from androgenetic embryos back into eggs in which a female pronucleus was retained, normal development to term was detected in 12−18 percent of cases, and both male and female progeny were obtained (Figure 13.11A). Therefore, androgenetic haploid embryos bearing either the X or Y chromosome had developed up to the 4-cell stage. However, when these nuclei were transplanted into eggs from which the female pronucleus was first removed and only a male pronucleus remained, development was not normal, but resembled that obtained with androgenetic eggs, in which embryonic development was extremely poor, but the trophoblast proliferated to about the same extent as in normal conceptuses (Barton et al. 1984).

These experiments were repeated using haploid nuclei from 2–16-cell gynogenetic haploid embryos (Figure 13.11B). When nuclei were transplanted back to fertilized eggs from which the female pronucleus was first removed, development to term was observed, and both male and female progeny were detected. However, when haploid nuclei from gynogenetic embryos were transplanted back to eggs in which the male pronucleus was first removed, development was typical of that obtained with gynogenetic eggs. Embryos implanted, and some small midgestation embryos were detected, but always with extremely sparse trophoblast and extraembryonic membranes (Surani et al. 1984).

Development to term has not been observed with nuclei from haploid gynogenetic embryos after the 16-cell stage. This could be because of a fundamental change in the differentiated state of nuclei from the equipotent 8-cell blastomeres (Kelly 1979) to the formation of distinct inner and outer groups of cells at the 16-cell stage (see Chapter 1). Alternatively, the asynchrony between the donor nucleus and the resident pronucleus may be too substantial after the 16-cell stage for complementation to occur. However, the results with transplantation of 8-cell nuclei showed that these donor nuclei were reprogrammed when transplanted to haploid eggs, because in the reconstituted eggs, activation of the genome and blastulation did not occur precociously.

The reason why eggs can develop to term after transfer of haploid 8–16-cell embryonic nuclei in the presence of an appropriate pronucleus, whereas diploid 8-cell nuclei fail to do so, is unclear at present. The presence of a pronucleus may control early development and allow donor nuclei greater time during early divisions to reprogram adequately and synchronize with the pronucleus. These findings are in accord with observations showing formation of apparently normal tetraploid blastocysts after transfer of morula and ICM nuclei into fertilized eggs (Modlinski 1978, 1981). Further studies are clearly needed to resolve many of these points. It will also be of interest to determine conditions under which haploid androgenetic eggs can develop beyond the 4-cell stage and whether or not this may occur in vivo (Tarkowski and Rossant 1976).

These initial results of our study demonstrate that the differential activities of parental genomes persist beyond the pronuclear stage and after activation of the embryonic genome. In addition, the information also persists after programming of donor nuclei by epigenetic factors in egg cytoplasm. Hence, the information concerning parental origin of the genomes is heritable and conserved, and it is not obliterated by changes in cytoplasmic factors. Genetic studies with paternal or maternal duplica-

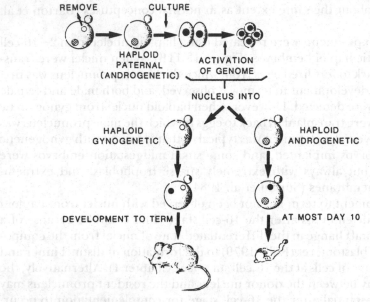

A

TRANSPLANTATION OF HAPLOID PATERNAL NUCLEI

REMOVE   CULTURE

HAPLOID
PATERNAL   ACTIVATION
(ANDROGENETIC)   OF GENOME

NUCLEUS INTO

HAPLOID   HAPLOID
GYNOGENETIC   ANDROGENETIC

DEVELOPMENT TO TERM   AT MOST DAY 10

◎ MALE PRONUCLEUS   ○ FEMALE PRONUCLEUS   ● MALE DONOR NUCLEUS

B

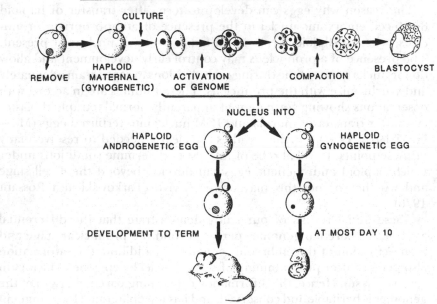

TRANSPLANTATION OF HAPLOID MATERNAL NUCLEI

CULTURE

REMOVE   HAPLOID   ACTIVATION   COMPACTION   BLASTOCYST
         MATERNAL   OF GENOME
         (GYNOGENETIC)

NUCLEUS INTO

HAPLOID   HAPLOID
ANDROGENETIC EGG   GYNOGENETIC EGG

DEVELOPMENT TO TERM   AT MOST DAY 10

○ FEMALE PRONUCLEUS   ◎ MALE PRONUCLEUS   ● FEMALE DONOR NUCLEUS

tions of certain regions of chromosomes 2 and 11 (Cattanach and Kirk 1985) also suggest that the parental origins of chromosomes can influence development at very advanced stages of embryogenesis. These studies therefore indicate that the molecular nature of the information and the process of its introduction should be compatible with the ability to transmit the information to later stages of development.

C. Reprogramming of maternal and paternal genomes

The mechanism of introduction and the nature of the information responsible for the differential roles of parental genomes during development and how it is acquired are unknown. Nevertheless, certain features of the process can be deduced from the available information. For example, it seems necessary that the information be stable and heritable so as to influence events at stages later in development. As for the timing of the process, it seems likely that reprogramming of the parental genomes occurs during gametogenesis, because the homologous chromosomes are spatially segregated at that time.

The epigenetic factors responsible for specific modifications of parental chromosomes presumably differ during spermatogenesis and oogenesis (see Chapters 6 and 7). There are indications to support this view, because the inactivated X-chromosome is reactivated during oogenesis, so that the XX chromosomes are active, but the XY bivalent is heterochromatic and inactive during spermatogenesis (Ohno, Kaplan, and Kinosita 1961; Lifschytz and Lindley 1972; Epstein et al. 1978; Johnston 1981). The paternal DNA is subsequently condensed as transcription ceases and remains inactive for a relatively long period of time following replacement of histones by protamines (Rodman et al. 1981; Rodman, Pruslin, and Allfrey 1982). By contrast, oocytes are transcriptionally active until late in their maturation, and they retain nucleosomal core histones H2b and H4 (Rodman et al. 1981). After fertilization, protamines are rapidly replaced by histones, and the more diffuse paternal chromatin replicates slightly ahead of more condensed maternal chromatin (Donahue 1972; Abramczuk and Sawicki 1975). No gross differences are visible by the 2-cell stage, when both parental alleles of β_2-microglobulin are expressed (Sawicki et al. 1981).

Substantial remodelling of chromatin apparently occurs during the formation of germ cells and gametogenesis. The process of preprogramming of parental genomes probably occurs in two stages: In the first stage, previous modifications are reversed; later, new information is introduced. The undermethylation of major and minor DNA satellite se-

Figure 13.11. Nuclear transplantation: development after transfer of 2–4-cell haploid paternal nuclei (A) or after transfer of 2–16-cell haploid maternal nuclei (B) to recipient haploid androgenetic or gynogenetic eggs. [Reproduced from Surani et al. (1986), with permission; copyright retained by *Cell*.]

424   M. Azim H. Surani

quences (Sturm and Taylor 1981; Ponzetto-Zimmerman and Wolgemuth 1984; Sanford et al. 1984) and the reactivation of the inactive X chromosome in oocytes (Ohno et al. 1961; Epstein et al. 1978; Johnston 1981; Monk 1981) probably reflect the first stage of the process. The second stage probably commences with meiosis, when parental chromosomes may be specifically modified, as judged by de novo methylation of DNA during the spermatogonial and primary spermatocyte stages (Groudine and Conkin 1985). There is evidence to show that some, but not all (Waalwijk and Flavell 1978; Rahe, Erickson, and Quinto 1983; Stein et al. 1983), structural genes in sperm are hypermethylated. Furthermore, some dispersed DNA sequences in oocytes are undermethylated, whereas the same sequences are hypermethylated in sperm DNA (Sanford et al. 1984).

The involvement of DNA methylation in gene expression during development has been considered in recent years (see review by Jähner and Jaenisch 1984), but the role of this modification is not yet clearly established. Demethylation of DNA and sensitivity of chromatin to deoxyribonuclease 1 (Elgin 1981; Groudine and Weintraub 1982) are associated with active genes, although not all genes that show these properties are active. However, these changes in chromatin are of particular interest, because the modifications have been demonstrated to be heritable through successive DNA replication cycles (Holliday and Pugh 1975; Elgin 1981; Wigler 1981). Hence, these modifications, if introduced into chromatin during gametogenesis, could be protected by protein binding to DNAse-1-hypersensitive sites (Groudine and Weintraub 1982) and transmitted to the embryo after fertilization. The modified DNA sequences could then be activated at particular stages of development by tissue-specific transacting proteins when they bind to DNA enhancer sequences that function in *cis* (Serpling, Jasin, and Schaffner 1985). Thus, there are possible mechanisms to explain programming of parental genomes; these provide a basis for further exploration and experimental verification.

IV. Wider implications of differences between parental origins of genomes

The purpose here is to briefly consider the extent to which the apparent differences between maternal and paternal genomes may be of relevance to some aspects of reproductive strategies, evolution of the species, and development in other animals and in humans.

A. Significance for reproductive strategies

One of the striking aspects of reproduction is the loss during vertebrate evolution of asexual or parthenogenetic modes of reproduction (Markert 1982). This is despite the fact that in many respects parthenogenesis is a less wasteful procedure for rapid multiplication of individual animals. The progeny derived by parthenogenesis are essentially clones of the parent, with little genetic variability between them. Because there are only a few surviving species in which this form of reproduction is exclusively used, it is fair to assume that sexual reproduction has advantages. It allows genetic variability to occur and thereby confers greater benefits on the species, enabling them to exploit new and changing environments. Parthenogenesis still survives in a strictly limited way in birds, especially in some breeds of turkeys, but only a few of them develop into normal adults (Harada and Buss 1981). In mammals as well there is a remnant of the process, expecially in the LT/Sv strain of mice, in which spontaneous parthenogenetic development occurs, followed by suppression of the second polar body, but all these embryos die around the time of implantation (Stevens 1975).

Perhaps, as the heterogametic mode of reproduction was introduced, the two gametes gradually acquired differential roles during development. This may be beneficial, because it allows greater control over events during embryogenesis, which is especially necessary in viviparous placental mammals. The acquisition of specialized functions by the two genomes probably led to the eventual elimination of asexual reproduction in mammals.

If parthenogenesis is undesirable in higher vertebrates, then inbreeding also is not beneficial, because it restricts the usefulness of genetic variability arising out of sexual reproduction. Indeed, severe inbreeding depression is encountered in mammals (Biemont 1974; Markert 1982). This observation is commonly attributed to the expression of recessive lethal genes. Perhaps there may also be a failure of appropriate gene expression because of the close relationship between the origins of these chromosomes (Biemont 1974). It will be of interest to compare the mechanism causing inbreeding depression with that in which development of embryos is affected by the absence of chromosomes of one parental origin.

The alternative problem encountered with the evolution of sexual reproduction is the need to restrict interspecies hybridization to some extent. There has been a substantial reduction in the incidence of cross-hybridization in mammals as compared with lower vertebrates (Wilson, Maxson, and Sarich 1974). One mechanism to prevent interspecies hybridization is suggested to be a regulatory one that causes allelic repression of hybrid genes (Wilson et al. 1974). For this reason it is of interest to note that there are major species differences especially in the newly synthesized proteins of eggs of several rodents (Norris, Barton, and Surani 1985). These differences presumably prevent appropriate activation of paternal genes from alien species. In any interspecies hybrids that do result, males are either absent or sterile (Haldane 1922; Jones and Barton 1985). The breakdown of chromosomal interaction is illustrated when a Y chromosome from *Mus domesticus* is introduced into a strain of *Mus musculus*. These XY progeny are infertile females or hermaphrodites, probably because the testis-bearing Y chromosome of *M. domesticus* fails to repress the appropriate X-linked locus of *M. musculus* (Eicher et al. 1982).

B. Relevance to development in humans

There are overt similarities in development of androgenetic eggs in the mouse and in the human (Surani 1985a, 1985b). In the human, androgenetic eggs occur spontaneously following fertilization by one spermatozoon, which undergoes diploidization, or by two spermatozoa, and in both cases the maternal genome is physically or functionally excluded from participating in embryogenesis (Jacobs, Wilson, and Sprenkle 1980). Such embryos give rise to complete hydatidiform moles that are either 46,XX or 46,XY (Szulman and Surti 1984). These moles usually are recognized at 4 weeks of pregnancy as consisting of trophoblastic hyperplasia, and in only two recorded instances were fetal remains found. This is very similar to the pattern of development of androgenetic mouse eggs. The development of human triploid eggs also shows similarities to the mouse (Szulman and Surti 1984). These fetuses usually are diandric, and they produce high levels of hCG, indicating hyperactivity of the trophoblast. Usually it is the syncytiotrophoblast that shows hyperplasia and vacuolization resembling the normal placenta. The fetus dies at 8–9 weeks of gestation. In the mouse, digynic triploids also develop with substantial trophoblast and with abnormal fetuses (Surani and Barton 1983). In rare instances in the human the trophoblast from such an embryo becomes malignant (Szulman and Surti 1984).

Because of the overt similarities in genotype and mode of development of mice and humans, it should be possible to deduce more precisely the causes of some fetal abnormalities in humans. Many different types of mouse eggs, such as diandric and digynic triploids, can be reconstituted and their development examined (Surani 1985b). Such studies may provide an experimental model for investigating the apparent association between abnormal fetal development, with hyperplasia of trophoblast, in the human and an excess of paternal over maternal genomes (A. E. Szulman, personal communication).

C. Relevance to animal breeding

Direct genetic manipulation of animal eggs by reconstitution to produce specific genotypes or by introduction of cloned genes may provide opportunities for directly altering their genetic characteristics (Markert and Seidel 1981). It is theoretically possible to rescue germ cells of nonviable embryos, such as parthenogenones, by making chimeras.

Further studies may indicate if parthenogenones can be rescued intact, perhaps by using different types of reconstituted blastocysts. Alternatively, further knowledge of chromosomal modifications during gametogenesis may, in the long term, provide means for interfering with the process. If so, replication of uniparental or biparental animals using only maternal or paternal genomes may become feasible.

V. Conclusions

The studies described here demonstrate differential roles for paternal and maternal genomes during mammalian embryogenesis. The former appears to be preferentially needed for the development of extra-embryonic tissues, and the latter for preimplantation development and embryogenesis. In addition, chromosomes of both parental origins are apparently needed within individual embryonic cells for development to proceed to term. Hence, specific interactions between homologous chromosomes and their alleles are also indicated. The presence of chromosomes from both parental origins is crucial for spatial organization and controlled growth of cells, tissues, and organs in embryos and their development to term. Genetic studies involving specific regions of chromosomes 2 and 11 demonstrate that the parental origin of these chromosomes influences the shape and size of neonatal mice. Embryonic cells that entirely lack a genetic contribution from either parental source have the potential for extensive proliferation and differentiation in ectopic sites, but this is not sufficient for embryonic development to term. Hence, this important aspect of the genetic control of development cannot be studied using embryo-derived cell lines outside the context of the embryo. Studies on chimeras reveal that the presence of some normal embryonic cells in developing fetuses is sufficient to provide such control.

The mechanism that leads to modifications of the parental chromosomes is unknown. However, it appears most likely that when homologous chromosomes are segregated during oogenesis and spermatogenesis, they are subjected to modifications, and these differences are apparently heritable and influence events throughout development. The exact time over which these differences persist in development is not known, but these modifications of homologous chromosomes may persist until the formation of new germ cells in the fetus.

Future studies to advance our understanding of this process can be achieved in several ways. It should be possible to analyze the consequences of genomic differences in greater detail, especially to gain an understanding of the process at the molecular level. Use can be made of mutants such as DDK and T^{hp}, particularly the latter, in which the chromosomal region involved in the mutation is known; this region can be mapped and cloned to provide the DNA probes to study gene expression in order to determine the influence of the parental origin of the chromosomal region on development. Alleles located within a number of other chromosomal regions identified by noncomplementation tests can be studied to establish the precise control of their expression during development. Eggs, such as gynogenones, in which a specific effect of the failure of trophoblast proliferation is observed, can also be studied in detail using both cellular and molecular approaches, with, for example, tissue-specific cDNA clones. Similarly, the role of the whole of the paternal and maternal genomes can be investigated by perturbing their functions by insertional mutagenesis to produce hemizygous dominant mutants, followed by eventual identification and study of the genes involved. Insertion of exogenous DNA, particularly into chromosomal regions already identified to function differentially, depending on their parental origin, could cause various abnormalities during development.

Better understanding of the consequences of the parental origin of chromosomes on development is likely to reveal the nature of the DNA modification itself and the mechanism of preprogramming that presumably occurs during gametogenesis. A link can them be established between structural and chemical modifications of chromatin and their transmission during development and eventual activation of specific genes by trans-acting factors. Another approach is to explore ways of experimentally altering or reversing information concerning the parental origin of chromosomes. This may be achieved by use of chemicals or by exposure of genomes to different cytoplasmic environments. If reversal of information concerning the parental origin of chromosomes can be achieved, this will result in production of uniparental animals and, more important, will provide biological models to explore mechanisms involved in modifications of chromosomes. In sum, considerable weight of evidence from embryologic and genetic studies demonstrates functional differences between parental genomes. Attempts to discover the molecular basis for these differences should provide one of the exciting challenges in developmental biology.

Acknowledgments

I am greatly indebted to my colleagues Mrs. Sheila Barton and Dr. Michael Norris for their collaboration in most of my studies described here. I am very grateful to Mr. Andrew Collick, Dr. Don Powell, and in particular Dr. Wolf Reik for stimulating discussion on a number of ideas described here and for comments on various parts of the manuscript.

References

Abramczuk, J., and Sawicki, W. (1975) Pronuclear synthesis of DNA in fertilized and parthenogenetically activated mouse eggs. *Exp. Cell Res.* 92, 361–72.

Anderegg, C., and Markert, C. L. (in press) Successful rescue of microsurgically produced homozygous, uniparental mouse embryos via production of aggregation chimeras. *Proc. Natl. Acad. Sci. U.S.A.*

Barton, S. C., Adams, C. A., Norris, M. L., and Surani, M. A. H. (1985) Development of gynogenetic and parthenogenetic inner cell mass and trophectoderm tissues in reconstituted blastocysts in the mouse. *J. Embryol. Exp. Morphol.* 90, 267–85.

Barton, S. C., and Surani, M. A. H. (1983) Microdissection of the mouse egg. *Exp. Cell Res.* 146, 187–91.

Barton, S. C., Surani, M. A. H., and Norris, M. L. (1984) Role of paternal and maternal genomes in mouse development. *Nature (Lond.)* 311, 374–6.

Bennett, D. (1975) The T-locus of the mouse. *Cell* 6, 441–54.

Bensaude, O., Babinet, C., Morange, M., and Jacob, F. (1983) Heat-shock proteins, first major products of zygotic gene activity in mouse embryo. *Nature (Lond.)* 305, 331–2.

Biemont, C. (1974) A chromosomal interaction system for the control of embryonic development. *Mech. Ageing Dev.* 3, 291–9.

Borsuk, E. (1982) Preimplantation development of gynogenetic diploid mouse embryos. *J. Embryol. Exp. Morphol.* 69, 215–22.

Braude, P., Pelham, H., Flach, G., and Lobatto, R. (1979) Post-transcriptional control in the early mouse embryo. *Nature (Lond.)* 282, 102–5.

Cattanach, B. M., and Kirk, M. (1985) Differential activity of maternally and paternally derived chromosome regions in mice. *Nature (Lond.)* 315, 496–8.

Chandra, H. S., and Brown, S. W. (1975) Chromosome imprinting and the mammalian X-chromosome. *Nature (Lond.)* 53, 165–8.

Chapman, V., Forrester, L., Sanford, J., Hastie, N., and Rossant, J. (1984) Cell lineage-specific undermethylation of mouse repetitive DNA. *Nature (Lond.)* 307, 284–6.

Donahue, R. P. (1972) Cytogenetic analysis of the first cleavage division in mouse embryos. *Proc. Natl. Acad. Sci. U.S.A.* 69, 74–7.

Eicher, E. M., Washburn, L. L., Whitney, J. B., and Morrow, K. E. (1982) *Mus poschiavinus* Y-chromosome in the C57BL/6J murine genome causes sex reversal. *Science* 217, 535–7.

Elgin, S. C. R. (1981) DNAse 1-hypersensitive sites of chromatin. *Cell* 27, 413–15.

Endo, S., and Takagi, N. (1981) A preliminary cytogenetic study of X chromosome inactivation in diploid parthenogenetic embryos from LT/Sv mice. *Jpn. J. Genet.* 56, 349–56.

Epstein, C. J. (1975) Gene expression and macromolecular synthesis during preimplantation embryonic development. *Biol. Reprod.* 12, 82–105.

Epstein, C. J., Smith, S., Travis, B., and Tucker, G. (1978) Both X chromosomes function before visible X-chromosome inactivation in female mouse embryos. *Nature (Lond.)* 274, 500–3.

Flach, G., Johnson, M. H., Braude, P. R., Taylor, R. A. S., and Bolton, V. N. (1982) The transition from maternal to embryonic control in the 2-cell mouse embryo. *EMBO Journal* 1, 681–6.

Gardner, R. L. (1968) Mouse chimeras obtained by the injection of cells into the blastocyst. *Nature (Lond.)* 220, 596–7.

—(1982) Investigation of cell lineage and differentiation in the extraembryonic endoderm of the mouse embryo. *J. Embryol. Exp. Morphol.* 68, 175–98.

Gardner, R. L., Papaioannou, V. E., and Barton, S. C. (1973) Origin of the ectoplacental cone and secondary giant cells in mouse blastocysts reconstituted from isolated trophoblast and inner cell mass. *J. Embryol. Exp. Morphol.* 30, 561–72.

Gardner, R. L., and Rossant, J. (1979) Investigations of the fate of 4.5 day post-coitum mouse inner cell mass cells by blastocyst injection. *J. Embryol. Exp. Morphol.* 52, 141–52.

Graham, C. F. (1971) Virus assisted fusion of embryonic cells. *Acta Endocrinol. (Kbh)* Suppl. 153, 154–65.

—(1974) The production of parthenogenetic mammalian embryos and their use in biological research. *Biol. Rev.* 49, 399–422.

Green, M. C., ed. (1981) *Genetic Variants and Strains of the Laboratory Mouse.* Stuttgart: Gustav Fischer.

Groudine, M., and Conkin, K. F. (1985) Chromatin structure and *de novo* methylation of sperm DNA: implications for activation of the paternal genome. *Science* 228, 1061–8.

Groudine, M., and Weintraub, H. (1982) Propagation of globin DNAse 1-hypersensitive sites in absence of factors required for induction: a possible mechanism for determination. *Cell* 30, 131–9.

Haldane, J. B. S. (1922) Sex ratio and unisexual sterility in hybrid animals. *J. Genet.* 12, 101–9.

Harada, K., and Buss, E. G. (1981) Cytogenetic studies of embryos developing parthenogenetically in turkeys. *Poultry Sci.* 60, 1362–4.

Harper, M. I., Fosten, M., and Monk, M. (1982) Preferential paternal X inactivation in extraembryonic tissue of early mouse embryo. *J. Embryol. Exp. Morphol.* 67, 127–35.

Holliday, R., and Pugh, J. E. (1975) DNA modification mechanisms and gene activity during development. *Science* 187, 226–32.

Hoppe, P. C., and Illmensee, K. (1977) Microsurgically produced homozygous-diploid uniparental mice. *Proc. Natl. Acad. Sci. U.S.A.* 74, 5657–61.

—(1982) Full-term development after transplantation of parthenogenetic embryonic nuclei into fertilized mouse eggs. *Proc. Natl. Acad. Sci. U.S.A.* 79, 1912–16.

Iles, S. A., McBurney, M. W., Bramwell, S. R., Deussen, S. R., and Graham, C. F. (1975) Development of parthenogenetic and fertilized mouse embryos in the uterus and in extra-uterine sites. *J. Embryol. Exp. Morphol.* 34, 387–405.

Illmensee, K., and Hoppe, P. C. (1981) Nuclear transplantation in *Mus musculus*: developmental potential of nuclei from preimplantation embryos. *Cell* 23, 9–18.

Jacobs, P. A., Wilson, C., Sprenkle, J. A., Rosenshein, N. B., and Migeon, B. R. (1980) Mechanism of origin of complete hydatidiform moles. *Nature (Lond.)* 286, 714–16.

Jähner, D., and Jaenisch, R. (1984) DNA methylation in early mammalian development. In *DNA Methylation, Biochemistry and Biological Significance*, eds. A. Razin, H. Cedar, and A. D. Riggs, pp. 189–219. New York: Springer-Verlag.

Johnson, D. R. (1974) Hairpin-tail: a case of post-reductional gene action in the mouse egg. *Genetics* 76, 795–805.

–(1975) Further observations on the hairpin-tail (T^{hp}) mutation in the mouse. *Genet. Res.* 24, 207–13.

Johnston, P. G. (1981) X chromosome activity in female germ cells of mice heterozygous for Searle's translocation T(X: 16)16H. *Genet. Res.* 37, 317.

Jones, T. S., and Barton, N. (1985) Sex chromosomes and evolution. *Nature (Lond.)* 314, 668–9.

Kaufman, M. H. (1978) Chromosome analysis of early postimplantation presumptive haploid parthenogenetic mouse embryos. *J. Embryol. Exp. Morphol.* 45, 85–91.

–(1981) Parthenogenesis: a system facilitating understanding of factors that influence early mammalian development. In *Progress in Anatomy*, Vol. 1, eds. R. J. Harrison and R. L. Holmes, pp. 1–34. Cambridge University Press.

Kaufman, M. H., Barton, S. C., and Surani, M. A. H. (1977) Normal postimplantation development of mouse parthenogenetic embryos to the forelimb bud stage. *Nature (Lond.)* 265, 53–5.

Kelly, S. J. (1979) Studies of the development potential of 4- and 8-cell stage mouse blastomeres. *J. Exp. Zool.* 200, 365–76.

Krietsch, W. K. G., Fundele, R., Kuntz, G. W. K., Fehlau, M., Burki, K., and Illmensee, K. (1982) The expression of X-linked phosphoglycerate kinase in the early mouse embryo. *Differentiation* 23, 141–4.

Lifschytz, E., and Lindsley, D. L. (1972) The role of X-chromosome inactivation during spermatogenesis. *Proc. Natl. Acad. Sci. U.S.A.* 69, 182–6.

Lin, T. P. (1971) Egg micromanipulation. In *Methods in Mammalian Embryology*, ed. J. C. Daniel, Jr., pp. 157–71. San Francisco: Freeman.

Lyon, M. F. (1983) The use of Robertsonian translocations for studies of nondisjunction. In *Radiation-induced Chromosome Damage in Man*, eds. T. Ishihara and M. S. Sasaki, pp. 327–46. New York: Alan R. Liss.

Lyon, M. F., and Glenister, P. (1977) Factors affecting the observed number of young resulting from adjacent-2 disfunction in mice carrying a translocation. *Genet. Res.* 29, 83–92.

Lyon, M. F., and Rastan, S. (1984) Parental source of chromosome imprinting and its relevance for X-chromosome inactivation. *Differentiation* 26, 63–7.

McGrath, J., and Solter, D. (1983) Nuclear transplantation in the mouse embryo by microsurgery and cell fusion. *Science*, 220, 1300–3.

–(1984a) Completion of mouse embryogenesis requires both the maternal and paternal genomes. *Cell* 37, 179–83.

–(1984b) Maternal T^{hp} lethality in the mouse is a nuclear, noncytoplasmic defect. *Nature (Lond.)* 308, 550–1.

—(1984c) Inability of mouse blastomere nuclei transferred to enucleated zygotes to support development *in vitro. Science* 226, 1317–19.

McLaren, A. (1976) *Mammalian Chimaeras.* Cambridge University Press.

—(1979) The impact of pre-fertilization events on post-fertilization development in mammals. In *Maternal Effects in Development,* eds. D. R. Newth and M. Balls, pp. 287–320. Cambridge University Press.

Manes, C., and Menzel, P. (1981) Demethylation of CpG sites in DNA of early rabbit trophoblast. *Nature (Lond.)* 293, 589–90.

Mann, J. R., and Lovell-Badge, R. H. (1984) Inviability of parthenogenones is determined by pronuclei, not egg cytoplasm. *Nature (Lond.)* 310, 66–7.

Markert, C. L. (1982) Parthenogenesis, homozygosity and cloning in mammals. *J. Hered.* 73, 390–7.

Markert, C. L., and Seidel, G. E. (1981) Parthenogenesis, identical twins and cloning in mammals. In *New Technologies in Animal Breeding,* eds. B. G. Brackett, G. E. Seidel, and S. M. Seidel, pp. 181–200. New York: Academic Press.

Modlinski, J. A. (1975) Haploid mouse embryos obtained by microsurgical removal of one pronucleus. *J. Embryol. Exp. Morphol.* 33, 897–905.

—(1978) Transfer of embryonic nuclei to fertilized mouse eggs and development of tetraploid blastocysts. *Nature (Lond.)* 273, 466–7.

—(1980) Preimplantation development of microsurgically obtained haploid and homozygous diploid mouse embryos and effects of pretreatment with cytochalasin B on enucleated eggs. *J. Embryol. Exp. Morphol.* 60, 153–61.

—(1981) The fate of inner cell mass and trophectoderm nuclei transplanted to fertilized mouse eggs. *Nature (Lond.)* 292, 342–3.

Monk, M. (1981) A stem-line model for cellular and chromosomal differentiation in early mouse development. *Differentiation* 19, 71–6.

Morris, J. (1968) The XO and OY chromosome constitution in the mouse. *Genet. Res.* 12, 125.

Müller, R., Verma, I. M., and Adamson, E. D. (1983) Expression of c-*onc* genes: c-*fos* transcripts accumulate to high levels during development of mouse placenta, yolk sac and amnion. *EMBO Journal* 2, 679–85.

Niemierko, A. (1975) Induction of triploidy in the mouse by cytochalasin B. *J. Embryol. Exp. Morphol.* 34, 279–89.

Norris, M. L., Barton, S. C., and Surani, M. A. H. (1985) A qualitative comparison of protein synthesis in the preimplantation embryos of four rodent species (mouse, rat, hamster and gerbil). *Gamete Res.* 12, 313–16.

Ohno, S., Kaplan, W. D., and Kinosita, R. (1961) X-chromosome behavior in germ somatic cells of *Rattus norvegicus. Exp. Cell Res.* 22, 535–44.

Papaioannou, V. E., and West, J. D. (1981) Relationship between the parental origin of the X-chromosomes, embryonic cell lineage and X-chromosome expression in mice. *Genet. Res.* 37, 183–97.

Petzoldt, U., and Hoppe, P. C. (1980) Spontaneous parthenogenesis in *Mus musculus*: comparison of protein synthesis in parthenogenetic and normal preimplantation embryos. *Mol. Gen. Genet.* 180, 547–52.

Petzoldt, U., Illmensee, G. R., Burki, K., Hoppe, P. C., and Illmensee, K. R. (1981) Protein synthesis in microsurgically produced androgenetic and gynogenetic mouse embryos. *Mol. Gen. Genet.* 184, 11–16.

Ponzetto-Zimmerman, C., and Wolgemuth, D. J. (1984) Methylation of satellite sequences in mouse spermatogenic and somatic DNAs. *Nucleic Acids Res.* 12, 2807–22.

Rahe, B., Erickson, R. P. and Quinto, M. (1983) Methylation of unique sequence DNA during spermatogenesis in mice. *Nucleic Acids Res.* 11, 7947–61.

Rastan, S., and Cattanach, B. M. (1983) Interaction between the Xce locus and imprinting of the paternal X-chromosome in mouse yolk sac endoderm. *Nature (Lond.)* 303, 635–7.

Rastan, S., Kaufman, M. H., Handyside, A. H., and Lyon, M. F. (1980) X-chromosome inactivation in extra-embryonic membranes of diploid parthenogenetic mouse embryos demonstrated by differential staining. *Nature (Lond.)* 288, 172–3.

Razin, A., Webb, C., Szyf, M., Yisraeli, Y., Rosenthal, A., Naveh-Many, T., Sciaky-Gallili, N., and Cedar, H. (1984) Variations in DNA methylation during mouse cell differentiation *in vivo* and *in vitro*. *Proc. Natl. Acad. Sci. U.S.A.* 81, 2275–9.

Rodman, T. C., Pruslin, F. H., and Allfrey, V. G. (1982) Mechanism of displacement of sperm basic nuclear proteins in mammals. An *in vitro* simulation of post-fertilization results. *J. Cell Sci.* 53, 227–44.

Rodman, T.C., Pruslin, F. H., Hoffman, H. P., and Allfrey, V. G. (1981) Turnover of basic chromosomal protein in fertilized eggs: a cytoimmunochemical study of events *in vivo*. *J. Cell Biol.* 90, 351–61.

Rossant, J., Vijh, M., Siracusa, L. D., and Chapman, V. M. (1983) Identification of embryonic cell lineages in histological sections of *M. musculus–M. caroli* chimaeras. *J. Embryol. Exp. Morphol.* 73, 179–91.

Sanford, J., Chapman, V. M., and Rossant, J. (1985) DNA methylation in extraembryonic lineage of mammals. *Trends in Genetics* 1, 89–93.

Sanford, J., Forrester, L., Chapman, V., Chandley, A., and Hastie, N. (1984) Methylation patterns of repetitive DNA sequences in germ cells of *Mus musculus*. *Nucleic Acids Res.* 12, 2823–36.

Sawicki, J. A., Magnuson, T., and Epstein, C. J. (1981) Evidence for expression of the paternal genome in the 2-cell mouse embryo. *Nature (Lond.)* 294, 450–1.

Searle, A. G., and Beechey, C. V. (1978) Complementation studies with mouse translocations. *Cytogenet. Cell Genet.* 20, 282–303.

–(1985) Non-complementation phenomena and their bearing on non-disjunctional effects. In *Aneuploidy, Aetiology and Mechanisms*, eds. V. L. Dellarco, P. E. Voytek, and A. Hollaender, pp. 363–76. New York: Plenum.

Serpling, E., Jasin, M., and Schaffner, W. (1985) Enhancers and eukaryotic gene transcription. *Trends in Genetics* 1, 224–30.

Sharman, G. B. (1971) Late DNA replication in the paternally derived X-chromosome of female kangaroos. *Nature (Lond.)* 230, 231–2.

Singer, J., Roberts-Ems, J., Luthardt, F. W., and Riggs, A. D. (1979) Methylation of DNA in mouse early embryos, teratocarcinoma cells and adult tissues of mouse and rabbit. *Nucleic Acids Res.* 7, 2369–85.

Stein, R., Sciaky-Gallili, N., Razin, A., and Cedar, H. (1983) Pattern of methylation of two genes coding for housekeeping functions. *Proc. Natl. Acad. Sci. U.S.A.* 80, 2422–6.

434   M. AZIM H. SURANI

Stevens. L. C. (1975) Teratocarcinogenesis and spontaneous parthenogenesis in mice. In *Developmental Biology of Reproduction,* eds. C. L. Markert and J. Papaconstantinou, pp. 13–106. New York: Academic Press.

–(1978) Totipotent cells of parthenogenetic origin in a chimaeric mouse. *Nature (Lond.)* 276, 266–7.

Stevens, L. C., Varnum, D. S., and Eicher, E. M. (1977) Viable chimaeras produced from normal and parthenogenetic mouse embryos. *Nature (Lond.)* 269, 515.

Sturm, K. S., and Taylor, J. H. (1981) Distribution of 5-methylcytosine in the DNA of somatic and germline cells from bovine tissues. *Nucleic Acids Res.* 9, 4537–46.

Surani, M. A. H. (1985a) Regulation of embryogenesis by maternal and paternal genomes in the mouse. In *Cold Spring Harbor Banbury Report 20,* eds. F. Costantini and R. Jaenisch, pp. 43–55. Cold Spring Harbor N.Y.: Cold Spring Harbor Laboratory.

–(1985b) Development of reconstituted mouse eggs and embryos provide genetic models for fetal development in the human. In *Implantation of the Human Embryo,* eds. R. G. Edwards, J. Purdy, and P. C. Steptoe, pp. 179–94. New York: Academic Press.

Surani, M. A. H., and Barton, S. C. (1983) Development of gynogenetic eggs in the mouse: implications for parthenogenetic embryos. *Science* 222, 1034–6.

Surani, M. A. H., Barton, S. C., and Kaufman, M. H. (1977) Development to term of chimaeras between diploid parthenogenetic and fertilized embryos. *Nature (Lond.)* 270, 601–2.

Surani, M. A. H., Barton, S. C., and Norris, M. L. (1984) Development of reconstituted mouse eggs suggests imprinting of the genome during gametogenesis. *Nature (Lond.)* 308, 548–50.

Surani, M. A. H., Barton, S. C., and Norris, M. L. (1986) Nuclear transplantation in the mouse: heritable differences between parental genomes after activation of the embryonic genome. *Cell* 45, 127–36.

Szulman, A. E., and Surti, V. (1984) Complete and partial hydatidiform moles: cytogenetic and morphological aspects. In *Human Trophoblast and Neoplasms,* eds. R. A. Pattillo and R. O. Hussa, pp. 135–46. New York: Plenum.

Takagi, N., and Sasaki, M. (1975) Preferential inactivation of the paternally derived X-chromosomes in the extraembryonic membranes of the mouse. *Nature (Lond.)* 256, 640–1.

Tarkowski, A. K. (1977) *In vitro* development of haploid mouse embryos produced by bisection of one-cell fertilized eggs. *J. Embryol. Exp. Morphol.* 38, 187–202.

Tarkowski, A. K., and Rossant, J. (1976) Haploid mouse blastocysts developed from bisected zygotes. *Nature (Lond.)* 259, 663–5.

Van Blerkom, J., and Brockway, G. O. (1975) Qualitative patterns of protein synthesis in the preimplantation mouse embryo. 1. Normal pregnancy. *Dev. Biol.* 44, 148–57.

Van Blerkom, J., Janzen, R., and Runner, M. N. (1982) The patterns of protein synthesis during foetal and neonatal organ development in the mouse are remarkably similar. *J. Embryol. Exp. Morphol.* 72, 97–116.

Waalwijk, C., and Flavell, R. A. (1978) DNA methylation at a CCGG sequence in the large intron of the rabbit β-globin gene: tissue-specific variations. *Nucleic Acids Res.* 5, 4631–41.

Wakasugi, N. (1974) A genetically determined incompatibility system between spermatozoa and eggs leading to embryonic death in mice. *J. Reprod. Fertil.* 41, 85–96.

West, J. D., Frels, W. I., Chapman, V. M., and Papaioannou, V. E. (1977) Preferential expression of the maternally derived X-chromosome in the mouse yolk sac. *Cell* 12, 873–82.

Whittingham, D. G. (1980) Parthenogenesis in mammals. In *Oxford Reviews of Reproductive Biology*, Vol. 2, ed. C. A. Finn, p. 205. Oxford, Clarendon Press.

Wigler, M. H. (1981) The inheritance of methylation patterns in vertebrates. *Cell* 24, 285–6.

Wilson, A. C., Maxson, L. R., and Sarich, V. M. (1974) Two types of molecular evolution. Evidence from studies of interspecific hybridization. *Proc. Natl. Acad. Sci. U.S.A.* 71, 2843–7.

Winking, H., and Silver, L. M. (1984) Characterization of a recombinant mouse *t* haplotype that expresses a dominant lethal maternal effect. *Genetics* 108, 1013–20.

Young, P. R., and Tilghman, S. M. (1984) Induction of α-fetoprotein synthesis in differentiating F9 teratocarcinoma cells is accompanied by a genome-wide loss of DNA methylation. *Mol. Cell Biol.* 4, 898–907.

14 Mutations and chromosomal abnormalities: How are they useful for studying genetic control of early mammalian development?

TERRY MAGNUSON

CONTENTS

I. Introduction

The overall picture of mammalian development is one of complex but integrated molecular activity. The result of such activity is that cells of the embryo become distinguishable from one another, and by expressing certain features of a differentiated phenotype, they cooperate in morphogenetic interactions to form tissues. There is no doubt that gene activity is necessary for this process to occur. Experimental evidence in the mouse clearly indicates that suppression of transcription or expression of zygotic lethal mutations can halt development at practically any time, even as early as the 2-cell stage, as reviewed by Johnson (1981) and Magnuson and Epstein (1981). Nevertheless, the complexity of genomic organization and the roles of specific genes concerned with mechanisms of mammalian development and differentiation are matters of paramount importance that have yet to be unraveled.

Recent spectacular advances in molecular genetics have opened new prospects for studying the structure, organization, and regulation of gene expression in eukaryotes. In organisms such as *Drosophila*, these techniques, coupled with classic genetics, have provided a means by which one can identify and study genes that are necessary for control of normal development and differentiation. For example, it has been possible to identify and clone the homeotic genes of *Drosophila*, in the absence of any biochemical information about their products, because of the fact that multiple alleles and appropriate deletions and/or inversions are available and provide recognizable chromosomal breakpoints surrounding the mutant gene. This information allows one to map the gene in question and to obtain DNA that spans the mutation of interest by chromosomal walking (Bender et al. 1983; Garber, Kuroiwa, and Gehring 1983; Scott et al. 1983). Thus, the combined approaches of the classic developmental geneticist and those of the molecular geneticist have helped to define the roles of the homeotic genes in specifying body pattern and segment identity in *Drosophila*. For an excellent review of homeosis, see Raff and Kaufman (1983).

More complex genomes, such as that of the mouse, are not readily amenable to many of the genetic manipulations that are available for *Drosophila* genetic studies. Consequently, geneticists working with mice have to explore new prospects for identifying and studying the structure, function, and organization of genes involved in the control of mammalian development. The purpose of this chapter is to discuss both the experimental means available and the recent progress that has been made to achieve this goal. I shall address development of the mouse embryo before organogenesis begins. This time period includes the first seven to eight days of gestation, up to and including primitive-streak formation. Basically, two experimental approaches will be considered. The first is the classic approach of asking what genetic defects tell us about development. We must ask whether or not particular mutant genes are involved in specific developmental processes such as compaction, formation of inner cell mass and trophectoderm, implantation, or mesoderm formation, all of which are fundamental to survival of the intact embryo, but not necessarily to the individual cells themselves (embryonic lethals). Alternatively, they may be involved in cellular processes and structures, such as energy metabolism or cytoskeletal and membrane proteins, all of which are fundamental to the survival of each cell independent of the intact embryo (cell lethals).

The second approach that will be considered is to ask what other means can be used to identify genes that are involved in the control of development and differentiation. Instead of depending on random mutations, can we create mutations at will for specific developmental stages and tissue types? Many recent technical advances in molecular genetics make this an exciting and interesting possibility.

II. Existing genetic abnormalities: What can they tell us about development?

In the mouse, there are numerous genetic abnormalities that result in death in the early embryo (Table 14.1); they have been reviewed elsewhere (McLaren 1976; Magnuson and Epstein 1981; Magnuson 1983). These defects include recessive lethal mutations, deletions that are lethal when homozygous, and conditions of chromosomal imbalance such as nullisomy, monosomy, and haploidy. Many of the recessive lethal mutations listed in Table 14.1 have heterozygous manifestations that are different from the homozygous lethal effects. It must be remembered that in most cases the relevant phenotypes represent various aspects of pleiotropic effects, and therefore both homozygotes and heterozygotes provide important information about the mechanisms by which a mutation produces deleterious effects. It is a general tenet that these naturally occurring or experimentally induced genetic variants, which presumably result in an alteration in the quantity or quality of one or more gene products, may provide valuable insight into the molecular activities of wild-type gene products and eventually lead to identification of the genes themselves. In fact, the classic approach of developmental geneticists has always been to exploit existing mutations in an attempt to proceed from the mutant phenotype toward the genome through analysis of the developmental steps disrupted in mutants.

In spite of numerous attempts and some important findings regarding the biological consequences of genetic deficiencies affecting the mouse embryo, not a single instance can be cited in which this approach has revealed information about the nature of a particular gene or its product. This inability to identify gene products and functions on the basis of phenotype is perhaps not surprising, given the number and complexity of steps that are likely to exist between the defective gene product and the detectable phenotypes in higher organisms. Nevertheless, if we are to understand the pathogenesis arising from these genetic defects, our goals clearly need to be directed toward not only understanding the lethal phenotype but also characterizing the affected genes and their products.

Although several genetic abnormalities are known to be lethal during the early stage of mouse development (listed in Table 14.1), only a few have been studied extensively. In the following paragraphs I shall discuss these particular abnormalities not to describe each lethal phenotype in detail but rather to present experimental approaches available for study

Table 14.1. *Early embryonic lethals*

Mutation or chromosomal abnormality	Chromosome location	Approximate time of death and appearance of phenotypic abnormalities
c^{25H}	7 (albino deletions)	2–6-cell stage; cessation of cell division, but able to live for 2 more days (Lewis 1978b; Nadijcka et al. 1979)
Nullisomy	14 or 15, and X	2–8-cell stage; no specific defect described (Morris 1968; Kaufman and Sachs 1975; Burgoyne and Biggers 1976; Luthardt 1976; Tarkowski 1977)
Monosomy	1–5, 10–12, 14–19	8-cell stage; implantation, no consistent defect other than general reduction in cell number per embryo (Epstein and Travis 1979; Magnuson et al. 1982; Baranov 1983; Magnuson et al. 1985)
t^{12}, t^{w32}	17 (t complex)	Unable to undergo morula-blastocyst transition; abnormal lipid synthesis at 2–4 cells (Smith 1956; Calarco and Brown 1968; Hillman et al. 1970; Hillman and Hillman 1975)
T^{hp} (hairpin)	17 (deletion)	Morula to blastocyst; reported to result in failure of morula-blastocyst transition (Babiarz 1983); however, our own data suggest that many embryos survive to late blastocyst stage (T. Magnuson and C. J. Epstein, unpublished results)
Haploidy	Missing full set	Early cleavage stage to implantation; no specific defect described (Kaufman and Gardner 1974; Kaufman and Sachs 1975; Modlinski 1975; Tarkowski and Rossant 1976; Tarkowski 1977)
om (ovum mutant)	Maternal effect	Morula stage to implantation; poor development of trophoblast, possible trophectoderm defect (Wakasugi et al. 1967; Wakasugi 1973)
Ts (tail-short)	11	Morula stage to implantation; reduction in cell number (Paterson 1980)
Os (oligosyndactyly)	8	Implantation; metaphase arrest beginning at early blastocyst stage (Paterson 1979; Magnuson and Epstein 1984)

Mutation	Chromosome	Description
Agouti locus	2	
a^x (lethal nonagouti)		Blastocyst to implantation; no tissue-specific defect (Papaioannou and Mardon 1983);
A^y (lethal yellow)		Implantation; possible trophectoderm defect (Robertson 1942a; Eaton and Green 1963; Papaioannou and Gardner 1979)
t^{wPa-1}	17 (t complex)	Implantation; general disorganization (Guenet et al. 1980)
t^{w73}	17 (t complex)	Implantation to egg cylinder; extraembryonic cell types most affected (Spiegelman et al. 1976; Babiarz et al. 1982)
l^0 (l^6)	17 (t complex)	Early egg cylinder; failure to differentiate into extraembryonic and embryonic ectoderm (Gluecksohn-Schoenheimer 1940; Dunn and Gluecksohn-Schoenheimer 1950; Nadijcka and Hillman 1975)
Ve (velvet coat)	15	Early egg cylinder; failure to differentiate into extraembryonic and embryonic ectoderm (cf. McLaren 1976)
T^{Orl} (Orleans)	17 (t complex)	Egg cylinder; lack of embryonic development; consists of parietal endoderm surrounded by giant cells (Erickson, Lewis, and Slusser 1978)
Bld (blind)	15	Late egg cylinder; retarded growth (Vankin 1956; cf. McLaren 1976)
T^{OR} (Oak Ridge)	17 (t complex)	Late egg cylinder; death of embryonic ectoderm (Bennett et al. 1975)
t^{w5}	17 (t complex)	Late egg cylinder; death of embryonic ectoderm (Bennett and Dunn 1958)
Dey (Dickie's small eye)	2	Late egg cylinder; death of embryonic ectoderm (cf. McLaren 1976)
Wc (waved coat)	14	Late egg cylinder; death of embryonic ectoderm (cf. McLaren 1976)
c^{6H}	7 (albino deletion)	Late egg cylinder; defective extraembryonic tissues (Lewis et al. 1976)
t^9	17 (t complex)	Primitive streak; failure to form mesodermal derivatives (Moser and Gluecksohn-Waelsch 1967; Spiegelman and Bennett 1974)
se (short-ear, deletion)	9	Primitive streak; absence of mesoderm, excess proliferation of extraembryonic ectoderm, overgrowth of trophoblast giant cells (Dunn 1972; cf. McLaren 1976)
Fu^{ki} (kinky)	17	Post primitive streak; multiple embryonic axes (Gluecksohn-Schoenheimer 1949)

of developmental mutations and their potential for analysis of mechanisms of development and differentiation.

A. Oligosyndactyly

Oligosyndactyly (*Os*) is a radiation-induced mutation located on mouse chromosome 8. It was first defined in heterozygous mice because of its effects on limb morphogenesis. The mutation causes fusion of the second, third, and sometimes the fourth digits in forelimbs and hindlimbs (Grüneberg 1956) that has been attributed to a reduction in size of the preaxial margin of the footplate beginning about day 10 of embryonic development (Grünebrg 1961). A second pleiotropic effect described in *Os* heterozygotes occurs in the feet, forearms, and lower legs and includes abnormal fusion of some muscles or tendons, as well as unusual insertions of certain muscles (Kadam 1962). Nephrogenic diabetes insipidus is a third pleiotropic effect that occurs in heterozygous mice; the severity of this condition is strain-dependent (Falconer, Latsyzewski, and Isaacson 1964).

Because of its heterozygous effects on limb morphogenesis, *Os* provides an interesting model for a class of human mutations that result in syndactyly (Temtamy and McKusick 1978) and for which mechanisms are unknown. Apparently, a reduction in size of the preaxial border of the footplate is detectable in *Os*/+ embryos as soon as the footplate becomes distinguishable as a separate structure (approximately day 10 of gestation) (Grüneberg 1961). The reduced size of the footplate results in crowding of digits II and III and an apparent reduction in the mesenchymal material available for skeletal development. This occurs before any membranous skeleton can be detected by standard histologic techniques, and it seems that the cartilaginous and osseous skeletons merely repeat a pattern established in the footplate. This reduction in size of the preaxial border has been reported to be due to a deficiency of cells in this area, possibly because of increased cell pyknosis (*cf.* Green 1966). Thus, the complex skeletal abnormalities of *Os* heterozygous adult limbs can be understood as consequences of a single initial disturbance.

In the homozygous state, *Os* is lethal at the time of implantation (Van Valen 1966). The mutation causes widespread mitotic arrest (Paterson 1979), suggesting that the primary effect of the *Os* mutation is on the formation of the mitotic apparatus during prophase or on chromosome movement during metaphase and/or anaphase. Furthermore, the mutation appears to result in cell autonomous lethality, because homozygous embryos cannot be rescued in aggregation chimeras (T. Magnuson, D. Yee, and C. J. Epstein, unpublished results).

Although there is no obvious link between the effects of *Os* in heterozygous mice and the mitotic arrest in homozygous embryos, two possibilities must be considered. First, it is possible that, as a radiation-induced mutation, *Os* is in fact a small deletion or rearrangement. If that were the case, homozygous and heterozygous manifestations could be due to independent effects of more than one mutant gene. An alternative possibility is that *Os* is a mutation of a single gene, the product of which is necessary for normal mitotic function. If that were the case, complete mitotic arrest would occur in homozygous embryos because of the lack of any wild-type gene product. The heterozygous defects could be explained as limitations induced by a 50 percent reduction in the wild-type gene product, which would be deleterious to those tissues undergoing extremely rapid expansion and proliferation.

By examining homozygous embryos for spindle organization, we have found that the cells arrest in metaphase, with intact, normal-appearing mitotic spindles (Figure 14.1). Although spindle ultrastructure has not yet been examined, we know that these mutant spindles are not hyperstable and that two important spindle-associated components, kinetochores and the pericentriolar material of centrosomes, are present (Magnuson and Epstein 1984). These results define the *Os* mutation as one that, in the homozygous state, blocks anaphase chromosome movement. This combination of metaphase arrest with intact spindles is unique among all mitotic-arrest mutations described in higher eukaryotes, as summarized in Table 1 of Magnuson and Epstein (1984).

Two principal possibilities must be considered to explain how an *Os/Os* embryo can divide apparently normally five to six times and then enter metaphase arrest. The first possibility is that the egg cytoplasm provides enough of the wild-type gene product required for mitosis up to day 3 of development, at which time the concentration of this maternally derived gene product drops below the required level. This product could be either a constituent of the mitotic apparatus or something required to effect chromosome movement and spindle disaggregation. Most maternal messenger RNA is degraded by the 2-cell stage; however, that may not be the case for maternally derived proteins, as reviewed by Magnuson and Epstein (1981) and Johnson (1981). For example, it is known that the activities of several maternally derived enzymes remain constant until the 8- to 16-cell stage, after which they drop acutely. Furthermore, a recent study (West and Green 1983) reported that maternally derived glucose phosphate isomerase is still present at the late blastocyst stage. The second possibility is that the structure or function of the mitotic spindle in day-3 embryos may be qualitatively different from that in early cleavage-stage embryos, and transcription of a new gene product (presumably the wild-type allele of *Os*) may be involved in this change. If that were the case, the time of onset of the mitotic block would coincide with the activation of the Os^+ locus or the need of its product.

In either of these two possible situations, the abnormalities could involve a structural component of the spindle, a specific process required

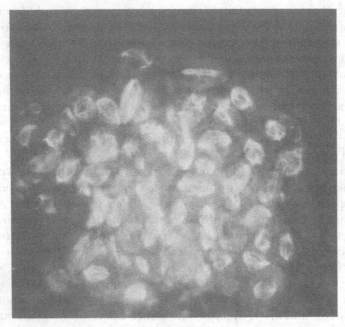

Figure 14.1. Fluorescent micrograph of an *Os/Os* embryo that has been labeled with a mixture of monoclonal antibodies that have specificities directed against α and β tubulin. Numerous mitotic spindles can be detected. Magnification X566. [Reprinted from Magnuson and Epstein (1984), with permission; copyright held by M.I.T. Press.]

for mitosis to proceed beyond metaphase, or a problem in some basic cellular process such as metabolism or energy production that would adversely affect many other cellular processes. Although the two possible types of *Os* defects have been framed in terms of absence of a gene product, it is also conceivable that they represent synthesis of an abnormal gene product that is unable to carry out its normal function in the homozygous state, whereas it can do so when mixed with normal gene products in heterozygous cells.

Clearly, the studies cited have shown that the *Os* wild-type gene probably is not involved in a developmental process such as specification of body pattern. Nevertheless, this does not by any means minimize the importance of the mutation. The homozygous phenotype of metaphase arrest is so distinct that a molecular analysis of mutant spindles is likely to produce information that will be important for understanding a basic cellular process such as mitosis. Furthermore, it must not be forgotten that in the heterozygous state the mutation results in malformations of the limb and kidney. We know little about the developmental mechanisms responsible for these genetically induced abnormalities, and a mutation such as *Os* is an important model for study of these congenital malformations.

B. Albino deletions

The genetic system in the mouse known as the albino-deletion complex offers an extremely promising model for studies dealing with genes that are known to be involved in control of development and differentiation. These deletions represent a series of overlapping chromosomal deficiencies that cover the region of the albino locus on mouse chromosome 7. Although several deletions were produced during the course of radiation experiments conducted at the Oak Ridge National Laboratory (W. L. Russell and L. B. Russell) and at the MRC Radiology Unit, Harwell, England (A. G. Searle), only six have been studied in detail, as reviewed by Gluecksohn-Waelsch (1979). Complementation tests have resulted in classification of these six deletions into four complementation groups (Figure 14.2). When homozygous, two of the deletions (c^{25H}, c^{62H}) result in death of the embryo early in development, whereas the remaining four deletions (c^{3H}, c^{112K}, c^{65K}, c^{14CoS}) are lethal at the time of birth.

The c^{25H} deletion represents about 7 percent of chromosome 7 (approximately 5 cM) (Miller et al. 1974) and causes cessation of cell division, which begins at the 2- to 6-cell stage, with death occurring 1–2 days after that time (Lewis 1978b; Nadijcka, Hillman, and Gluecksohn-Waelsch 1979). The only obvious ultrastructural abnormality that has been found is aberrantly shaped nuclei. This phenotype is quite distinct from that observed for other preimplantation lethals (Magnuson and Epstein 1981).

Embryos homozygous for the c^{6H} deletion (approximately 2 cM in length) are morphologically abnormal by 6.5 to 7.0 days of gestation (Lewis, Turchin, and Gluecksohn-Waelsch 1976). At that time, cells of the ectoplacental cone as well as the extraembryonic ectoderm stop dividing and begin to show signs of degeneration. In addition, the parietal yolk sac endoderm extends markedly into the decidual tissue. The embryonic ectoderm, however, appears structurally normal. By 7.5 days of gestation, the extraembryonic ectoderm is extremely disorganized and pyknotic in appearance. The embryonic ectoderm is reduced in size but still remains organized normally. Death and resorption finally occur by day 8.0. Lewis and associates (1976) have proposed that the c^{6H} deletion, when homozygous, interferes with normal differentiation of the parietal endoderm, ectoplacental cone, and extraembryonic ectoderm and that reduction in embryonic ectoderm growth and failure of primitive-streak formation are secondary abnormalities attributed to a dying embryo.

A region of nonoverlap, distal to the *Mod-2* locus, between c^{3H} (a perinatal lethal) and c^{6H} (a prenatal lethal) (Figure 14.2), is assumed to include the gene or genes whose absence is responsible for the c^{6H}

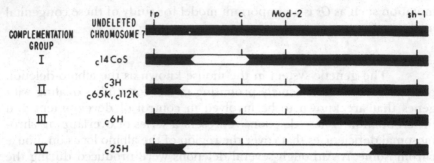

Figure 14.2. Complementation map of the albino deletions: *tp* (taupe), *c* (albino), *Mod-2* (mitochondrial malic enzyme), *sh-1* (shaker 1). [Reprinted from Gluecksohn-Waelsch (1979), with permission; copyright held by M.I.T. Press.]

lethal phenotype. Although the possible function of this gene is as yet completely unknown, the unique situation of the availability of overlapping deletions in the mouse is promising, because they provide a means by which specific DNA segments can be localized. In fact, the feasibility of this approach has recently been demonstrated by Disteche and Adler (1984), who constructed a genomic library from chromosomes enriched 10-fold, by flow sorting, for Cattanach's X^7 translocation chromosome. In this rearrangement, the middle portion of chromosome 7, which covers the albino deletions, is translocated in an inverted fashion into the middle of the X chromosome. Female mice homozygous for the relevant albino-deletion sequences can be rescued, to varying degrees, by making them heterozygous for the X^7 translocation chromosome (Gluecksohn-Waelsch et al. 1980). Therefore, the genomic library constructed by Disteche and Adler (1984) should contain chromosome-7 DNA sequences that are missing from the albino-deletion chromosomes. These investigators analyzed 48 independently isolated clones from this library and found that 57 percent of the inserts contained repetitive DNA, 25 percent contained single-copy sequences, and 19 percent were nonrecombinant. One of the single-copy inserts was mapped to the region of overlap common to the c^{3H} and c^{6H} deletions. These results are very encouraging because they clearly show the possibility of locating specific DNA fragments to the region of the albino deletions. Toward this goal, we are attempting to produce c^{6H} homozygous embryonic stem cell lines (T. Magnuson and D. Yee, unpublished results). Such cells would be a valuable source of homozygous DNA, particularly because the embryos die very early in development. This DNA could then be used in conjunction with c^{3H}/c^{3H} DNA for localizing segments to the area of chromosomal nonoverlap.

The remaining four deletions (c^{3H}, c^{112K}, c^{65K}, c^{14CoS}) that have been studied in detail are very interesting, because they represent one of the first systems known to be involved in the control of specific developmental processes. Studies of homozygous fetuses or newborns have revealed defects specific to the differentiation of liver parenchymal cells as well as cells of the proximal convoluted tubules of the kidney, as reviewed by Gleucksohn-Waelsch (1979). The c^{3H}, c^{112K}, and c^{65K} deletions are approximately 3 cM in length, whereas the c^{14CoS} deletion is about 1.5 cM in length. These deletions do not complement one another, and homozygous fetuses show dramatic reductions in both blood glucose levels and activities of some liver-specific enzymes (phosphoenolpyruvate carboxylase, glucose-6-phosphatase, tyrosine aminotransferase, serine dehydratase, glutamine synthetase, UDP glucuronosyltransferase), but not of all enzymes tested. In addition to enzyme deficiencies, three plasma proteins (albumin, transferrin, and α-fetoprotein) are also affected, with their synthesis and secretion being reduced to about 20 percent of normal (Erickson, Pickart, and Thaler 1976; Garland et al. 1976). Ultrastructural abnormalities include a vesiculated rough endoplasmic reticulum that has lost a majority of its membrane-bound ribosomes, as well as swollen and dilated Golgi apparatus and nuclear membrane (Trigg and Glueksohn-Waelsch 1973). Binding studies of insulin (Goldfeld et al. 1981), epidermal growth factor and glucagon (Shaw and Glueksohn-Waelsch 1983), and glucocorticoids (Goldfeld et al. 1983) have also revealed receptor deficiencies to about 25 percent of normal.

Although it is difficult to find some common denominator to explain these liver- and kidney-specific defects, experimental studies have revealed that they are not due to deletions of respective structural genes. For example, the structural genes for several of the affected enzymes (Cori et al. 1981, 1983), the glucocorticoid receptor (Francke and Gehring 1980), and the plasma proteins (cf. Roderick and Davisson 1984) have been mapped to other chromosomes. Furthermore, it must be emphasized that a specific combination of defects is associated with these deletions and that death is not due, for example, to a total lack of liver differentiation, because most other aspects of liver function, including the total protein machinery, appear to be normal (Glueksohn-Waelsch 1979).

Taken together, these data strongly suggest that the albino deletions include either a single gene or possibly a series of controlling or regulatory genes that interact with a specific set of structural genes located on other chromosomes (Glueksohn-Waelsch 1979). Although these regulatory genes have yet to be identified on a molecular level, some of the structural genes that encode the affected enzymes have been cloned, and in the case of tyrosine aminotransferase, steady-state mRNA either is not detectable or is dramatically reduced in livers from homozygous newborns (Schmid et al. 1985). These data suggest that either transcription or message processing is being affected by the deletions.

In sum, the albino deletions represent one of the few systems in the mouse in which specific developmental defects are associated with a series of overlapping chromosomal deficiencies. Characterized largely through the pioneering work of Gluecksohn-Waelsch and her colleagues, the albino-deletion complex should provide opportunities for rapid progress at the molecular and cellular levels.

C. Lethal yellow

Numerous alleles of the agouti locus, located on mouse chromosome 2, have been described, ranging in manifestation from extreme nonagouti (a^e), which produces black coat color in homozygous mice, to yellow (A^y), which produces yellow hair in heterozygotes. The locus apparently determines whether melanocytes will synthesize phaeomelanin (yellow) or eumelanin (black or brown) and appears to operate through the environment of the hair bulb rather than the melanocyte itself (Silvers 1961; Mayer and Fishbane 1972; Poole, 1980). No biochemical basis for this effect has yet been demonstrated. However, the available genetic evidence suggests that the agouti locus is a complex, multigenic locus (Wallace 1965; Phillips 1966). In addition to having yellow coat color, A^y heterozygotes also become obese and develop mild glucose intolerance and hyperinsulinemia. An enhanced susceptibility to tumors to which the strain or tissue is already susceptible is also associated with A^y (Wolff and Pitot 1973). The mechanisms responsible for these diverse phenotypic effects are unknown.

Two alleles of the agouti locus, lethal nonagouti (a^x) and yellow (A^y), cause early embryonic death of homozygotes, as reviewed by Pedersen and Spindle (1981) and Magnuson (1983). The time and mode of action of the a^x gene were studied by Papaioannou and Mardon (1983). They found, by histologic analysis, that presumptive a^x/a^x embryos first appear abnormal at the midblastocyst stage (about 4.5 days post coitum). Nevertheless, these embryos are able to implant and form trophoblastic giant cells and, in fact, are able to form a primitive endoderm layer. However, by day 7.5 to 8.5, the putative homozygotes persist only as disorganized clumps of embryonic tissue. No tissue-specific defect has been observed.

For A^y, it has been shown that homozygotes die sometime after implantation (Robertson 1942a). Based on histologic observations, Eaton and Green (1963) have attributed this death to defective differentiation of trophoblastic giant cells. This conjecture has been supported by chimera experiments suggesting that putative A^y/A^y cells of the inner cell mass are able to survive when combined with normal trophectoderm (Papaioannou and Gardner 1979). However, other factors may also contribute to the lethal phenotype. For example, Robertson (1942b) demonstrated that when ovaries from heterozygous yellow mice were transplanted to homozygous agouti mice, the homozygous yellow embryos survived significantly longer, indicating that uterine environment is important.

Any further work on either the a^x or A^y lethal phenotype has been hindered by the problem of identifying mutant embryos prior to the appearance of gross abnormalities. The inability to identify mutants necessitates the somewhat circular process of inferring from appearance which embryos are mutant and then using these embryos as the basis for studies of the mutant state. Although that is probably a satisfactory procedure for morphological studies, it does not really permit biochemical and immunological investigations, particularly before the time at which gross morphological abnormalities become apparent. For A^y, it seemed that this difficulty might be circumvented by the exciting finding of Copeland, Jenkins, and Lee (1983b). These investigators demonstrated that the mutant locus segregated concordantly with an integrated ecotropic murine leukemia provirus. This suggested that retroviral integration may have resulted in the A^y mutation by causing a mutagenic event either by interrupting normal gene expression or by activating genes flanking the insertional site. However, more recent preliminary evidence suggests that this might not be the case. When two independent strains of mice that were not used in the original experiments, but both of which carried the A^y mutation, were probed with the ecotropic-specific envelope probe, no evidence was found for the presence of the provirus. Experiments are now in progress to attempt to separate the A^y mutation from the proviral insertion by a crossover event in the strains originally studied (N. Jenkins, personal communication).

D. *T* complex

The *T* complex encompasses a region of mouse chromosome 17 that begins at the locus of *T* and extends beyond the *H-2* complex (Artzt, McCormick, and Bennett 1982a). Most wild populations are polymorphic for different *t* haplotypes, each of which carries variant alleles of a number of genes that affect embryonic development, sperm formation, and function (Bennett 1975). Some of the phenotypic effects associated with each *t* haplotype include interaction with the dominant mutation *T* to produce taillessness in *T/t* animals, homozygous lethality or semilethality, transmission-ratio distortion in *+/t* or *T/t* males, and sterility in males homozygous for a semilethal *t* haplotype or doubly heterozygous for two different lethal *t* haplotypes. In addition to these effects, suppression of meiotic recombination has also been observed to occur in *+/t* heterozygotes. Normally, one would expect a $10-15$ percent recombination frequency, but at most only $0.1-0.2$ percent is typically observed. It was L. C. Dunn who realized the importance of the *T* complex by recognizing its role in early development. All of the phenotypic effects associated with the *T* complex have been reviewed in great detail (Glueksohn-Waelsch and Erickson 1970; Bennett 1975; Klein and Hammerberg 1977; Sherman and Wudl 1977; Bennett 1980; Lyon 1981; Magnuson and Epstein

1981; Silver 1981; Magnuson 1983; Silver, Garrels, and Lehrach 1984) and therefore will not be covered here. Instead, I shall deal primarily with the most recent advances concerning the molecular genetics of the T complex, particularly those aspects that are directly relevant to future studies on the lethal mutations.

Perhaps the single most important advancement in understanding the T complex has been that of unraveling its genetic structure, and the most significant contribution in this regard comes from work by Silver and Artzt (1981). These investigators found that t chromatin is fully capable of undergoing meiotic recombination when paired with other t chromatin in double heterozygotes, and because of this discovery it has become possible to map some of the t lethal genes relative to one another. In doing so, it has become clear that at least some of the lethal genes are nonallelic (Artzt et al. 1982a; Artzt 1984). This is an important concept, because explanations for the disparate t-associated effects no longer need be based on abnormalities within a single gene.

During the progress of these mapping studies, it was also discovered that the H-2 complex occupies an anomalous position in t chromosomes as compared with wild-type chromosomes (Artzt, Shin, and Bennett 1982b). Using a combination of classic and molecular genetics, as well as serology, it was later found that the whole H-2 complex is included in a simple inversion, suggesting that a chromosomal rearrangement has occurred in t haplotypes and that this rearrangement may be responsible for some of the recombination suppression between t chromatin and wild-type chromatin (Shin et al. 1983, Shin, Bennett, Artzt, 1984). The extent of the inversion outside the H-2 complex is not known; however, it probably does not cover the whole length of the region of T-complex recombination suppression, because no cytological abnormalities have been seen, and the relative positions of the allelic genes for T and *quaking* do not appear to be altered (cf. Shin et al. 1984). Interestingly, during the course of these experiments, it was discovered that some of the t lethal genes (t^{12}, t^{w32}, t^{w18}, t^{Lub-1}, t^{w5}) map very near to (<1 cM) or are intermingled with genes of the H-2 complex. The significance of this close relationship is unclear, but as Shin and associates (1984) pointed out, their genetic proximity provides a possible beginning for a chromosomal walk in attempts to clone embryonic t lethal genes.

Another approach to molecular analysis of the T complex that is being pursued is that of identifying and mapping restriction-fragment length polymorphisms between wild-type and t-haplotype DNA (Fox et al. 1984a; Rohme et al. 1984). In an elegant series of experiments, these investigators microdissected fragments of the proximal half of a wild-type chromosome 17, isolated DNA from these, and then established a library of genomic clones (Rohme et al. 1984). Using these clones as probes, many were found to hybridize to sequences located on chromosome 17, and when screened against wild-type DNA and *t*-containing DNA, several restriction-fragment length polymorphisms were detected. Four of these genomic clones have since been analyzed in greater detail, and a series of *t*-haplotype-specific restriction fragments was defined (Fox et al. 1984a). By examining partial *t* haplotypes for the presence or absence of these *t*-specific fragments, a genetic map has been constructed showing that these fragments map to different regions within the proximal portion of the *T* complex and that the proximal-to-distal order of these DNA fragments is inverted when compared with wild-type chromosomes (Herrmann et al. 1986). Thus, this newly described proximal inversion and the previously described distal inversion (Shin et al. 1984) provide an explanation for recombination suppression in the *T* complex.

These microdissected clones have failed to detect any polymorphisms among the eight *t* haplotypes thus far analyzed. This result is similar to that obtained with other *T*-complex DNA probes, such as the *H-2* genes (Silver 1982; Shin et al. 1982) and an α-globin pseudogene (Fox, Silver, and Martin 1984b). This is unusual, because polymorphisms are detectable with these probes in DNA from different wild-type strains, and that suggests that all complete *t* haplotypes are very closely related to one another, at least at the present level of analysis, and that they share a common ancestry.

Even though a significant amount has been learned in the past five years about the organization and molecular structure of the *T* complex on the DNA level, we are still as ignorant as ever about the nature of the developmental defects attributed to the *t* lethal genes, as reviewed in extensive detail by Bennett (1975), Sherman and Wudl (1977), and Magnuson (1983). Cloning of DNA sequences within the *T* complex has not yet yielded any information about the possible nature of the gene products that are absent or altered by the *t* lethal genes. Although it is anticipated that the *H-2* genes and the microdissected clones will provide important entry points needed for cloning *t* lethal genes by chromosome walking, it will undoubtedly be difficult. In addition to the obvious technical problems, a clear-cut assay, short of rescue by microinjection techniques, does not exist for identifying these genes on a molecular level. Thus, the next few years will be critical for determining if this approach will yield any information about the *t* lethal genes and their products.

E. Insertional mutagenesis

Another approach to the study of genetic control of early mammalian development is to induce mutations by insertion of foreign genetic material into the mouse genome. The insertional event can result in mutant phenotypes either by altering transcription of genes flanking the insertion site or by physically disrupting transcription by integration into an active gene. Insertional mutagenesis by mobile genetic elements has been known to occur as a spontaneous event in maize (McClintock 1956), yeast (Roeder and Fink 1980), and *Drosophila* (Kidwell, Kidwell, and Sved 1977). In fact, cloned transposable elements in *Drosophila* now serve as vectors in gene-transfer experiments (Rubin and Spradling 1982; Spradling and Rubin 1982). Although comparable factors have not yet been identified with complete certainty in mouse, spontaneous integration of viral DNA into the mouse genome is known to occur, and in the case of the *dilute* locus, such an event seems to have caused a recessive mutation (Jenkins et al. 1981). Another approach has been to induce mutations by experimentally introducing foreign DNA (both viral and nonviral cloned DNAs) into the embryo (Jaenisch et al. 1983; Schnieke, Harbers, and Jaenisch 1983; Wagner et al. 1983; Palmiter et al. 1984). In each case, interesting results have been obtained, and they will be discussed in this section. The critical difference between mutations caused by experimental insertions and those that result from spontaneous events or conventional mutagenesis experiments is that the inserted foreign DNA provides a molecular probe that can be used subsequently to isolate, clone, and characterize the interrupted gene and its flanking areas. This information has been, in certain cases, extremely valuable for identifying the gene product and its function.

i. Endogenous retroviruses. It has been estimated that as much as 0.05 percent of the mouse genome consists of endogenous retrovirus-related sequences (Callahan and Todaro 1978; Phillips et al. 1982), and because more than 700 mutant alleles have been identified in the mouse (Green 1981), one might predict that some of the spontaneous mutations are the result of virus integrations. Evidence that this is in fact the case has been obtained, as described by Jenkins and associates (1981). These investigators found that the single endogenous ecotropic murine leukemia virus (MuLV) of DBA/2J mice segregated concordantly with the *dilute* (*d*) mutation in 53 of 53 recombinant inbred lines, as well as with seven other strains that carry the mutation. Furthermore, DNA from five spontaneous *dilute* revertants was found to lack most, but not all, of the ecotropic virus-specific sequences, suggesting that the mutation had indeed been caused by virus integration (Jenkins et al. 1981; Copeland, Hutchison, and Jenkins 1983a). Molecular analysis of two of the *dilute* revertants suggested that reversion was due to virus excision by homologous recombination involving the viral long terminal repeats (LTR) and leaving exactly one LTR sequence in the DNA. So far, no identification of a mutant or wild-type gene product has been reported. These investigators have suggested that perhaps other mutations should be screened for endogenous retrovirus-related sequences and that this approach may be of more general use than originally thought.

ii. The *Mov* loci. Jaenisch and associates have experimentally inserted Moloney murine leukemia virus (M-MuLV) into the mouse germ line either by exposing mouse embryos to the virus (Jaenisch 1976; Jaenisch et al. 1981; Jähner and Jaenisch 1980) or by microinjecting cloned DNA into zygotes (Harbers, Jähner, and Jaenisch 1981). In this manner, 14 substrains (*Mov-1 − 14*) have been isolated. On analysis, the *Mov-1 − 13* substrains were found to carry one copy of the provirus, each of which is inherited as a single gene according to Mendelian rules. The *Mov-14* substrain carries a tandem array of multiple proviral genomes (Stewart et al. 1983). Breeding experiments have shown that virus integration at the *Mov 1 − 12* and *-14* loci does not lead to any recognizable mutant effect in the respective homozygous animals. However, virus insertion at the *Mov-13* locus resulted in a recessive lethal mutation that in homozygotes caused developmental arrest during midgestation, between days 11 and 12, with death occurring 1 or 2 days later (Jaenisch et al. 1983). Integration of the proviral genome occurred at the 5′ end of the first intron of the $\alpha 1(I)$ collagen gene, resulting in complete transcriptional block (Schnieke et al. 1983; Harbers et al. 1984). The viral insertion prevents the developmentally regulated appearance of a transcription-associated hypersensitive site (Breindl, Harbers, and Jaenisch 1984). The phenotypic abnormalities associated with the developmental arrest include necrosis of mesenchymal cells and of hemopoietic cells of the fetal liver. Sudden death finally results from rupture of major blood vessels (Lohler, Timpl, and Jaenisch 1984). Normally, abundant transcription of the type-1 collagen gene does not occur until day 12, which coincides perfectly with the appearance of phenotypic abnormalities observed in *Mov-13* homozygotes (Schnieke et al. 1983).

iii. Transgenic mice. Introduction of cloned genes into mice by microinjection techniques has been used by many groups to achieve integration of defined genes into the germ line, thereby producing transgenic mice (see Chapter 16 for a detailed discussion of this approach). The purpose of these experiments has been to elucidate the DNA sequences involved in tissue and stage-specific gene regulation, and this has been done by attempting to achieve developmentally appropriate expression of the foreign genes by correlating specific modifications in gene design with corresponding modifications in expression. Another major goal has to do with the possibility of gene therapy by introducing foreign genes. Even though most of the work with transgenic mice has been done with these goals in mind, one might also expect that, depending on the insertion site, the injected DNA could act as a mutagen, thereby producing both dominant and recessive mutations. The ability to detect these mutations will depend on an extensive screening program, and recessive lethal mutations will essentially go unnoticed unless attempts are made to generate mice homozygous for the insertion. A preliminary estimate of the frequency of recessive lethal mutations produced to date by microinjection of cloned DNAs is between 1 and 10 percent (E. Lacy, personal communication). More work is needed before any correlations can be made between the molecular nature of the lesions produced by the insertional event and the resulting genetic consequences.

So far, three reports have been published regarding the production of recessive mutations by microinjected DNAs. Wagner and associates (1983) have produced six transgenic substrains, each of which carries human growth-hormone sequences at a unique site. Four of the substrains are able to produce apparently healthy homozygotes. However, in two of the substrains, no postnatal homozygotes have been found, and litter sizes are abnormally small. In one of these substrains (HUGH/3), approximately 25 percent of the embryos obtained from heterozygous parents died shortly after implantation (day 4–5), and an additional 34 percent died sometime between days 9 and 12 of development (Covarrubius, Nishida, and Mintz 1985). This phenomenon of "second-stage" death remains unexplained. In the second substrain (HUGH/4), early postimplantation lethality of homozygotes also occurred, and the lethal phenotype was similar to the early lethality detected for HUGH/3 homozygotes, even though both lines are attributable to integration events occurring on different chromosomes (Covarrubius et al. 1985). A preliminary molecular analysis of DNA flanking the insertion site has been done for both substrains, and the available evidence indicates that wild-type mouse sequences were rearranged during the integration of the foreign DNA (Covarrubius et al. 1985). This, of course, will make it difficult to identify on the molecular level the disrupted genes responsible for lethality, and it remains to be determined if DNA rearrangements and/or deletions will be common events in the production of transgenic mice.

A second instance of insertional mutagenesis has occurred in one of seven transgenic substrains of mice, each of which carries a metallothionein thymidine kinase fusion gene. In this particular case, Myk103, the foreign DNA construct, has become integrated into the germ line in low copy number and is stably transmitted by females, but not by males, even though they are fertile (Palmiter et al. 1984). Apparently, about half of the sperm carrying the insert fail to reach the cauda epididymis and vas deferens, and those that do are incapable of fertilization. Thus, the foreign DNA insert appears to have disrupted a gene that is expressed during the haploid stages of spermatogenesis, resulting in a male transmission-ratio distortion (see Chapter 6 for further discussion of haploid gene expression during spermatogenesis). These results imply that postmeiotic transcription allows for expression of genes important in the development of fertile sperm.

A third instance of insertional mutagenesis has been reported by Woychik and associates (1985). These investigators have created several strains of transgenic mice by injecting into fertilized mouse eggs a fusion construct of the mouse mammary tumor virus long terminal repeat and the mouse c-*myc* gene. These experiments were designed to test various hypotheses concerning the action of c-*myc*. However, in one particular line, 21 percent of the offspring from heterozygous parents possessed specific skeletal defects in both forelimbs and hindlimbs and paws. Thus, it appears that this particular insertional event has created a recessive mutation (ld^{Hd}) that leads to severe defects in limb formation. Furthermore, it turns out that this mutation is phenotypically identical and noncomplementary to two previously defined limb-deformity mutations: one that was described at Jackson Labs (ld^J) and one that was described at Oak Ridge (ld^{OR}). All three mutations have been mapped to chromosome 2. These data indicate that ld^{Hd} is allelic to ld^J nd ld^{OR}, and hopefully the inserted element will provide the necessary marker for molecular characterization of a gene involved in pattern formation of the limb in the mammalian embryo.

The *Mov-13* locus represents one of the few instances in which a lethal phenotype has been correlated with a molecular characterization of the mutagenized gene. This elegant example of the power of molecular genetics should be repeatable for any mutant caused by an insertional event provided that gross DNA rearrangements or deletions do not occur. It remains to be determined if insertional mutagenesis will be of any more use in identifying genes involved in specific developmental programs or processes than methods of generating mutations by conventional means. In both cases, the investigator has no control over particular genes to be mutagenized. Nevertheless, the real advantage of insertional mutagenesis lies in the fact that the mutagenized gene can be recognized at the molecular level.

F. Autosomal monosomy

In general, gross chromosomal imbalance in mammals is not compatible with postnatal survival, and autosomal aneuploidy is no exception. In fact, most aneuploids die sometime prior to term, as detailed in the excellent discussion of aneuploidy by Epstein (1986). However, unlike autosomal trisomics, which live at least until midgestation, autosomal monosomic embryos die sometime during the preimplantation or periimplantation period (see Table 14.1 for survival of mouse autosomal monosomics) (Ford and Evans 1973; Gropp, Giers, and Kolbus 1974; Gropp, Kolbus, and Giers 1975; Dyban and Baranov 1978; Epstein and Travis 1979; Magnuson, Smith, and Epstein 1982; Magnuson et al. 1985; Baranov 1983). In the mouse, only rare survivors have been detected after implantation, and these have always been morphologically abnormal and retarded in development. This early lethality of monosomic embryos also appears to be true in humans (Boué, Boué, and Lazar 1975; Hassold et al. 1978). Although we do not understand the reasons for autosomal monosomy being lethal so much earlier in development than autosomal trisomy, the universal lethality of monosomies suggests that large numbers of loci scattered over all of the autosomes are involved in dosage-dependent processes, so that a 50 percent reduction is able to produce serious consequences very early in development.

The actual time of onset of lethality in mice varies considerably for each of the different monosomies. For example, monosomies 1, 2, 5, 15, and possibly 11 and 16 begin to cause death at least by the early blastocyst stage, but for monosomies 14 and 19 the lethal period does not begin until the late blastocyst stage, and for monosomies 4, 10, 12, 17, and 18 lethality occurs sometime after the late blastocyst stage (Baranov 1983; Magnuson et al. 1985). Even though monosomies can be roughly grouped according to the time when lethality is first detected, the lethal period often extends over several days. For example, some embryos monosomic for chromosome 1 die at the 8- to 16-cell stage, whereas others do not die until after the late blastocyst stage. Thus, lethality does not seem to be restricted to any stage-specific event, and therefore one might conclude that basic cellular processes or structures are being affected. Our earlier studies have excluded as the cause of death, at least for monosomy 19, expression of recessive lethal genes (Epstein and Travis 1979), as well as generalized derangement in the expression of the early embryonic genome (Magnuson et al. 1982). In fact, the only consistent defect that has been detected among the different monosomies is a marked decrease in cell number, suggesting that cell proliferation may be affected (Magnuson et al. 1982, 1985).

Obviously, genetic background may play an important role in determining exactly when a particular monosomy will result in embryonic death (Magnuson et al. 1985). The effect of genetic constitution on the phenotype is not unique to the monosomies. Strain differences in both time of death and nature of malformations have also been observed with trisomic mouse fetuses (Evans, Brown, and Burtenshaw 1980; Gropp, Gropp, and Winking 1981; Gropp and Grohé 1981).

In an earlier study (Magnuson et al. 1982), we found that monosomy-19 embryos could be rescued in aggregation chimeras until at least day 9. However, the monosomic contribution to the resulting chimera was always less than 50 percent. When examined at a later stage (day 14), no monosomy-19 chimeric fetuses were found, suggesting that cells with monosomy 19 did not survive (S. Debrot, T. Magnuson, and C. J. Epstein, unpublished results). The same negative result was found for chimeras involving monosomy 14 when examined even as early as day 9 (S. Debrot and C. J. Epstein, unpublished results). We have interpreted these data to mean either that monosomic cells are dying or that they are at a proliferative disadvantage and are overgrown by diploid cells. Thus, it appears that autosomal monosomy is not generally compatible with survival to term, even in the form of an aggregation chimera (Magnuson et al. 1985). This conjecture is supported by our other studies on chromosomes 15, 16, and 17, in which 149 term or live-born aneuploid chimeras were examined, and only 1 monosomy chimera was found (Epstein et al. 1982; Cox et al. 1984; Epstein, Smith, and Cox 1984).

It is important to note that other studies of reciprocal translocations (Searle and Beechey 1978) and partial deletions (McGrath and Solter 1984) have identified several chromosomal regions that display differential activity depending on whether they are maternally or paternally derived (see Chapter 13 for further discussion). In all of the chimera studies described earlier, the monosomy embryos used to generate the chimeras contained only the maternally derived chromosome in question. Thus, in order to rule out the possibility that lethality was due to an absolute need for the paternally derived chromosome, these experiments should be repeated in a reciprocal fashion such that the monosomic embryos will be missing the maternally derived chromosome rather than the paternally derived chromosome.

We do not know the reasons for the early lethality of embryos with autosomal monosomy. Because a direct consequence of any aneuploid state is a dosage-dependent alteration in the synthesis of gene products of the unbalanced chromosome (Epstein et al. 1981; Epstein 1985), it is likely that a 50 percent decrease in the synthesis of at least one, but more likely several, gene products may be responsible for death of the embryo. Although one would like to believe that specific phenotypic effects are associated with each of the monosomies, and that these effects are due to imbalance of specific genes on the unbalanced chromosomes, the early lethality of the monosomies makes it difficult to demonstrate this possibility. Actual identification of the responsible loci and the processes controlled by them represents a fascinating but formidable challenge for future experiments. Some possible approaches include the production of partial aneuploids, injection of suspect genes to increase gene dosage, and the use of techniques such as anti-sense RNA (to be discussed later) in order to decrease gene dosage.

III. Genes that control development: How can we identity them?

The classic approach of utilizing mutations that result from spontaneous events or from conventional mutagenesis experiments (in the hope of identifying developmental genes) encounters a series of problems. For example, it is time-consuming and expensive to attempt to generate and screen for new lethal mutations in mice, particularly if one is interested in a specific stage of development or a specific tissue type within the embryo. In addition, it is practically impossible to generate multiple alleles or temperature-sensitive alleles of existing mutations, two approaches that have been extremely beneficial in other systems for studying the biological consequences of mutations affecting development. Furthermore, it is not easy to make attempts at precise mapping or chromosomal walking for cloning experiments because it is not possible to generate multiple chromosomal breakpoints surrounding the mutations of interest. Because of these limitations, it has been very difficult to transcend the gap between phenotype, gene, and gene product. All of these factors have frustrated mammalian developmental geneticists for years. It is essential to find means for identifying and characterizing important stage-specific genes and to determine the effects on the embryo when they are not correctly expressed. The remainder of this chapter will deal with the evidence suggesting that this is a realistic possibility.

A. Neutralization of specific gene products

i. Anti-sense RNA. One possible way to examine the molecular details of gene activity at specific stages is to interfere with the production or function of gene products. In this regard, exciting recent experiments have demonstrated that it is possible to prevent gene expression by blocking translation of specific messenger RNAs with complementary sequences. Izant and Weintraub (1984) have shown that thymidine kinase (TK) gene expression can be reduced fourfold to fivefold when DNA plasmids containing the TK gene are injected into TK-deficient mouse L cells together with a 100- to 200-fold excess of plasmids that direct the reverse-polarity (anti-sense) transcription of the TK gene. These experiments have demonstrated that gene plasmids designed to produce anti-sense RNA can diminish the appearance of a specific gene product.

A similar approach was used by Rubenstein, Nicolas, and Jacob (1984). These investigators cotransfected into 3T6 mouse fibroblasts the *Escherichia coli LacZ* gene, which had been cloned into a eukaryotic expression vector, along with a 2,566-bp 5′ fragment of the *LacZ* gene inserted into the expression vector in an anti-sense orientation. When equal amounts of both constructs were cotransfected, β-galactosidase activity was found to be decreased approximately 10-fold. Thus, not only can the anti-sense copy result in a significant reduction in expression of the *LacZ* gene, but this reduction requires only the 5′ portion of the gene.

An alternative method involving direct injection of in vitro synthesized anti-sense RNA has recently been reported (Melton 1985). Using the frog oocyte system, it was demonstrated that translation of injected globin messenger RNA can be specifically and completely blocked if a 50-fold excess of anti-sense globin RNA is injected. Melton found that an effective block would occur only if the anti-sense RNA was injected prior to the globin mRNA or if the two were coinjected. If the globin mRNA was injected first, followed by the anti-sense RNA, a low level of globin protein synthesis could still be detected. These results suggest that it may not be possible to inhibit protein synthesis from a message that is already being translated. Additional experiments confirmed that the critical region for blocking mRNA translation with the anti-sense RNA is the 5' region of the message and showed that the block to translation appears to be the result of RNA-RNA duplex formation.

The potential for use of anti-sense RNA to dissect the activities of genes involved in *Drosophila* development has recently been demonstrated by Rosenberg and associates (1985) studying the *Krüppel (Kr)* mutation. The *Kr* gene has an early zygotic function that is involved in the segmentation process (Nusslein-Volhard and Wieschaus 1980; Wieschaus, Nusslein-Volhard, and Kluding 1984; Preiss et al. 1985). Embryos homozygous for the mutation die before hatching and show a unique lethal phenotype that is characterized by a deletion of adjacent thoracic and anterior abdominal segments. In amorphic alleles, the missing segments are replaced by a mirror-image duplication of parts of the normal posterior abdomen. In weaker alleles, progressively fewer segments are deleted. The Kr^+ gene results in the production of a rare 2.5-kb poly(A)$^+$ RNA transcript, most of which has been recovered in a 2.5-kb cDNA clone. Rosenberg and associates (1985) found that when RNA transcribed from this clone in an anti-sense orientation was injected into wild-type embryos, phenocopies of *Kr* homozygous lethal phenotype could be produced. However, none of the injected embryos showed the extreme, amorphic *Kr* phenotype, suggesting that even though an excess of the anti-sense RNA was injected, the wild-type gene activity was not completely abolished. This could perhaps be due to a problem with stability of the anti-sense RNA or with the kinetic parameters of duplex formation in vivo. Neither of these factors has been examined.

The value of anti-sense transcription as a versatile tool for genetic analysis of development lies in its potential use for analyzing the cellular functions associated with DNA sequences expressed at specific times and in discrete regions of the embryo. The goal will be to determine if specific mRNAs are involved in stage- and tissue-specific events. Whether or not this can be achieved by causing constitutive anti-sense transcription to interfere with activation of endogenous genes is still unknown.

ii. Antibodies. Another possibility to be explored is neutralization of specific gene products by introduction of antibodies. This approach was used successfully in early amphibian embryos in which antibodies to a gap-junction protein were found to disrupt junction-mediated communication selectively (Warner, Guthrie, and Gilula 1984). Interestingly, when injected into a specific cell of an 8-cell embryo, this disruption led to defects much later in development. The most frequent defect observed involved varying degrees of right/left asymmetry. A small percentage of the embryos failed to form brains and eyes on both sides. These results suggest that intercellular communication has a pronounced influence on embryonic development. A similar approach has been used by Izant, Weatherbee, and McIntosh (1983), who produced a monoclonal antibody unique to a microtubule-associated protein of the mitotic spindle. When injected into cells prior to the onset of anaphase, the antibody was found to inhibit proper chromosome movement.

The use of antibodies to study early mouse development will, of course, require prior preparation and characterization of a target protein and may be applicable only during the early stages of development. This approach has been used to study the role of cell surface molecules in compaction. It was shown that an antiembryonal carcinoma serum could prevent compaction or could reversibly trigger decompaction of mouse embryos. The antigen has been characterized as an extracellular molecule, known variously as uvomorulin, cadherin, or cell-adhesion molecule; it is stable at the cell surface and is involved in cell-cell recognition during early embryogenesis (Kemler, Babinet, and Jacob 1977; Hayafil et al. 1980; Hyafil, Babinet, and Jacob 1981; Peyrieras et al. 1983).

When considering the potential for using techniques that allow for neutralization of specific gene products, there is no doubt that murine developmental genetics will advance during the next few years. Among the formidable challenges will be that of determining the normal function of these gene products within the cell.

B. Homologous genes: the homeo box

Analysis of genetic control of morphogenesis in *Drosophila* has advanced rapidly in recent years with the identification and molecular cloning of genes that are involved in the control of segment pattern formation. These genes can be categorized basically into two groups. One group includes those genes that control the number and polarity of segments (Nusslein-Volhard and Wieschaus 1980). For example, embryos lacking function of the wild-type allele of fushi tarazu (*ftz*) form half the number of expected segments (Wakimoto and Kaufman 1984). The second group includes the homeotic genes that specify segment diversity and identity, but not segmentation itself (Garcia-Bellido 1975; Lewis 1978a). For example, inactivation of bithorax (*bx*), a homeotic gene, causes the anterior half of the third thoracic segment to transform from a haltere to a wing structure.

In a recent series of reports (McGinnis et al. 1984a, 1984c; Scott and

Weiner 1984), a highly conserved nucleotide sequence, named the homeo box, was found associated with several, but not all, of the *Drosophila* homeotic and segmentation genes. In fact, the presence of this conserved sequence has allowed identification of previously unknown homeotic genes. This domain was also found in the genomes of a wide spectrum of animals, including annelid worms, frogs, birds, mice, and humans (Carrasco et al. 1984; Levine, Rubin, and Tjion 1984; McGinnis et al. 1984a, 1984b; Joyner et al. 1985a; Rabin et al. 1985). Needless to say, this discovery has created much excitement, particularly because it is tempting to expect a similar set of conserved functions for homeo-box-associated genes in vertebrate organisms. If so, the highly conserved homeo-box sequence should then provide the molecular tag needed for identifying these genes in higher organisms.

Some notable features associated with the homeo-box domain include its consistent size of about 180 nucleotides, the fact that it generally lies within an extended open reading frame, and the fact that in most cases it is highly conserved (Gehring 1985). This extensive homology strongly suggests that the homeo-box domain is subject to selective pressure. It is interesting to point out that a low but nevertheless significant homology exists with the amino acid sequences encoded by the a1 and α2 yeast mating-type genes (Sheperd et al. 1984), and also with regulatory proteins in prokaryotic systems (Laughon and Scott 1984; Shepherd et al. 1984). Another consistent feature of the homeo-box domain is its large content of basic amino acids (approximately 30 percent of the residues are either arginine or lysine). This amino acid composition is compatible with the idea that the homeo-box-containing proteins may be DNA-binding proteins. This conjecture is supported by immunofluorescence data indicating that the Ultrabithorax homeotic proteins are localized in cell nuclei (White and Wilcox 1984).

The segmentation and homeotic genes are all essential for normal segmental development in the fly. The absence of any of them results in distorted embryonic development. They have also been shown to be expressed in very discrete, coherent groups of cells, the fates of which they apparently determine in concert with other pattern-formation genes (Akam 1983; Levine et al. 1984; Hafen, Kuroiwa, and Gehring 1984a; Hafen, Levine, and Gehring 1984b; Kuroiwa, Hafen, and Gehring 1984). Like the *Drosophila* homeo-box-containing genes, the homologous *Xenopus* genes (*AC1* and *MM3*) are expressed during development (Carrasco et al. 1984; Müller, Carrasco, and DeRobertis 1984). One transcript of the *AC1* is first detected at late gastrulation; it increases during neurulation and then declines by the tadpole stage. Another transcript is switched on at late gastrulation; its amount continues to increase and remains high even in swimming tadpoles. A third transcript is not detected until the tail-bud stage. A transcript of the *MM3* gene is first detected in oocytes (Müller et al. 1984) and then declines during early development; it reappears by gastrulation and then remains until the swimming-tadpole stage. The functional significance of these transcripts is not known.

Sequences homologous to the *Drosophila* homeo box are present in at least 10 copies in both mice and humans (Levine et al. 1984; McGinnis et al. 1984a). One of the homeo-box-containing mouse genomic clones (Mo-10) has been mapped to mouse chromosome 6 (McGinnis et al. 1984b), and another (Mo-en.1) has been mapped to chromosome 1 (Joyner et al. 1985b). At least five other homeo-box-loci have been mapped to mouse chromosome 11 (Hart et al. 1985; Jackson et al. 1985; Joyner et al. 1985a; Rabin et al. 1985). In one study, two of the chromosome-11 homeo-box loci were discovered by using two human single-copy homeo-box flanking probes that map to human chromosome 17 (Joyner et al. 1985a). This is interesting, because several genes that map to human chromosome 17 also map to mouse chromosome 11, suggesting that a region of mouse chromosome 11 may be syntenic with human chromosome 17 (Joyner et al. 1985a; Rabin et al. 1985).

These genetic mapping studies have led to speculation regarding possible allelism of the homeo-box loci with genomic loci known to affect mouse development. For example, McGinnis and associates (1984b) pointed out that there are approximately 50 murine mutations that have been mapped and are known to affect morphogenesis. On mouse chromosome 6 alone there are five such genes (hypodactyly, truncate, postaxial hemimelia, microphthalmia, and crooked). However, regional mapping of the Mo-10 homeo-box loci distal to the kappa-immunoglobulin locus rules out allelism with everything but hypodactyly. Hypodactyly is known to cause a reduction in the phalanges in both forefeet and hindfeet of homozygotes. In heterozygotes, the malformation is restricted to the hindfeet. On mouse chromosome 11 there are two known genes (tail-short and vestigial tail) that affect morphogenesis (cf. Joyner et al. 1985b; Rabin et al. 1985). Homozygotes for tail-short die during implantation (Paterson 1980), whereas heterozygotes have skeletal abnormalities that include vertebral fusion, asymmetry of limb length, triphalangy, and an additional pair of ribs. Vestigial-tail is a viable recessive mutation that also affects the number and form of vertebrae. It remains to be determined if any of these genes contain homeo boxes.

The homeo-box loci mapped to chromosome 11 have been shown to encode multiple transcripts, some of which are expressed as early as day 7.5 of embryogenesis (Jackson, Schofield, and Hogan 1985). Maximum expression, however, does not occur until days 11.5−12.5 (Hart et al. 1985; Hauser et al. 1985; Jackson et al. 1985), and, in the one case tested so far, a transcript was found to be enriched in the embryonic spinal cord and brain (Jackson et al. 1985). The only adult tissue that expresses this particular homeo-box gene in high levels is the kidney. The chromosome-1 homeo-box-containing gene Mo-en.1 has been shown to be expressed both in differentiated teratocarcinomas and in mid- to late-gestation fetuses (Joyner et al., 1985b). This particular mouse homeo box shows strong homolgy with the *Drosophila engrailed* homeo box and, even more interesting, for 63 nucleotides immediately 3' to the homeo box itself (Joyner et al. 1985b). Another mouse homeo-box region that has not yet been mapped to a mouse chromosome has been shown to contain at least three homeo-box loci (Colberg-Poley et al. 1985a, 1985b), two of which are expressed in F9 teratocarcinoma cells and mid-gestation fetuses. One of these genes (*M6*) was found to be expressed in a variety of adult tissues (Colberg-Poley et al. 1985b).

The high degree of homology among the homeo boxes of flies, frogs, mice, and humans is intriguing. It is not known if the homeo boxes of vertebrates are associated with genes similar to the segmentation and homeotic genes of flies. It is possible that the function of the homeo domain has been conserved, whereas the function of the gene itself has changed. In any case, more work is needed before the significance of either the homeo-box domain or the homeo-box loci can be understood.

IV. Summary

A considerable body of biochemical and genetic evidence clearly indicates that the mammalian embryonic genome is functional during early development and that it plays an essential role even as early as the 2-cell stage. It is therefore of major interest and importance to determine the transcriptional and posttranscriptional control mechanisms involved in regulating this gene activity. Traditionally, this has been approached by studying mutations that result in abnormal development. It is important to distinguish mutant genes causing metabolic or structural errors within the cell from genes associated with developmental programs or their regulation. Only in a few cases has the effect of a mutant been defined on a cellular or molecular level (e.g., *Os* affects the process of cell division; *Mov-13* affects the collagen-1 gene). Systems that offer the most promise for molecular approaches include the albino deletions, the *t* complex, and the insertional mutations.

There has been much recent interest in the possibility of using a more directed approach for identifying genes involved in specific developmental processes, rather than simply relying on randomly generated mutations. One possibility involves elimination of specific gene products and their functions by use of complementary nucleotides or antibodies. In theory, this would allow the production of phenocopies of mutations for any gene that has been cloned. In addition to this approach, increasing attention has recently been focused on identifying homologous genes between species, especially with the *Drosophila* homeotic genes, and some of this information may be directly applicable to mammalian systems.

Acknowledgments

I am deeply indebted to Ms. Della Yee for her never-ending support, enthusiasm, and constant discussion. I would like to thank Drs. Salome G. Waelsch and Wendy Golden and Mr. Dan Odom for critically reading this manuscript and for their helpful suggestions. I would also like to express my sincere gratitude to Dr. Salome G. Waelsch for encouraging my interests in the albino deletions. The work originating from this laboratory was supported by NIH grant HD-19892 and by the Pew Memorial Trust. The author is a Pew Scholar in the Biomedical Sciences.

References

Akam, M. (1983). The location of *Ultrabithorax* transcripts in *Drosophila* tissue sections. *EMBO Journal* 2, 2075–84.

Artzt, K. (1984) Gene mapping within T/t complex of the mouse. III. t-lethal genes are arranged in three clusters on chromosome 17. *Cell* 39, 565–72.

Artzt, K., McCormick, P., and Bennett, D. (1982a) Gene mapping within the T/t complex of the mouse. 1. t-lethal genes are nonallelic. *Cell* 28 463–70.

Artzt, K., Shin, H., and Bennett D. (1982b) Gene mapping within the T/t complex of the mouse. II. Anomalous position of the *H-2* complex in t haplotypes. *Cell* 28, 471–76.

Babiarz, B. (1983) Deletion mapping of the T/t complex: evidence for a second region of critical embryonic genes. *Dev. Biol.* 95, 342–51.

Babiarz, B., Garrisi, G. J., and Bennett, D. (1982) Genetic analysis of the t^{w73} haplotype of the mouse using deletion mutations: evidence for a parasitic lethal mutation. *Genet. Res.* 39, 111–20.

Baranov, V. S. (1983) Chromosomal control of early embryonic development in mice. I. Experiments on embryos with autosomal monosomy. *Genetica* 61, 165–77.

Bender, W., Akam, M. A., Beachy, P. A., Karch, F., Peifer, M., Lewis, E. B., and Hogness, D. S. (1983) Molecular genetics of the bithorax complex in *Drosophila melanogaster*. *Science* 221, 23–9.

Bennett, D. (1975) The T locus of the mouse. *Cell* 6, 441–54.

– (1980) The T-complex in the mouse: an assessment after 50 years of study. *Harvey Lect.* 74, 1–21.

Bennett, D., and Dunn, L. C. (1958) Effects on embryonic development of a group of genetically similar lethal alleles derived from different populations of wild house mice. *J. Morphol.* 103, 135–57.

Bennett, D., Dunn, L. C., Spiegelman, M., Artzt K., Cookingham, J., and Schermerhorn, E. (1975) Observations on a set of radiation-induced dominant T-like mutations in the mouse. *Genet. Res.* 26, 95–108.

Boúe, J., Boúe, A., and Lazar, P. (1975) Retrospective and prospective epidemiological studies of 1500 karyotyped spontaneous human abortions. *Teratology* , 12, 11−26.

Breindl, M., Harbers, K., and Jaenisch, R. (1984) Retrovirus-induced lethal mutation in collagen 1 gene is associated with altered chromatin structure. *Cell* 38, 9−16.

Burgoyne, P. S., and Biggers, J. D. (1976). The consequences of X-dosage deficiency in the germ line: impaired development in vitro of preimplantation embryos from XO mice. *Dev. Biol.* 51, 109−17.

Calarco, P. G., and Brown, E. H. (1968) Cytological and ultrastructural comparisons of t^{12}/t ^{12} and normal mouse morulae. *J. Exp. Zool.* 168, 169−86.

Callahan, R., and Todaro, G. J. (1978) Four major endogenous retrovirus classes each genetically transmitted in various species of *Mus*. In *Origins of Inbred Mice*, ed. H. C. Morse III, pp. 689−713. New York: Academic Press.

Carrasco, A. E., McGinnis, W., Gehring, W. J., and DeRobertis, E. M. (1984) A homologous protein-coding sequence in *Drosophila* homeotic genes and its conservation in other metazoans. *Cell* 37, 409−14.

Colberg-Poley, A. M., Voss, S. D., Chowdhury, K., and Gruss, P. (1985a) Structural analysis of murine genes containing homeo box sequences and their expression in embryonal carcinoma cells. *Nature* 314, 713−18.

Colberg-Poley, A. M., Voss, S. D., Chowdhury, K., Stewart, C. L., Wagner, E. F., and Gruss, P. (1985b) Clustered homeo boxes are differentially expressed during murine development. *Cell* 43, 39−45.

Copeland, N. G., Hutchison, K. W., and Jenkins, N. A. (1983a) Excision of the DBA ecotropic provirus in dilute coat-color revertants of mice occurs by homologous recombination involving the viral LTRs. *Cell* 33, 379−87.

Copeland, N. G., Jenkins, N. A., and Lee, B. K. (1983b) Association of the lethal yellow (A^y) coat color mutation with an ecotropic murine leukemia virus genome. *Proc. Natl. Acad. Sci. U.S.A.* 80, 247−49.

Cori, C. F., Gluecksohn-Waelsch, S., Klinger H. P., Pick, L., Schlagman, S. L., Teicher, L. S., and Chang, H. W. (1981) Complementation of gene deletions by cell hybridization. *Proc. Natl. Acad. Sci. U.S.A.* 78, 479−83.

Cori, C. F., Gluecksohn-Waelsch, S., Shaw, P. A., and Robinson, C. (1983) Correction of a genetically caused enzyme defect by somatic cell hybridization. *Proc. Natl. Acad. Sci. U.S.A.* 80, 6611−14.

Covarrubius, L., Nishida, Y., and Mintz, B. (1985) Early developmental mutations due to DNA rearrangements in transgenic mouse embryos. *Cold Spring Harbor Symp. Quant. Biol.* 50, 447−52.

Cox, D. R., Smith, S. A., Epstein, L. B., and Epstein, C. J. (1984) Mouse trisomy 16 as an animal model of human trisomy 21 (Down syndrome): production of viable trisomy 16↔ diploid mouse chimeras. *Dev. Biol.* 101, 416−24.

Disteche, C. M., and Adler, D. (1984) Localization of cloned mouse chromosome 7-specific DNA to lethal albino deletions. *Somatic Cell Mol. Genet.* 10, 211−15.

Dunn, G. R. (1972) Embryological effects of a minute deficiency in linkage group II of the mouse. *J. Embryol. Exp. Morphol.* 27, 147−54.

Dunn, L. C., and Gluecksohn-Schoenheimer, S. (1950) Repeated mutations in one area of a mouse chromosome. *Prot. Natl. Acad. Sci. U.S.A.* 36, 233−7.

Dyban, A. P., and Baranov, V. S. (1978) *The Cytogenetics of Mammalian Embryogenesis.* Moscow: Nauka.

Eaton, G. J., and Green, M. M. (1963) Giant cell differentiation and lethality of homozygous yellow mouse embryos. *Genetica* 34, 155–61.

Epstein, C. J. (1986) *The Consequences of Chromosome Imbalance: Principles, Mechanisms, and Models.* Cambridge University Press.

Epstein, C J., Epstein, L. B., Cox, D. R., and Weil, J. (1981) Functional implications of gene dosage effects in trisomy 21. In *Trisomy* 21, eds. R. Burgio, M. Fraccaro, L. Tiepolo, and U. Wolf, pp. 155–72. Berlin: Springer-Verlag.

Epstein, C. J., Smith, S., and Cox, D. R. (1984) Production and properties of mouse trisomy 15 ↔ diploid chimeras. *Dev. Genet.* 4, 159–65.

Epstein, C. J., Smith, S. A., Zamora, T., Sawicki, J. A., Magnuson, T. R., and Cox, D. R. (1982) Production of viable adult trisomy 17 ↔ diploid mouse chimeras. *Proc. Natl. Acad. Sci. U.S.A.* 79, 4376–80.

Epstein, C. J., and Travis, B. (1979) Preimplantation lethality of monosomy for mouse chromosome 19. *Nature* 280, 144–5.

Erickson, R. P., Lewis, S. E., and Slusser, K. S. (1978) Deletion mapping of the *t* complex of chromosome 17 of the mouse. *Nature* 274, 163–4.

Erickson, R. P., Pickart, L., and Thaler, M. M. 1976) Serum protein synthesis in mutant mice with abnormal hepatic endoplasmic reticulum. *Cytobios* 15, 49–56.

Evans, E. P., Brown, B. B., and Burtenshaw, M. D. (1980). Personal communication. *Mouse News Letter* No. 63, 30.

Falconer, D. S., Latsyzewski, M., and Isaacson, J. H. (1964) Diabetes insipidus associated with oligosyndactyly in the mouse. *Genet. Res.* 5, 473–88.

Ford, C. E., and Evans, E. P. (1973) Non-expression of genome unbalance in haplophase and early diplophase of the mouse and incidence of karyotypic abnormality in post-implantation embryos. In *Proceedings of the Symposium on Chromosomal Errors in Relation to Reproductive Failure,* eds. A. Boúe and C. Thibault, pp. 271–285. Paris: INSERM.

Fox, H. S., Martin, G. R., Lyon, M. F., Herrmann, B., Frischauf, A. M., Lehrach, H., and Silver, L. M. (1984a) Molecular probes define different regions of the mouse *t* complex. *Cell* 40, 63–9.

Fox, H. S., Silver, L. M., and Martin, G. R. (1984b) An alpha globin pseudogene is located within the mouse *t* complex. *Immunogenetics* 19, 125–30.

Francke, U., and Gehring, U. (1980) Chromosome assignment of a murine glucocorticoid receptor gene (*Grl-1*) using intraspecies somatic cell hybrids. *Cell* 22, 657–64.

Garber, R. L., Kuroiwa, A., and Gehring, W. J. (1983) Genomic and cDNA clones of the homeotic locus *Antennapedia* in *Drosophila. EMBO Journal* 2, 2027–36.

Garcio-Bellido, A. (1975) Genetic control of wing disc development in Drosophila. In *Cell Patterning,* Ciba Foundation Symposium, pp. 161–78. Amsterdam: Associated Scientific.

Garland, R. C., Satrustegui, J., Gluecksohn-Waelsch, S., and Cori, C. F. (1976) Deficiency of plasma protein synthesis caused by X-ray-induced lethal albino alleles in mouse. *Proc. Natl. Acad. Sci. U.S.A.* 73, 3376–80.

Gehring, W. J. (1985) The homeo box: a key to the understanding of development. *Cell* 40, 3–5.

Gluecksohn-Schoenheimer, S. (1940) The effect of an early lethal (t^o) in the house mouse. *Genetics* 25, 391–400.

– (1949) The effects of a lethal mutation responsible for duplications and twinning in mouse embryo. *J. Exp. Zool.* 110, 47–76.

Gluecksohn-Waelsch, S. (1979) Genetic control of morphogenetic and biochemical differentiation: lethal albino deletions in the mouse. *Cell* 16, 225–37.

Gluecksohn-Waelsch, S., and Erickson, R. P. (1970) The *T* locus of the mouse: implications for mechanisms of development. *Curr. Top. Dev. Biol.* 5, 281–316.

Gluecksohn-Waelsch, S., Teicher, L. S., Pick, L., and Cori, C. F. (1980) Genetic rescue of lethal genotypes in the mouse. *Dev. Genet.* 1, 219–28.

Goldfeld, A. E., Firestone, G. L., Shaw, P.A., and Glueckson-Waelsch, S. (1983) Recessive lethal deletion on mouse chromosome 7 affects glucocorticoid receptor binding activities. *Proc. Natl. Acad. Sci. U.S.A.* 80, 1431–4.

Goldfeld, A. E., Rubin, C. S., Siegel, T. W., Shaw, P. A., Schiffer, S. G., and Gluecksohn-Waelsch, S. (1981) Genetic control of insulin receptors. *Proc. Natl. Acad. Sci. U.S.A.* 78, 6359–61.

Green, E. L. (1966) *Biology of the Laboratory Mouse.* New York: McGraw-Hill.

– (1981) *Genetic Variants and Strains of the Laboratory Mouse.* London: Oxford University Press.

Gropp, A., Giers, D., and Kolbus, U. (1974) Trisomy in fetal backcross progeny of male and female metacentric heterozygotes of the mouse. I. *Cytogenet. Cell Genet.* 13, 511–35.

Gropp, A., and Grohé, G. (1981) Strain background dependence of expression of chromosome triplication in the mouse embryo. *Hereditas* 94, 7–8.

Gropp, A., Kolbus, U., and Giers, D. (1975) Systematic approach to the study of trisomy in the mouse. II. *Cytogenet. Cell Genet.* 14, 42–62.

Gropp, D., Gropp, A., and Winking, H. (1981) Personal communication. *Mouse News Letter* No. 64, 70.

Grüneberg, H. (1956) Genetical studies on the skeleton of the mouse. XVIII. Three genes for syndactylism. *J. Genet.* 54, 113–45.

– (1961) Genetical studies on the skeleton of the mouse. XXVII. The development of oliogosyndactylism. *Genet. Res.* 2, 33–42.

Guenet, J., Condamine, H., Gaillard, J., and Jacob, F. (1980) t^{wPa-1}, t^{wPa-2}, t^{wPa-3}: three new *t*-haplotypes in the mouse. *Genet. Res.* 36, 211–17.

Hafen, E., Kuroiwa, A., and Gehring, W. J. (1984a) Spatial distribution of transcripts from the segmentation gene *fushi tarazu* during Drosophila embryonic development. *Cell* 37, 833–41.

Hafen, E., Levine, M., and Gehring, W. J. (1984b) Regulation of *Antennapedia* transcript by the bithorax complex in *Drosophila. Nature* 307, 287–9.

Harbers, K., Jähner, D., and Jaenisch, R. (1981) Microinjection of cloned retroviral genomes into mouse zygotes: integration and expression in the animal. *Nature* 293, 540–2.

Harbers, K., Kuehn, M., Delius, H., and Jaenisch, R. (1984) Insertion of retrovirus into the first intron of α1(I) collagen gene leads to embryonic lethal mutation in mice. *Proc. Natl. Acad. Sci. U.S.A.* 81, 1504–8.

Hart, C.P., Awgulewitsch, A., Fainsed, A., McGinnis, W., and Ruddle, F. H. (1985) Homeo box gene complex on mouse chromosome 11: molecular cloning, expression in embryogenesis, and homology to a human homeo box locus. *Cell* 43, 9–18.

Hassold, T. J., Matsuymama, A., Newlands, J. M., Matsuura, J. S., Jacobs, P. A., Manuel, B., and Tsuei, J. (1978) A cytogenetic study of spontaneous abortions in Hawaii. *Ann. Hum. Genet.* 41, 443–54.

Hauser, C. A., Joyner, A. L., Klein, R. D., Learned, T. K., Martin, G. R., and Tjian, R. (1985) Expression of homologous homeo-box-containing genes in differentiated human teratocarcinoma cells and mouse embryos. *Cell* 43, 19–28.

Herrmann, B. A., Bucan, M., Mains, P. E., Frischauf, A. M., Silver, L. M., and Lehrach, H. (1986) Genetic analysis of the proximal portion of the mouse *t* complex: Evidence for a second inversion within *t* haplotypes. *Cell* 44, 469–75.

Hillman, N., and Hillman, R. (1975) Ultrastructural studies of t^{w32}/t^{w32} mouse embryos. *J. Embryol. Exp. Morphol.* 33, 685–95.

Hillman, N., Hillman, R., and Wileman, G. (1970) Ultrastructural studies of cleavage stage t^{12}/t^{12} mouse embryos. *J. Reprod. Fertil.* 33, 501–6.

Hyafil, F., Babinet, C., and Jacob, F. (1981) Cell-cell interactions in early embryogenesis: a molecular approach to the role of calcium. *Cell* 26, 447–54.

Hyafil, F., Morello, D., Babinet, C., and Jacob, F. (1980) A cell surface glycoprotein involved in the compaction of embryonal carcinoma cells and cleavage stage embryos. *Cell* 21, 927–34.

Izant, J. G., Weatherbee, J. A., and McIntosh, J. R. (1983) A microtubule-associated protein antigen unique to mitotic spindle microtubules in PtK1 cells. *J. Cell Biol.* 96, 424–34.

Izant, J. G., and Weintraub, H. (1984) Inhibition of thymidine kinase gene expression by anti-sense RNA: a molecular approach to genetic analysis. *Cell* 36, 1007–15.

Jackson, I., Schofield, P., and Hogan, B. (1985) A mouse homeo box gene is expressed during embryogenesis and in the adult kidney. *Nature* 317, 745–8.

Jaenisch, R. (1976) Germ line integration and Mendelian transmission of the exogenous Moloney leukemia virus. *Proc. Natl. Acad. Sci. U.S.A.* 73, 1260–6.

Jaenisch, R., Harbers, K., Schnieke, A., Lohler, J., Chumakov, I., Jähner, D., Grotkopp, D., and Hoffman, E. (1983) Germline integration of Moloney murine leukemia virus at the *Mov-13* locus leads to recessive lethal mutation and early embryonic death. *Cell* 32, 209–16.

Jaenisch, R., Jähner, D., Nobis, P., Simon, I., Lohler, J., Harbers, K., and Grotkopp, D. (1981) Chromosomal position and activation of retroviral genomes inserted into the germ line of mice. *Cell* 24, 519–24.

Jähner, D., and Jaenisch, R. (1980) Integration of Moloney leukemia virus into the germ line of mice: correlation between site of integration and virus activation. *Nature* 287, 456–8.

Jenkins, N. A., Copeland, N. G., Taylor, B. A., and Lee, B. K. (1981) *Dilute (d)* coat colour mutation of DBA/2J mice is associated with the site of integration of an ecotropic MuLV genome. *Nature* 293, 370–4.

Johnson, M. H. (1981) The molecular and cellular basis of preimplantation mouse development. *Biol. Rev.* 56, 463–98.

Joyner, A. L., Kornberg, T., Coleman, K. G., Cox, D. R., and Martin, G. R. (1985a) Expression during embryogenesis of a mouse gene with sequence homology to the Drosophila *engrailed* gene. *Cell* 43, 29–37.

Joyner, A. L., Lebo, R. V., Kan, Y. W., Tjian, R., Cox, D. R., and Martin, G. R. (1985b) Comparative chromosome mapping of a conserved homeo box region in mouse and human. *Nature* 314, 173–5.

Kadam, K. M. (1962) Genetical studies on the skeleton of the mouse. XXXI. The muscular anatomy of syndactylism and oligosyndactylism. *Genet. Res.* 3, 139–56.

Kaufman, M. H., and Gardner, R. L. (1974) Diploid and haploid mouse parthenogenetic development following in vitro activation and embryo transfer. *J. Embryol. Exp. Morphol.* 31, 635–42.

Kaufman, M. H., and Sachs, L. (1975) The early development of haploid and aneuploid parthenogenetic embryos. *J. Embryol. Exp. Morphol.* 34, 645–55.

Kemler, R., Babinet, C., and Jacob, F. (1977) Surface antigen in early differentiation. *Proc. Natl. Acad. Sci. U.S.A.* 74, 4449–52.

Kidwell, M. G., Kidwell, J. F., and Sved, J. A. (1977) Hybrid dysgenesis in *Drosophila melanogaster:* a syndrome of aberrant traits including mutation, sterility and male recombination. *Genetics* 86, 813–33.

Klein, J., and Hammerberg, C. (1977) The control of differentiation by the *T* complex. *Immunol. Rev.* 33, 70–104.

Kuroiwa, A., Hafen, E., and Gehring, J. (1984) Cloning and transcriptional analysis of the segmentation gene *fushi tarazu* of Drosophila. *Cell* 37, 825–31.

Laughon, A., and Scott, M. P. (1984) Sequence of a *Drosophila* segmentation gene: protein structure homology with DNA-binding proteins. *Nature* 310, 25–31.

Levine, M., Hafen, E., Garber, R. L., and Gehring, W. J. (1983) Spatial distribution of *Antennapedia* transcripts during *Drosophila* development. *EMBO Journal* 2, 2037–46.

Levine, M., Rubin, G. M., and Tjian, R. (1984) Human DNA sequences homologous to a protein coding region conserved between homeotic genes of Drosophila. *Cell* 38, 667–73.

Lewis, E. B. (1978a) A gene complex controlling segmentation in *Drosophila*. *Nature* 276, 565–70.

Lewis, S. E. (1978b). Developmental analysis of lethal effects of homozygosity for the c^{25H} deletion in the mouse. *Dev. Biol.* 65, 553–7.

Lewis, S. E., Turchin, H. A., and Gluecksohn-Waelsch, S. (1976) The developmental analysis of an embryonic lethal (c^{6H}) in the mouse. *J. Embryol. Exp. Morphol.* 36, 363–71.

Lohler, J., Timpl, R., and Jaenisch, R. (1984) Embryonic lethal mutation in mouse collagen 1 gene causes rupture of blood vessels and is associated with erythropoietic and mesenchymal cell death. *Cell* 38, 597–607.

Luthardt, F. W. (1976) Cytogenetic analysis of oocytes and early preimplantation embryos from XO mice. *Dev. Biol.* 54, 73–81.

Lyon, M. F. (1981) The t-complex and the genetical control of development. *Symp. Zool. Soc. Lond.* 47, 455–77.

McClintock, B. (1956) Controlling elements and the gene. *Cold Spring Harbor Symp. Quant. Biol.* 21, 197–216.

McGinnis, W., Garber, R. L., Wirz, J., Kuroiwa, A., and Gehring, W. J. (1984a) A homologous protein-coding sequence in *Drosophila* homeotic genes and its conservation in other metazoans. *Cell* 37, 403–8.

McGinnis, W., Hart, C. P., Gehring, W. J., and Ruddle, F. H. (1984b) Molecular cloning and chromosome mapping of a mouse DNA sequence homologous to homeotic genes of *Drosophila*. *Cell* 38, 675–80.

McGinnis, W., Levine, M., Hafen, E., Kuroiwa, A., and Gehring, W. J. (1984c) A conserved DNA sequence in homeotic genes of the *Drosophila* Antennapedia and bithorax complexes. *Nature* 308, 428–33.

McGrath, J., and Solter, D. (1984) Maternal T^{hp} lethality in the mouse is a nuclear, not cytoplasmic, defect. *Nature* 308, 550–1.

McLaren, A. (1976) Genetics of the early mouse embryo. *Annu. Rev. Genet.* 10, 361–88.

Magnuson, T. (1983) Genetic abnormalities and early mammalian development. In *Development in Mammals*, Vol. 5, ed. M. H. Johnson, pp. 209–49. Amsterdam: Elsevier.

Magnuson, T., Debrot, S., Dimpfl, J., Zweig, A., Zamora, T., and Epstein, C. J. (1985) The early lethality of autosomal monosomy in the mouse. *J. Exp. Zool.* 236, 353–60.

Magnuson, T., and Epstein, C. J. (1981) Genetic control of very early mammalian development. *Biol. Rev.* 56, 369–408.

– (1984) Oligosyndactyly: a lethal mutation in the mouse that results in mitotic arrest very early in development. *Cell* 38, 823–33.

Magnuson, T., Smith, S., and Epstein, C. J. (1982) The development of monosomy 19 mouse embryos. *J. Embryol. Exp. Morphol.* 69, 223–36.

Mayer, T. C., and Fishbane, J. L. (1972) Mesoderm-ectoderm interaction in the production of the agouti pigmentation pattern in mice. *Genetics* 71, 297–303.

Melton, D. (1985) Injected anti-sense RNAs specifically block messenger RNA translation in vivo. *Proc. Natl. Acad. Sci. U.S.A.* 82, 144–8.

Miller, D. A., Dev, V. G., Tantravahi, R., Miller, O. J., Schiffman, M. B., Yates, R. A., and Gluecksohn-Waelsch, S. (1974) Cytological detection of the c^{25H} deletion involving the albino (c) locus on chromosome 7 in the mouse. *Genetics* 78, 905–10.

Modlinski, J. A. (1975) Haploid mouse embryos obtained by microsurgical removal of one pronucleus. *J. Embryol. Exp. Morphol.* 33, 897–905.

Morris, T. (1968) The XO and OY chromosome constitution in the mouse. *Genet. Res.* 12, 125–37.

Moser, G. C., and Gluecksohn-Waelsch, S. (1967) Developmental genetics of a recessive allele at the complex T-locus in the mouse. *Dev. Biol.* 16, 564–76.

Muller, M. M., Carrasco, A. E., and DeRobertis, E. M. (1984) A homeo-box-containing gene expressed during oogenesis in Xenopus. *Cell* 39, 157–62.

Nadijcka, M. D., and Hillman, N. (1975) Autoradiographic studies of t^n/t^n mouse embryo. *J. Embryol. Exp. Morphol.* 33, 725–30.

Nadijcka, M. D., Hillman, N., and Gluecksohn-Waelsch, S. (1979) Ultrastructural studies of lethal c^{25H}/c^{25H} mouse embryos. *J. Embryol. Exp. Morphol.* 52, 1–11.

Nusslein-Volhard, C., and Wieschaus, E. (1980) Mutations affecting segment number and polarity in *Drosophila. Nature* 287, 795–801.

Palmiter, R. D., Wilkie, T. M., Chen, H. Y., and Brinster, R. L. (1984) Transmission distortion and mosaicism in an unusual trangenic mouse pedigree. *Cell* 36, 869–77.

Papaioannou, V. E., and Gardner, R. L. (1979). Investigation of the lethal yellow A^y/A^y embryo using mouse chimaeras. *J. Embryol. Exp. Morphol.* 52, 153–63.

Papaioannou, V. E., and Mardon, H. (1983) Lethal nonagouti (a^x): description of a second embryonic lethal at the agouti locus. *Dev. Genet.* 4, 21–9.

Paterson, H. F. (1979). In vivo and in vitro studies on the early embryonic lethal oligosyndactylism (*Os*) in the mouse. *J. Embryol. Exp. Morphol.* 52, 115–25.

– (1980) In vivo and in vitro studies on the early embryonic lethal tail-short (*Ts*) in the mouse. *J. Exp. Zool.* 211, 247–56.

Pedersen, R. A., and Spindle, A. I. (1981) Cellular and genetic analysis of mouse blastocyst development. In *Cellular and Molecular Aspects of Implantation*, eds. S. R. Glasser and D. W. Bullock, pp. 91–108. New York: Plenum.

Peyrieras, N., Hyafil, F., Louvard, D., Ploegh, H. L., and Jacob, F. (1983) Uvomorulin: a non-integral membrane protein of early mouse embryo. *Proc. Natl. Acad. Sci. U.S.A.* 80, 6274–7.

Phillips, R. J. S. (1966) A cis-trans position effect at the *A* locus of the house mouse. *Genetics* 54, 485–95.

Phillips, S. J., Birkenmeier, E. H., Callahan, R., and Eicher, E. M. (1982) Male and female mouse DNAs can be discriminated using retroviral probes. *Nature* 297, 1241–3.

Poole, T. W. (1980) Dermal-epidermal interactions and the action of alleles at the agouti locus in the mouse. II. The viable yellow (A^{vy}) and mottled agouti (a^m) alleles. *Dev. Biol.* 80, 495–500.

Preiss, A., Rosenberg, U. B., Kienlin, A., Seifert, E., and Jackle, K. (1985) Molecular genetics of *Krüppel*, a gene required for segmentation of the *Drosophila* embryo. *Nature* 313, 27–32.

Rabin, M., Hart, C. P., Ferguson-Smith, A., McGinnis, W., Levine, M., and Ruddle, F. H. (1985) Two homeo box loci mapped in evolutionarily related mouse and human chromosomes. *Nature* 314, 175–8.

Raff, R. A., and Kaufman, T. C. (1983) *Embryos, Genes, and Evolution.* New York: Macmillan.

Robertson, G. G. (1942a) An analysis of the development of homozygous yellow mouse embryos. *J. Exp. Zool.* 89, 197–231.

– (1942b) Increased viability of homozygous yellow mouse embryos in new uterine environments. *Genetics* 27, 166–7.

Roderick, T. H., and Davisson, M. T. (1984) Personal communication. *Mouse News Letter* 71, 9–10.

Roeder, G. S., and Fink, G. R. (1980) DNA rearrangements associated with a transposable element in yeast. *Cell* 21, 239–49.

Rohme, D., Fox, H. S., Herrmann, B., Frischauf, A. A., Edstrom, J. E., Mains, P., Silver, L. M., and Lehrach, H. (1984) Molecular clones of the mouse *t* complex derived from microdissected metaphase chromosomes. *Cell* 36, 783–8.

Rosenberg, U. B., Preiss, A., Seifert, E., Jackle, H., and Knipple, D. C. (1985) Production of phenocopies by *Krüppel* antisense RNA injection into *Drosophila* embryos. *Nature* 313, 703–706.

Rubenstein, J. R., Nicolas, J. R., and Jacob, F. (1984) Non-sense RNA (nsRNA): a tool to specifically inhibit gene expression in vivo. *C. R. Acad. Sci.* [D] *(Paris)* 299, 271–4.

Rubin, G. M., and Spradling, A. C. (1982) Genetic transformation of *Drosophila* with transposable element vectors. *Science* 218, 348–53.

Schmid, W., Muller, G., Schutz, G., and Gluecksohn-Waelsch, S. (1985) Deletions near the albino locus on chromsome 7 of the mouse affect the level of tyrosine aminotransferase mRNA. *Proc. Natl. Acad. Sci. U.S.A.* 82, 2866–9.

Schnieke, A., Harbers, K., and Jaenisch, R. (1983) Embryonic lethal mutation in mice induced by retrovirus insertion into the $\alpha 1(I)$ collagen gene. *Nature* 304, 315–20.

Scott, M. P., and Weiner, A. J. (1984) Structural relationships among genes that control development: sequence homology between the Antennapedia, Ultrabithorax, and fushi tarazu loci of *Drosophila. Proc. Natl. Acad. Sci. U.S.A.* 81, 4115–19.

Scott, M. P., Weiner, A. J., Hazelrigg, T. I., Polisky, B. A., Pirrotta, V., Scalenghe, F., and Kaufman, T. C. (1983) The molecular organization of the Antennapedia locus in *Drosophila. Cell* 35, 763–76.

Searle, A. G., and Beechey, C. V. (1978) Complementation studies with mouse translocations. *Cytogenet. Cell Genet.* 20, 282–303.

Shaw, P. A., and Gluecksohn-Waelsch, S. (1983) Epidermal growth factor and glucagon receptors in mice homozygous for a lethal chromosomal deletion. *Proc. Natl. Acad. Sci. U.S.A.* 80, 5379–82.

Shepherd, J. C. W., McGinnis, W., Carrasco, A. E., DeRobertis, E. M., and Gehring, W. J. (1984) Fly and frog homeo domains show homologies with yeast mating type regulatory proteins. *Nature* 310 70–1.

Sherman, M. I., and Wudl, L. R. (1977). The mouse *T/t* complex. In *Concepts in Mammalian Embryogenesis*, ed. M. I. Sherman, pp. 136–234. Cambridge: M.I.T. Press.

Shin, H. S., Bennett, D., and Artzt, K. (1984) Gene mapping within the *T/t* complex of the mouse. IV. The inverted MHC is intermingled with several *t*-lethal genes. *Cell* 39, 573–8.

Shin, H. S., Flaherty, L., Artzt, K., Bennett, D., and Ravetch, J. (1983) Inversion in the *H-2* complex of *t*-haplotypes in mice. *Nature* 306, 380–3.

Shin, H. S., Stavenezer, J., Artzt, K., and Bennett, D. (1982) The genetic structure and origin of *t*-haplotypes of mice, analyzed with *H-2* c-DNA probes. *Cell* 29, 969–76.

Silver, L. M. (1981) Genetic organization of the mouse *t* complex. *Cell* 27, 239–40.

– (1982) Genomic analysis of the *H-2* complex region associated with mouse *t* haplotypes. *Cell* 29, 961–8.

Silver, L. M., and Artzt, K. (1981) Recombination suppression of mouse *t*-haplotypes due to chromatin mismatching. *Nature* 290, 68–71.

Silver, L. M., Garrels, J. I., and Lehrach, H. (1984) Molecular studies of mouse chromosome 17 and the *t* complex. In *Genetic Engineering—Principles and Methods*, Vol. 6, eds. J. K. Setlow and A. Hollaender, pp. 141–56. New York: Plenum.

Silvers, W. K. (1961) Genes and the pigment cells of mammals. *Science* 134, 368–73.

Smith, L. J. (1956) A morphological and histochemical investigation of a pre-implantation lethal (t^{12}) in the house mouse. *J. Exp. Zool.* 132, 51–83.

Spiegelman, M., Artzt, K., and Bennett, D. (1976) Embryological study of a T/t locus mutation (t^{w73}) affecting trophectoderm development. *J. Embryol. Exp. Morphol.* 36, 373–81.

Spiegelman, M., and Bennett, D. (1974) Fine structural study of cell migration in the early mesoderm of normal and mutant mouse embryos (T-locus: t^9/t^9). *J. Embryol. Exp. Morphol.* 32, 723–38.

Spradling, A. C., and Rubin, G. M. (1982) Transposition of cloned P elements into *Drosophila* germ line chromosomes. *Science* 218, 341–7.

Stewart, C., Harbers, K., Jahner, D., and Jaenisch, R. (1983) X chromosome-linked transmission and expression of retrovirus genomes microinjected into mouse zygotes. *Science* 221, 760–2.

Tarkowski, A. K. (1977) In vitro development of haploid mouse embryos produced by bisection of one-cell fertilized eggs. *J. Embryol. Exp. Morphol.* 38, 187–202.

Tarkowski, A. K., and Rossant, J. (1976) Haploid mouse blastocysts developed from bisected zygotes. *Nature* 259, 663–5.

Temtamy, S., and McKusick, V. (1978) *The Genetics of Hand Malformations.* New York: Alan R. Liss.

Trigg, M. J., and Gluecksohn-Waelsch, S. (1973) Ultrastructural basis of bio-chemical effects in a series of lethal alleles in the mouse. Neonatal and developmental studies. *J. Cell Biol.* 58, 549–63.

Vankin, L. (1956) The embryonic effects of "Blind," a new early lethal mutation in mice. *Anat. Rec.* 125, 648 (abstract).

Van Valen, P. (1966) Oligosyndactylism, an early embryonic lethal in the mouse. *J. Embryol. Exp. Morphol.* 15, 119–24.

Wagner, E. F., Covarrubias, L., Stewart, T. A., and Mintz, B. (1983) Prenatal lethalities in mice homozygous for human growth hormome gene sequences integrated in the germ line. *Cell* 35, 647–55.

Wakasugi, N. (1973) Studies on fertility of DDK mice: reciprocal crosses between DDK and C57BL/6J strains and experimental transplantation of the ovary. *J. Reprod. Fertil.* 33, 283–91.

Wakasugi, N., Tomita, T., and Kondo, K. (1967) Differences of fertility in reciprocal crosses between inbred strains of mice: DDK, KK and NC. *J. Reprod. Fertil.* 13, 41–50.

Wakimoto, B. T., and Kaufman, T. C. (1984) Defects in embryogenesis in mutants associated with the Antennapedia gene complex in *Drosophila melanogaster*. *Dev. Biol.* 102, 147–72.

474   TERRY MAGNUSON

Wallace, M. E. (1965) Pseudoallelism at the agouti locus in the mouse. *J. Hered.* 56, 267–71.

Warner, A. E., Guthrie, S. C., and Gilula, N. B. (1984) Antibodies to gap-junctional protein selectively disrupt junctional communication in the early amphibian embryo. *Nature* 311, 127–31.

West, J. D., and Green, J. F. (1983) The transition from oocyte-coded to embryo-coded glucose phosphate isomerase in the early mouse embryo. *J. Embryol. Exp. Morphol.* 78, 127–40.

White, R. A. H., and Wilcox, M. (1984) Protein products of the bithorax complex in *Drosophila*. *Cell* 39, 163–71.

Wieschaus, E., Nusslein-Volhard, C., and Kluding, H. (1984) *Krüppel*, a gene whose activity is required early in the zygotic genome for normal embryonic segmentation. *Dev. Biol.* 104, 172–86.

Wolff, G. L., and Pitot, H. C. (1973) Influence of background genome on enzymatic characteristics of yellow (*A/-, A/-*) mice. *Genetics* 73, 109–23.

Woychik, R. P., Stewart, T. A., Davis, L. G., D'Eustachio, P., and Leder, P. (1985) An inherited limb deformity created by insertional mutagenesis in a transgenic mouse. *Nature* 318, 36–40.

15 Production of permanent cell lines from early embryos and their use in studying developmental problems

ELIZABETH J. ROBERTSON AND ALLAN BRADLEY

CONTENTS

I. Introduction

The first visible differentiation of the mammalian embryo occurs with formation of the blastocyst. At this stage the embryo can be seen to have two distinct cell types: the trophectoderm cells, which have limited developmental capacity that restricts them to the extraembryonic lineages, and the pluripotent inner cell mass (ICM) cells, which give rise to all of the differentiated cell types of the adult, including the germ line (see Chapters 4 and 5). As the ICM cells proliferate, groups of cells become committed to specific developmental pathways. By 7.5 days of development, proliferating pluripotent cells are restricted to the embryonic ectoderm of the egg cylinder, and it is these cells that give rise to the tissues of the adult mouse. The methods for rescuing the undetermined pluripotent cells present during the first 8 days of embryonic development and the establishment of permanent tissue culture lines from them are the subjects of this chapter.

Isolating cell lines from embryos into tissue culture provides an opportunity to perpetuate a given cell phenotype indefinitely. Permanent cell lines established under in vitro conditions provide large quantities of homogeneous material that can be used for biochemical analysis of the patterns of gene expression. Embryo-derived cell lines are becoming increasingly important as sources of material to facilitate molecular investigation of the differentiation processes occurring in the embryo.

Cell lines from embryos can be classified as being of either a differentiated or an undifferentiated stem cell phenotype. Differentiated cell lines are relatively easy to define, both phenotypically and by expression of particular biochemical markers (Sherman 1975; Evans 1981).

Undifferentiated cell lines fall into two classes: (1) "determined" stem cells, capable of self-renewal, but with restricted developmental capacity (e.g., hematopoietic stem cells), and (2) "undetermined" stem cells, so-called pluripotent stem cells that are also capable of self-renewal, but with much wider differentiative capacity, and that give rise to the determined stem cells. Both types of stem cells clearly have their experimental applications, although the latter type may have a wider variety of uses for studying differentiation in early development.

It has been realized for many years that pluripotent stem cells are an important component of teratocarcinoma tumors. Spontaneous teratocarcinomas are gonadal tumors, typically composed of a variety of well-differentiated cell and tissue types in addition to an undifferentiated population of stem cells termed embryonal carcinoma (EC) cells. The technique of single-cell cloning has demonstrated that the differentiated tissues within such a tumor are derived by proliferation and differentiation of the EC stem cells (Kleinsmith and Pierce 1964).

Teratoma tumors were first discovered as spontaneous tumors in the testes of strain-129 mice (Stevens and Little 1954). These tumors were demonstrated to be derived from germ cells during fetal development by experimental transplantation of 11–12-day fetal genital ridges to adult host testes. Histologic examination revealed that transplanted male genital ridges formed fetal testes in which a proportion of the germ cells proliferated and eventually developed with an EC cell morphology (Stevens 1964).

The teratocarcinoma tumors that develop in strain-129 mice are derivatives of premeiotic germ cells, and as such their precise relationship with the pluripotent cells of the early embryo might be called into question. However, the spontaneous teratocarcinomas that develop in the ovaries of LT mice (Stevens and Varnum 1974) have a more direct relationship with the embryo. These tumors grow from oocytes that initiate development spontaneously (i.e., parthenogenetically) in the ovary. These postmeiotic oocytes (Eppig et al. 1977) develop as fairly normal preimplantation embryos, but subsequent growth in this ectopic site leads to abortive embryogenesis and results in the development of a teratoma. In rare cases the stem cells proliferate to produce a progressive teratocarcinoma (Stevens 1983).

The apparent relationship among teratocarcinoma cells, embryos, and germ cells has led to experimental induction of teratocarcinoma from embryos grafted into ectopic sites. The derivation of EC cell lines from these tumors is a low-frequency event and may involve a "malignant-transformation" event. This poses questions about the exact relationship between EC cells and the pluripotent cells of the early embryo.

EC cells have many phenotypic characteristics that are shared by normal embryo cells (Graham 1977; Martin 1980; Evans 1981), not least of which is the ability of some EC cells to participate in normal embryogenesis. The most critical test of homology is whether or not pluripotent cells, similar to EC cells, can be derived from the embryo in the absence of tumor formation. Recently, pluripotent stem cell lines have indeed been isolated directly from the mouse embryo into tissue culture. These stem cell lines were originally termed EK cells by Evans and Kaufman (1981) and later embryonic stem (ES) cells by Martin (1981). ES cells present many possibilities for studying developmental problems and have many advantages over the use of tumor-derived counterparts (Bradley and Robertson 1985). In this chapter we shall compare the properties of EC and ES cell lines and discuss their relationship to cells of the normal embryo.

II. Establishing cell lines from embryos

A. Experimental production of tumors from embryos

Survival and growth of embryos can be supported if they are grafted to a nonuterine site in a suitable recipient animal. In the majority of grafts the morphology of the embryos rapidly becomes abnormal, and the embryo forms a benign mass or teratoma composed of a disorderly array of differentiated adult and embryonic tissues. A progressively growing malignant tumor, or teratocarcinoma, will form at the site of the graft if the tumor contains undifferentiated EC stem cells. These tumors can be maintained and propagated by repeatedly transferring portions either subcutaneously or into the peritoneal cavities of successive syngeneic hosts. EC cells can be propagated indefinitely within the tumor environment, but it is a more difficult technique to recover the stem cells into culture. The primary tumors are slow-growing, and a tumor will typically take several months to reach a substantial size (1–2 cm). For successful isolation of EC cells into culture it is normally necessary to passage tumors through animals for several generations to select for masses containing an elevated proportion of more rapidly dividing stem cells. Few EC cell lines have been obtained from primary tumors (Mintz and Cronmiller 1981; McBurney and Rogers 1982).

Three main factors determine whether grafted embryos give rise to teratomas or to teratocarcinomas: the developmental stage of the embryo, the portion of the embryo grafted, and the genetic backgrounds of the embryo and host animal (Solter, Dominis, and Damjanov 1979, 1980, 1981; Stevens 1983). Transfer of early stage embryos is inefficient for generating teratocarcinomas. On average, less than 5 percent of grafts made using preimplantation-stage embryos will give rise to transplantable tumors (Stevens 1968, 1970). The efficiency of teratocarcinoma formation increases with the age of the embryo, reaching a maximum (up to 50%) at 8.5 days of development (Damjanov, Solter, and Skreb 1971; Iles 1977). After that time, teratocarcinoma production falls off abruptly, and embryos give rise solely to teratomas (Damjanov et al. 1971). This pattern of teratocarcinoma formation is explained in part by the reduced survival rates of the early preimplantation stages after grafting, compared with those of the longer established and rapidly growing postimplantation stages. The inability of embryos older than 8.5 days to form teratocarcinomas presumably reflects loss of pluripotent stem cell progenitors by that time. By transfer of more specifically defined regions of the embryo, obtained from carefully dissected egg-cylinder-stage embryos, the capacity for teratocarcinoma formation can be shown to be a property that is restricted to the embryonic ectoderm cells. Grafts that consist solely of extraembryonic ectoderm and endoderm portions will form only completely differentiated teratoma-like structures (Solter and Damjanov 1973; Diwan and Stevens 1976; Solter et al. 1980).

Embryo grafting experiments using different mouse strains have shown that teratocarcinoma formation is also affected by genetic factors. The genotype of the host animal is crucial for proliferation, as opposed to differentiation, of EC cells (Solter et al. 1981). Some inbred strains of mice (e.g., C3H and BALB/c) appear to be highly permissive, with up to 50–70 percent of transferred embryos developing into tumors containing EC cells (Damjanov and Solter 1982). In contrast, transplantation of embryos from other inbred strains (e.g., AKR and C57Bl) rarely results in formation of progressively growing tumors, and these strains are said to be nonpermissive. The efficiency of teratocarcinoma formation by nonpermissive strains can be substantially increased by allowing the embryos to grow in an F_1 host. Additionally, there appears to be a maternally trans-mitted teratocarcinoma permissiveness in both host and graft (Damjanov and Solter 1982).

B. Derivation of pluripotent embryo cells in culture

Fertilized and parthenogenetically derived embryos can be grown successfully in culture up to and beyond the period at which they would normally implant within the uterus. For growth to continue after the first four days of independent embryonic development it is necessary that the blastocyst mimic implantation by first hatching from the zona pellucida and then attaching to a surface. Attachment to a two-dimensional sub-strate disturbs the morphological development of the embryo. The outer encasing layer of trophectoderm cells spreads to form a monolayer exposing the ICM cells to the culture environment. The ICM will con-tinue to grow and may differentiate to form an abnormal egg-cylinder-like structure that is apparently devoid of stem cells. It is possible to derive permanent cell lines of differentiated phenotypes from such cul-tured blastocyst outgrowths (Sherman 1975). These studies show that undisturbed growth of the explanted embryo is not a sufficiently strong selection procedure to allow for continued proliferation of stem cells, and all of the stem cells are induced to differentiate. This is presum-ably in response to the culture environment or to developmental signals that persist within the embryonic clump. In order to prevent the sequence of differentiation events from taking place, while encouraging the cells to proliferate, it is necessary to mechanically manipulate the embryo.

The culture techniques discussed here use, for convenience, embryos retrieved during the preimplantation phase of development. Typically, 3.5-day post caitum (p.c.) expanded blastocysts are recovered from the reproductive tract and placed in highly supplemented tissue-culture me-dium. The embryos attach and are allowed to proceed undisturbed to a later stage, approximately equivalent to 6.5 to 7.5 days of development, prior to manipulation of the embryonic outgrowth (Evans and Kaufman, 1981; Axelrod and Lader, 1983; Robertson et al. 1983b).

The ICM component, which normally forms a discrete cellular mass, is separated from the trophectoderm layer and enzymatically and me-chanically dissociated into a number of smaller multicellular aggregates. These are then placed onto a feeder layer of mitotically inactive fibroblast cells. The majority of cell clumps appear to survive successfully and attach. Normally a single ICM-derived clump will give up to 10 primary colonies. These are easily distinguished from the feeder cells within 48 h of plating, and in successful cultures a disaggregated ICM clump will give rise directly to between 1 and 3 primary colonies that can be classified as having a stem cell phenotype. The remaining colonies, although some

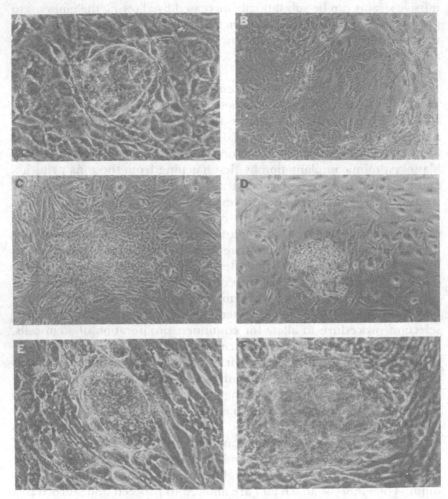

Figure 15.1. Cell colonies obtained after plating disaggregated ICMs onto feeder cells. A and B: Trophoblast-like cells. (A) Appearance 2 days after plating. (B) Appearance 4 days after plating. (C) Fibroblast-like cell colony. (D) Endoderm-like cell colony. E and F: Colonies of embryonic stem cells. (E) Appearance 2 days after plating. (F) Appearance 4 days after plating.

may transiently resemble stem cells, are formed from a variety of differentiated cell phenotypes, including trophoblast giant cells, endoderm cells, and epithelial cells. The appearances of the various ICM-derived colonies are shown in Figure 15.1. Similar colonies were described by Axelrod and Lader (1983). These non-stem-cell colonies initially grow vigorously but do not passage well. EC-like colonies are picked approximately 7 days following the first disaggregation, dissociated into a single-cell suspension in drops of tryspin, and transferred to fresh feeder layers.

Normally, all of the putative stem cell colonies give rise to cultures composed of a homogeneous population of undifferentiated cells. Permanent cell lines can then be established by further rounds of trypsinization at 4–5 day intervals.

A different experimental technique has been described by Martin (1981) in which the embryos are manipulated at the blastocyst stage, with ICMs being isolated from the trophectoderm by immunosurgery. Isolated ICMs are then cultured on feeder cells in medium conditioned by prior exposure to a culture of undifferentiated EC cells. A proportion of such ICMs (4/30) were reported to give rise directly to colonies of stem cells, apparently without the necessity for any further manipulation. No information is available on the fate of the remaining ICMs regarding whether, for example, they failed to grow or gave rise to colonies of differentiated cells.

In a comparative study, Axelrod and Lader (1983) showed that the efficiencies of isolation of stem cells from groups of embryos grown in normal medium and conditioned medium were similar (13% and 12%, respectively). They also noted that conditioned medium seemed to have an inhibiting effect on the ability of embryos to form initial outgrowths from the ICM. We have observed that conditioned medium enhances the proliferation of primary colonies, but the majority are of a "pretrophoblast" type and ultimately differentiate to trophoblast-like cells.

What other factors might influence success rates? The most critical factor appears to be the time at which the ICM-derived outgrowth is first disturbed. We find that normal fertilized blastocysts give best results if they are cultured for a four-day period. The ability to recover undifferentiated stem cells seems to be closely correlated with the extent of morphological differentiation exhibited by the ICM outgrowths at four days. There is a marked degree of variability between embryos even if they have been recovered from the same animal. ICM clumps in which there is extensive endoderm formation or a relatively rapid progression to an overtly multilayered egg-cylinder-like structure tend to have a reduced chance of containing pluripotent stem cells, whereas embryos in which there is growth in the absence of substantial overt differentiation fare better.

Using these techniques, we routinely expect that at least 10 percent of embryos will give rise to lines, although there may be noticeable variation in success rates between groups of embryos cultured on different occasions.

The use of blastocysts that have been subjected to a period of implantational delay also appears to increase the efficiency of recovering stem cells. There is a variety of explanations for this observation. Implantational delay, brought about by removal of the ovarian tissue 2.5 days after fertilization, together with simultaneous administration of progesterone, prevents the embryos from implanting in the uterine wall. One effect of delay is to selectively increase the number of cells in the ICM (Evans and Kaufman 1983). The delaying procedure also artificially prolongs the period the embryo spends in the blastocyst conformation, during which time the ICM is sequestered within the layer of trophectodermal cells. Additionally, the embryo ceases growth and enters a state of dipause. This reduction in the rate of cell division and differentiation may be the most important factor in the isolation of a cell line. The effect on growth rate is evidenced by the necessity to shift the timing of the first disaggregation. For normal embryos this culture period is four days, but using delayed embryos the highest isolation frequencies coincide with a six-day period of culture. Through the use of delayed blastocysts, the efficiency of isolation of permanent stem cell lines can be as high as 30 percent.

There is no evidence for any maternal uterine influence affecting the ability of an embryo to give rise to stem cells. Pluripotent cells have been derived from embryos retained in culture from the 1-cell state onward. Indeed, stem cell lines can be derived from cultured parthenogenetically activated oocytes (Kaufman et al. 1982). There also is no evidence for strain restrictions similar to those seen to affect teratocarcinoma formation. In one study, efficiency rates were compared in parallel groups of blastocysts collected from the inbred 129Sv strain and from matings between C3H females and Rm males. In both groups, approximately one of three embryos gave pluripotent cell lines. We have, to date, no indication that specific strains are nonpermissive. A review of ES cell lines (Robertson et al. 1983b; Silver, Martin, and Strickland 1983) does not reveal any obvious genetic restrictions, with lines being isolated from inbred strains, outbred strains, and various F_1 crosses. This is not an unexpected finding. From tumor-induction studies it appears that the most important factor is the genetic background of the maternal environment, not that of the embryo graft. Variation in the success of isolating stem cells might arise as a consequence of the noticeable differences in relative growth rate of early embryonic stages between different strains. It may be necessary to compensate for any such differences by making appropriate alterations in the timing of the first disaggregation.

C. Characteristics of EC cell lines

A larger number of EC cell lines that have their origins in embryo-derived tumors have been described. This apparent wealth of cell lines is misleading, because all EC cell lines can be traced back to only some 16 independently derived tumors, and of these only 12 have been well described in the literature. A smaller number of EC cell lines have been isolated from spontaneously occurring teratocarcinomas. The majority of the available EC cell lines are listed in Appendix 1 of Silver and associates (1983).

The different EC cell lines, although superficially resembling one another in morphology and in the expression of specific cell surface antigens, compose a very heterogeneous set of tissue-culture lines. Particularly well documented are differences in growth characteristics and requirements (e.g., feeder-cell dependence, cloning efficiency), chromosome constitution, and differentiation ability.

Cell lines differ in the ease with which differentiation can be elicited in culture. Some lines differentiate readily following suspension culture of cellular aggregates. These structures, embryoid bodies, enter a synchronous and reproducible sequence of differentiation. Endoderm cells form within 48 h on the outer surfaces of the structures; longer-term culture gives structures containing a range of cell and tissue types (Martin, Wiley, and Damjanov 1977). Other EC cell lines show little tendency to form well-differentiated embryoid-body structures, but will form a range of cell types if the cultures continue to be refed after being allowed to reach confluence (Nicolas et al. 1976). Other EC cell lines are reluctant to differentiate unless exogenous promoters are added to the culture medium. In these instances, differentiation may be more restricted; for example, F9 cells form either endodermal cells or neural cells (Strickland and Mahdavi 1978; Kuff and Fewell 1980). The relative differentiation ability in culture is generally correlated with both the range and extent of tissue formation seen if the cells are induced to reform teratocarcinomas and with the ability of EC cells to respond correctly to the regulatory signals of the normal embryonic environment.

EC cell lines are also heterogeneous with respect to chromosome complement. Unlike murine somatic cell lines, EC cell lines exhibit a narrow karyotypic distribution, with a mode equal or close to the normal diploid complement of 40, although few are described as being euploid. Presumably, retention of a stem cell phenotype must of necessity be associated with possession of a reasonably balanced chromosome complement.

What are the factors responsible for this observed heterogeneity between EC cell lines, and how can these differences be rationalized if it is assumed that they are all derived from the same source of embryonic progenitor cell? Divergence of EC cells away from the embryonic phenotype occurs both during the time spent in the tumor form and as a consequence of the methods used for routine maintenance in tissue culture.

Primary embryo-derived tumor masses typically are slow-growing and are composed of a wide range of differentiated adult somatic tissues. Cytogenetic studies show that for the majority of these tumors the modal chromosome count is 40. Longer-term propagation of either solid tumors or ascites embryoid bodies results in two well-documented effects: restrictions in number and extent of the differentiated tissues formed, and acquisition of predominant karyotypic alterations. This can occur extremely rapidly; indeed, Iles and Evans (1977) noted changes in differentiative capacity and karyotype in four transplantable tumors within the first few transplant generations. One tumor was grossly karyotypically abnormal when first analyzed at the third transplant. One interesting observation from this study was that sublines maintained as ascites transplants altered more rapidly than cells maintained as a solid tumor.

The effect of long-term propagation is particularly well exemplified by cell lines derived from two independently induced 129Sv tumors. These are detailed in Table 15.1. EC cells from the OTT6050 tumor maintained as in vivo ascites cultures of simple embryoid bodies are reportedly both euploid (although 80 percent of cells were recorded as having trisomy 8) and totipotent (Cronmiller and Mintz 1978). Different EC cell lines derived from portions of serially transplanted tumor, as feeder-independent lines, are very different from one another. They show a wide range of differentiation abilities in culture and in embryos and differ in karyotype. In comparison, EC cell lines derived from the OTT5568 tumor and maintained on feeder layers all retain high differentiation ability, as assessed by the range of differentiation induced following suspension culture of cellular aggregates and on reformation of teratocarcinomas. Interestingly, for this tumor, all primary cell lines and sublines share the same karyotype.

EC cell lines derived from the spontaneous ovarian tumor LT72484 provide another example. Sublines have been described as euploid and totipotent when maintained by subcutaneous transfer (Cronmiller and Mintz 1978). A subline isolated from a 34th transplant generation by Martin and associates was found to be XX trisomy 1 (Martin et al. 1978). Another LT 72484 derivative has been described as nondifferentiating (Nicolas, Jakob, and Jacob, 1978).

Published information on karyotype analysis of primary cell lines and tumors is summarized in Table 15.2, and as noted earlier, most have a modal number of 40. Analysis of sex chromosomes in 12 independently derived tumors shows that 5 are XX, 4 are XY, and 3 are XO. This high percentage of XO cell lines suggests that the sex-chromosome complement is unstable, but whether this arises through loss of the second X chromosome or loss of the Y chromosome is debatable. Results from OTT6050 lines suggest that the Y chromosome is unstable, with only a single subline apparently retaining the Y chromosome (Cronmiller and Mintz 1978). McBurney and Strutt (1980) have suggested, from a study of the P10 cell line, that one of the X chromosomes is unstable. Similarly, results from cytogenetic analysis of embryo-derived stem cell lines made from parthenogenetic oocytes (see Section IID) showed that all lines were unstable with respect to the X chromosomes.

Only 5 completely euploid EC cell lines have been established from the 12 primary teratocarcinomas described in Table 15.2, namely OTT6050 maintained in vivo (transient) (Cronmiller and Mintz 1978), C145b (Papaioannou et al. 1979), P10 (McBurney and Strutt, 1980), P19 (McBurney and Rogers 1982), and METT-1 (Mintz and Cronmiller 1981). It is

Table 15.1. *EC cell lines derived from 129Sv embryo-induced teratocarcinomas*

Tumor	Cell line	Karyotype	Characteristics	References[a]
OTT6050 (6.5-day embryo)	6050 in vivo	XY : XO, trisomy 8	Rate of chimera formation 20%; 2 germ-line animals	1, 2
	PCC3	XO, modal no. 40	Differentiates extensively in culture after reaching confluence	3
	PCC4	n.d.	As for PCC3	3
	PC13	XO, modal no. 41	Restricted differentiation; mainly neural tissue formed	4
	OC15	XO, trisomy 11	Differentiation restricted to endodermal, muscle, and neural tissue	5
	F9	XO, modal no. variable	Differentiates to endoderm cells on addition of retinoic acid	6, 7
OTT5568 (3.5-day embryo)	PSA1 and subclones	XO, trisomy 6	Differentiates via embryoid-body formation; chimera formation 50% at midterm	8, 9, 10
	PSA4TG	XO, trisomy 6	Differentiation as for PSA1 and subclones	11, 12
	NG-2	XO, trisomy 6	Differentiation as for PSA1 and subclones	9, 10, 12
	PSMB	XO, trisomy 6	Differentiation in culture extensive, via formation of cystic embryoid bodies	13

Note:

[a](1) Illmensee and Mintz (1976), (2) Cronmiller and Mintz (1978), (3) Nicolas et al. (1976), (4) Hooper and Slack (1977), (5) McBurney (1976), (6) Bernstine et al. (1973), (7) Strickland and Mahdavi (1978), (8) Martin et al. (1977), (9) Dewey et al. (1977), (10) Fujii and Martin (1983), (11) Slack et al. (1978), (12) Stewart (1982), (13) Magrane (1982).

Table 15.2. *Karyotype analysis of embryo-derived teratocarcinomas*

Tumor	Strain	Sex chromosome	Chromosomal abnormalities	References[a]
OTT6050	129Sv (6.5 days)	XY	Ascites form, modal no. 39/40; trisomy 8 in 80% of cells	1
OTT5568	129Sv (3.5 days)	XO	All EC cell lines have modal no. 40, trisomy 6	2
C86	C3H/He (7.5 days)	XX	Tumor initially modal no. 40; abnormalities evident by 4th transplant	3, 4
C145b	C3H/He (6.5 days)	XX	Primary tumor modal no. 40; abnormalities evident by 3rd transplant	3, 4
C17	C3H/He (6.5 days)	XX	Primary tumor model no. 40; abnormalities evident by 3rd transplant	3, 4
C106	C3H (3.5 days)	XO	Primary tumor XO, with additional autosomal abnormalities	3, 4
P10	C3H/He (7.5 days)	XX	EC cells in culture euploid	5
P19	C3H (7.5 days)	XY	EC cells in culture euploid	6
5'	C3H/HeHa (7.5 days)	XY	Modal no. 41, trisomy 11	6
7'	C3H/HeHa (7.5 days)	XY	Modal no. 41, trisomy 11	6
METT-1	129Sv (6 days)	XX	EC cells in culture euploid	7
ARK	AKR	XO	EC cells in culture XO, tetrasomy 8	8

Note:

[a](1) Cronmiller and Mintz (1978), (2) Martin et al. (1977), (3) Iles (1977), (4) Iles and Evans (1977), (5) McBurney and Strutt (1980), (6) McBurney and Rogers (1982), (7) Mintz and Cronmiller (1981), (8) Martin et al. (1978).

pertinent to the present discussion to note that two of these culture cell lines, P19 and METT-1, were isolated from the primary embryo-derived tumor mass without the necessity for serial transplantation. With one exception, all euploid EC lines have been reported to integrate into host embryos with high efficiency (Papaioannou and Rossant 1983), and two of the four (OTT6050 and METT-1) have given rise to germ-line animals.

Studies of EC cell lines monitored during successive culture generations also indicate that cell lines will alter in culture. Hogan (1976) showed that growth of OTT6050-derived clonal lines in the absence of feeder cells for a four-month period dramatically reduced the ability of mass cultures to form embryoid bodies. A similar study by Magrane (1982) on an OTT5568 feeder-dependent cell line corroborated the observation made by Hogan. It was shown that denying the EC cells feeder layers caused a rapid change in the population. This was seen as an increased rate of cell division, loss of differentiation during in vivo and in vitro assays, and an increase in the modal number from 40 to 43 within 70 passage generations. The same cell line maintained in parallel on feeder layers for in excess of 12 months continuous culture retained a very stable differentiation capacity and karyotype.

In summary, it would seem that the abnormalities documented in established EC cell lines do not preexist in the tumor progenitor-cell population. The fact that the tumor masses remain undisturbed for long periods of time should favor differentiation of more competent (and, by implication, more normal) pluripotent cells and preferentially select for more actively proliferating cells. The developmentally abnormal phenotype of EC cell lines can be further accentuated by prolonged growth in tissue culture, which generates heterogeneity in the cell population, although the stability of the pluripotent phenotype may be prolonged through the use of feeder-cell layers.

D. Characteristics of pluripotent cell lines from embryos

Stem cell lines obtained directly following dissociation of the normal embryo have characteristics that give them a number of advantages over tumor-derived EC cell lines. The most important of these, in the experimental context, is the behavior of ES cells when they are returned to the embryonic environment. This will be considered at length in Section III.

There is no restriction regarding the mouse strains from which ES cell lines can be isolated. Within the few years that the techniques have been available, a range of new genetically marked stem cell lines have been generated. These include lines that have been designed to carry isozymal variants (Bradley et al. 1984), lines with isozymally marked X chromosome (Martin and Lock 1983), lines with prominent translocation markers, and lines carrying a variety of coat-color phenotypes (Robertson et al. 1983b). Perhaps the most valuable use of this technique is to derive cell lines from developmental lethal mutations (Magnuson et al. 1982). Production of EC cell lines from such a wide variety of genetic variants has not been possible because of the necessity for in vivo growth within a compatible animal. Tumors containing EC cells fail to grow in immunosuppressed or nude mice (Solter and Damjanov 1979). Growth of ES lines directly in culture removes such constraints.

A universal feature of ES cells is a high differentiation ability both in culture and in tumor masses. We find that the majority of cell lines from embryos possess, at least initially, a normal euploid chromosome complement. These two features are probably interrelated and result from the very rapid isolation process that has subsequently been followed by a relatively short period in culture. By virtue of the isolation technique, these cell lines represent primary cultures and have not been subjected to the deleterious selection pressures encountered in the derivation of EC cell lines.

A cytogenetic analysis of 35 independently derived stem cell lines has shown that the majority were euploid, as judged by G-banding techniques. The karyotype analysis is summarized in Table 15.3. Only 6 cell lines possessed autosomal abnormalities. In 4 lines, derived from different mouse strains, the abnormality was attributable to trisomy 11. This correlates with the finding of McBurney (1976), who also found chromosome 11 to be unstable in EC cell lines.

In our laboratory, the majority of ES cell lines have been found to be derived from male embryos, with some 25 of the 35 lines possessing a Y chromosome. Only 5 XX lines have been isolated. The remaining cell lines were XO when first analyzed. If these results prove significant, it suggests both that female embryos are less efficient at giving stem cell lines using this isolation procedure and that the sex-chromosome complement is unstable. Are these observations correlated? It may be that female embryos in which the undifferentiated cells carry two active X chromosomes are discriminated against by some feature of the isolation process. The XO lines may therefore arise from XX embryos. There are three lines of evidence that support this view. The first, which is largely circumstantial, is that the Y chromosome is inherited very stably in XY cell lines. In contrast, XO cells can be detected among the cells of predominantly XX cell lines. The best evidence comes from cytogenetic analysis of stem cell lines derived from parthogenetic embryos. The isolation procedure detailed in Section IIB has been extended for use with blastocysts derived from ethanol-activated oocytes (Kaufman et al. 1982). Fifteen cell lines derived from one pronuclear haploid and three lines from two pronuclear diploid embryos were isolated, from both inbred and F_1 embryos, and established as permanent cell lines. All cell lines showed an alteration

Table 15.3. *Karyotype analysis of stem cell lines derived from fertilized blastocysts*

		XY	XX	XO
Number of lines analyzed[a]	35	24	5	5
Number of lines euploid	27	23	4	0
Number of lines aneuploid	8	2	2	5
Abnormalities affecting sex chromosomes only	2	1[b]	1[c]	0
Abnormalities affecting autosomal chromosomes	2	1	1	0
Abnormalities involving sex chromosomes and autosomes	4[d]	0	0	4

Note:
[a]Lines analyzed by G-banding of metaphase spreads.
[b]XYY line.
[c]XX[del] line.
[d]Four independently derived lines characterized by trisomy 11.

in the expected XX constitution. This was seen as either loss of the entire second X chromosome or a partial deletion involving a variable region of a single X chromosome (Robertson, Evans, and Kaufman 1983a). The loss of the second X chromosome may represent some form of compensatory mechanism, because both X chromosomes are active in parthenogenetically derived ES cell lines (Rastan and Robertson 1985).

III. Behavior of embryonic stem cells on reintroduction into blastocysts

A. EC cells returned to the embryonic environment
The pluripotent nature of EC cell lines has been demonstrated by their ability to form chimeras when combined with genetically dissimilar normal preimplantation mouse embryos (Brinster 1974; Mintz and Illmensee 1975; Papaioannou et al. 1975). The majority of studies have used a blastocyst injection technique, although recently techniques involving aggregation of cleavage-stage embryos with EC cell clumps have been used successfully (Stewart 1982; Fujii and Martin 1983).

The variation between EC cell lines discussed in Section IIC is highlighted by chimera-formation studies showing that EC cells can have a variety of fates in the host embryo. Some EC cell lines fail to show either normal or neoplastic growth in the embryonic environment (R5/3, F1/9, and 1009) (Table 15.4). Some cell lines, such as C145b (Papaioannou et al. 1979), show contributions to extraembryonic lineages only. Lines such as C86, C17, and P19 will form chimeras only with tumors. If the

Table 15.4. *Summary of results of EC-embryo combinations prenatally and postnatally*

Tumor cell line	Strain	Sex chromosomes	Midterm			Term			References[a]
			No. analyzed	No. normal	No. abnormal	No. analyzed	No. normal	No. abnormal	
OTT6050[b]	129	XO : XY	179	8	6	279	32[c]	2	1–8
OTT5568	129	XO	272	63	68	168	28	1	9–12
METT-1	129	XX	0	0	0	324+	46+[d]	1+	13, 14
LT72484	LT	XX	30	2	10	74	8[c]	3	9, 15
C17	C3H	XO	0	0	0	77	5	8	5, 12
C86	C3H	XO	13	3	0	71	0	6	5, 12
C145b	C3H	XO	404	5	0	201	0	0	16
P19	C3H	XY	422	36	73	62	0	10	8. 17
P10	C3H	XX	9	3	0	58	31	0	8, 18
F1/9	F1	XY	71	0	0	18	0	0	6
1009	129	XY	15	0	0	137	0	0	6
			1,415	120 (8.4%)	157 (11.1%)	1,469	150 (10.2%)	31 (2.1%)	

Note: EC lines grouped under original tumor for clarity (see Tables 15.1 and 15.2).

[a] (1) Brinster (1974), (2) Illmensee and Mintz (1976), (3) Mintz and Cronmiller (1981), (4) Mintz and Illmensee (1975), (5) Papaioannou et al. (1978), (6) Papaioannou et al. (1979), (7) Illmensee (1978), (8) Papaioannou and Rossant (1983), (9) Fujii and Martin (1983), (10) Dewey et al. (1977), (11) Stewart (1982), (12) Papaioannou et al. (1975), (13) Stewart and Mintz (1981), (14) Stewart and Mintz (1982), (15) Cronmiller and Mintz (1978), (16) Papaioannou et al. (1979), (17) Rossant and McBurney (1982), (18) Rossant and McBurney (1983).

[b] Excluding cell hybrids.   [c] Two germ-line chimeras reported.   [d] Two germ-line chimeras reported.   [e] One germ-line chimera reported.

behavior of EC cells is compared critically with the behavior of normal ICM cells injected in the same way, then only two EC cell lines have been described that differentiate reasonably normally following introduction into the embryonic environment. These are P10 and METT-1, both euploid XX lines.

Table 15.4 summarizes the chimera-forming efficiency of EC cell lines. Overall, EC cell lines form live-born chimeras with an efficiency in the range of 10 percent. If conceptuses are analyzed at midterm, the proportion of chimeras is substantially higher (19.6%). Interestingly, over half of chimeric conceptuses are abnormal, whereas the remainder are morphologically normal at this stage. The majority of abnormal chimeras do not survive to term, and this is reflected in the percentage of chimeras at birth. Of the live-born chimeras, one in five develops tumors.

GPI isoenzyme variants and coat-color markers have been used to score for the presence of EC-derived cells in a variety of tissues in live-born and midgestation embryos. The results indicate that EC contributions to chimeras are patchy. Occasional individuals are found to have contributions to all tissues, but such animals are the exception rather than the rule (Illmensee and Mintz 1976; Illmensee 1978; Stewart and Mintz 1981). This restricted pattern of mixing by EC cells might be a contributory factor to the high rates of fetal loss seen in chimeras. Presumably, if the cells remain clumped, EC contributions to specific organs and tissues may be substantial, and the developing embryo will be subjected to high contributions of what are likely to be aneuploid cells to these tissues. In the absence of a process to regulate this contribution, the embryo might not survive to term.

B. Formation of functional germ cells by EC cells

At the outset of the EC cell line studies it was hoped that cells from culture would populate the germ line with high efficiency. This is not the case, and on reflection not surprising, because many of the chimera studies have used aneuploid cells. Just as the contributions by EC cells to the organs are noticeably sporadic, the chances of the culture-derived cells populating the genital ridges are also reduced. Contributions by EC cells to the germ line have proved to be rare events. Only 5 germ-line individuals have been reported out of 147 live-born mice (3.4%, Table 15.4). Only 2 of these animals were derived from a cell line that had been cultured entirely in vitro (Stewart and Mintz 1981, 1982). The extent of contribution to the germ cells in these particular female animals was low (6.3% in one individual and 6.7% in the second, which had a $W^v/+$ genotype to be described later). It is relevant to note that the cell line used, METT-1, has a euploid XX karyotype and a short passage history. However, even this cell line showed many general EC characteristics: Chimera-forming efficiency was low (13%), chimeras generally had a low EC-derived component, mosaicism was sporadic, and only 10 percent ot tne chimeras had contributions to the gonads, as assayed by GPI separation and germ-line transmission.

It is rather surprising that the P10 EC cell line has not demonstrated any contributions to the germ line (Rossant and McBurney 1983). P10 forms chimeras with a greatly increased efficiency of 53 percent, compared with some 13 percent for METT-1, and the animals typically have a uniform and high contribution from the EC cells in the absence of tumor formation. It is likely that the use of XX lines in this context necessitates a careful choice of host embryo strain. These should preferably carry mutations that restrict the fertility of the host component if germ-line contributions are to be detected (see Section IIID).

C. Chimera formation by ES cell lines

We have performed a large study of the chimera-forming ability of 17 independently derived ES cell lines. We have been encouraged to find that, without exception, all have proliferated and integrated normally (Evans, Bradley, and Robertson 1985). In fact, the chimeras obtained have many similarities to chimeras formed by injection of ICM cells into host blastocysts (Papaioannou and Rossant 1983).

Chimeras formed from ES cell lines exhibited the following features:

1.   The proportion of chimeras generated usually was very high (average 35% overall), without significant prenatal loss (70% came to term).
2.   The level of contribution to individuals, as assayed by GPI and coat color, was often in excess of 50 percent.
3.   The chimerism was extensive (involving all the organs of the chimeric mice), and the cells integrated evenly to give animals with a widespread and fine-grained pattern of mosaicism. This feature is illustrated in Figure 15.2.

An interesting and important result follows from the use of some of the ES lines derived from parthenogenetically activated embryos (Kaufman et al. 1982) in some of our chimera experiments. These lines appear to repopulate the embryo with high efficiency. GPI analysis has indicated that the descendants of diploid parthenogenetic lines are able to differentiate along any developmental pathway. Their success in chimeras contrasts with the developmental failure of the intact parthenote (see Chapter 13).

D. Formation of germ cells by ES cells

Test breeding of male chimeras obtained by blastocyst injection of XY stem cells has shown that the cultured cells will routinely populate the germ line (Bradley et al. 1984). Three independently derived lines were used in this particular study. Over 50 percent of the resulting

Figure 15.2. Chimeric male generated following injection of 10 XY ES cells derived from a pigmented-strain donor embryo into an albino host blastocyst. High levels of donor contribution and fine-grained mosaicism are evident.

live-born animals were mosaics (Table 15.5, series II). In the chimeric population, we observed a marked sex-distortion effect toward males. Of 78 phenotypic males that were set up for test breeding, 31 failed to breed; 15 of the 47 fertile males showed transmission of the culture-derived genome. Analysis of the breeding data from these animals demonstrated two classes of chimeras. Eight animals transmitted their culture-derived genome to only a small proportion of their progeny, whereas the remaining 7 transmitted only sperm derived from the culture-derived cells.

The reason for this observation is that injection of XY stem cells into random-host embryos is equally likely to involve an XX host embryo (inappropriate combination) or an XY (appropriate combination) host embryo. Because the Y chromosome is responsible for sex determination in mammals, the former combination (XY → XX) produces an embryo that contains cells of both sexes, and as a result the normal pattern of sexual development of the XX host might be disturbed. Embryo aggregation experiments between two different morula-stage embryos have shown that the resulting ratio of the phenotypic sexes in chimeras is dependent on strain. Certain strain combinations give a 1 : 1 ratio of males to fe-

Table 15.5. *Chimera-forming efficiency and sex-ratio distortion following injection of embryonic stem cells into host embryos*

Cell line	Strain	Sex chromosomes	No. injected	No. born	No. chimeric	Chimeras Male	Female
Series I[a]							
B2B2	129	XY	66	40	14	2	7
CP2	129	XY	123	94	18	11	4
CP3	129	XY	156	111	43	17	15
A13	129	XX	160	109	36	15	18
CL6	LT	XY	161	103	21	7	11
			666	457 (68.7%)	132 (28.9%)	52	55
Series II[b]							
CP1	129	XY	246	161	70	40	27
CC1.1	129	XY	83	61	37	24	13
CC1.2	129	XY	321	205	127	97	29
			650	427 (65.7%)	234 (54.8%)	161 (70.0%)	69 (30.0%)

Note:
[a]Series I: 3–5 cells injected into each blastocyst.
[b]Series II: 10–15 cells injected into each blastocyst.

males, whereas others show a sex-distortion effect if the strains used are very similar (Mullen and Whitten 1971; Ohno 1978).

In the ES cell study described earlier, the strain combination was constant, but between chimeras there was variation in the number of viable stem cells that survived the injection process and take part in embryogenesis. In our experiments, an average of 40 percent of the XX embryos were converted to phenotypic males (the sex ratio among live-born chimeras was 70% male to 30% female). All of the chimeras that have undergone sex conversion would be expected to transmit the ES genotype in their germ lines, because functional sperm can be derived only from the XY component of a male chimera. Only an estimated one in three of the sexual mosaics proved to be fertile, and, as predicted, all the functional sperm were derived from the cultured XY cells. The remaining sexual mosaics were sterile, and at autopsy they proved to be hermaphrodites and showed a range of morphological abnormalities affecting the reproductive system.

Integration of XY stem cells into a male host embryo results in a normal phenotypic male. One in four of the fertile chimeric males of this class were shown to be germ-line chimeras. The contribution to the sperm was also relatively low, ranging from 3 to 0.3 percent in the eight individuals. This level of transmission is substantially lower than we see in the contribution to the gonads, as assessed by quantitative GPI analysis. There are two possible explanations: Either the general level of incorporation is high, but only a subset of the introduced stem cells is able to form functional sperm, or the mitotic rates of the germ cells from the inbred 129 strain (ES component) and the outbred MF1 host differ significantly. We believe that the latter explanation is more plausible. Clearly, a complete explanation must await a full examination of the fertility of aggregation chimeras between 129 (ES strain) and MF1 host embryos.

There is evidence that XX → XY females are occasionally able to ovulate XY oocytes (Evans, Ford, and Lyon 1977), but we have not bred chimeric females to assay for this event. We have, however, examined how the functional contribution of cultured stem cells in the germ line can be increased by using host embryos carrying mutations that affect their fertility.

Alleles at the dominant white spotting (*W*) locus appear to reduce the proliferation of migratory stem cells in a cell autonomous manner (Silvers 1979). Principally germ cells, neural-crest-derived melanocytes, and primordial-blood stem cells are affected. We have utilized three alleles at this locus. The phenotypes of these alleles are described in Figure 15.3.

By intercrossing animals carrying these alleles, either host embryos carrying a single *W* allele or embryos that are a compound of two different alleles have been generated for blastocyst injection. Breeding analysis of seven phenotypic male chimeras that were generated in this series has shown five to be germ-line ES chimeras. Of these, one is a completely germ-line animal, whereas the remaining four transmit sperm from both the host and culture-derived genotype. In this series we have observed an increase in the efficiency of transmission of the introduced cells from the less than 3 percent transmission seen in the previous series to a transmission rate between 9.3 and 80.3 percent. The effects of host blastocyst genotype on transmission rates are shown in Table 15.6. The increase in transmission rate might be attributable to the similar genetic backgrounds of the host embryo and introduced cells (both backcrossed to the 129 strain). However, the difference in transmission between the homozygous *W* host embryos and the heterozygous *W* embryos indicates that the *W* locus is having an effect.

It may be possible to use mutations to bias the incorporation of cultured stem cells into different cell populations, allowing rescue of defective embryos. The *W* locus probably will be an important asset where transmission of XX cell lines to the next generation is required, particularly when an XY cell line is not available (e.g., lines derived from parthenogenetically activated embryos).

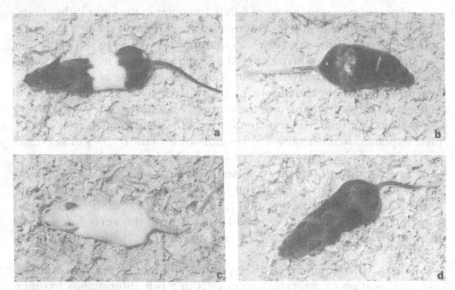

Figure 15.3. Alleles at the W locus: (a) $W^{sh}/+$ heterozygote; adults are fertile; (b) $W^e/+$ heterozygote (W^e/W^e homozygotes are prenatal lethals); (c) W^{sh}/W^{sh} homozygote (adults are fertile, black-eyed, and white-coated); (d) $W^v/+$ heterozygote (W^v/W^v homozygotes are viable but infertile).

Table 15.6. *Transmission of culture-derived progeny from chimeras using host embryos of different genotypes*

Germ-line Chimera	Host blastocyst Strain	Genotype	No. host progeny	No. F_1 EK progeny	Percentage culture-derived progeny
CP1.3	MF1	+/+	227	1	0.4
CP1.5	MF1	+/+	244	7	2.8
CP1.11	MF1	+/+	91	2	2.1
CC1.1.5	MF1	+/+	103	2	1.9
CC1.1.24	MF1	+/+	67	1	1.5
CC1.2.32	MF1	+/+	137	1	0.7
CC1.2.33	MF1	+/+	126	1	0.8
CC1.2.40	MF1	+/+	100	1	1.0
WCC1.2.5	129	$W^e/+$	39	4	9.3
WCC1.2.T8.1	129	W^v/W^{sh}	10	41	80.3
WCC1.2.T17.1	129	W^v/W^{sh}	29	4	12.1
WCC1.2.T17.3	129	$W^v/+$	14	2	12.5

IV. Relationship of pluripotent stem cells to cells of normal embryos

The fundamental difference between cultured stem cells and the cells of the normal mouse embryo is that cultured cells are apparently capable of proliferating indefinitely with an undifferentiated and pluripotent state. In the embryo, this property is lost around the eighth day of development, when all of the component cells of the conceptus can be considered as being restricted to follow various pathways of differentiation. From which embryonic cell type are EC and embryo-derived stem cells derived? Embryo grafting experiments indicate that pluripotent cells are present up to the eighth day of development and that transfer of embryonic ectoderm alone is sufficient to give a teratocarcinoma. Embryo manipulation in culture suggests that there is a well-defined period during which it is possible to isolate stem cells. This falls between four days and six days of culture subsequent to the blastocyst stage. These observations, taken together, suggest that the EC progenitor cells are present in the definitive embryonic ectoderm. These embryonic ectoderm cells must be developmentally labile, because various features of EC cells and ES cells suggest that they are most closely equivalent to the cells of the 5.5-day primitive ectoderm (Martin 1980; Evans, 1981). This is indicated by a number of features, including routine formation of both parietal and visceral endoderm cells by many EC and ES cell lines. EC and ES cells share patterns of expression of cell surface antigens with 5.5-day primitive ectoderm cells. For example, the Forssman antigenic determinant is present on cultured stem cells, late morula cells, and ICM cells, but is not present on the cell surface in 6.5-day embryonic ectoderm cells (Evans et al. 1979). Cultured pluripotent cell lines cannot, however, be considered as being homologous to early ICM cells. For example, clumps of cultured cells will not form a recognizable embryo if microsurgically transferred to a trophoblast vesicle (Rossant and Papaioannou 1985).

How is it possible to derive permanent cell lines from normal embryonic cells? There are two possible explanations. First, the non-uterine culture of embryos may disrupt the normal pattern of development sufficiently to allow pluripotent cells to be diverted from their normal fate and adopt a program of continued proliferation. These cells may have tumorigenic potential that is suppressed by factors in the normal embryonic environment. These cells can be retrieved from embryo masses and will continue their growth pattern in culture while retaining the ability to differentiate in an appropriate way in response to developmental signals. Alternatively, it is possible that tumor formation and long-term propagation of stem cells are possible only after some kind of "transformation" event, so that pluripotent cell lines, although initially derived from normal embryonic stem cells, become progressively dissimilar from their progenitors. It can be argued that ES cell lines fall into the former category. This is suggested both by the high efficiency and rapidity of the isolation procedure and by the lack of any obvious genetic constraints. With few exceptions, tumor-derived EC cell lines appear to fall into the second category.

Pluripotent cell lines, from whatever source, will behave as tumor cells if introduced into a somatic location in a histocompatible host animal. However, this malignant phenotype may be reversed following transfer into the embryonic environment. EC and ES cell lines can be clearly distinguished by the degree to which the malignant properties are controlled or suppressed by a host embryo. Whereas ES cell lines have yet to show uncontrolled proliferation or differentiation on return to the embryo, EC cells can, in general, be viewed as being no longer competent to respond appropriately to regulatory signals provided by the embryonic environment. In the majority of cases, EC cell lines have proved to show restricted developmental phenotype and an uncontrollable and unregulated pattern of development in the embryonic environment, proliferating indiscriminantly to form tumors and/or cause fetal death. The malignancy of EC cells may be partly attributable to the unbalanced karyotype displayed by the majority of these lines.

However, tumor-derived cells placed in an embryonic environment sometimes are capable of reverting to a cell type that behaves in many respects like a normal embryo cell (Brinster 1974; Mintz and Illmensee 1975; Illmensee and Mintz 1976). Furthermore, this intriguing observation is not restricted to EC cells. Neoplastic regulation has been demonstrated using other embryonic tumors. C1300 neuroblastoma cells are regulated following injection into the somite region of 8-day embryos followed by transplantation of this somite in vivo (Pierce, Podesta, and Wells 1983). Similar regulation has also proved to be the case with leukemia cells in combination with day-10 (but not day-11) embryos (Gootwine, Webb, and Sachs 1982). These observations have stimulated studies aimed at understanding the factors important in the embryonic environment that might restrict and control the malignant phenotype.

This type of study has consisted in introducing EC cells into the embryonic environment at the blastocyst stage, followed by an assessment of proliferation of the EC component in vivo or in vitro (Pierce et al. 1979).

This type of study must be interpreted with a degree of caution, particularly when the experimental procedure requires assessment of embryonic regulation by growth in a nonuterine site, namely ectopic sites or in vitro. Additionally, the cell lines used in these experiments have not been tested for their chimera-forming potential, and their behavior in an embryonic environment might always be unregulated. This reservation has recently been vindicated with in vitro studies using P10 and subclones from the P19 EC lines in an embryonic environment in vitro (Rossant and Papaioannou 1985).

The demonstration that rampantly malignant cell lines, such as P19 (Rossant and McBurney 1982), can also form chimeras with EC-derived normal tissue is an important observation. The apparent unregulated tumorigenicity of this line might be used in a comparison with a less malignant EC cell line or embryo-derived stem cells to determine differences between these cell types.

V. Use of pluripotent stem cells for genetic analysis of development

A. Study of developmental mutants

One approach to the study of development in the mouse embryo is through investigation of existing mutations that are known to interfere in developmental processes. Several developmentally lethal mutations have been described, and the corresponding genetic loci well defined. Failure of embryogenesis in developmental mutants may be a result of generalized cell lethal effects or may be attributed to defects in specific cell lineages, as, for example, in the *Steel* mutation and alleles of the *W* locus in which there are well-documented effects on the blood cell, pigment cell, and germ cell lineages (Silvers 1979).

Biochemical and molecular approaches to the study of these mutations and identification of the gene product or products involved are difficult. One reason for this is that the mutations are of necessity maintained in the heterozygous form, and it is difficult to obtain the appropriate quantities of the homozygous material. The possibility of establishing permanent tissue-culture lines from embryos homozygous for developmentally acting alleles would solve this problem. The availability in culture of such stem cell lines offers attractive possibilities for characterizing the lethal effect. It is possible to examine homozygote failure by analysis of mutant embryos, but comparing the differentiative capacities of mutant and wild-type cell lines in vitro may be more informative. The consequence of aberrant expression may not be seen as having a detectable effect on differentiation, but might affect the ability of the cell to respond correctly to developmental signals. This could better be approached by monitoring the behavior of cells when they are returned to the embryonic environment. This, of course, can be achieved by aggregating defective embryos with normal carrier embryos. Blastocyst injection of cultured stem cells offers two advantages in that, first, it is possible to manipulate the relative contributions of the normal and mutant genotype cells to the chimeric conceptus and, second, all injected ES cells are of the mutant genotype, whereas only one in four of embryos obtained by mating heterozygotes will contain cells homozygous for the mutation.

The potential for use of the system has already been illustrated by the study of Magnuson and associates (1982). Through the use of appropriate chromosomal markers they were able to identify a stem cell line that was homogzygous for the t^{w5} mutation (a lethal haplotype of the t complex). Because the cell line could be both maintained in culture and induced to differentiate into a variety of cell types, it can be concluded that the mutation does not act via some generalized cell lethal effect, nor is it due to a detectable block to differentiation. Interestingly, embryo aggregation experiments (Magnuson et al. 1983) have shown that it is not feasible to rescue chimeric embryos, and those chimeras in which the homozygous t^{w5} cells are present in a high proportion die within eight days. This is an important finding, because an in vivo functional assay would be a prerequisite for detecting restoration of wild-type function in homozygous cell lines transformed by cloned t-complex genes.

B. Analysis of growth and differentiation of cells in culture

The use of cultured pluripotent cells, which are undoubtedly similar in phenotype to embryo cells and whose differentiation sequence closely resembles that of the normal embryo, provides a unique opportunity for identification and characterization of developmentally important molecules. These may be involved in maintenance of the undifferentiated proliferating state or associated with acquisition of a differentiated phenotype.

The stability of the undifferentiated cell state must be ensured by the expression of a particular combination of gene products, many of which will be present in other cell types. This is illustrated by comparisons of patterns of total protein synthesis between undifferentiated stem cells and differentiating populations (Failly-Crepin and Martin 1979; Lovell-Badge and Evans 1980). Proteins that show qualitative differences in expression may be potentially important in the control of growth and differentiation. Cultured stem cells provide a valuable source of material from which it is possible to identify these developmentally important markers. These can then be applied to the normal developing embryo. The types of experimental strategies that have been adopted include the preparation and screening of cDNA libraries from stem cells to indentify genes that are differentially expressed according to cell phenotype (Stacey and Evans 1984). Alternatively, EC cells can be used to generate immunologic reagents that can be used subsequently to identify cell surface determinants and intracellular proteins important to the developing embryo (Brulet, Condamine, and Jacob, 1983).

The availability of undifferentiated cells that can be maintained and monitored under increasingly carefully defined conditions enables a more accurate description of the growth requirements of the cell. Studies on the responses of EC and ES cells to exogenously added growth factors will help in defining the roles that growth factors and receptors play in normal development (Heath 1983).

Pluripotent stem cells will differentiate in culture in response to alterations in the growth environment. It is not possible to control the differentiation of the majority of EC cell lines and embryo-derived stem cell lines, although the cells will differentiate reproducibly via embryoid-body formation. The use of EC cell lines that have a more restricted pattern of differentiation has been invaluable for differentiation studies. The F9 line fails to differentiate under normal conditions of culture. Addition of low levels of retinoic acid to F9 cultures has a dramatic effect, causing rapid and synchronous differentiation of the entire population to cells of an endodermal phenotype (Strickland and Mahdavi 1978). By further alteration of the culture conditions the cells can be induced to follow one of two different pathways to give either parietal or visceral endoderm (Hogan, Barlow, and Tilly 1983). These culture-derived endodermal cells appear to be homologous with the cells of the extra-embryonic lineages in terms of the quantitative and qualitative expressions of sets of protein markers. Although the cells may be subjected to an embryologically inappropriate signal, the controlled differentiation response enables the differential expressions of well-characterized genes to be studied in a tissue-culture model system. The results from this study can be extended to consider the mechanisms that are responsible for cell commitment in the normal embryo. The P19 EC cell line may also be potentially very useful. The differentiated cell type formed by P19 cells can be controlled to give cardiac muscle cells or nerve cells (McBurney et al. 1982).

C. Introduction of new mutants or new genetic material into embryos

In recent years, improved methods for transformation of cell lines in culture have been devised and have provided a number of potentially interesting ways in which stem cells can be used to study the molecular biology of developing embryos. In particular, the development of dominant selection systems, based on the uptake and expression of bacterial genes that confer drug resistance (Mulligan and Berg 1981), and the construction of highly infective viral vectors (Cepko, Roberts, and Mulligan, 1984; King et al. 1985), allows routine and efficient introduction of exogenous DNA into tissue-culture cells. These techniques should allow stable integration of specific gene constructs into undifferentiated cells. Providing the transformation procedure does not alter the developmental capabilities of the cells, it should then be possible to monitor the expression of the introduced genes as the cells differentiate in culture. More important, this will provide a system whereby differentiation can be achieved in the embryonic context under the control of a host embryo. For this type of study to be practicable, it is obviously preferable to use stem cells that are karyotypically normal and developmentally competent. Studies that have used EC cell lines have been of limited value, because with the exception of the P10 and METT-1 lines, EC cell lines differentiate poorly and unpredictably on return to the embryo. Embryo-derived ES cells offer a much improved model system in this regard. It should be possible to use transformation techniques in conjunction with developmentally normal cell lines to examine the control of gene activity during normal somatic differentiation. Two classes of genes might be useful in such study, namely, those that are known to be temporally and spatially regulated during embryogenesis, and those that might have a controlling influence on differentiation. These genes can be placed in carefully designed gene constructs that control the expression of the gene of interest. Genes could be introduced with the aim of examining the pattern of expression under controlling influence provided by the host cell genome. An alternative strategy would be to place the gene of interest under the control of promoters that give constitutive expression or, alternatively, under the control of tissue-specific or inducible promoters. Following incorporation of cells into the embryo, it should then become possible to monitor the consequences of inappropriate expression of the gene product on the differentiation of the cell and on the host embryo.

Transformation of cells in culture by constitutively expressed genes may enable stem cells to be phenotypically marked. The cells may then be distinguished from host cells and should allow the fate of the cultured cells during development to be followed with some accuracy.

If ES cells that have been genetically modified in culture can reproducibly and efficiently colonize the functional germ cell population, this opens up further possibilities. Stem cells may be selected in culture for the loss of a specific biochemical function. Providing the selection procedure is not deleterious, it may be possible to stably transfer this mutation into the mouse genome following formation of functional germ cells by the defective stem cells.

Incorporation of transformed cells into the germ line would allow stable integration and transmission of gene constructs in a way similar to that described for transgenic mice obtained by injection of zygotes with DNA. Test breeding of such animals would also allow detection of novel mutations created by insertion of exogenous DNA into transcriptionally active regions of the genome.

It is thus becoming clear that ES cells offer a unique experimental tool that will facilitate novel genetic approaches to mouse embryo development.

Acknowledgments

A.B. is a Beit Memorial Fellow. The original research described here has been funded by the Medical Research Council and the Cancer Research Campaign. We would like to thank Lesley Cooke and Pam Fletcher for their contributions to this work.

References

Axelrod, H. R., and Lader, E. (1983) A simplified method for obtaining embryonic stem cell lines from blastocysts. In *Cold Spring Harbor Conference on Cell Proliferation*, Vol. 10, eds. L. M. Silver, G. R. Martin, and S. Strickland, pp. 665–70. Cold Spring Harbor, N.Y.: Cold Spring Harbor Laboratory.

Bernstine, E. G., Hooper, M. L., Grandchamp, S., and Ephrussi, B. (1973) Alkaline phosphatase activity in mouse teratocarcinoma. *Proc. Natl. Acad. Sci. U.S.A.* 70, 3899–903.

Bradley, A., Evans, M., Kaufman, M. H., and Robertson, E. (1984) Formation of germline chimeras from embryo-derived teratocarcinoma cell lines. *Nature* 309, 255–6.

Bradley, A. and Robertson, E. J. (1985) Embryonic stem cells: a tool for elucidating the developmental genetics of the mouse. In *Current Topics in Developmental Biology*, Vol. 20, eds. T. S. Okada and A. A. Moscona. New York: Academic Press.

Brinster, R. L. (1974) The effects of cells transferred into the blastocyst on subsequent development. *J. Exp. Med.* 140, 1049–56.

Brulet, P., Condamine, H., and Jacob, F. (1983) Murine teratocarcinoma cells and the identification of developmentally relevant molecules. *Cancer Surveys* 2, 93–113.

Cepko, C. L., Roberts, B. E., and Mulligan, R. C. (1984) Construction and applications of a highly transmissible murine retrovirus shuttle vector. *Cell* 37, 1053–62.

Cronmiller, C., and Mintz, B. (1978) Karyotypic normalcy and quasi-normalcy of developmentally totipotent mouse teratocarcinoma cells. *Div. Biol.* 67, 465–77.

Damjanov, I., and Solter, D. (1982) Maternally transmitted factors modify development and malignancy of teratomas in mice. *Nature* 296, 95–7.

Damjanov, I., Solter D., and Skreb, N. (1971) Teratocarcinogenesis as related to the age of embryos grafted under the kidney capsule. *Wilhelm Roux Arch.* 167, 288–90.

Dewey, M. J., Martin, D. W., Martin, G. R., and Mintz, B. (1977) Mosaic mice with teratocarcinoma derived mutant cells deficient in hypoxanthine phosphoribosyl transferase. *Proc. Natl. Acad. Sci. U.S.A.* 74, 5564–8.

Diwan, S., and Stevens, L. C. (1976) The development of teratomas from endoderm of mouse egg cylinders. *J. Natl. Cancer Inst.* 57, 937–42.

Eppig, J. J., Kozak, L. P., Eicher, E. M., and Stevens, L. C. (1977) Ovarian teratomas in mice are derived from oocytes that have completed the 1st meiotic division. *Nature* 269, 517–18.

Evans, E. P., Ford, C. E., and Lyon, M. F. (1977) Direct evidence of the capacity of the XY germ cell in the mouse to become an oocyte. *Nature* 267, 430–1.

Evans, M. (1981) Origin of mouse embryonal carcinoma cells and the possibility of their direct isolation into tissue culture. *J. Reprod. Fertil.* 62, 625–31.

Evans, M., Bradley, A., and Robertson, E. J. (1985) EK contribution to chimaeric mice: from tissue culture to sperm. In *Genetic Manipulation of the Mammalian Ovum and Early Embryos*, Banbury report. Cold Spring Harbor, N.Y.: Cold Spring Harbor Laboratory.

Evans, M. J., and Kaufman, M. H. (1981) Establishment in culture of pluripotential cells from mouse embryos. *Nature* 292, 154–5.

— (1983)Pluripotential cells grown directly from normal mouse embryos. *Cancer Surveys* 2, 186–207.

Evans, M. J., Lovell-Badge, R. H., Stern, P. L., and Stinnakre, M. G. (1979) Cell lineages of the mouse embryo and embryonal carcinoma cells; Forssman antigen distribution and patterns of protein synthesis. In *Cell Lineage, Stem Cells, and Cell Determination*, ed. N. Le Douarin, pp. 115–29. Amsterdam: Elsevier/North Holland.

Failly-Crepin, C., and Martin, G. R. (1979) Protein synthesis and differentiation in a clonal line of teratocarcinoma and in pre-implantation mouse embryos. *Cell Differ.* 8, 61–73.

Fujii, J. T., and Martin, G. R. (1983) Developmental potential of teratocarcinoma cells in utero following aggregation with cleavage stage mouse embryos. *J. Embryol. Exp. Morphol.* 74, 79–90.

Gootwine, E., Webb, C. G., and Sachs, L. (1982) Participation of myeloid leukaemic cells integrated into embryos in haematopoietic differentiation in adult mice. *Nature* 299, 63–5.

Graham, C. F. (1977) Teratocarcinoma cells and normal mouse embryogenesis. In *Concepts in Mammalian Embryogenesis*, ed. M. I. Sherman, pp. 315–99. Cambridge: M.I.T. Press.

Heath, J. K. (1983) Regulation of murine embryonal carcinoma cell proliferation and differentiation. *Cancer Surveys* 2, 141–64.

Hogan, B. L. M. (1976) Changes in the behaviour of teratocarcinoma cells cultured *in vitro*. *Nature* 263, 136–7.

Hogan, B. L. M., Barlow, D. P., and Tilly, R. (1983) F9 teratocarcinoma cells as a model for the differentiation of parietal and visceral endoderm in the mouse embryo. *Cancer Surveys* 2, 115–40.

Hooper, M. L., and Slack, C. (1977) Metabolic cooperation in HGPRT$^+$ and HGPRT$^-$ embryonal carcinoma cells. *Dev. Biol.* 55, 271–84.

Iles, S. A. (1977) Mouse teratomas and embryoid bodies: their induction and differentiation. *J. Embryol. Exp. Morphol.* 38, 63–75.

Iles, S. A., and Evans, E. P. (1977) Karyotype analysis of teratocarcinomas and embryoid bodies in C3H mice. *J. Embryol. Exp. Morphol.* 38, 77–92.

Illmensee, K. (1978) Reversion of malignancy and normalized differentiation of teratocarcinoma cells in chimaeric mice. In *Genetic Mosaics and Chimaeras in Mammals*, ed. L. B. Russell, pp. 3–25. New York: Plenum.

Illmensee, K., and Mintz, B. (1976) Totipotency and normal differentiation of single teratocarcinoma cells cloned by injection into blastocysts. *Proc. Natl. Acad. Sci. U.S.A.* 73, 549–53.

Kaufman, M. H., Robertson, E. J., Handyside, A. H., and Evans, M. J. (1982) Establishment of pluripotent cell lines from haploid mouse embryos. *J. Embryol. Exp. Morphol.* 73, 249–61.

King, W., Patel, M. D., Lobel, L. I., Goff, S. P., and Nguyen-Huu, M. L. (1985) Insertional mutagenesis of embryonal carcinoma cells by retroviruses. *Science* 228, 554–8.

Kleinsmith, L. J., and Pierce, G. B. (1964) Multipotentiality of single embryonal carcinoma cells. *Cancer Res.* 24, 1544–52.

Kuff, E. D., and Fewell, J. W. (1980) Induction of neural-like cells and acetylcholinesterase activity in cultures of F9 teratocarcinoma cells treated with retinoic acid and dibutyl cyclic adenosine monophosphate. *Dev. Biol.* 77, 103–15.

Lovell-Badge, R. H., and Evans, M. J. (1980) Changes in protein synthesis during differentiation of embryonal carcinoma cells and a comparison with embryo cells. *J. Embryol. Exp. Morphol.* 59, 187–206.

McBurney, M. W. (1976) Clonal lines of teratocarcinoma cells *in vitro*: differentiation and cytogenetic characteristics. *J. Cell. Physiol.* 89, 441–55.

McBurney, M., Jones-Villeneuve, E., Edwards, M., and Anderson, P. (1982) Control of muscle and neuronal differentiation in a cultured embryonal carcinoma cell line. *Nature* 299, 165–7.

McBurney, M. W., and Rogers, B. J. (1982) Isolation of male embryonal carcinoma cell lines and their chromosome replication patterns. *Dev. Biol.* 89, 503–8.

McBurney, M. W., and Strutt, B. J. (1980) Genetic activity of X-chromosomes in pluripotential female teratocarcinoma cells and their differentiated progeny. *Cell* 21, 357–64.

Magnuson, T., Epstein, C. J., Silver, L. M., and Martin, G. R. (1982) Pluripotent embryonic stem cells can be derived from t^{w5}/t^{w5} blastocysts. *Nature* 298, 750–2.

Magnuson, T., Martin, G. R., Silver, L. M., and Epstein, C. J. (1983) Studies on the viability of t^{w5}/t^{w5} embryonic cells *in vitro* and *in vivo*. In *Cold Spring Harbor Conferences on Cell Proliferation*, Vol. 10, eds. L. M. Silver, G. R. Martin, and S. Strickland, pp. 671–81. Cold Spring Harbor, N.Y.: Cold Spring Harbor Laboratory.

Magrane, G. G. (1982) A comparative study of human and mouse teratocarcinomas. Ph.D. thesis, University of London.

Martin, G. R. (1980) Teratocarcinomas and mammalian embryogenesis. *Science* 209, 678–76.

–   (1981) Isolation of a pluripotent cell line from early mouse embryos cultured in medium conditioned by teratocarcinoma stem cells. *Proc. Natl. Acad. Sci. U.S.A.* 78, 7634–8.

Martin, G. R., Epstein, C. J., Travis, B., Tucker, G., Yatziv, S., Martin, D. W., Jr., Clift, S., and Cohen, S. (1978) X-chromosome inactivation during differentiation of female teratocarcinoma stem cells *in vitro*. *Nature* 271, 329–33.

Martin, G. R., and Lock, L. F. (1983) Pluripotent cell lines derived from early mouse embryos cultured in medium conditioned by teratocarcinoma stem cells. In *Cold Spring Harbor Conference on Cell Proliferation*, Vol. 10, eds. L. M. Silver, G. R. Martin, and S. Strickland, pp. 635–46. Cold Spring Harbor, N.Y.: Cold Spring Harbor Laboratory.

Martin, G. R., Wiley, L. M., and Damjanov, I. (1977) The development of cystic embryoid bodies *in vitro* from clonal teratocarcinoma stem cells. *Dev. Biol.* 61, 230–44.

Mintz, B., and Cronmiller, C. (1981) METT-1: a karyotypically normal in vitro line of developmentally totipotent mouse teratocarcinoma cells. *Somatic Cell Genetics*, 7, 489–505.

Mintz, B., and Illmensee, K. (1975) Normal genetically mosaic mice produced from malignant teratocarcinoma cells. *Proc. Natl. Acad. Sci. U.S.A.* 72, 3585–9.

Mullen, R. J., and Whitten, W. K. (1971) Relationship of genotype and degree of coat colour to sex ratios and gametogenesis in chimaeric mice. *J. Exp. Zool.* 178, 165–76.

Mulligan, R. C., and Berg, P. (1981) Selection for animal cells that express the *Escherichia coli* gene coding for xanthine-guanine phosphoribosyl transferase. *Proc. Natl. Acad. Sci. U.S.A.* 78, 2072–6.

Nicolas, J. F., Avner, P., Gaillard, J., Guenet, J. L., Jakob, H., and Jacob, F. (1976) Cell lines derived from teratocarcinomas. *Cancer Res.* 36, 4224–31.

Nicolas, J. F., Jakob, H., and Jacob, F. (1978) Metabolic cooperation between mouse EC cells and their differentiated derivatives. *Proc. Natl. Acad. Sci. U.S.A.* 72, 3292–6.

Ohno, S. (1978) Why not androgynes among animals? In *Genetic Mosaics and Chimeras in Mammals*, ed. L. B. Russell, pp. 165–85. New York: Plenum.

Papaioannou, V. E., Evans, E. P., Gardner, R. L., and Graham, C. F. (1979) Growth and differentiation of an embryonal carcinoma cell line (C145b). *J. Embryol. Exp. Morphol.* 54, 277–95.

Papaioannou, V. E., Gardner, R. L., McBurney, M. W., Babinet, C., and Evans, M. J. (1978) Participation of cultured teratocarcinoma cells in mouse embryogenesis. *J. Embryol. Exp. Morphol.* 44, 93–104.

Papaioannou, V. E., McBurney, M. W., Gardner, R. L., and Evans, M. J. (1975) Fate of teratocarcinoma cells injected into early mouse embryos. *Nature* 258, 70–3.

Papaioannou, V. E., and Rossant, J. (1983) Effects of the embryonic environment on proliferation and differentiation of embryonal carcinoma cells. *Cancer Surveys* 2, 165–83.

Pierce, G. B., Lewis, S. H., Miller, G. J., Moritz, E., and Miller, P. (1979) Tumorigenicity of embryonal carcinoma as an assay to study control of malignancy by the murine blastocyst. *Proc. Natl. Acad. Sci. U.S.A.* 76, 6699–51.

Pierce, G. B., Podesta, A., and Wells, R. S. (1983) Malignancy and differentiation: the role of the blastocyst in control of colony formation in teratocarcinoma stem cells. In *Cold Spring Harbor Conference on Cell Proliferation*, Vol. 10, eds. L. M. Silver, G. R. Martin, and S. Strickland, pp. 15–22. Cold Spring Harbor, N.Y.: Cold Spring Harbor Laboratory.

Rastan, S., and Robertson, E. J. (1985) X-chromosome deletions in embryo-derived (EK) cell lines associated with lack of X-chromosome inactivation. *J. Embryol. Exp. Morphol.* 90, 379–88.

Robertson, E. J., Evans, M. J., and Kaufman, M. H. (1983a) X-chromosome instability in pluripotent stem cell lines derived from parthenogenetic embryos. *J. Embryol. Exp. Morphol.* 74, 297–309.

Robertson, E. J., Kaufman, M. H., Bradley, A., and Evans, M. J. (1983b) Isolation, properties and karyotype analysis of pluripotent (EK) cell lines from normal and parthenogenetic embryos. In *Cold Spring Harbor Conference on Cell Proliferation*, Vol. 10, eds. L. M. Silver, G. R. Martin, and S. Strickland, pp. 647–63. Cold Spring Harbor, N.Y.: Cold Spring Harbor Laboratory.

Rossant, J., and McBurney, M. W. (1982) The developmental potential of an euploid male teratocarcinoma cell line after blastocyst injection. *J. Embryol. Exp. Morphol.* 70, 99–112.

– (1983) Diploid teratocarcinoma cell lines differ in their ability to differentiate normally after blastocyst injection. In *Cold Spring Harbor Conference on Cell Proliferation*, Vol. 10, eds. L. M. Silver, G. R. Martin, and S. Strickland, pp. 625–33. Cold Spring Harbor, N.Y.: Cold Spring Harbor Laboratory.

Rossant, J., and Papaioannou, V. E. (1985) Outgrowth of embryonal carcinoma cells from injected blastocysts *in vitro* correlates with abnormal chimera development *in vivo*. *Exp. Cell Res.* 156, 213–20.

Sherman, M. I. (1975) The culture of cells derived from mouse blastocysts. *Cell*, 5, 343–9.

Silver, L. M., Martin, G. R., and Strickland, S., eds. (1983) *Cold Spring Harbor Conference on Cell Proliferation*, Vol. 10, Cold Spring Harbor, N.Y.: Cold Spring Harbor Laboratory.

Silvers, W. K. (1979) *The Coat Colours of Mice*. Berlin: Springer-Verlag.

Slack, C., Morgan, R. H. M., and Hooper, M. L. (1978) Isolation of metabolic cooperation defective variants from mouse embryonal carcinoma cells. *Exp. Cell Res.* 117, 195–205.

Solter, D., and Damjanov, I. (1973) Explantation of extra-embryonic parts of 7 day old mouse egg cylinder. *Experientia* 29, 701–5.

– (1979) Teratocarcinomas rarely develop from embryos transplanted into athymic mice. *Nature* 278, 554–6.

Solter, D., Dominis, M., and Damjanov, I. (1979) Embryo-derived teratocarcinoma. I. The role of strain and gender in the control of teratocarcinogenesis. *Int. J. Cancer* 24, 770–2.

– (1980) Embryo-derived teratocarcinoma. II. Teratocarcinogenesis depends on the type of embryonic graft. *Int. J. Cancer* 25, 341–9.

– (1981) Embryo-derived teratocarcinoma. III. Development of tumors from teratocarcinoma-permissive and nonpermissive strain embryos transplanted to F₁ hybrids. *Int. J. Cancer* 28, 479–85.

Stacey, A. J., and Evans, M. J. (1984) A gene sequence expressed only in undifferentiated EC, EK cells and testes. *EMBO Journal* 3, 2279–85.

Stevens, L. C. (1964) Experimental production of testicular teratomas in mice. *Proc. Natl. Acad. Sci. U.S.A.* 52, 654–61.

– (1968) The development of teratomas from intratesticular grafts of tubal mouse eggs. *J. Embryol. Exp. Morphol.* 20, 329–41.

– (1970) The development of transplantable teratocarcinomas from intratesticular grafts of pre- and post-implantation mouse embryos. *Dev. Biol.* 21, 364–82.

 – (1983) Testicular, ovarian, and embryo-derived teratomas. *Cancer Surveys* 2, 75–91.

Stevens, L. C., and Little, C. C. (1954) Spontaneous testicular tumors in an inbred strain of mice. *Proc. Natl. Acad. Sci. U.S.A.* 40, 1080–7.

Stevens, L. C., and Varnum, D. S. (1974) The development of teratomas from parthenogenetically activated ovarian mouse eggs. *Dev. Biol.*, 37, 369–80.

Stewart, C. L. (1982) Formation of viable chimaeras by aggregation between teratocarcinomas and preimplantation mouse embryos. *J. Embryol. Exp. Morphol.* 67, 167–79.

Stewart, T. A., and Mintz, B. (1981) Successive generations of mice produced from an established culture of euploid teratocarcinoma cells. *Proc. Natl. Acad. Sci. U.S.A.* 78, 6314–18.

 – (1982) Recurrent germ line transmission of the teratocarcinoma genome from the METT-1 culture line to progeny *in vivo. J. Exp. Zool.* 224, 465–71.

Strickland, S., and Mahdavi, V. (1978) The induction of differentiation in teratocarcinoma stem cells by retinoic acid. *Cell,* 15, 393–403.

16 Integration and expression of genes introduced into mouse embryos

ERWIN F. WAGNER AND COLIN L. STEWART

CONTENTS

I. Introduction

Intimately controlled, complex biological networks are responsible for the functioning of a multicellular organism. These controls operate at various levels, including cell-cell communication and coordination

509

as well as intracellular regulation. Regulation also must occur at the level of the gene. The reductionist approach suggests that understanding gene regulation will help to understand yet unsolved biological problems such as morphogenesis, cell differentiation, and cell growth in mammalian development.

The advent of gene cloning as a routine method has facilitated analysis of eukaryotic gene regulation through powerful techniques such as introduction of viruses or cloned genes into mammalian tissue-culture cells, as reviewed by Pellicer et al. (1980a). Because of the limitations of such in vitro studies for analyzing tissue-specific regulation of gene expression, attempts were first made to introduce viruses as model genes, and then cloned genes, into the mammalian embryo. The major rationale underlying this approach is that regulation and function of the gene can be studied in vivo, where it will be exposed from the onset of development to all possible regulatory factors. This should permit rigorous analysis of gene expression in various somatic cells of the adult organism. If the gene is also present in the germ cells, the stability and inheritance of expression can be studied in subsequent generations of animals carrying the gene.

With the development of successful methods for introducing genes into mouse embryos, other areas of research have also been opened up. Thus, to date, the major uses of gene transfer into mouse embryos and embryonic stem cells have been (1) to define genetic elements controlling specific gene expression in vivo, (2) to study the functions of individual genes in cell differentiation and development, (3) to analyze the consequences of expression on the organism and of DNA integration (insertional mutagenesis), and (4) to explore the possibility of genetically manipulating a mammal either to improve it for agricultural purposes or to correct a genetic defect through gene therapy; for recent reviews, see Gordon and Ruddle (1985), Palmiter and Brinster (1985), Wagner, Rüther, and Stewart 1984, 1986).

In this review we shall summarize the studies that have addressed the points raised earlier. We shall confine ourselves to experiments dealing with stable introduction of genetic material into mouse embryos and embryonic stem cells. After briefly describing the techniques for introducing genes into mice, we shall review the results obtained from injecting cloned genes into mouse eggs, from using viruses as model genetic elements, and from gene transfer to embryonic stem cells.

II. Methods for introducing genes into mouse embryos

In this section, the methods available for introducing exogenous genetic information into mouse embryos will be briefly reviewed, because detailed descriptions have already been published.

A. DNA injection into fertilized eggs

The three techniques used for introducing genes into mouse embryos are shown diagrammatically in Figure 16.1. Currently, the one that is most frequently used is injection of recombinant DNA into the pronuclei of fertilized mouse eggs; see Gordon and Ruddle (1983) and Brinster and associates (1985) for a detailed description of this technique. Briefly, this method involves isolation of fertilized eggs that are then cultured until the pronuclei are visible; the DNA solution is microinjected directly into one of the pronuclei. Successful injection is evidenced by pronuclear swelling (Figure 16.2A). The injected eggs are then transferred to a pseudopregnant female for development to term. Once the mice are born, they are analyzed as outlined in Figure 16.2B. Identification of the mice carrying the injected DNA sequences is necessary because the average frequency of positive mice that develop from the injected eggs is approximately 10–40 percent. Further analysis of such transgenic animals is carried out to determine if the genes are transmitted to the offspring and if expression occurs (Wagner et al. 1984; Palmiter and Brinster 1985).

B. Infection of embryos with viral vectors

The second technique for introducing genes into mouse embryos uses viruses, particularly cloned retroviruses, as vectors for exogenous DNA. An obligate step in the life cycle of a retrovirus is its stable integration into the infected cell's chromosomes (Weiss et al. 1982). There are several features that make cloned retroviral vectors particularly versatile: (1) A wide variety of cells can be infected using the same vectors, (2) the efficiency of gene transfer to the infected cells can reach 100 percent, (3) genes up to 6–8 kb can be incorporated into these vectors, and (4) expression of a gene from a retroviral vector can be 50-fold greater than for the same gene transferred into cultured cells using DNA transfection (Hwang and Gilboa 1984). It has been shown that wild-type retroviruses can efficiently infect preimplantation embryos and integrate into the germ line of mice (Jaenisch et al. 1981). One of the attractions of this system is its simplicity. In order to introduce recombinant retroviral vectors into mouse embryos, all that is necessary is to culture preimplantation embryos (denuded of their zona pellucida, because it blocks virus entry) on cells that are producing the vector (Jaenisch, Fan, and Croker 1975). The infected embryos are then returned to a foster mother, and when born they are analyzed exactly as outlined in Figure 16.2B. One of the disadvantages of using these vectors has been the inefficient expression of the viral genome in embryonic cells (Jähner et al. 1981; Stewart et al. 1981a). This problem and its possible solutions will be discussed subsequently.

ROUTES FOR INTRODUCING GENES INTO MICE

1) MICROINJECTION OF CLONED
   DNA INTO ZYGOTES

2) INFECTION OF PRE-
   AND POSTIMPLANTATION
   EMBRYOS WITH
   RECOMBINANT RETROVIRUS

3) TRANSFECTION AND INFECTION
   OF EC/ES CELLS WITH CLONED
   DNA

SELECTION, CHARACTERIZATION

CHIMAERA FORMATION

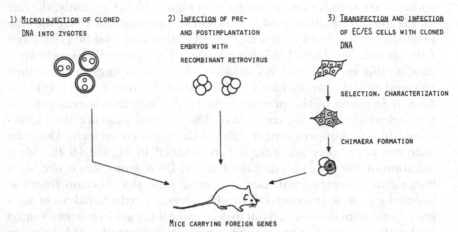

MICE CARRYING FOREIGN GENES

Figure 16.1. Three routes for introduction of exogenous DNA into mice.

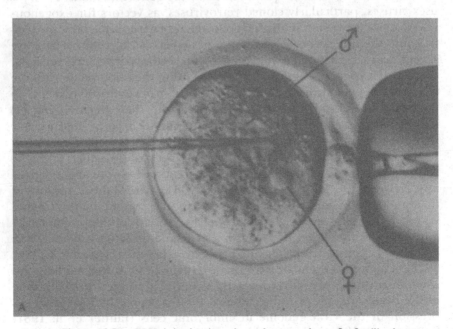

Figure 16.2A. DNA injection into the male pronucleus of a fertilized mouse egg (Nomarski optics, ×1200).

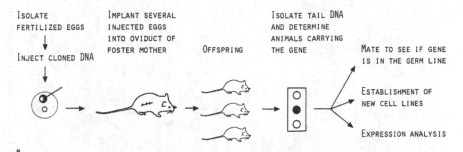

Figure 16.2B. Scheme for production and analysis of transgenic mice from DNA-injected fertilized eggs.

C. Gene transfer using embryonic stem cells

A third technique for introducing foreign genetic material into embryos has been to use the stem cells of teratocarcinomas, or embryonal carcinomas (EC cells). These cells, derived from germ cells or embryos, have been widely used as an in vitro model for mouse embryogenesis (see Chapter 15). Interest in this approach was stimulated by the discovery that EC cells were capable of forming functional gametes in chimeric mice (Mintz, Illmensee, and Gearhart 1975; Stewart and Mintz 1981; Bradley et al. 1984). This offered the possibility of manipulating the mouse genome by first isolating a clone of stem cells in vitro that had been selected to carry a particular alteration in its genome. This alteration could then be introduced into the mouse germ line by reintroducing the EC cells into blastocysts by microinjection, as reviewed by Papaioannou and Dierterlen-Lievre (1984), or into morulae by aggregation (Figures 16.3 and 16.4) (Stewart 1982, Fujii and Martin 1983). Although chimeras have been produced from cells carrying exogenous DNA or selected mutations in their genomes, there has been no report, even with the recently isolated embryonic stem (ES) cell lines, that descendants of the altered cells are able to form viable gametes. Thus, the potential of this approach has yet to be realized. (See note in proof on page 540.)

III. DNA injection into fertilized mouse eggs

A. Integration and germ-line transmission of exogenous DNA

In experiments with a large number of genes (>40) that have been injected into the male pronucleus of the fertilized egg, the foreign gene(s) is usually found in both the somatic and germ cells of the trans-genic mouse. Integration of the DNA probably occurs very early in

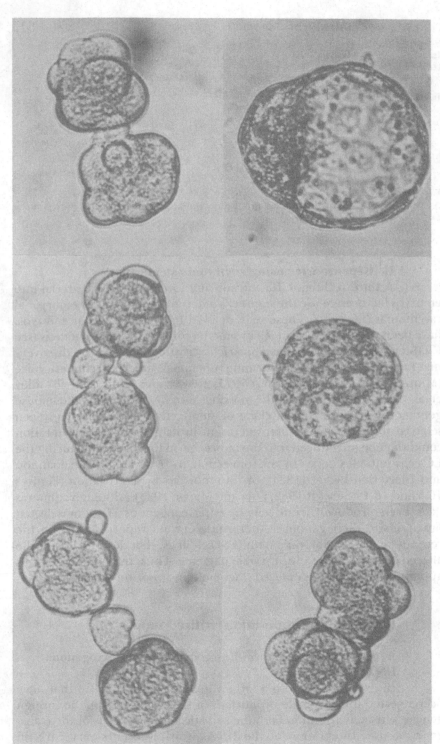

Figure 16.3. Two 8-cell-stage mouse embryos are shown aggregating with a group of neomycin-resistant EC cells, with subsequent formation of a single blastocyst.

Figure 16.4. Adult chimera derived from embryo aggregation with neomycin-resistant EC cells. The pigmented region of the coat shows that the selected EC cell clone is still able to undergo differentiation, and subsequent analysis showed extensive internal chimerism, with all chimeric tissues expressing the introduced gene.

development, perhaps even during the first cell division, and in most cases before the germ cell population is formed. In cases in which the genes are transmitted through the germ line, they behave as Mendelian traits, thus forming a new strain of mice. From subsequent breeding of heterozygous founder animals, homozygous offspring can be obtained. The foreign DNA usually is found randomly localized on one or two chromosomes of the diploid set in the adult mouse. The integration event appears to be nonspecific; no evidence of site-directed integration of DNA has been reported. The number of copies of the gene can range from 1 to over 100, usually found in a head-to-tail tandem array, with most copies intact. The first experiments that led to these conclusions were performed with cloned genes such as thymidine kinase of herpes simplex (HSV-TK) and the β-globin genes of human and rabbit origin (see Table 16.1 for a list of genes that have been studied with this approach).

Once it was shown that genes could be stably integrated in the somatic and germ cells of adult mice, the next step was to determine if these genes were expressed and correctly regulated. A report by Wagner, Stewart,

Table 16.1. *Expression of injected genes in transgenic mice*

DNA injected	Mode of expression			References
	None	Nonspecific	Specific	
Viral genes				
SV40	ND			Jaenisch and Mintz (1974)
HSV-TK[a]	ND			Gordon et al. (1980)
HSV-TK		+		Wagner et al. (1981a)
MLV[b]		+		Harbers et al. (1981a)
MLV		+		Stewart et al. (1983a)
Tissue-specific genes				
Growth hormone, human	+			Wagner et al. (1983)
Transferrin, chicken			+)	McKnight et al. (1983)
Elastase I, rat			+	Swift et al. (1984)
Myosin light-chain 2, rat			+	Shani (1985)
AFP[c] minigene, mouse			+	Krumlauf et al. (1985)
β-globin, human	+			Stewart et al. (1982b)
β-globin, rabbit	ND			Costantini and Lacy (1981)
β-globin, rabbit		+		Lacy et al. (1983)
β-globin, rabbit			?	Wagner et al. (1981b)
β-globin, hybrid mouse/human			+	Chada et al. (1985)
β-globin, hybrid mouse/human			+	Magram et al. (1985)
β-globin, human			+	Townes et al. (1985)
α-globin, mouse		+		Rusconi (1984)
Igκ, mouse[d]		+		Brinster et al. (1983)
Igκ, mouse			+	Storb et al. (1984)
Igμ, mouse			+	Grosschedl et al. (1984)
Igκ, Igμ, mouse			+	Rusconi and Köhler (1985)
SLA, porcine			+	Frels et al. (1985)
E_α, mouse			+	Le Meur et al. (1985)

Gene	Expression	Reference
E_α, mouse	+	Pinkert et al. (1985)
E_α, mouse	+	Yamamura et al. (1985)
Fusion genes		
MT-TK[e]	+	Brinster et al. (1981)
MT-TK	+	Palmiter et al. (1982b)
MMTV-TK[f]	(+)	Ross and Solter (1985)
MT-GH[g], rat	+	Palmiter et al. (1982a)
MT-GH, human	+	Palmiter et al. (1983)
MT-GRF[h], human	+	Hammer et al. (1985a)
MT-somatostatin, rat	(+)	Low et al. (1985)
MT-HPRT[i], human	+	Stout et al. (1985)
Elastase-GH	+	Ornitz et al. (1985)
Crystallin-CAT	+	Overbeck et al. (1985)
Collagen α2(I)-CAT		Westphal et al. (1985)
RSV-CAT[j]	+	Westphal et al. (1985)
Oncogenes		
SV40 T antigen	(+)	Brinster et al. (1984)
SV40 T antigen	(+)	Palmiter et al. (1985)
SV40 T antigen	+	Messing et al. (1985)
MMTV-myc	(+)	Stewart et al. (1984)
SV40 and v-myc	(+)	Small et al. (1985)
Insulin-SV40 T antigen	+	Hanahan (1985)
Hepatitis-B surface antigen	(+)	Babinet et al. (1985)
Hepatitis-B surface antigen	+	Chisari et al. (1985)

Note:

Abbreviations: [a]HSV-TK, herpes simplex virus thymidine kinase; [b]MLV, Moloney murine leukemia virus; [c]AFP, alphafetoprotein; [d]Ig, immunoglobulin; [e]MT-TK, mouse metallothionein-I-thymidine kinase fusion gene; [f]MMTV-TK, mouse mammary tumor virus-TK; [g]MT-GH, mouse metallothionein-I-growth hormone; [h]MT-GRF, mouse metallothionein-I-growth-hormone-releasing factor; [i]MT-HPRT, mouse metallothionein-I-hypoxanthine phosphoribosyl transferase; [j]RSV-CAT, Rous sarcoma virus-chloramphenicol acetyltransferase; ND, not determined; + only preference for specific expression.

and Mintz (1981a) first showed the potential of obtaining expression from exogenous DNA introduced into fertilized eggs, finding HSV-TK activity in late gestation fetuses. The data now available on the expression of exogenous DNA in mice will be divided into two parts. First we shall discuss the results obtained with genes containing their own controlling elements (promoters); then we shall summarize data obtained with fusion genes, which contain heterologous and/or inducible promoter elements.

B. Expression of unmodified genes in transgenic mice

    i. Tissue-specific expression of genes. Following a detailed report on the somatic expression of an HSV-TK gene in mice by Brinster and associates (1981), the door was open to seriously test the tissue-specific or regulated expression of various foreign genes. One question was whether or not the genes themselves and/or the immediate sequences surrounding them carried the information directing tissue-specific expression.

    The first indications of tissue-specific expression were obtained from analysis of transgenic mice carrying the chicken transferrin gene (McKnight et al. 1983). Preferential expression of the foreign gene was consistently observed in mouse liver, the organ in which transferrin is specifically expressed in chickens and mice, although some transcription was also observed in several other tissues. Quantitatively, the level of transcription in the transgenic mice was low, possibly because of inefficient recognition of the chicken promoter or because of integration of the foreign gene into a less active region of the mouse chromosome. However, the observed expression was stably transmitted to offspring, leading to the conclusion that the intact chicken transferrin gene or sequences in its vicinity directed expression in the liver. It should be noted that for quantitative studies dealing with the control of gene expression, the arbitrary copy number of integrated genes is a disadvantage, because a comparative analysis of expression between different genes or within a set of genes subjected to deletion analysis is difficult.

    Evidence for both qualitatively and quantitatively specific expression was obtained with transgenic mice expressing the rat elastase-I gene in the pancreas (Swift et al. 1984). The injected DNA, 23 kb in size, contained 7 kb of upstream and 5 kb of downstream flanking sequences and directed correct expression of the foreign gene independent of chromosomal location in four of five transgenic mice. The level of rat elastase-I mRNA in the pancreas was equal to or greater than the normal rat level and correlated with the number of integrated gene copies. This pancreas-specific element was further characterized in two additional series of experiments employing fusion genes, as discussed in Section IIIC.

    Another rat gene, myosin light-chain 2, was examined, and skeletal muscle-specific expression was found in two of three transgenic mice tested (Shani 1985). However, the levels of expression of the inserted gene were 100-fold different in the two mice examined; when compared with a rat myogenic cell line, expression in one mouse was reduced to 10 percent, and in the second it was 5–10 times greater than in the cultured cells. Another example of tissue-specific and temporally correct expression of introduced genes was obtained from studies of transgenic mice carrying modified copies of the mouse α-fetoprotein gene (Krumlauf et al. 1985). The α-fetoprotein gene is normally expressed in embryonic development in tissues such as visceral endoderm, yolk sac, fetal liver and gut; the gene is not expressed in the adult animal. Approximately half of the mice carrying integrated copies of a modified "mini-gene" expressed the introduced gene, and in each of these expression occurred only in the three tissues that express the authentic α-fetoprotein gene; expression was extinguished after birth. Thus, modified genes carrying either 7 or 14 kb of flanking DNA sequences could direct both tissue-specific and temporally correct expression, although the level of expression was highly variable. The precise cis-element sequence requirement for quantitatively correct expression of the α-fetoprotein gene is not yet known, but should be determined from further experiments. From the results of these quantitative studies one can conclude that the genes and their adjacent sequences are sufficient for tissue-specific expression, but also that the local chromosomal environment at the site of integration has a positional effect, and additional sequences or other unknown factors may limit the expression of the genes. One such factor is the presence of plasmid vector sequences in the injected DNA, as will be discussed subsequently.

ii. Developmental regulation of globin genes. The globin genes of different species have been particularly useful in analysis of gene regulation in vivo, because the differentiation, cell biology, and abnormalities of the hemopoietic system are exceptionally well characterized. At different stages of mammalian development, distinct embryonic, fetal, and adult hemoglobins are synthesized in erythroid cells, and several of these genes have already been cloned and sequenced. The molecular mechanisms controlling globin gene activation and globin switching during erythroid cell differentiation are unknown, and transgenic mice are expected to be instrumental in defining cis-acting regulatory sequences involved in these processes.

Data obtained from initial experiments using the human β-globin or a rabbit β-globin gene were rather disappointing and showed only a low level of expression that was not limited to the appropriate tissues (Table 16.1). One exception was reported by Wagner and associates (1981b) using a rabbit β-globin gene that appeared to be expressed in erythrocytes of adult mice and their offspring. Unfortunately, only a preliminary characterization of blood cells in these transgenic mice was described, and the report did not include an analysis of any other tissues; thus, it is impossible to clarify the discrepancy between these findings and those of Lacy and associates (1983), who found nonspecific expression of rabbit β-globin genes.

Convincing data on specific β-globin gene expression in transgenic mice were obtained by Chada and associates (1985), who used a hybrid mouse/human-adult β-globin gene. Correct tissue-specific expression was found in several transgenic lines that were produced by injection of a purified fragment without prokaryotic vector sequences. The level of expression per gene copy was 2−4 percent of the normal level. Interestingly, this hybrid β-globin gene was expressed at the correct stage of erythroid cell differentiation: Like the endogenous adult β-globin genes, it was inactive in yolk-sac-derived embryonic erythroid cells and was expressed for the first time in fetal liver erythroid cells (Magram, Chada, and Costantini 1985). These data confirm that cis-acting regulatory elements closely linked to a gene (in this case, adult β-globin) can confer tissue-specific and temporal regulation of expression and imply that distinct trans-acting regulatory factors modulate expression in the different cell types.

A definitive study relating to erythroid-specific expression of human β-globin genes that also supports the conclusions drawn earlier has been reported by Townes and associates (1985). Several constructs carrying between 48 and 4,300 bp of 5′ flanking sequence and 1,700 bp of 3′ flanking sequence were used to produce transgenic mice, of which most (15 to 20) expressed the human β-globin genes only in blood cells. In parallel with the endogenous mouse β-globin gene, the level of expression in some mice was comparable to that from endogenous β-globin genes, and expression was first detected between 11 and 14 days of development. The foreign genes synthesized correct mRNAs, and the human β-globin protein was also detectable in mature erythrocytes, even in the progeny of one of those mice. This study also included 16 mouse lines derived from injections with DNA containing the prokaryotic vector sequences. In almost all these lines (15 of 16) no expression of the human β-globin gene was detected in erythroid cells. This confirms the observations made by Chada and associates (1985) and Tilghman and associates (unpublished observations) that vector sequences impair expression. These observations may partly explain the initial lack of expression found in several transgenic mice (Table 16.1). However, it is a common finding that some transgenic mice do not express their foreign genes, most likely because of integration into a region of chromatin that is incompatible with expression. Further work is needed to understand the molecular mechanisms underlying hemoglobin switching, the role and properties of trans-acting factors, and the controls operating in regulating the balanced/unbalanced expression of globin synthesis.

iii. Transgenic mice with altered immume systems. It is possible to perturb the immune system and study the control of immunoglobulin (Ig) gene expression through introduction of exogenous Ig genes into mice. Recent reports have analyzed (1) the correct cell-type-specific expression of the exogenous Ig genes and (2) the consequences of an additional foreign Ig gene for the rearrangements of the endogenous genes.

In the studies of Brinster and associates (1983) and Storb and associates (1984), a functional immunoglobulin x light-chain gene was used to produce transgenic mice. All these mice expressed the gene at similar levels in lymphoid tissues, with a marked specificity for B lymphocytes. No correlation between the copy number of the foreign gene and the amount of x-specific mRNA was found, suggesting that this gene is activated by B-cell-specific signals regardless of the integration site. B-cell hybridomas that produced the foreign x chains together with an endogenous heavy chain had endogenous x genes that were not rearranged. Apparently the complete Ig molecule, in an unknown way, inhibits the rearrangement process. In another study (Grosschedl et al. 1984), a rearranged immunoglobulin μ heavy-chain gene introduced into the germ line of mice was expressed only in lymphoid tissues. However, the gene was transcriptionally active in both purified B and T cells, where it is normally not expressed. This suggests that control of Ig gene rearrangement might be the mechanism that determines the specificity of heavy-chain gene expression within the lymphoid cell lineage. To what extent the presence of the transgene influences the rearrangement and expression of the endogenous gene was studied in Abelson virus transformants and in hybridomas (Weaver et al. 1985). It was found that the transgenic μ gene somehow prevented rearrangement of endogenous Ig genes, because in 40 percent of the lines a germ-line J_H allele was found. Work by Rusconi and Köhler (1985) showed that μ and x genes introduced together into the mouse germ line could be expressed specifically in B cells. Here again, B-cell hybridomas were produced, and the data show that a negative-feedback inhibition of μ and x transgenic products on the endogenous heavy-chain rearrangement is possible, although the interpretation of these results is rather difficult. Taken together, these reports illustrate important aspects of Ig gene regulation and will facilitate a deeper understanding of the complex interactions operating within the immune system.

A different way to introduce changes in the immune system with transgenic animals is introduction of major histocompatibility complex (MHC) genes, such as class-I genes or "immune-response" genes coding for the Ia antigens (class-II genes). A study by Frels and associates (1985) used a porcine class-I MHC gene (SLA) that was introduced into the germ line of a C57BL/10 mouse. Various levels of expression of SLA antigen were found in the transgenic mice in various tissues. Skin graft rejections by normal C57BL/10 mice suggested that the foreign SLA antigen could be recognized by T cells as a functional transplantation antigen. Whether or not the porcine class-I molecule can also function as a restriction element in T-cell-mediated antigen recognition in the transgenic mouse remains to be investigated. Using this approach it should be possible to study the mechanisms mediating self-tolerance, T-cell specificities, and the T-cell repertoire for antigen recognition.

A related question, namely, whether or not class-II antigens encoded by genes introduced into transgenic mice can function as T-cell restriction elements, has now been answered in elegant experiments carried out by three groups (Le Meur et al. 1985; Pinkert et al. 1985; Yamamura et al. 1985). All three groups showed that a new immune response can be established by a single-copy insertion of a cloned E_α gene into eggs from mice incapable of expressing their endogenous E_α genes. The resulting transgenic mice express the E_α gene in a tissue-specific manner; Le Meur and associates and Pinkert and associates found elevated levels of E_α in response to interferon stimulation in peritoneal macrophages. A small DNA fragment with 2 kb 5' and 1.4 kb 3' flanking DNA was sufficient to confer tissue specificity and inducibility. In contrast, Yamamura and associates (1985) found that the introduced E_α gene and the endogenous E_β gene are constitutively expressed in macrophages. The reasons for the discrepancy are not clear. Aside from defining the control elements responsible for regulated expression of an E_α gene, the engineering of a new immune response by a single gene insertion permits further studies of Ia function in the context of the whole organism.

C. Expression of fusion genes in transgenic mice

The aim of experiments with transgenic mice produced from DNA injections with fusion genes is to obtain inducible or tissue-specific expression of the injected gene. Here, the prime interest is not only to analyze elements responsible for tissue specificity but also to develop an experimental system that can be manipulated to produce certain gene products.

i. Mouse MT-I fusion genes. Extensive studies on several lines of transgenic mice were performed by Brinster and associates (1981) and Palmiter and associates (1982b) with a herpes simplex virus thymidine kinase (HSV-TK) gene fused to the mouse metallothionein I (MT) promoter. Initially, animals were identified that expressed HSV-TK at significant levels in liver tissue after administration of cadmium at a concentration known to induce the endogenous MT gene. Significant expression of the fusion gene was also found in kidney, and less in brain, which corresponded to the expression pattern of the endogenous MT gene. No correlation between the amount of enzyme and the number of integrated copies of the fusion gene was observed. In subsequent studies, transgenic mice with the MT-TK genes were analyzed for inheritance of expression following germ-line transmission to offspring and also for a possible correlation of DNA methylation with expression of the trans-ferred genes. Both issues were rather complex. Expression in offspring was analyzed in two lines in great detail; it was extremely variable, being extinguished, diminished, or enhanced relative to that of the parent. The foreign sequences were extensively methylated at a variety of restriction sites but still expressed the HSV-TK gene; a correlation between changes in DNA methylation and expression of these genes was found in only a few animals. Most important, these data showed that achieving gene expression from microinjected genes did not ensure stable inheritance of that expression in all cases.

Transgenic mice carrying the MT promoter fused to the rat or human growth-hormone gene showed a clear alteration in growth-hormone syn-thesis (Palmiter et al. 1982a; Palmiter et al. 1983). As a consequence, the majority of these mice grew significantly larger than their littermates. Several mice had extraordinarily high levels of the fusion mRNA in their liver and contained 100–800-fold more growth hormone in their sera than normal mice. The rapid growth of the animals was not regulated solely by the heavy metals used as external inducers of expression of the MT fusion gene, because one animal, raised without heavy-metal induc-tion, still grew more rapidly than the metal-induced ones. However, in some cases, heavy-metal stimulation did increase expression, as expected. It is interesting to note that in one of these transgenic mouse lines the fusion gene was expressed in a subset of neurons (Swanson et al. 1985). The results with an MT-human-growth-hormone fusion gene were al-most identical (Palmiter et al. 1983). A detailed analysis revealed that the fusion genes were expressed in a variety of tissues, but the ratio of expression of human growth-hormone mRNA to endogenous MT mRNA varied markedly among the tissues, suggesting an influence on expression by the site of integration and tissue environment. In this study, the physiological consequences of excess human growth-hormone production were also examined. It was found that these animals had increased levels of insulin-like growth factor I in their sera, and the histology of their pituitaries suggested a dysfunction of the cells that normally synthesize growth hormone. The accelerated growth trait of these mice was inherited by the offspring, although female mice express-ing human or rat growth hormone were generally infertile. In a related set of experiments, Hammer, Palmiter, and Brinster (1984) used the rat MT growth-hormone fusion gene to inject eggs of the homozygous mu-tant, little (*lit/lit*), which is a mouse model for a human hereditary dis-order: isolated growth-hormone deficiency type I (Eicher and Beamer

1976). Expression of the fusion gene in transgenic mice restored growth to the dwarf mutants, changing them from dwarfs to "giants." This phenotype was also transmitted to their offspring. Chronic growth-hormone production in male transgenic animals improved their fertility, but it had a deleterious effect on female fertility, as in wild-type females expressing exogenous growth-hormone genes. The mechanism by which growth-hormone levels influence fertility in females is unknown.

In normal mice, growth-hormone-releasing factor (GRF) positively regulates growth-hormone synthesis and secretion, whereas somatostatin inhibits the release of growth hormone. When transgenic mice were produced by injecting an MT-human-GRF fusion gene, increased somatic growth was observed with animals that expressed human GRF, and consequently they had increased levels of mouse growth hormone in their plasma (Hammer et al. 1985a). Surprisingly, female transgenic mice carrying the MT-GRF fusion gene were then fertile, in contrast to female transgenic mice expressing human or rat growth hormone. In experiments using an MT-rat-somatostatin-cDNA fusion gene, transgenic mice expressing and correctly processing the somatostatin peptide did not show accelerated growth rates (Low et al. 1985). Expression of the foreign gene was most active in the anterior pituitary. Efficient processing of the presomatostatin peptide to the biologically active peptides occurred in the anterior pituitary, a tissue that does not normally synthesize somatostatin-like peptides. From these data one can conclude that the processing machinery is not specific to tissues that normally express somatostatin. Recently, another cDNA coding for human hypoxanthine phosphoribosyltransferase (HPRT) fused to the MT promoter was used to produce transgenic mice, which preferentially expressed the fusion gene in tissues of the central nervous system (Stout et al. 1985). Based on some of these initial experiments, further studies are under way on the biological consequences of growth-hormone production, as well as application of these and other fusion genes in various livestock species, including rabbits, sheep, and pigs (Hammer et al. 1985b).

ii. Other enhancer/promoter fusion genes. The study of tissue-specific gene expression in vivo can also be addressed systematically and definitively by using fusion genes containing a tissue-specific element with an easily assayable marker gene. The rat elastase-I gene is an excellent candidate, because this gene is expressed in a strictly tissue-specific manner. Ornitz and associates (1985) set out to identify the cis-acting DNA elements required for pancreas-specific expression by joining various parts of the 5' elastase flanking region to the human growth-hormone structural gene as a marker. Detailed deletion analysis in combination with production of many transgenic lines showed that the sequence between 200 and 80 bp 5' of the cap site is necessary and sufficient for pancreas-specific expression, which is regulated at the transcriptional level. This is the first identification of a mammalian tissue-specific element using the stringent criteria of the in vivo transgenic test system.

In another study by Westphal and associates (1985), the structural gene for the easily assayable bacterial enzyme chloramphenicol acetyltransferase (CAT) was joined to 5' control sequences of the murine αA-crystallin gene, the murine α2(I) collagen gene, or the 3' long terminal repeat (LTR) of avian sarcoma virus (RSV). Transgenic mice carrying these three fusion genes were produced and assayed for specific expression in different tissues. The α-crystallin fusion gene was expressed only in the lens (Overbeek et al. 1985), whereas the α2(I)-collagen construct was active in those organs that contained high levels of α2(I)-collagen mRNA, such as the tail. Transgenic mice carrying the RSV-CAT gene showed preferential expression in connective tissues. These results show that using marker genes joined to tissue-specific elements is a powerful approach in defining the cis-acting elements responsible for differential cell-type-specific gene expression.

iii. Oncogenes and transgenic mice. The study of oncogenes is directed toward identifying the functions of their products and dissecting the mechanisms underlying oncogene activation and tissue specificity of tumor formation. Several recent reports using SV40 T-antigen genes, *myc* fusion genes, and insulin/SV40 T-antigen genes indicate that this approach can provide new information on oncogene function and the development of tumors.

A report by Brinster and associates (1984) described the introduction of DNA containing the SV40 early region gene (the T antigen) and an MT fusion gene (either the TK or growth hormone) into the germ line. About 25 percent of the transgenic mice developed tumors of the choroid plexus, a layer of epithelial cells that lines the brain and forms the blood-brain barrier. This phenotype was heritable and was transmitted concordantly with the T-antigen gene, which is readily detected as mRNA and protein in affected tissues and cell lines derived from them. However, SV40 T-antigen expression was barely detectable in unaffected tissues, suggesting that tumorigenesis depended on activation of the SV40 genes. Furthermore, the MT fusion genes were rarely expressed in most of these mice. One could speculate that they had been inactivated early in development by an unknown mechanism. This series of experiments has recently been extended by Palmiter and associates (1985), who reported that the large T antigen alone is sufficient for tumorigenesis (the MT fusion gene is dispensable) and that the SV40 enhancer has an important role in limiting tumors to the choroid plexus. A deletion of the SV40 enhancer region led to a drastic change in the tissues affected, with the animals developing peripheral neuropathies, hepatocellular carcinomas, and islet cell adenomas (Messing et al. 1985). In studies by Jaenisch and Mintz (1974) and Williamson and associates (1983), who introduced SV40 DNA at the blastocyst stage, no choroid plexus tumors were detected, which is somewhat surprising. It is also worth noting that several of the choroid plexus tumors and their cell lines contained amplified SV40 sequences, suggesting that DNA amplification and rearrangements may be one mechanism for SV40 gene activation in the transgenic mice.

Stewart, Pattengale, and Leder (1984) used the inducible mouse mammary tumor virus (MMTV) promoter linked to various portions of the mouse c-*myc* proto-oncogene for injection into the embryos. This resulted in transgenic mice that developed tissue-specific mammary adenocarcinomas after several pregnancies. This promoter element, which was also shown to function when fused to the TK gene (Ross and Solter 1985), is normally responsive to prolactogenic stimuli and glucocorticoids and is the probable reason for pregnancy-associated tumorigenesis. Tumorigenesis was heritable in one of the females. Expression of the fusion gene was highly variable in different tissues, and penetrance of tumor formation was incomplete, because it did not arise in every mammary gland and occurred in only a fraction of the animals carrying the fusion gene. This implied that some other secondary event was necessary to elicit tumor formation.

A recent study by Hanahan (1985) attempted to examine tissue-specific expression of an oncogene and the consequences of this specific expression for the organism. Furthermore, the possibility of establishing stable cell lines from rare cell types was suggested on the basis of the expression of oncogenes in these cells. Transgenic mice were produced that carried the SV40 T antigen under the control of the 5' flanking region of the rat insulin gene. They developed normally but died prematurely at 9–12 weeks of age. T-antigen expression was confined to the insulin-producing β cells of the endocrine pancreas in the islets of Langerhans, indicating that the upstream sequences of the insulin gene are sufficient for cell-type-specific expression in vivo. In young mice, the islets of Langerhans were of normal size, with all β cells expressing T antigens; however, with increasing age the islets became disorganized and hyperplastic, and solid β-cell tumors arose. Although tumor formation was frequent (every mouse carrying this fusion gene developed tumors), it was not absolute, because only a small fraction of the hyperplastic islets became tumorigenic. Together, these and other studies (Small et al. 1985) support the view that naturally occurring neoplasias arise by multiple steps, because the presence of an oncogene in the germ line is not sufficient to produce a tumorigenic phenotype in all cells expressing the gene.

The wide application of transgenic mice is nicely documented by recent experiments reported by two groups using cloned hepatitis-B virus (HBV) (Babinet et al. 1985; Chisari et al. 1985). In both studies, transgenic mice were obtained that carried sequences coding for the surface antigen (HBsAg). Babinet and associates found the foreign gene to be expressed specifically in the liver of two transgenic lines (males expressed 5–10 times more than females), whereas Chisari and associates observed only nonspecific expression at a low level. The discrepancy of these results is not clear and may be resolved when more transgenic lines are analyzed. These studies may provide a laboratory animal model to study several aspects of the pathology of HBV, which normally infects only humans and some primates. These recent findings should provide encouragement that similar approaches will help to unravel the mechanisms of tumorigenesis in vivo.

D. Consequences of DNA injection: integrated sequences as insertional mutagens

Detailed genetic analysis of transgenic mice shows that integration of exogenous DNA can disrupt important chromosomal regions through DNA rearrangements or deletions, leading to developmental defects or sterility. Such "insertional" mutations, when present in the germ line, can be recognized in the homozygous progeny; in some cases they are lethal. In addition, the foreign DNA can serve as a tag for isolating the surrounding host DNA and defining the affected gene. The possibility of producing developmental mutants in order to find the genes that control mammalian development is an exciting prospect and is one of the major aims of the developmental geneticist (see Chapter 14 for further discussion of this approach). The first report describing recessive embryonic lethal mutations as a result of experimental insertion of DNA was published by Wagner and associates (1983). Two lines carrying the human growth-hormone gene failed to yield viable homozygous offspring. In these instances, lethality of homozygous embryos occurred at 4–5 days of gestation (E.F. Wagner, and C.L. Stewart, unpublished data). Because of the complex integration pattern of the foreign DNA, the affected host DNA sequences have not yet been characterized in order to isolate the affected host genes, which may play a role in early mouse development.

Palmiter and associates (1984) described an unusual transgenic mouse carrying two copies of an MT-TK fusion gene that was stably transmitted only by females through several generations. Males never transmitted the insert, although they were fertile. These authors proposed that the foreign DNA disrupted a gene required during haploid stages of spermatogenesis and that cloning the gene might reveal an essential function in this important developmental pathway (see Chapter 6 for further discussion of haploid gene expression). Westphal and associates (1985) described a

RSV-CAT transgenic line that is characterized by a semidominant trait of embryonic lethality and distortion of transmission.

Most recently, Woychik and associates (1985) reported the occurrence of a limb-deforming mutation produced by insertional mutagenesis in a transgenic line that carried the MMTV-myc fusion gene. The mutant gene, called limb deformity (ld^{Hd}), is probably allelic to a spontaneous mutant gene and distorts pattern formation in all four developing limb buds of the embryo: Single bones replaced the radius and ulna, and more distal bones of each limb fused. A DNA segment was cloned from the border of the interrupted region of mouse chromosome 2, and it segregates with the ld^{Hd} phenotype as an autosomal recessive. These and other mutants may thus provide a key to the actions of pattern-specific genes and their products.

III. Gene expression studies with viruses

Viral genes were the first to be used to study how the state of differentiation of a cell influences gene expression. This study of virus/host cell relationships has been a fruitful approach to mammalian development, because for some viruses, in particularly SV40, polyoma, and Moloney murine leukemia virus (MLV), their ability to replicate or express at least parts of the genome is influenced by the state of differentiation. The aim behind these studies is to determine the mechanisms that underlie the control of endogenous gene expression in mammalian development.

These three viruses are generally unable to replicate efficiently in embryos or embryonic stem cells of the mouse. However, in terminally differentiated derivatives of these stem cells they do replicate and express their genomes. These observations suggested that the control of gene expression in the totipotent or pluripotent cells of the embryo may differ from that in differentiated cells. The first part of this section will deal with the expression of these three viruses in embryos. The latter part will deal with the expression in murine teratocarcinoma cells in vitro. More extensive reviews covering studies performed on other viruses are available (Maltzman and Levine 1981; Kelly and Condamine 1982; Levine 1982).

A. Viral interactions with embryos

i. SV40 virus. This DNA virus undergoes complete replication and expression of its genome in monkey kidney cells (Tooze 1981). In mouse cells this does not occur; only the early part of the SV40 genome is expressed, resulting in appearance of the T antigens. As a consequence, SV40 infection of mouse cells leads to transformation (Tooze 1981).

However, the ability of this virus to express its genome in preimplantation mouse embryos is unclear, although replication can occur. Initial studies claimed that infection of preimplantation embryos (morulae) with high amounts of virus (4×10^4 plaque-forming units per milliliter) (Abramczuk et al. 1978) resulted in death of the infected embryo, and that implied that replication of the virus was occurring in these cells. It was also demonstrated that T and V antigens (a product of the late region) could be detected in the nuclei of trophectoderm, but not in the inner cell mass (Abramczuk et al. 1978). In a separate series of experiments, Jaenisch reported that replication of SV40 DNA in infected preimplantation embryos could not be observed (Jaenisch 1974). However, relatively high levels of [^{3}H]thymidine were used, and the infected embryos were incubated over a relatively long period of time (30 h) in order to detect replication of SV40 DNA. Mouse preimplantation embryos are extremely sensitive to [^{3}H]thymidine, with concentrations as low as 0.05 μCi/ml resulting in cell death (Snow 1973). Thus, it is possible that failure to detect SV40 replication was due to the very high concentrations of [^{3}H]thymidine.

Evidence that SV40 virus can replicate in embryos has come from experiments in which viral DNA was introduced into embryos by blastocyst injection or by morulae infection (Jaenisch and Mintz 1974; Willison et al. 1983). The adult mice were tested for the presence of SV40 sequences, and these were found in varying amounts in different organs. Willison and associates (1983) showed that the SV40 DNA existed in the episomal form (i.e., unintegrated closed circular molecules); a few animals also contained chromosomally integrated SV40 sequences. Both studies reported that no obvious cytopathological effects were detected, and whether or not the viral genome was expressed in any way was not described (see Section IIICiii). It is possible that viral replication was occurring in the absence of gene expression in the embryos, because infectious virus was recovered from a number of tissues (Willison et al. 1983). In addition, a recent study has shown that plasmids containing the SV40 genome do replicate when injected into fertilized eggs or embryos (Wirak et al. 1985). The only other report addressing the question whether or not SV40 can be expressed has suggested that cells taken from 9–10-day postimplantation embryos and infected in vitro express SV40 T antigen; cells from embryos prior to this stage are not permissive (Kelly and Condamine 1982).

ii. Polyoma virus. Polyoma is another DNA virus whose genome organization is very similar to that of SV40 (Tooze 1981). Polyoma, in contrast to SV40, does undergo a complete life cycle in mouse cells, resulting in lysis of the infected cells. Abramczuk and associates (1978) reported that infected preimplantation embryos did express polyoma V antigen in trophoblast nuclei, presumably indicating that the virus had replicated its genome; however, the embryos did not die, as did SV40 infected embryos. Jaenisch reported that replication of the polyoma genome could not be detected after labeling infected embryos with [³H]-thymidine (Jaenisch and Berns 1977), but the same criticisms raised to the SV40 studies would also apply here.

More recent studies have shown that polyoma-containing plasmids can replicate in fertilized eggs and early embryos (Wirak et al. 1985). In postimplantation embryos, cells become permissive for V antigen expression on the ninth day of gestation. Cells derived from embryos prior to this stage are apparently resistant to polyoma expression (Kelly and Condamine 1982).

iii. **Moloney murine leukemia virus.** The most extensive studies of the interactions between viruses and embryos have been performed with the murine retrovirus MLV. This virus differs from the two previously discussed DNA viruses in that an obligatory step in its life cycle is integration of its genome into the infected cell's chromosomes. This integration is extremely stable, and the integrated or proviral copy of the virus effectively becomes another chromosomal gene (Weiss et al. 1982).

Productive infection of cultured mouse fibroblasts results in the cells synthesizing new copies of the virus, which bud from the cell surface and can then infect other cells. Preimplantation embryos are not permissive for MLV expression (Jaenisch et al. 1975), even though virus can integrate at these stages (Jähner et al. 1982). This resistance to productive infection disappears at or around eight to nine days of gestation, when virus injected into embryos leads to productive infection in many different cell types (Jähner et al. 1982). Infection of postimplantation embryos rarely resulted in MLV being transmitted through the germ line. However, this occurred more frequently when preimplantation embryos were infected (Jaenisch et al. 1981). Using both of these techniques, Jaenisch established 14 *Mov* substrains of mice. Each contained a single proviral copy at an autosomal locus in its genome. The exception was *Mov-14*, which was derived by microinjection of cloned retroviral DNA into a zygote pronucleus and contained multiple tandemly repeated copies present on the X chromosome (Stewart et al. 1983a).

The salient points that have emerged from studies of these *Mov*-1−14 substrains can be summarized as follows: Integration of a MLV proviral copy into the mouse's germ line is stable, with the proviral genome being transmitted to all offspring as an autosomal locus in a Mendelian manner (Jaenisch et al. 1983). In some of these *Mov* substrains (1, 2, 3, 9, 13, and 14) the proviral genomes were activated, which resulted in synthesis and detection of new viral particles (viremia) in the mice. The timing of activation varied between these substrains, and it has been suggested that this was due to some developmentally regulated event initiating the expression of the proviral genome (Jaenisch et al. 1981). Whether activation occurs in all cells of a particular tissue or whether it is occurring randomly in some undefined group of cells is yet to be determined. With the other *Mov* substrains, no evidence has been found for activation of the proviral genome. However, in many of these substrains the proviral copy was found by recloning to be defective because of the occurrence of some mutation (Chumakov et al. 1982).

In one substrain (*Mov-13*) it was shown that the provirus had integrated into the α1(I) collagen gene, thus preventing its transcription (Harbers et al. 1984). When mice were made homozygous for the proviral integration, they died at or around 12 days of gestation (Schnieke, Harbers, and Jaenisch 1983), owing to rupture of the major blood vessels (Löhler, Timpl, and Jaenisch 1984). These results also demonstrate the usefulness of retroviral insertion into the mouse germ line as a potential method for generating insertional mutants (Jenkins et al. 1981; Copeland, Jenkins, and Lee 1983) (see Chapter 14). In addition, recent reports have described the introduction of retroviral vectors into preimplantation and postimplantation embryos. Van der Putten and associates (1985), Jähner and associates (1985), and Huszar and associates (1985) showed that the vectors could be transmitted to offspring, although the frequency was low and expression either did not occur or occurred infrequently. In contrast, similar vectors injected into postimplantation embryos were expressed in some tissues, when helper virus was present (Stuhlmann et al. 1984). However, we have recently obtained evidence that embryos infected with retroviral vectors carrying genes under the control of an internal thymidine kinase promoter express these genes stably in all tissues of the adult mouse and over many generations (C. Stewart et al. unpublished observations).

In summary, all three viruses described earlier fail to replicate or express their genes efficiently in early embryos. Embryonic cells would appear to be permissive for virus expression only when extensive differentiation has occurred either at or around the ninth day of gestation.

B. Analysis of viral gene expression in teratocarcinoma cell lines

In order to investigate the block to viral gene expression that exists in early embryonic cells, it was necessary to use teratocarcinoma cells. As in embryos, the three viruses replicate poorly in EC stem cells; however, they are expressed and undergo replication in the differentiated derivatives of the EC cells. The results from investigations with the three viruses will be briefly summarized, because extensive reviews on this issue have recently been published (Kelly and Condamine 1982; Levine 1982; Stewart 1984; Sleigh 1985).

i. **SV40 virus.** SV40 can readily penetrate and uncoat itself in EC cells (Swartzendruber and Lehman 1975). Although SV40 DNA can persist in the infected cell for about 7–10 days, there is no evidence that it can replicate (Friedrich and Lehman 1981). Similarly, no T antigen was detected in the stem cells, although SV40 is perfectly capable of infecting and expressing both large-T and small-t antigens in the differentiated derivatives (Swartzendruber and Lehman 1975; Huebner et al. 1983). At what level this inhibition of T-antigen expression occurs is controversial. Accumulation of SV40 early transcripts is greatly reduced in EC cells as compared with their differentiated derivatives. It has been suggested that this failure to express T antigen is due to unspliced or incorrectly spliced mRNA (Segal, Levine, and Khoury 1980). However, others have reported that they detected correctly spliced RNA, although at lower amounts than found in differentiated cells, and in an altered ratio of large-T mRNA to small-t mRNA transcripts (Linnenbach, Huebner, and Croce 1980, 1981; Huebner et al. 1983). These results suggested that the failure to express T antigen in stem cells was related to the lower level of transcripts present, but whether this was due to increased turnover of RNA or to a lower rate of transcription is not clear (Levine 1982). The lower levels of mRNA appear to have been correctly spliced; thus, the lack of expression may be due to additional controls that influence transcription and at present remain undefined.

ii. **Polyoma virus.** With few exceptions, polyoma virus is unable to replicate in EC cells (Dandolo, Aghion, and Blangy 1984). Transcription and expression of the polyoma genome in the stem cells are generally repressed by 1,000-fold as compared with infected mouse fibroblasts, although the transcripts are apparently correctly spliced (Fujimura et al. 1981a; Dandolo, Blangy, and Kamen 1983). However, polyoma mutants have been isolated from which expression of the T antigen can occur in infected EC cells. In all cases the mutations were mapped to the enhancer/promoter region 5′ to the start of the early region genes (Vasseur et al. 1980; Fujimura et al. 1981b; Sekikawa and Levine 1981; Levine 1982). The structure of these mutants is variable, because they involve point mutations, deletions, insertions, and duplications of segments within this region (Levine 1982). Some of these mutants function only in the line of EC cells from which they were isolated; most appeared to function by increasing the levels of stable transcripts that were correctly spliced and could accumulate in the infected cells (Dandolo et al. 1983). Precisely how these mutations overcome the block to expression is unclear, because they appear to affect not only transcription but also replication of the polyoma genome in the infected cells (Fujimura and Linney 1982).

iii. **Moloney murine leukemia virus.** MLV has been shown to efficiently infect and integrate into EC cells (Jähner et al. 1982; Stewart et al. 1982a; Gautsch and Wilson 1983; Niwa et al. 1983). In the infected EC cells, the accumulation of the proviral mRNA was 100-fold lower than levels in infected fibroblasts and differentiated derivatives of EC cells (Teisch et al. 1977; Stewart et al. 1982a; Gautsch and Wilson 1983). With these barely detectable levels, there was no evidence for production of infectious virus by EC cells (Teich et al. 1977; Stewart et al. 1982a). Differentiation of the infected EC cells did not result in significant virus expression, which is similar to the situation for infected preimplantation embryos that were allowed to undergo further development (Jähner et al. 1982; Stewart et al. 1982a). However, infection of differentiated EC derivatives (in analogy with postimplantation embryos) resulted in efficient virus production (Jähner et al. 1982; Stewart et al. 1982a). Thus, integration of the provirus into EC cell or preimplantation embryo chromosomes imposes some block to proviral DNA expression, and this block can be maintained when the infected cells are allowed to differentiate. This block is not confined to the pluripotent or totipotent cells of the mouse embryo, because PYS-2 cells (a differentiated cell line with characteristics in common with the parietal yolk sac endoderm) are likewise inefficient for proviral gene expression (Teich et al. 1977; Stewart et al. 1983b).

The nature of this block to expression is controversial. Stewart and associates (1982a) reported that the loss of expression of the provirus in the F9 line of EC cells could be correlated with de novo methylation of the proviral genomes. Similarly, all proviral copies in mice derived from infected preimplantation embryos were found to be highly methylated (Jähner et al. 1982). In experiments involving other EC cell lines it was reported that the onset of methylation was delayed when compared with the time of provirus inactivation (Gautsch and Wilson 1983; Niwa et al. 1983). All these experiments suggested that the absence of provirus expression was due to some inhibitory mechanisms, rather than lack of factors necessary for expression. This has been confirmed by transfecting genomic DNA from infected F9 cells into NIH 3T3 cells. The proviral genomes were not irreversibly inactivated, because treating the 3T3 fibroblasts with 5-azacytidine or iododeoxythymidine prior to or after transfection resulted in virus synthesis in the transfected cells (Stewart et al. 1982a; Gautsch and Wilson 1983; C. Stewart, unpublished observations). It remains to be determined if other inhibitory mechanisms also prevent expression of the proviral genomes in EC cells (Niwa et al. 1983) and what relationship, if any, exists between such inhibition and DNA methylation. Gorman, Rigby, and Lane (1985) recently reported that trans-acting regulatory factors may also inhibit transcription from unintegrated LTRs.

Other studies have shown that, in general, the LTR of wild-type MLV cannot function as an efficient promoter in EC cells. However, a hybrid MLV, containing mutant polyoma virus sequences in its LTR, was able to productively infect fibroblasts and mice (Linney et al. 1984; Davis, Linney, and Fan 1985). EC cells infected with the hybrid did not produce virus, although some viral proteins apparently were synthesized (Linney et al. 1984). When the hybrid LTR was linked to a gene (CAT) whose transcription could be easily measured, the hybrid LTR was able to function in transient expression studies in EC cells, whereas the wild-type LTR did not (Linney et al. 1984). The failure to obtain infectious virus in EC cells from an intact virus carrying a hybrid LTR suggested that translational or other posttranscriptional controls might also operate in the cells.

Under certain circumstances the wild-type 5' LTR can function in EC cells. Infection of EC cells with a vector containing the neomycin resistance gene, followed by selection, resulted in isolation, at a low frequency, of EC cell clones (in 10^{-5} to 10^{-3} cells, compared with an efficiency of 100% in fibroblasts). These clones contained a single copy of the vector, with the selectable gene being inefficiently transcribed from the LTR (Sorge et al. 1984; Wagner et al. 1984; Stewart, Vanek, and Wagner 1985; Taketo et al. 1985). Thus, the regulatory elements of the LTR were capable of functioning in EC cells, although only after selection was applied. This would therefore suggest that the block to retroviral expression can be overcome, possibly through the chromosomal site at which the vector integrates and thus through some cis-acting mechanism, although a mutation in the LTR cannot be excluded. It has also been shown recently that in retroviral vectors that carry an internal promoter sequence between the LTRs, the gene under its control can be expressed in most EC cells (Rubenstein, Nicolas, and Jacob 1984; Wagner et al. 1985b). Thus, inhibition of expression would appear to extend only to genes or sequences under the control of the LTR, although this has yet to be shown in the absence of selection.

C. Perspectives

Infections of early mouse embryos and EC cells with the three viruses described have demonstrated that the mechanisms regulating virus expression in early mouse embryonic cells differ from those existing in terminally differentiated cells. With all three viruses, accumulation of stable transcripts from their genomes is reduced in EC cells. In polyoma virus and also in MLV this would appear to be due to insufficient levels of transcription being initiated from the promoter/enhancer sequences, although with polyoma a further complicating factor is the intimate relationship between expression of the early part of the genome and replication of the virus. In cells infected with SV40 virus, accumulation of stable mRNA is also diminished. Whether this is due to instability of the mRNAs, insufficient transcription, or additional posttranscriptional as well as translational controls is not yet clear.

One of the more revealing features of these studies is that posttranscriptional controls may be equally as important as transcription in obtaining correct expression of a gene within these cells. This aspect should be particularly emphasized, because studies on adenovirus-infected (Ad5 strain) F9 cells showed striking evidence for controls other than transcription affecting expression (Cheng and Praskier 1982). Whether or not such controls also operate on the expression of endogenous genes during differentiation is not known. When cloned copies of the mouse H-2K antigen, alpha-fetoprotein (AFP), and chicken δ-crystallin genes were introduced into EC cells (Kondoh, Takahashi, and Okada 1984; Rosenthal et al. 1984; Scott et al. 1984), expression occurred infrequently, but coincided with the onset of differentiation.

A further intriguing result from infecting embryos, especially with MLV, has been the discovery of the de novo methylation activity in these cells. The role of an efficient de novo methylation mechanism that usually completely methylates proviral DNA, and in some situations DNA injected into embryos, remains a mystery, although some provocation theories have been proposed (Jaenisch and Jähner 1984; Jähner and Jaenisch 1985). The observation that methylation occurred in pluripotent or totipotent cells of the embryo suggested that it might be a characteristic of the developmental state of these cells. Jaenisch and Jähner (1984) proposed that de novo methylation is involved in regulating gene expression, especially that of proviruses, because all *Mov* strains that lacked expression of the provirus had high levels of methylation of the viral genome. However, clones derived from the P19 EC cell line carrying a single vector copy express retroviral sequences despite being highly methylated in the LTR as well as in gene coding sequences (C. Stewart and E. Wagner, unpublished observations). In addition, the discovery that the PYS-2 differentiated cell line could also carry out de novo methylation of proviral copies showed that de novo methylation is a more general property of early embryonic cells. Thus, a direct role for methylation in controlling the expression of provirus during differentiation has not been conclusively demonstrated.

Finally, another result that has emerged from these studies is the efficient block to expression imposed by the early embryonic cells. This is particularly striking for MLV and SV40, because the inhibition persists throughout differentiation of the infected cells and also into the adult life of the mouse (Friedrich and Lehman 1981; Jaenisch et al. 1981; Willison et al. 1983). In both cases, infectious virus has been recovered from the silent copies, thereby showing that the block to expression is not irreversible (Friedrich and Lehman 1981; Harbers, Jähner, and Jaenisch 1981b; Willison et al. 1983). In some *Mov* substrains, virus activation occurred spontaneously, perhaps in a small subpopulation of cells, such as the circulating lymphocytes (Jaenisch and Berns 1977). The reason why activation occurs at all is not clear, and further studies on the inactivation mechanism should be useful for investigating such important developmental events as the switching off of genes during cell differentiation and proliferation and in X-chromosome inactivation.

V. Manipulation of mouse genome using embryonic stem cells

The recent demonstration that cell lines closely resembling EC cells could be readily isolated from explanted blastocysts has rekindled interest in using embryonic cells as a potential means to manipulate the mouse genome. These cells, called embryonic stem (ES) cells (Evans and Kaufman 1981; Martin 1981), have been shown to colonize the mouse germ line at a high frequency (Bradley et al. 1984; Wagner et al. 1985a). This is in contrast to the results obtained from established EC cell lines, in which the frequency is very low (Mintz et al. 1975, Stewart and Mintz 1981). This section will briefly review the approaches to manipulating the mouse genome that are offered by these EC and ES cells. A more detailed description of their isolation, culture, and cellular characteristics can be found in Chapter 15.

The interest aroused by these cells is still largely due to their potential as an alternative means of altering the mouse genome, compared with DNA injection or virus infection of embryos. The potential advantage of this system is that these cells can be cultured and exposed to selective conditions while maintaining their pluripotentiality. Any method open to somatic cell genetics for altering the mouse genome could be applied to the cells, providing, of course, that altered cells are still able to colonize the germ line. No one has yet reported isolation of a new strain of mice derived from a genetically altered EC or ES cell line.

Some of the methods that are available to manipulate the genome in these cells in vitro are shown in Figure 16.5. However, only a few of these methods have been used to isolate genetically altered cells in vitro with attempts to form chimeras. These will be briefly outlined here; a more detailed review has recently been published (Stewart 1984).

A. Mutant EC cell lines

EC cells defective in certain pathways of nucleotide metabolism have been isolated, namely, those with deficiency in HPRT activity (Dewey et al. 1977) and in TK activity. The aim of these experiments was to generate mice that would also lack these enzymes and thus would serve as useful models for genetically inherited diseases in humans, such as Lesch-Nyhan syndrome in patients lacking HPRT (Kelley and Wyngaar 1978). Chimeric animals have been produced in a number of laboratories; however, none of the chimeras transmitted the mutant genotype to their offspring (Dewey et al. 1977; Illmensee, Hoppe, and Croce 1978; Stewart 1982; Fujii and Martin 1983). Another report described a chloramphenicol-

Methods Available for Altering Mouse Genome in Vitro

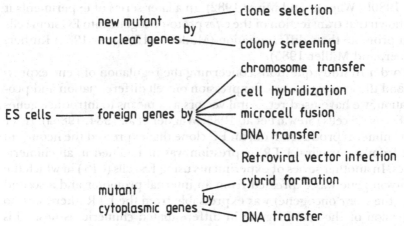

Figure 16.5. Diagram of the possibilities available for altering the ES genome in vitro.

resistant EC cell line that was used to form chimeras, but no evidence was presented that the chimeric tissues had retained the chloramphenicol-resistant phenotype (Watanabe, Dewey, and Mintz 1978). A mouse carrying such a mitochondrial trait would be useful in studying cytoplasmic inheritance.

B. Gene transfer into EC cell lines

Illmensee and associates investigated the feasibility of introducing new genes into EC cells by chromosome-mediated gene transfer. An initial report claimed that a human chromosome was transferred into an EC cell line that was able to form chimeras; it was claimed that a weak expression of the human galactokinase enzyme was detectable in these mice (Illmensee et al. 1978). Subsequent experiments were performed using mouse × rat hybrid cells, and evidence was presented that some rat proteins (albumin and glycerol phosphate dehydrogenase) that were not expressed in stem cells were expressed in the liver in the chimeras (Illmensee and Croce 1978; Duboule et al. 1982a, 1982b).

Isolation and development of dominant selectable markers such as the guanine phosphoribosyltransferase gene (*Ecogpt*) (Mulligan and Berg 1981) and the neomycin resistance gene (Southern and Berg 1982) have greatly simplified the methods by which new genetic information can be transferred into cells by DNA-mediated gene transfer (Pellicer et al. 1980a) by chromosome-mediated gene transfer (Tunnacliffe et al. 1983), or by retroviral vectors (Mann, Mulligan, and Baltimore 1983; Wagner et al. 1985b). Early studies on DNA-mediated gene transfer with EC cells showed that genes could be introduced and expressed in these cells. Expression of the exogenous genes was also maintained during differentiation of these cells in vitro and during tumor formation in vivo (Pellicer et al. 1980b; Wagner and Mintz 1982). In a later series of experiments it was shown that transfection of the c-*fos* proto-oncogene into EC stem cells could promote their differentiation (Müller and Wagner 1984; Rüther, Wagner, and Müller 1985).

In order to study questions concerning the regulation of gene expression and the consequences of expression on cell differentiation and proliferation, we have used retroviral vectors as a means to introduce genes into EC or ES cells (Stewart et al. 1985; Wagner et al. 1984, 1985a). In 11 adult chimeras produced from an EC clone that expressed the neomycin gene from a retroviral LTR, expression was maintained in all chimeric tissues. In another series of experiments using ES cells (CP1) in which the neomycin gene was expressed from an internal promoter and a second gene (the v-*myc* oncogene) was expressable from the LTR, there was no expression of the second gene in differentiated chimeric tissues. This suggested that efficient activation of a gene under the control of an LTR generally does not occur during differentiation.

These experiments demonstrated that one can isolate embryonic stem cell clones and characterize them with respect to copy number and expression of the introduced vector. They also showed that selection did not eliminate the ability of the cells to form chimeras, nor their ability to colonize a wide variety of different tissues (Stewart et al. 1985). None of the chimeras produced offspring carrying the vector. This may be due to the clones having an abnormal karyotype or some other lesion prior to infection.

VI. Conclusions

The study of developmental genetics in the mouse is still descriptive, and molecular analysis is at an early stage. In the *Drosophila* system, for example, there are large numbers of developmental mutants and an efficient means of introducing genes into the germ line, based on the P-element transposon (Rubin and Spradling 1982). Genes introduced into the fly are expressed not only correctly in a tissue- and temporal-specific manner but also in a quantitatively correct way (Goldberg, Posakony, and Maniotis 1983). With this system it is now possible to isolate and clone developmental genes, mutate or alter them in vitro, and then reintroduce them back into the fly, where their effects on development can be studied (Bonner et al. 1984). This is proving to be an extremely powerful approach for unraveling the control elements of genes, their functions, and how they interact with each other to coordinate development.

In mammalian systems, by contrast, there are relatively few mutants known that affect early development, and it is not clear if they play any role in regulating developmental decisions (by analogy with the bithorax complex in *Drosophila* or the lineage mutants in the nematode *C. elegans*). Furthermore, the majority of the mutants that have been identified have not yet been cloned in the mouse (see Chapter 14 for further discussion of this subject). One exception to this situation in which transgenic mice may have an impact is in manipulation of the immune system. Here, the genetic diversity, together with the well-characterized polymorphisms of MHC molecules and immunoglobulins, can be exploited in transgenic mice to further unravel complex cellular interactions.

Transgenic mice produced by DNA injection have also been important in identifying cis-acting DNA sequences that are involved in regulating correct tissue-specific and temporal expression of a few genes, most notably for pancreatic elastase. The use of fusion genes combining the structural elements of one gene with the regulatory elements of another has been important, as in investigating the roles that oncogenes or transforming genes play in initiating malignancy.

Viruses have been used extensively to study regulation of gene expression in early development. With this approach it has been shown that for at least three well-studied viruses (SV40, polyoma, and MLV) there is an apparent lack of expression in the stem cells as compared with their differentiated derivatives. It has, however, proved to be difficult to determine the molecular basis of this difference in expression because of the diversity of controls that are operating. Whether this is a true reflection of how cells control their own genes during differentiation or is only relevant to the viruses is not yet clear. This should not, however, completely detract from viruses, because they have been instrumental in uncovering some features of gene regulation for which a proper understanding is still lacking. For instance, we have yet to determine the role of de novo methylation in regulating gene expression as well as the replication of SV40 and polyoma as episomes in embryonic cells. A better understanding of MLV expression in embryonic cells will also help in the design of efficient retroviral vectors. These can be used either to insert single-copy genes into the mouse germ line and somatic tissues or as possible insertional mutagens that tag the affected gene.

Embryonic stem cells provide an alternative and useful complement to the approaches discussed earlier for analyzing gene expression in vitro and in vivo. These cells can be used for isolating mutants or well-characterized clones carrying exogenous DNA. Whether or not these induced changes can be introduced into the germ line of the mouse remains to be seen. If homologous recombination could be achieved for the purpose of correct insertion (Smithies et al. 1985) or inactivation of genes (King et al. 1985), it would also be a powerful tool for studying gene organization and regulation in higher eukaryotic cells. In addition, these cells should be useful to determine the molecular organization of the totipotent genome and how it changes when the cells are stimulated to differentiate along different developmental pathways.

Note in proof. A recent report has indicated that the neomycin-resistance gene can be stably transmitted into the germ-line after introduction of neo-resistant ES cells into chimeras by blastocyst injection (Gossler, Deotschman, Korn, Serfling, and Kemler in press).

Acknowledgments

We thank Ulrich Rüther for helpful criticism and Ines Benner for her enormous speed and efficiency in helping us to prepare the manuscript, which otherwise would never have been finished on time.

References

Abramczuk, J., Vorbrodt, A., Solter, D., and Koprowski, H. (1978) Infection of mouse preimplantation embryos with simian virus 40 and polyoma virus. *Proc. Natl. Acad. Sci. U.S.A.* 75, 999–1003.

Babinet, C., Farza, H., Morello, D., Hadchouel, M., and Pourcel, C. (1985) Specific expression of hepatitis B surface antigen (HBsAg) in transgenic mice. *Science* 230, 1160–3.

Bonner, J. J., Parks, C., Parker-Thornburg J., Martin, M. A., and Pelham, H. R. B. (1984) The use of promoter fusions in *Drosophila* genetics: isolation of mutations affecting the heat shock response. *Cell* 37, 979–91.

Bradley, A., Evans, M., Kaufman, M. H., and Robertson, E. (1984) Formation of germ line chimaeras from embryo derived teratocarcinoma cell lines. *Nature* 309, 255–6.

Brinster, R. L., Chen, H. Y., Messing, A., Van Dyke, T., Levine, A. J., and Palmiter, R. D. (1984) Transgenic mice harboring SV40 T-antigen genes develop characteristic brain tumors. *Cell* 37, 367–79.

Brinster, R. L., Chen, H. Y., Trumbauer, M., Denear, A. W., Warren, R., and Palmiter, R. D. (1981) Somatic expression of herpes thymidine kinase in mice following injection of a fusion gene into eggs. *Cell* 27, 223–31.

Brinster, R. L., Chen, H. Y., Trumbauer, M. E., Yagle, M. K., and Palmiter, R. D. (1985) Factors affecting the efficiency of introducing foreign DNA into mice by microinjecting eggs. *Proc. Natl. Acad. Sci. U.S.A.* 82, 4438–42.

Brinster, R. L., Ritchie, K. A., Hammer, R. E., O'Brien, R. L., Arp, B., and Storb, U. (1983) Expression of a microinjected immunoglobin gene in the spleen of transgenic mice. *Nature* 306, 332–6.

Chada, K., Magram, J., Raphael, K., Radice, G., Lacy, E., and Costantini, F. (1985) Specific expression of a foreign β-globin gene in erythroid cells of transgenic mice. *Nature* 314, 377–80.

Cheng, C., and Praskier, J. (1982) Regulation of type 5 adenovirus replication in murine teratocarcinoma cell lines. *Virology* 123, 45–59.

Chisari, F. V., Pinkert, C. A., Milich, D. R., Filippi, P., McLachlan, A., Palmiter, R. D., and Brinster, R. L. (1985) A transgenic mouse model of the chronic hepatitis B surface antigen carrier state. *Science* 230, 1157−60.

Chumakov, I., Stuhlmann, H., Harbers, K., and Jaenisch, R. (1982) Cloning of two genetically transmitted Moloney leukaemia proviral genomes: correlation between biological activity of the cloned DNA and viral genome activation in the animal. *J. Virol.* 42, 1088−98.

Copeland, N. G., Jenkins, N. A., and Lee, B. K. (1983) Association of the lethal yellow (A^y) coat color mutation with an ecotropic murine leukaemia virus genome. *Proc. Natl. Acad. Sci. U.S.A.* 80, 247−9.

Costantini, F., and Lacy, E. (1981) Introduction of a rabbit β-globin gene into the mouse germ line. *Nature* 294, 92−4.

Dandolo, L., Aghion, J., and Blangy, D. (1984) T-antigen-independent replication of polyomavirus DNA in murine embryonal carcinoma cells. *Mol. Cell. Biol.* 4, 317−23.

Dandolo, L., Blangy, D., and Kamen, R. (1983) Regulation of polyoma virus transcription in murine embryonal carcinoma cells. *J. Virol.* 47, 55−64.

Davis, B., Linney, E., and Fan, H. (1985) Suppression of leukaemia virus pathogenicity by polyoma virus enhancer. *Nature* 314, 550−3.

Dewey, M. J., Martin, D. W., Martin, G. R., and Mintz, B. (1977) Mosaic mice with teratocarcinoma derived mutant cells deficient in hypoxanthine phosphoribosyl transferase. *Proc. Natl. Acad. Sci. U.S.A.* 74, 5564−8.

Duboule, D., Croce, C. M., and Illmensee, K. (1982a) Tissue preference and differentiation of malignant rat × mouse hybrid cells in chimaeric mouse fetuses. *EMBO Journal* 1, 1595−603.

Duboule, D., Petzoldt, U., Illmensee, G. R., Croce, C. M., and Illmensee, K. (1982b) Protein synthesis in hybrid cells derived from fetal rat × mouse chimaeric organs. *Differentiation* 23, 145−52.

Eicher, E. M., and Beamer, W. G. (1976) Inherited ateliotic dwarfism in mice: characteristics of the mutation, *little*, on chromosome 6. *J. Hered.* 67, 87−91.

Evans, M. J., and Kaufman, M. H. (1981) Establishment in culture of pluripotential cells from mouse embryos. *Nature* 292, 154−6.

Frels, W. I., Bluestone, J. A., Hodes, R. J., Capecchi, M. R., and Singer, D. S. (1985) Expression of a microinjected porcine class I major histocompatibility complex gene in transgenic mice. *Science* 228, 577−80.

Friedrich, T. D., and Lehman, J. M. (1981) The state of simian virus 40 DNA in the embryonal carcinoma cells of the murine teratocarcinoma. *Virology* 110, 159−66.

Fujii, J. T., and Martin, G. R. (1983) Developmental potential of teratocarcinoma stem cells in utero following aggregation with cleavage stage embryos. *J. Embryol. Exp. Morphol.* 74, 79−86.

Fujimura, F. K., Deininger, P. L., Friedmann, T., and Linney, E. (1981a) Mutation near the polyoma DNA replication origin permits productive infection of F9 embryonal carcinoma cells. *Cell* 23, 809−14.

Fujimura, F. K., Gilbert, P. E., Eckhart, W., and Linney, E. (1981b) Polyoma virus infection of retinoic acid-induced differentiated teratocarcinoma cells. *J. Virol.* 39, 36−12.

Fujimura, F., and Linney, E. (1982) Polyoma mutants that productively infect

F9 embryonal carcinoma cells do not rescue wild-type polyoma in F9 cells. *Proc. Natl. Acad. Sci. U.S.A.* 79, 1479–83.

Gautsch, J. W., and Wilson, M. C. (1983) Delayed de novo methylation in teratocarcinoma cells suggests additional tissue specific mechanisms for controlling gene expression. *Nature* 301, 32–5.

Goldberg, D. A., Posakony, J. W., and Maniatis, T. (1983) Correct developmental expression of a cloned alcohol dehydrogenase gene transduced into the *Drosophila* germ line. *Cell* 34, 59–73.

Gordon, J. W., and Ruddle, F. H. (1983) Gene transfer into mouse embryos: production of transgenic mice by pronuclear injection. *Methods Enzymol.* 101, 411–33.

– (1985) DNA-mediated genetic transformation of mouse embryos and bone marrow – a review. *Gene* 33, 121–36.

Gordon, J. W., Scangos, G. A., Plotkin, D. J., Barbosa, J. A., and Ruddle, F. H. (1980) Genetic transformation of mouse embryos by microinjection of purified DNA. *Proc. Natl. Acad. Sci. U.S.A.* 77, 7380–4.

Gorman, C. M., Rigby, P. W. J., and Lane, D. P. (1985) Negative regulation of viral enhancers in undifferentiated embryonic stem cells. *Cell* 42, 519–26.

Gossler, A., Doetschman, T., Korn, R., Serfling, E., and Kemler, R. (in press) *Proc. Natl. Acad. Sci. U.S.A.*

Grosschedl, R., Weaver, D., Baltimore, D., and Costantini, F. (1984) Introduction of a μ-immunoglobulin gene into the mouse germ line: specific expression in lymphoid cells and synthesis of functional antibody. *Cell* 38, 647–58.

Hammer, R. E., Brinster, R. L., Rosenfeld, M. G., Evans, R. M., and Mayo, K. E. (1985a) Expression of human growth factor in transgenic mice results in increased somatic growth. *Nature* 315, 413–16.

Hammer, R. E., Palmiter, R. D., and Brinster, R. L. (1984) Partial correction of murine hereditary growth disorder by germ line incorporation of a new gene. *Nature* 311, 65–7.

Hammer, R., Pursel, V. G., Rexroad, C. E., Jr., Wall, R. J., Bolt, D. J., Ebert, K. M., Palmiter, R. D., and Brinster, R. L. (1985b) Production of transgenic rabbits, sheep and pigs by microinjection. *Nature* 315, 680–3.

Hanahan, D. (1985) Heritable formation of pancreatic β-cell tumors in transgenic mice expressing recombinant insulin/SV40 oncogenes. *Nature* 315, 115–22.

Harbers, K., Jähner, D., and Jaenisch, R. (1981a) Microinjection of cloned retroviral genomes into mouse zygotes: integration and expression in the animal. *Nature* 293, 540–2.

Harbers, K., Kuehn, M., Delius, H., and Jaenisch, R. (1984) Insertion of retrovirus into the first intron of α1(I) collagen gene leads to embryonic lethal mutation in mice. *Proc. Natl. Acad. Sci. U.S.A.* 81, 1504–8.

Harbers, K., Schnieke, A., Stuhlman, H., Jähner, D., and Jaenisch, R. (1981b) DNA methylation and gene expression: endogenous retroviral genome becomes infectious after molecular cloning. *Proc. Natl. Acad. Sci. U.S.A.* 78, 7609–13.

Huebner, K., Linnenbach, A., Ghosh, P. K., Rushdi, A., Romanczuk, H., Tsuchida, N., and Croce, C. M. (1983) Tumor virus genomes in DNA-transformed F9 cells. In *Cold Spring Harbor Conference on Cell Proliferation*,

Vol. 10, eds. L. M. Silver, G. R. Martin, and S. Strickland, pp. 343–61. Cold Spring Harbor, N.Y.: Cold Spring Harbor Laboratory.

Huszar, D., Balling, R., Kothary, R., Magli, M. C., Hozumi, N., Rossant, J., and Bernstein, A. (1985) Insertion of a bacterial gene into the mouse germline using an infectious retrovirus vector. *Proc. Natl. Acad. Sci. U.S.A.* 82, 8587–91.

Hwang, L., and Gilboa, E. (1984) Expression of genes introduced into cells by retroviral infection is more efficient than that of genes introduced into cells by DNA transfection. *J. Virol.* 50, 417–24.

Illmensee, K., and Croce, C. M. (1979) Xenogenic gene expression in chimaeric mice derived from rat-mouse hybrid cells. *Proc. Natl. Acad. Sci. U.S.A.* 76, 879–83.

Illmensee, K., Hoppe, P. C., and Croce, C. M. (1978) Chimaeric mice derived from human-mouse hybrid cells. *Proc. Natl. Acad. Sci. U.S.A.* 75, 1914–8.

Jaenisch, R. (1974) Infection of mouse blastocysts with SV40 DNA: normal development of the infected embryos and persistence of SV40 specific DNA sequences in the adult animals. *Cold Spring Harbor Symp. Quant. Biol.* 39, 375–80.

Jaenisch, R., and Berns, A. (1977) Tumour virus expression during mammalian embryogenesis. In *Concepts in Mammalian Embryogenesis,* ed. M. I. Sherman, pp. 267–314. Cambridge: M.I.T. Press.

Jaenisch, R., Fan, H., and Croker, B. (1975) Infection of preimplantation mouse embryos and of newborn mice with leukaemia virus: tissue distribution of viral DNA and RNA and leukemogenesis in the adult animal. *Proc. Natl. Acad. Sci. U.S.A.* 72, 4008–12.

Jaenisch, R , Harbers, K., Schnieke, A., Löhler, J., Chumakov, I., Jähner, D., Grotkopp, D., and Hoffman, E. (1983) Germline integration of Moloney murine leukaemia virus at the *Mov13* locus leads to recessive lethal mutation and early embryonic death. *Cell* 32, 209–16.

Jaenisch, R., and Jähner, D. (1984) Methylation, expression and chromosomal position of genes in mammals. *Biochim. Biophys. Acta* 782, 1–9.

Jaenisch, R., Jähner, D., Nobis, P., Simon, I., Löhler, J., Harbers, K., and Grotkopp, D. (1981) Chromosomal position and activation of retroviral genomes inserted into the germ line of mice. *Cell* 24, 519–29.

Jaenisch, R., and Mintz, B. (1974) Simian virus 40 DNA sequences in DNA of healthy adult mice derived from preimplantation blastocysts injected with viral DNA. *Proc. Natl. Acad. Sci. U.S.A.* 71, 1250–4.

Jähner, D., Haase, K., Mulligan, R., and Jaenisch, R. (1985) Insertion of the bacterial *gpt* gene into the germ line of mice by retroviral infection. *Proc. Natl. Acad. Sci. U.S.A.* 82, 6927–31.

Jähner, D., and Jaenisch, R. (1985) Chromosomal position and specific demethylation in enhancer sequences of germ line-transmitted retroviral genomes during mouse development. *Mol. Cell. Biol.* 5, 2212–20.

Jähner, D., Stuhlmann, H., Stewart, C. L., Harbers, K., Löhler, J., Simon, I., and Jaenisch, R. (1982) De novo methylation and expression of retroviral genomes during mouse embryogenesis. *Nature* 298, 623–8.

Jenkins, N. A., Copeland, N. G., Taylor, B. A., and Lee, B. K. (1981) Dilute (*d*) coat color mutation of DBA/2J mice is associated with the site of integration of an ectotropic MuLV genome. *Nature* 293, 370–4.

Kelly, F., and Condamine, H. (1982) Tumor viruses and early mouse embryos. *Biochim. Biophys. Acta* 651, 105–41.

Kelley, W. N., and Wyngaarden, J. B. (1978) The Lesch-Nyhan syndrome. In *The Metabolic Basis of Inherited Disease*, Vol. 4, eds. J. B. Stanbury, J. B. Wyngaarden, and D. S. Fredrickson, pp. 141–66. New York: McGraw-Hill.

King, W., Patel, M. D., Lobel, L. I., Goff, S. P., and Nguyen-Huu, M. C. (1985) Insertion mutagenesis of embryonal carcinoma cells by retroviruses. *Science* 228, 554–8.

Kondoh, H., Takahashi, Y., and Okada, T. S. (1984) Differentiation dependent expression of the chicken δ crystallin gene introduced into mouse teratocarcinoma stem cells. *EMBO Journal* 9, 2009–14.

Krumlauf, R., Hammer, R., Tilghman, S., and Brinster, R. L. (1985) Developmental regulation of alphafoetoprotein genes in transgenic mice. *Mol. Cell. Biol.* 5, 1639–48.

Lacy, E., Roberts, S., Evans, E. P., Burtenshaw, M. D., and Costantini, F. (1983) A foreign β-globin gene in transgenic mice: integration at abnormal chromosomal positions and expression in inappropriate tissues. *Cell* 34, 343–58.

LeMeur, M., Gerlinger, P., Benoist, C., and Mathis, D. (1985) Correcting an immune response deficiency by creating E_α gene transgenic mice. *Nature* 316, 38–42.

Levine, A. J. (1982) The nature of the host range restriction of SV40 and polyoma viruses in embryonal carcinoma cells. *Curr. Top. Microbiol. Immunol.* 101, 1–30.

Linnenbach, A., Huebner, K., and Croce, C. M. (1980) DNA transformed murine teratocarcinoma cells: regulation of expression of simian virus 40 tumor antigen in stem versus differentiated cells. *Proc. Natl. Acad. Sci. U.S.A.* 77, 4875–9.

– (1981) Transcription of the simian virus 40 genome in DNA-transformed murine teratocarcinoma stem cells. *Proc. Natl. Acad. Sci. U.S.A.* 78, 6386–90.

Linney, E., David, B., Overhauser, J., Chao, E., and Fan, H. (1984) Nonfunction of a Moloney murine leukaemia virus regulatory sequence in F9 embryonal carcinoma cells. *Nature* 308, 470–2.

Löhler, J., Timpl, R., and Jaenisch, R. (1984) Embryonic lethal mutation in mouse collagen I gene causes rupture of blood vessels and is associated with erythropoietic and mesenchymal cell death. *Cell* 38, 597–607.

Low, M. J., Hammer, R. E., Goodman, R. H., Habener, J. F., Palmiter, R. D., and Brinster, R. L. (1985) Tissue-specific posttranslational processing of pre-prosomatostatin encoded by a metallothionein-somatostatin fusion gene in transgenic mice. *Cell* 41, 211–19.

McKnight, G. S., Hammer, R. E., Kuenzel, E. A., and Brinster, R. L. (1983) Expression of the chicken transferrin gene in transgenic mice. *Cell* 34, 335–41.

Magram, J., Chada, K., and Costantini, F. (1985) Developmental regulation of a cloned adult β-globin gene in transgenic mice. *Nature* 315, 338–40.

Maltzman, W., and Levine, A. J. (1981) Viruses as probes for development and differentiation. *Adv. Virus. Res.* 26, 65–117.

Mann, R., Mulligan, R. C., and Baltimore, D. (1983) Construction of a retrovirus packaging mutant and its use to produce helper-free defective retrovirus. *Cell* 33, 153–9.

Martin, G. R. (1981) Isolation of a pluripotent cell line from early mouse embryos cultured in medium conditioned by teratocarcinoma stem cells. *Proc. Natl. Acad. Sci. U.S.A.* 78, 7634–6.

Messing, A., Chen, H. Y., Palmiter, R. D., and Brinster, R. L. (1985) Peripheral neuropathies, hepatocellular carcinomas, and islet cell adenomas in transgenic mice. *Nature* 316, 461–3.

Mintz, B., Illmensee, K., and Gearhart, J. D. (1975) Developmental and experimental potentialities of mouse teratocarcinoma cells from embryoid body cores. In *Teratomas and Differentiation*, eds. M. I. Sherman and D. Solter, pp. 59–82. New York: Academic Press.

Müller, R., and Wagner, E. F. (1984) Differentiation of F9 teratocarcinoma stem cells after transfer of c-*fos* proto-oncogenes. *Nature* 311, 438–42.

Mulligan, R. C., and Berg, P. (1981) Selection for animal cells that express the *Escherichia coli* gene coding for xanthine-guanine phosphoribosyl transferase. *Proc. Natl. Acad. Sci. U.S.A.* 78, 2072–6.

Niwa, O., Yokota, Y., Ishida, H., and Sugahara, T. (1983) Independent mechanisms involved in suppression of the Moloney leukaemia virus genome during differentiation of murine teratocarcinoma cells. *Cell* 22, 1105–13.

Ornitz, D. M., Palmiter, R. D., Hammer, R. E., Brinster, R. L., Swift, G. H., and McDonald, J. R. (1985) Specific expression of an elastase–human growth hormone fusion gene in pancreatic acinar cells of transgenic mice. *Nature* 313, 600–2.

Overbeek, P. A., Chepelinsky, A., Khillan, J. S., Piatigorsky, J., and Westphal, H. (1985) Lens-specific expression and developmental regulation of the bacterial chloramphenicol acetyltransferase gene driven by the murine αA-crystallin promoter in transgenic mice. *Proc. Natl. Acad. Sci. U.S.A.* 82, 7815–19.

Palmiter, R. D., and Brinster, R. L. (1985) Transgenic mice. *Cell* 14, 343–5.

Palmiter, R. D., Brinster, R. L., Hammer, R. E., Trumbauer, M.E., Rosenfeld, M. G., Birnberg, N. C. and Evans, R. M. (1982a) Dramatic growth of mice that develop from eggs microinjected with metallothionein-growth hormone fusion genes. *Nature* 300, 611–15.

Palmiter, R. D., Chen, H. Y., and Brinster, R. H. (1982b) Differential regulation of methallothionein–thymidine kinase fusion gene in transgenic mice and their offspring. *Cell* 29, 701–10.

Palmiter, R. D., Chen, H. Y., Messing, A., and Brinster, R. L. (1985) SV40 enhancer and large-T antigen are instrumental in development of choroid plexus tumours in transgenic mice. *Nature* 36, 457–60.

Palmiter, R. D., Norstedt, G., Gelinas, R. E., Hammer, R. E., and Brinster, R. L. (1983) Metallothionein–human GH fusion genes stimulate growth of mice. *Science* 222, 809–14.

Palmiter, R. D., Wilkie, T. M., Chen, H. Y., and Brinster, R. L. (1984) Transmission distortion and mosaicism in an unusual transgenic mouse pedigree. *Cell* 36, 869–77.

Papaioannou, V. E., and Dieterlen-Lievre, F. (1984) Making chimaeras. In *Chimaeras in Developmental Biology*, eds. N. Le Douarin and A. McLaren, pp. 3–37. New York: Academic Press.

Pellicer, A., Robins, D., Wold, B., Sweet, R., Jackson, J., Lowy, I., Roberts, J. M., Sim, G. K., Silverstein, S., and Axel, R. (1980a) Altering genotype and phenotype by DNA-mediated gene transfer. *Science* 209, 1414–22.

Pellicer, A., Wagner, E. F., El Kareh, A., Dewey, M. J., Reuser, A. J., Silverstein, S., Axel, R., and Mintz, B. (1980b) Introduction of a viral thymidine kinase gene and the human β-globin gene into developmentally multipotential mouse teratocarcinoma cells. *Proc. Natl. Acad. Sci. U.S.A.* 77, 2098–102.

Pinkert, C. A., Widera, G., Cowing, C., Heber-Katz, E., Palmiter, R. D., Flavell, R. A., and Brinster, R. L. (1985) Tissue-specific, inducible and functional expression of the E $_\alpha^d$ MHC class II gene in transgenic mice. *EMBO Journal* 4, 2225–30.

Rosenthal, A., Wright, S., Cedar, H., Flavell, R., and Grossveld, F. (1984) Regulated expression of an introduced MHC H-2K^{bml} gene in murine embryonal carcinoma cells. *Nature* 310, 415–17.

Ross, S. R., and Solter, D. (1985) Glucocorticoid regulation of mouse mammary tumor virus sequences in transgenic mice. *Proc. Natl. Acad. Sci. U.S.A.* 82, 5880–4.

Rubenstein, J. L. R., Nicolas, J.-F., and Jacob, F. (1984) Construction of a retrovirus capable of transducing and expressing genes in multipotential embryonic cells. *Proc. Natl. Acad. Sci. U.S.A.* 81, 7137–40.

Rubin, G. M., and Spradling, A. C. (1982) Transposition of cloned P elements into *Drosophila* germ line chromosomes. *Science* 218, 348–53.

Rusconi, S. (1984) Gene transfer in living organisms. In *The Impact of Gene Transfer Techniques in Eucaryotic Cell Biology*, eds. J. S. Schell and P. Starlinger, pp. 134–52. Berlin: Springer-Verlag.

Rusconi, S., and Köhler, G. (1985) Transmission and expression of a specific pair of rearranged immunoglobin μ and κ genes in a transgenic mouse line. *Nature* 314, 330–4.

Rüther, U., Wagner, E. F., and Müller, R. (1985) Analysis of the differentiation-promoting potential of inducible c-*fos* genes introduced into embryonal carcinoma cells. *EMBO Journal* 4, 1775–81.

Schnieke, A., Harbers, K., and Jaenisch, R. (1983) Embryonic lethal mutation in mice introduced by retrovirus insertion into the α1(I) collagen gene. *Nature* 304, 315–20.

Scott, R. W., Vogt, T. F., Croke, M. E., and Tilghman, S. M. (1984) Tissue-specific activation of a cloned α-fetoprotein gene during differentiation of a transfected embryonal carcinoma cell line. *Nature* 310, 562–7.

Segal, S., Levine, A. J., and Khoury, G. (1980) Evidence for non-spliced SV40 RNA in undifferentiated murine teratocarcinoma stem cells. *Nature* 20, 335–7.

Sekikawa, K., and Levine, A. J. (1981) Isolation and characterization of polyoma host range mutants that replicate in nulli-potential embryonal carcinoma cells. *Proc. Natl. Acad. Sci. U.S.A.* 78, 1100–4.

Shani, M. (1985) Tissue-specific expression of rat myosin light-chain 2 gene in transgenic mice. *Nature* 314, 283–6.

Sleigh, M. J. (1985) Virus expression as a probe of regulatory events in early mouse embryogenesis. *Trends Genet.* 1, 17–21.

Small, J. A., Blair, D. G., Showalter, S. D., and Scangos G. A. (1985) Analysis of a transgenic mouse containing SV40 and v-*myc* sequences. *Mol. Cell. Biol.* 5, 642–8.

Smithies, O., Gregg, R. G., Boggs, S. S., Koralewski, M. A., and Kucherlapati, R. S. (1985) Insertion of DNA sequence into the human chromosomal β-globin locus by homologous recombination. *Nature* 317, 230–4.

Snow, M. H. L. (1973) Abnormal development of pre-implantation mouse embryos grown *in vitro* with ^{3}H-thymidine. *J. Embryol. Exp. Morphol.* 29, 601–15.

Sorge, J., Cutting, A. E., Erdman, V. D., and Gautsch, J. W. (1984) Integration-specific retrovirus expression in embryonal carcinoma cells. *Proc. Natl. Acad. Sci. U.S.A.* 81, 6627–32.

Southern, P. J., and Berg, P. (1982) Transformation of mammalian cells to antibiotic resistance with a bacterial gene under control of the SV40 early region promoter. *J. Mol. Appl. Genet.* 1, 327–41.

Stewart, C. L. (1982) Formation of viable chimaeras by aggregation between teratocarcinomas and preimplantation mouse embryos. *J. Embryol. Exp. Morphol.* 67, 167–79.

− (1984) Teratocarcinoma chimaeras and gene expression. In *Chimaeras and Developmental Biology,* eds. N. Le Douarin and A. McLaren, pp. 409–27. New York: Academic Press.

Stewart, C. L., Harbers, K., Jähner, D., and Jaenisch, R. (1983a) X chromosome-linked transmission and expression of retroviral genomes microinjected into mouse zygotes. *Science* 221, 760–2.

Stewart, C. L., Jähner, D., Stuhlmann, H., and Jaenisch, R. (1983b) Retrovirus as probes for studying gene expression in mouse embryogenesis. In *Cold Spring Harbor Conferences on Cell Proliferation,* Vol. 10, eds. L. M. Silver, G. R. Martin, and S. Strickland, pp. 379–85. Cold Spring Harbor, N.Y.: Cold Spring Harbor Laboratory.

Stewart, C. L., Stuhlmann, H., Jähner, D., and Jaenisch, R. (1982a) De novo methylation, expression and infectivity of retroviral genomes introduced into embryonal carcinoma cells. *Proc. Natl. Acad. Sci. U.S.A.* 79, 4098–102.

Stewart, C. L., Vanek, M., and Wagner, E. F. (1985) Expression of foreign genes from retroviral vectors in mouse teratocarcinoma chimaeras. *EMBO Journal* 4, 3701–9.

Stewart, T. A., and Mintz, B. (1981) Successive generations of mice produced from an established culture line of euploid teratocarcinoma cells. *Proc. Natl. Acad. Sci. U.S.A.* 78, 6314–17.

Stewart, T. A., Pattengale, P. K., and Leder, P. (1984) Spontaneous mammary adenocarcinomas in transgenic mice that carry and express MTV/*myc* fusion genes. *Cell* 38, 627–37.

Stewart, T. A., Wagner, E. F., and Mintz, B. (1982b) Human β-globin gene sequences injected into mouse eggs, retained in adults, and transmitted to progeny. *Science* 217, 1046–8.

Storb, U., O'Brien, R. L., McMullen, M. D., Gollahon, K. A. and Brinster, R. L. (1984) High expression of cloned immunoglobulin x gene in transgenic mice is restricted to B lymphocytes. *Nature* 310, 238–41.

Stout, J. T., Chen, H. Y., Brennand, J., Caskey, T. C., and Brinster, R. L. (1985) Expression of human HPRT in the central nervous system of transgenic mice. *Nature* 317, 250–2.

Stuhlmann, H., Cone, R., Mulligan, R. C., and Jaenisch, R. (1984) Introduction of a selectable gene into different animal tissues by a retrovirus recombinant vector. *Proc. Natl. Acad. Sci. U.S.A.* 81, 7151–5.

Swanson, L. W., Simmons, D. A., Arriza, J., Hammer, R., Brinster, R., Rosenfeld, M. G., and Evans, R. M. (1985) Novel developmental specificity in the nervous system of transgenic animals expressing growth hormone fusion genes. *Nature* 317, 363–6.

Swartzendruber, E. C., and Lehman, J. M. (1975) Neoplastic differentiation: interaction of SV40 and polyoma virus with murine teratocarcinoma cells *in vitro. J. Cell Physiol.* 85, 179–80.

Swift, G. H., Hammer, R. E., McDonald, R. J., and Brinster, R. L. (1984) Tissue-specific expression of the rat pancreatic elastase I gene in transgenic mice. *Cell* 38, 639–46.

Taketo, M., Gilboa, E., and Sherman, M. I. (1985) Isolation of embryonal carcinoma cell lines that express integrated recombinant genes flanked by the Moloney murine leukaemia virus long terminal repeat. *Proc. Natl. Acad. Sci. U.S.A.* 82, 2422–6.

Teich, N. M., Weiss, R. A., Martin, G. R., and Lowy, D. R. (1977) Virus infection of murine teratocarcinoma stem cell lines. *Cell* 12, 973–82.

Tooze, J. (1981) *DNA tumour viruses. Molecular Biology of Tumour Viruses,* 2nd ed., part 2. Cold Spring Harbor, N.Y.: Cold Spring Harbor Laboratory.

Townes, T. M., Lingrel, J. B., Chen, H. Y., Brinster, R. L., and Palmiter, R. D. (1985) Erythroid specific expression of human β-globin genes in transgenic mice. *EMBO Journal* 4, 1715–23.

Tunnacliffe, A., Parkar, M., Povey, S., Bengtsson, B., Stanley, K., Solomon, E., and Goodfellow, P. (1983) Integration of *Ecogpt* and SV40 early region sequences into human chromosome 17: a dominant selection system in whole cell and microcell human-mouse hybrids. *EMBO Journal* 2, 1577–4.

Van der Putten, H., Botteri, F. M., Miller, A. D., Rosenfeld, M. G., Fan, H., Evans, R. M., and Verma, I. M. (1985) Efficient insertion of genes into the mouse germ line via retroviral vectors. *Proc. Natl. Acad. Sci. U.S.A.* 82, 6148–52.

Vasseur, M., Kress, C., Montreau, N., and Blangy, D. (1980) Isolation and characterization of polyoma virus mutants able to develop in multipotent murine embryonal carcinoma cells. *Proc. Natl. Acad. Sci. U.S.A.* 77, 1068–72.

Wagner, E. F., Covarrubias, L., Stewart, T. A., and Mintz, B. (1983) Prenatal lethalities in mice homozygous for human growth hormone gene sequences integrated in the germ line. *Cell* 35, 647–55.

Wagner, E. F., Keller, G., Gilboa, E., Rüther, U., and Stewart, C. L. (1985a) Gene transfer into murine stem cells and mice using retroviral vectors. *Cold Spring Harbor Symp. Quant. Biol.* 50, 691–700.

Wagner, E. F., and Mintz, B. (1982) Transfer of nonselectable genes into mouse teratocarcinoma cells and transcription of the transferred human β-globin gene. *Mol. Cell. Biol.* 2, 190–8.

Wagner, E. F., Rüther, U. and Stewart, C. L. (1984) Introducing genes into mice and into embryonal carcinoma stem cells. In *The Impact of Gene Transfer Techniques in Eucaryotic Cell Biology*, eds. J. S. Schell and P. Starlinger, pp. 127–33. Berlin: Springer-Verlag.

— (1986) Gene transfer into mouse stem cells. In *Biotechnology: Potentials and Limitations*, Dahlem Workshop Report, Life Sciences Report 35, ed. S. Silver, pp. 185–96. Berlin: Springer-Verlag.

Wagner, E. F., Stewart, T. A., and Mintz, B. (1981a) The human β-globin gene and a functional viral thymidine kinase gene in developing mice. *Proc. Natl. Acad. Sci. U.S.A.* 78, 5016–20.

Wagner, E. F., Vanek, M., and Vennström, B. (1985b) Transfer of genes into embryonal carcinoma cells by retrovirus infection: efficient expression from an internal promoter. *EMBO Journal* 4, 663–9.

Wagner, T. E., Hoppe, P. C., Jollick, J. D., Scholl, D. R., Hodinka, R. L., and Gault, J. B. (1981b) Microinjection of a rabbit β-globin gene into zygotes and its subsequent expression in adult mice and their offspring. *Proc. Natl. Acad. Sci. U.S.A.* 78, 6376–80.

Watanabe, T., Dewey, M. J., and Mintz, B. (1978) Teratocarcinoma cells as vehicles for introducing specific mutant mitochondrial genes into mice. *Proc. Natl. Acad. Sci. U.S.A.* 75, 5113–17.

Weaver, D., Costantini, F., Imanishi-Kari, T., and Baltimore, D. (1985) A transgenic immunoglobulin Mu gene prevents rearrangement of endogenous genes. *Cell* 42, 117–27.

Weiss, R., Teich, N., Varmus, H., and Coffin, J. (1982) *RNA Tumor Viruses*, 2nd ed. Cold Spring Harbor, N.Y.: Cold Spring Harbor Laboratory.

Westphal, H., Overbeek, P. A., Khillan, J. S., Chepelinsky, A.B., Schmidt, A., Mahon, K. A., Bernstein, K. E., Piatigorsky, J., and de Crombrugghe, B. (1985) Promoter sequences of murine αA crystalline, murine α2(1) collagen or avian sarcoma virus genes linked to the bacterial CAT gene direct tissue specific patterns of CAT expression in transgenic mice. *Cold Spring Harbor Symp. Quant. Biol.* 50, 411–16.

Willison, K., Babinet, C., Boccara, M., and Kelly, F. (1983) Infection of pre-implantation mouse embryos with simian virus 40. In *Cold Spring Harbor Conferences on Cell Proliferation*, Vol. 10, eds. L. M. Silver, G. R. Martin, and S. Strickland, pp. 307–17. Cold Spring Harbor, N.Y.: Cold Spring Harbor Laboratory.

Wirak, D. O., Chalifour, L. E., Wassarman, P. M., Muller, W. J., Hassell, J. A., and De Pamphilis, M. L. (1985) Sequence-dependent DNA replication in preimplantation mouse embryos. *Mol. Cell. Biol.* 5, 2924–35.

Woychik, R. P., Stewart, T. A., Davis, L. G., D'Eustachio, P., and Leder, P. (1985) An inherited limb deformity created by insertional mutagenesis in a transgenic mouse. *Nature* 318, 36–40.

Yamamura, K., Kikutani, H., Folson, V., Clayton, L. K., Kimoto, M., Akira, S., Kashiwamura, S., Tonegawa, S., and Kishimoto, R. (1985) Functional expression of a microinjected E_α^d gene in C57BL/6 transgenic mice. *Nature* 316, 67–9.

Index

558 INDEX

Printed in the United States
By Bookmasters